Electrical & Electronics Engineering

Electrical & Electronics Engineering

DR. RAJ KUMAR BANSAL
BE, ME (BITS, Pilani), Ph.D. (IIT Kanpur)
Director
Surendera Group of Institutions
Sri Ganganagar

ABHISHEK SHARMA
B.Tech, M.Tech (Hons)
Reader & Head
Deptt. of Electronics and Communication Engg.
S.M.C.E.T., Jaipur

VISHWAS TANEJA
B.Tech (Hons)
Lecturer
Surendera Group of Institutions
Sri Ganganagar

(AN ISO 9001:2008 Certified Company)
VAYU EDUCATION OF INDIA
2/25, Ansari Road, Darya Ganj, New Delhi-110 002

Electrical & Electronics Engineering

ISBN: 978-93-80097-93-0

First Edition: 2011

Rs. 275/-

Distributed by:

(AN ISO 9001:2008 CERTIFIED COMPANY)
JBC PUBLISHERS & DISTRIBUTERS
22-A, Street No.4, Near B.L. Complex, Tonk Phatak, Jaipur-302015
Phone; 9269664791, 0141-3079287
Fax: 0141-3079286 E-mail: jbcpd.jaipur@gmail.com
Website: www.jbcpd.info

Published by:

(AN ISO 9001:2008 CERTIFIED COMPANY)
VAYU EDUCATION OF INDIA
2/25, Ansari Road, Darya Ganj, New Delhi-110 002
Ph.: 91-11-47236600, 41564445
Jaipur Ph : 0141-3079287
E-mail: vayueducation@rediffmail.com, vayueducation@gmail.com
Website: www.veiindia.com

To

Our

Parents & Teachers

Preface

This book caters to the needs of first year students of all branches of Engineering of Rajasthan technical university RTU, Kota

This book will facilitate the students to prepare for the university exams as well as help them in understanding the basic concepts of electrical and electronics engg in a systematic manner. A large number of solved examples have been given to illustrate the various concepts.

The author does not claim the originality of the subject matter given in this book and help has been taken from various sources like books, manuscripts, internet, lecture notes etc.

It has the following salient features:

1.. At the end of each chapter summary, short questions and highlights have been given.
2. The subject matter has been explained in a simple and easy to understand method..
3. A large number of numerical problems has been included.

The authors will feel satisfied if the book meets the requirements and aspirations of all sections of students i.e. weak as well as intelligent.

Errors might have crept in despite utmost care to avoid them. Authors shall be grateful if these are pointed out. Suggestions and criticism for improvement of the book are welcome and will be highly appreciated.

Feb. 2011 **–Authors**

Electrical & Electronics Engineering

Unit 1 : **DC Networks :** Kirchoff's Laws, Node Voltage and Mesh Current Analysis; Delta-Star and Star-Delta Transformation, Source Conversion. Classification of Network Elements, Superposition Theorem, Thevenin's Theorem.

Unit 2 : **Single Phase AC Circuits :** Generation of Single Phase AC Voltage, EMF Equation, Average, RMS and Effective Values. RLC Series, Parallel and Series-Parallel Circuits, Complex Representation of Impedances. Phasor Diagram, Power and Power Factor.

Three Phase A.C. Circuits : Generation of Three-Phase AC Voltage, Delta and Star-Connection, Line & Phase Quantities, 3-Phase Balanced Circuits, Phasor Diagram, Measurement of Power in Three Phase Balanced Circuits.

Unit 3 : **Transformer :** Faraday's Law of Electromagnetic Induction, Construction and Operation of Single Phase Transformer, EMF Equation, Voltage & Current Relationship and Phasor Diagram of Ideal Transformer.

Electrical DC Machine : Principle of DC Machines, Types, Different Parts of DC Machines.

Unit 4 : **Transistor :** Bipolar Junction Transistor, Transistor Current Components, Characteristics of CE, CB and CC Transistor Amplifiers.

Thyristors : The four layer diode, Bi-directional thyristors, the uni-junction transistor and its application in thyristor circuits.

Unit 5 : **Communication System :** Introduction to modulation (AM, FM & PM),demodulation, multiplexing. Superhetrodyne radio receiver, television.Elementary concepts of optical, satellite & mobile communication.

Contents

UNIT-1

UNIT-2

3. Single Phase A.C. Circuits 135-262

UNIT-3

5. Electromagnetic Induction and Transformations .298-344

UNIT-1

Chapter 1

Introduction to Electrical Engineering

Chapter 2

Networks Analysis and Theorems

Chapter 1

Introduction to Electrical Engineering

OBJECTIVE

- To define basic electrical quantities and their associated units
- To understand the concepts of charge, current, voltage, potential, emf.
- Ability to work with the sign conventions of voltage and current
- To Introduce dependent and independent voltage and current sources
- To understand the behavior of the resistance , inductance and capacitance
- To understand Ohm's law, Kirchhoff's current law, Kirchhoff's voltage law, Short circuit and Open circuit concepts
- To learn circuit simplification techniques-series circuit, parallel circuit, series-parallel circuit, redundant branches, Star-Delta transformation

1.1 INTRODUCTION

In electrical power plants, energy is converted from various sources to electrical form. Electrical distribution systems transport the energy to industries, homes and business houses, where it is converted to useful forms, such as mechanical energy, heat and light

The electronics components of an average automobile is growing rapidly. Electronics technology leads to increased functionality at a lower cost. Many common household appliances contain keypads for operator control, sensors, electronics displays and computer chips . Electronic systems have become so much integrated with mechanical systems that a new name, Mechatronics, is being used for this combination.

In this chapter , basic concepts of electrical engineering i.e. systems of units , several basic definitions and conventions, basic network elements, Ohm's law, circuit simplification techniques and star-delta transformation etc have been explained.

1.2 UNITS

All engineers handle measurable quantities which have to be communicated in a standard language. In order to state the value of any measurable quantity, we must give it both a number and a unit.

The standard adopted by the National Bureau of Standards in 1964; is used by all professional engineering societies. This is the International System of units (abbreviated SI). The SI is built upon seven basic units: the meter, kilogram, second, ampere, Kelvin, mole, and candela (Table 1.1).

Table 1.1 : SI Base Units

Base Quantity	Name of unit	Symbol
Length	meter	m
Mass	kilogram	kg
Time	second	s
Electric current	ampere	A
Thermodynamic temperature	kelvin	K
Amount of substance	mole	mol
Luminous intensity	candela	cd

This is a "**metric system,**" which is in common use in most of the countries of the world, although it is not yet widely used in the United States.

Units for other quantities such as volume, force, energy, etc., are derived from these seven base units.

The SI uses the decimal system to relate larger and smaller units to the basic unit and employs prefixes to signify the various powers (multiples and sub multiples)of 10. A list of prefixes and their symbols is given in Table 1.2:

Table 1.2 : SI Prefixes

Factor	Name	Symbol	Factor	Name	Symbol
10^{-24}	Yocto	y	10^{24}	Yotta	Y
10^{-21}	Zepto	z	10^{21}	Zetta	Z
10^{-18}	Atto	a	10^{18}	Exa	E
10^{-15}	Femto	f	10^{15}	Peta	P
10^{-12}	Pico	p	10^{12}	Tera	T
10^{-9}	Nano	n	10^{9}	Giga	G
10^{-6}	Micro	μ	10^{6}	Mega	M
10^{-3}	Milli	m	10^{3}	Kilo	k
10^{-2}	Centi	c	10^{2}	Hecto	h
10^{-1}	Deci	d	10^{1}	Deka	da

Combinations of several prefixes, such as the milli microsecond, are unacceptable. It is fairly common to express numbers in "engineering units." In engineering notation, a quantity is represented by a number between 1 and 999 and an appropriate metric unit using a power divisible by 3. For example, it is preferable to express the quantity 0.048 W as 48 mW i.e.48 $\times 10^{-3}$ watts, instead of 48,000 μW.

1.3 Electrons Theory

The laws governing the behaviour of electrons is called electron theory. This theory helps the scientists in exploring the nature of the universe and life. As per this theory, all electrical and electronic effects are caused either by the movement of electrons or due to the reason that too many or too few electrons exist in a particular place at a particular time.

1.3.1 Atomic Structure

Everything that we see, feel or touch is made of **matter**. There are five forms of matter in the universe: solid, liquid, gas, plasma, and Einstein-Bose. The simplest form of any matter is called an **element** in which the unique characteristics of the substance can be identified.

The smallest piece of an element that has all the characteristics of that element is the **atom**. These atoms are of different sizes, degrees of structural complexity and weight. The atoms have three main parts: **electrons, protons** and **neutrons**. Protons and neutrons form the central body of the atom, called the **nucleus**. Electrons move in approximately elliptical orbits called **shells** around the nucleus.

Smaller sub particles of the protons and neutrons are called **quarks**.

Electron has a negative charge, proton has a positive charge and neutron has no charge and is neutral. It means that a nucleus is positively charged. The number of protons is equal to the number of electrons. The number of protons is also equal to the atomic number of an element.

The electrons and protons have two other characteristics i.e. size and mass. An electron is about three times larger than the proton but is much lighter (1/1836).

Mass of electron = 9.1×10^{-31} kg. Mass of proton = 1.66×10^{-27} kg.

Electrons and protons have lines of forces. The electrons lines of force flow inward toward the electron. The lines of force of proton flow outward in all directions, Fig1.1 (a)

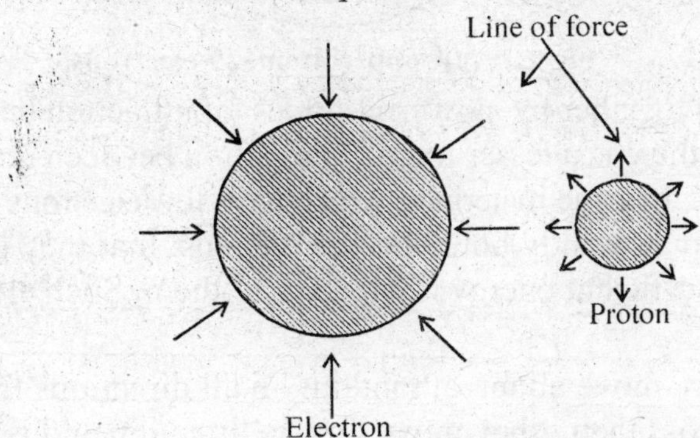

Fig 1.1 : (a) lines of Force of Electrons and Protons

Work has to be done to separate the electrons and protons. Unlike charges attract and like charges repel each other , Fig 1.1 (b)

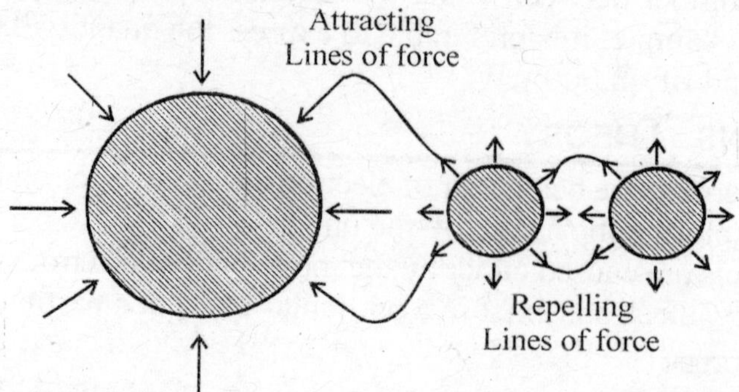

Fig 1.1: (b)Attracting and Repelling lines of Force

There are a number of orbital rings around a nucleus that can hold electrons. The maximum number of electrons possible in each orbital ring is given by the formula $2N^2$, where N is ring's number, subject to the condition that maximum number of electrons in any orbit will not exceed 32. Outermost ring is called the **valence ring** and the electrons orbiting in the outermost orbit are called **valence electrons.**

Copper atom contains 29 electrons with 2, 8, 18 and 1 electrons in the first, second , third and fourth orbits respectively, Fig 1.1 (c)

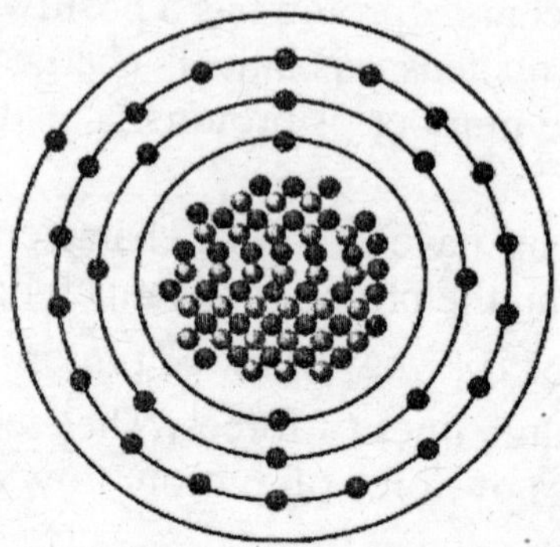

Fig 1.1: (c)Copper atom, 29 electrons

All atoms are bound together by powerful forces of attraction between the nucleus and the electrons. However , there is a lesser force of attraction between the electrons in the outer most orbit and the nucleus. In some materials , which have few electrons in the outer orbit ,these outer electrons are so much weakly bound to the nucleus, that they can be easily freed or forced away from it , if sufficient energy is supplied to them. Such electrons are called **free electrons**

These **free electrons** move about at random, in all directions from atom to atom ,but when emf is applied across them ,they move in one direction and is known as **current.**

Each orbit has an **energy level** associated with it. The electrons which are in the innermost shell require maximum amount of energy for their extraction from the atom. The electrons which are in the outermost shell require minimum amount of energy for their extraction from the valence shell.

Depending upon the number of electrons in the valence ring, materials can be classified as **conductors, insulators** and **semi conductors**. Conductors have only one or two electrons in the valence ring.

The body which contains unequal number of electrons and protons is said to be **electrically charged**.

If an atom loses an electron, it loses the negative charge, it is therefore left as positively charged. This positively charged body is called a **positive ion**. In liquids and gases ions can move and contribute to the flow of current.

All materials are made up of different combinations of atoms to form molecules of the materials. There are about 100 different kinds of atoms and the human body is made up of only 15 elements and only six of them are prominent i.e. carbon, hydrogen, oxygen, nitrogen, calcium, phosphorus etc.

1.3.2 CHARGE

Charge is the most basic quantity of electricity and one of the fundamental concepts in circuit analysis is that of charge conservation. There are two types of charges: positive (corresponding to a proton), and negative (corresponding to an electron).

There are many devices , such as batteries, diodes etc. in which positive charge motion is important to understand their operation, but external to the device the electron flow through the connecting wires is important. Charge in motion is called a **current.**

A quantity of charge that does not change with time is represented by upper case, Q. The instantaneous amount of charge that changes with time, is commonly represented by q(t), or simply 'q'.

Current through the conductors is due to the motion of negatively charged electrons .

The unit of charge is 'coulomb'. One electron has charge, $Q = \dfrac{1}{6.24\times 10^{18}} = 1.6\times 10^{-19}$ coulombs and one coulomb of charge has 6.24×10^{18} electrons i.e. one coulomb of charge is carried by 6.24×10^{18} electrons.

1.4 FUNDAMENTAL CONCEPTS OF ELECTRICAL CIRCUITS

In order to understand the behaviour of circuits , we will first study the concepts of Potential , Potential difference, Electric flux,voltage and current along with voltage and current sources.

1.4.1 Concept of Potential, Potential difference, Electric flux

(i) Potential : The capacity of two oppositely charged bodies to produce a flow of current between them may be considered as a potential for the production or flow of electric current. It indicates the electrical state of a body. Earth is assumed to be at zero potential.

An electrically charged particle sets up an electric field around it. If the particle is stationary, the field is called **Electrostatic Field.** The electric field lines are radial and are directed away from the positive charge and are directed towards the negative charge. The force of attraction or repulsion between two charges q_1 and q_2 is governed by coulomb's law and is given by:

$$\boxed{F = \frac{q_1 q_2}{4\pi\varepsilon R^2}} \qquad (1.1)$$

Where F is the force in Newtons, q_1 and q_2 are the two charges in coulombs, R is the distance between the charges in meters ε is the permittivity of the medium in which charges are placed, in farads/meter

The intensity of an electric field at any point in the field is the force experienced by a unit positive charge placed at that point. Hence **Electric field intensity, E** of the field set up by a charge 'q' is given as:

$$\boxed{E = \frac{q}{4\pi\varepsilon R^2}\ volts/m} \qquad (1.2)$$

If this unit positive charge is to be moved from one point to another in the electric field, work has to be done against the force.

The work done in moving a unit positive charge from infinity up to a point in an electric field is called the **potential** at that point expressed in terms of volts or joules/coulomb

A positively charged body has a **Positive Potential** and a negatively charged body has a **Negative Potential.**

Any charged conductor, which when connected to earth , can send electricity to the earth, is said to be at **Positive Potential** and any conductor , which receives electricity from earth, when connected to it , is said to be at **Negative Potential.**

(ii) Potential difference : Potential difference between any two points is defined as the difference between the electric potentials at those two points.

In Fig 1.2(a) , point A has a potential of V_A Volts and point B has a potential of V_B Volts, then the potential difference between points A and B is:

$V_{AB} = (V_A - V_B)$ volts

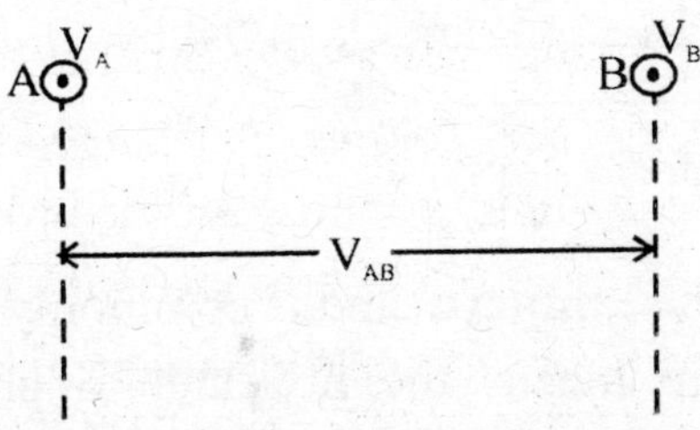

Fig.1.2 : (a) Potential difference

If the potential difference between the two points is zero, then no current will flow between them.

There is always a potential difference between a negatively charged body and a positively charged body. When two bodies having a potential difference are joined by a metallic wire, electrons flow from a negatively charged body to positively charged body.

The work done in moving a unit positive charge from one point to another point in an electric field is called the **Potential difference** in volts.

It is the difference between the electrical pressure or potentials at any two points of a conductor. Electric current flows from a point of higher potential to a point of lower potential , if these two points are connected to each other by a conductor. Magnitude of the electric current depends upon the potential difference. Batteries or generators are used to create the desired potential difference for the flow of current.

(iii) Electrical Flux (ψ) : The electrical flux is assumed to stream away symmetrically from a point charge and is coincident with the stream lines corresponding to the electric field. It is denoted by 'ψ' and is measured in 'coulombs'. A body which has a charge of 'q' coulombs, emits a total electric flux of 'q' coulombs. Electrical Flux (ψ) = q

1.4.2 Current

Electrons flow from a negatively charged body to a positively charged body whenever oppositely charged bodies come in contact with each other. Electrons also flow from a negatively charged body to a body which has a lower negative charge or which does not have any charge. The flow or movement of electrons is called flow of **electric current**.

The idea of "charge in motion" is important because, in moving a charge from one place to another place, energy may be transferred from one point to another. The power transmission line is an example of a device which transfers energy.

The current in a metallic wire, has both a numerical value and a direction. It is a measure of the rate at which charge is moving past a given reference point in a specified direction.

Let q(t) be the total charge that has passed the reference point since an arbitrary time t = 0, moving in the defined direction.

The current is defined at a specific point as the instantaneous rate at which net positive charge is moving past that point in the specified direction.

Current is symbolized by I or i, and is given by:

$$\boxed{i \text{ or } I = \frac{dq}{dt}} \tag{1.3}$$

The unit of current is the '**ampere**' and is commonly abbreviated as "**amp**".

A current flow of 1 amp. means the movement of charge equal to one coulomb per second.

1.4.2.1 Types of Current

There are several different types of current. A current that remains constant over all time is termed a **Direct Current**, or when the flow of electric charge has a constant rate in one direction or when the current flows continuously in one direction, it is called direct current or simply DC, and is shown by Fig. 1.2(b).

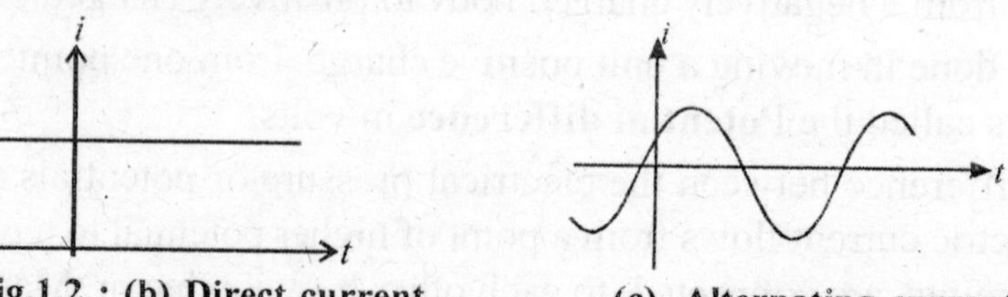

Fig.1.2 : (b) Direct current **(c) Alternating current**

There are also currents that vary sinusoidally with time, Fig1.2(c). Such currents are present in household circuits. Such a current is often referred to as **Alternating Current**, or AC.

Exponential currents, Fig 1.2(d) and **Damped Sinusoidal Currents** , Fig 1.2(e) are also encountered frequently.

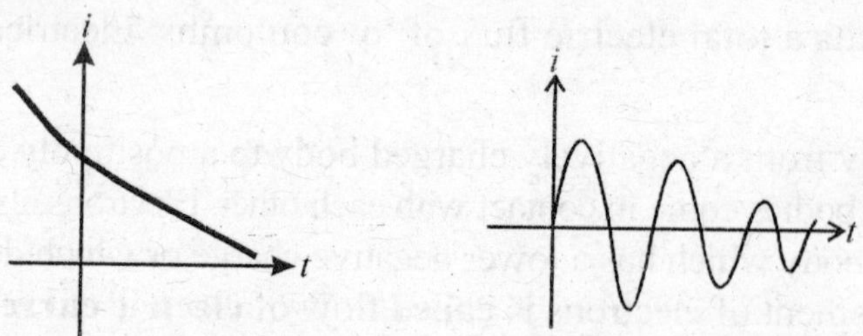

Fig.1.2 : (d) Exponential currents **(e) Damped sinusoidal currents**

If a current is varying but has the same polarity as shown in Fig.1.2(f), it is known as **Unidirectional current**.

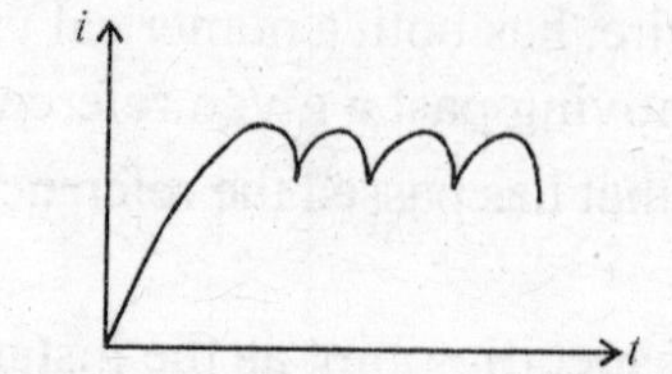

Fig 1.2 : (f) Unidirectional current.

1.4.2.2 Direction Of Current

In Fig1.3 (a), the direction of the arrow and the value 3 amperes indicate that either a current of 3 amps is moving to the right or that a current of –3 amps is moving to the left each second.

Fig.1.3 (b) indicates that :

Either –3 amp. is flowing to the left side or +3 amp. is flowing to the right side.

Fig.1.3 : Representation of current

The current arrow does not indicate the "actual" direction of current flow but is simply part of a convention. To indicate the value of a current $i_1(t)$ without specifying the arrow is incorrect . For example, Fig 1.3(c) is not proper representation of $i_1(t)$, whereas Fig 1.3 (d) is the proper representation.

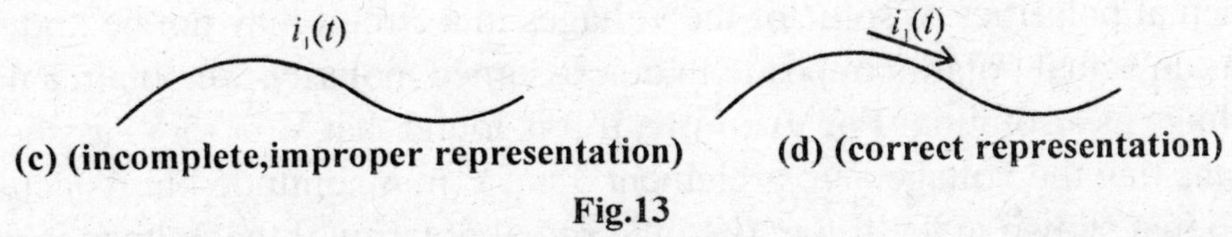

(c) (incomplete,improper representation) **(d) (correct representation)**

Fig.13

1.4.2.3 Double Subscript Notation For Current

Another way to indicate the direction of current for a circuit element is to label the ends of the element and use double subscripts to define the reference direction for the current. For example, consider a resistance shown in Figure1.4 ,where ends have been labeled as 'a' and 'b'.

The current denoted by i_{ab} is the current through the element with its direction pointing from a to b. Similarly, i_{ba} is the current with its direction pointing from 'b' to 'a'.

Of course, i_{ab} and i_{ba} have the same magnitude but are opposite in sign.

Thus, we have , $i_{ab} = -i_{ba}$

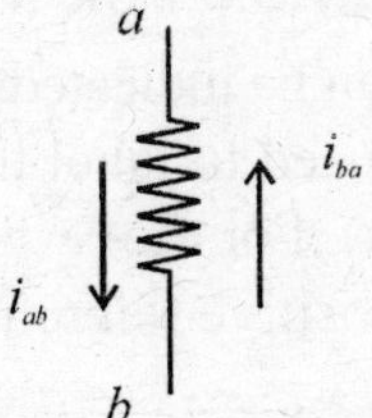

Fig. 1.4 : Double subscript notation for current variables.

1.4.3 Voltage

It may be thought of as a type of electric pressure. When charge moves through circuit elements, energy can be transferred. The voltage associated with a circuit element is the energy transferred per unit of charge that flows through the element. The units of voltage are Volts (V), which are equivalent to Joules per Coulomb (J/C). It is measured across any two points.

For example, suppose the voltage across the terminals of a battery is 12 Volt. This means that an energy of 12 Joules is transferred to or from the battery for each coulomb that flows through it.

1.4.3.1 Types of Voltages

Voltages can remain constant at every instant of time or can vary with time. Constant voltages are called **DC voltages**. Voltages that change in magnitude and alternate in polarity with time are said to be **AC Voltages.**

For example, voltage $v_1(t) = 10$ V, is a DC voltage. It has the same constant magnitude and polarity for all time. However, voltage $v_2(t) = 10 \sin(100\pi t)$, is an AC voltage that varies in magnitude and polarity with change in time.

1.4.3.2 Reference Polarities

The actual polarities of some of the voltages in a circuit may not be known. If a given voltage has an actual polarity opposite to our assumed polarity, we obtain a negative value for the voltage as a solution. **For example,** if it is found that $V_3 = -5$ V as shown in Figure 1.5, it means that the voltage across element 3 is 5V in magnitude but its actual polarity is opposite to that shown in the figure (i.e., the actual polarity of the voltage is positive at the bottom end of element 3 and negative at the top end).

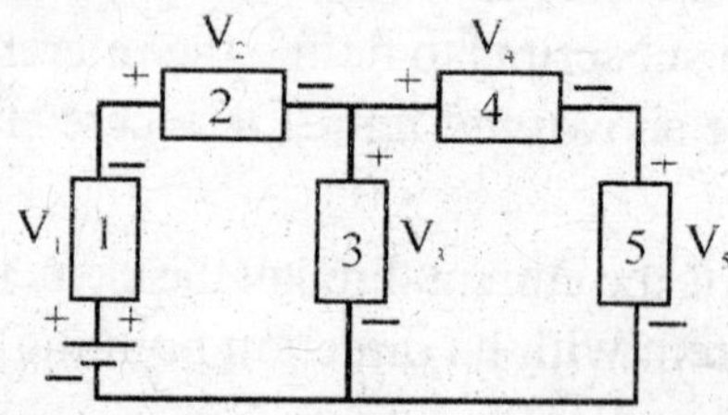

Fig.1.5 : Reference Polarities of voltages

1.4.3.3 Double-Subscript Notation For Voltage

Reference polarity of a voltage can be indicated by using double subscripts on the voltage variable. Letters or numbers can be used to label the terminals between which the voltage appears, as illustrated in Figure 1.6(a). For the resistance shown, V_{ab} represents the voltage between points 'a' and 'b' with the positive reference at point 'a'.

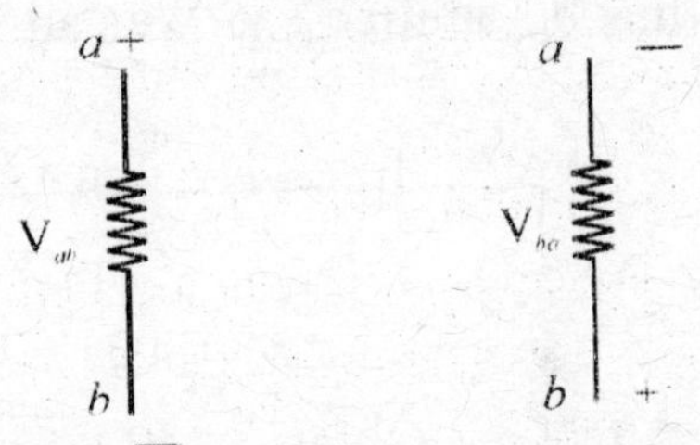

Fig 1.6(a) Double-subscript notation for voltages

The two subscripts identify the points across which the voltage appears, and the first subscript indicates the positive reference. Similarly, V_{ba} is the voltage between 'a' and 'b' with the positive reference at point 'b', i.e. $V_{ab} = -V_{ba}$

Reference polarity of voltage can also be indicated by an arrow, as shown in Figure1.6(b) The arrows are drawn such that head of the arrow corresponds to the positively defined terminal. Here terminal '1' is positive w.r.t. terminal'2'.

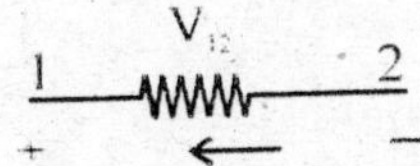

Fig 1.6(b) Reference Polarity of Voltage

A voltage can exist between a pair of electrical terminals whether a current is flowing or not. An automobile battery, for example, has a voltage of 12 V across its terminals even if nothing is connected to the terminals.

1.4.4 Electromotive Force

The electric force that causes the flow of current in a conductor is known as electromotive force i.e. EMF.

It is the external worth done per unit of charge to produce an electric potential difference across two open circuited terminals. The electric potential difference is created by separating positive and negative charges there by generating an electric field. The created electrical potential difference drives the current flow if a circuit is attached to the source of EMF.

1.4.5 Circuit Ground

The term **'ground'** appears frequently in electrical engineering applications . A reference voltage is usually maintained which is called as "ground voltage" with reference to which , all other voltages are referred. This reference voltage is at ground potential or zero potential and other terminal voltages are measured with respect to ground potential . Some terminals in the circuit will have voltages above it (positive) and some terminals will have voltages below it (negative) .

Consider the circuit shown in Fig.1.7 (a) where node D is selected as ground (or reference) node. When the voltages at different nodes are referred to this ground (or reference) point, the second subscript may be omitted e.g. We can write V_{ED} as V_E, V_{AD} as V_A and so on.

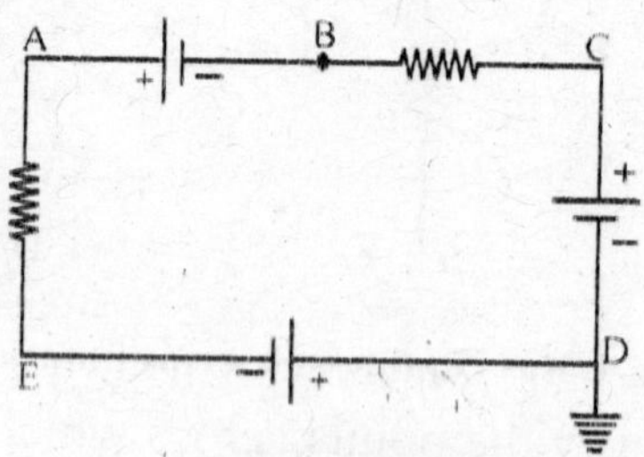

Fig 1.7(a) Circuit Ground

In many electronic circuits, the common ground point may be obtained by a metal frame/ chassis with in the device which may or may not be connected to the power. Different types of grounding symbols are shown in Fig 1.7(b)

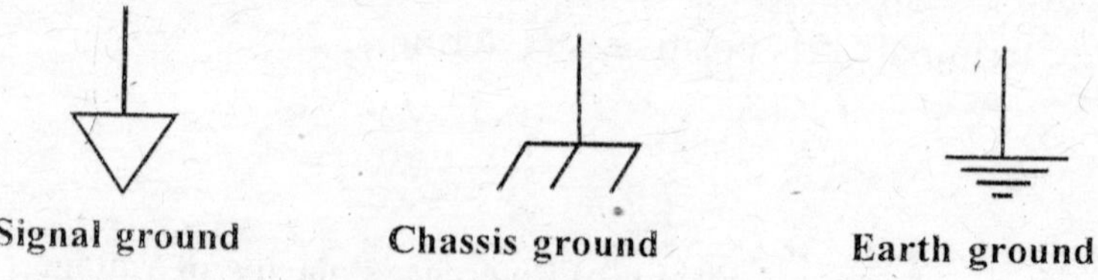

Fig.1.7 : (b) Different types of grounding symbols

1.5 Sources of Electric Energy

All electric circuits are driven by sources that supply energy. Sources are of two types: (i) voltage sources and current sources. Independent and Dependent voltage and (ii) current sources are capable of delivering power to external devices.

1.5.1 Voltage Sources

A **Voltage Source** is assumed to deliver energy with a potential difference of a given magnitude if it is a steady voltage source or with a potential difference of prescribed waveform , if the voltage changes with time, independently of the magnitude and direction of current flowing through it . A general voltage source which is a function of time is represented as a two-terminal device as shown in Figure 1.8 (a).The plus sign indicates the reference polarity of the voltage source.

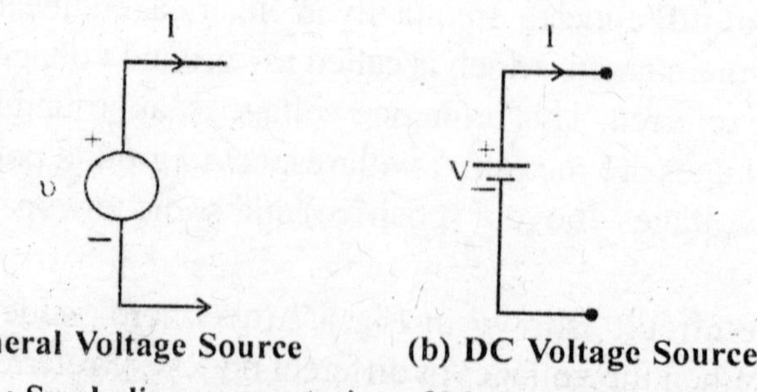

(a) General Voltage Source (b) DC Voltage Source

Fig. 1.8 : Symbolic representation of Ideal voltage sources

When it is a DC voltage source , it maintains a constant potential difference across its terminals and is represented by a battery, as shown in Figure 1.8 (b). Positive terminal is shown next to the wider of the two lines.

Plus sign at the upper end of the symbol, does not necessarily mean that the upper terminal is numerically positive with respect to the lower terminal. Instead, it means that the upper terminal is V volts positive with respect to the lower terminal at a particular instant of time. If at some instant , V happens to be negative, then it means that the upper terminal is actually negative with respect to the lower terminal at that instant.

1.5.1.1 Independent Voltage Sources

The sources which maintain a voltage across their terminals fixed or time varying ,completely independent of any other quantity i.e.all other voltages and currents in the circuit are termed Independent Voltage Sources

The independent voltage source is an ideal source and does not represent exactly any real physical device, because the ideal source could theoretically deliver an infinite amount of energy from its terminals.

An independent DC voltage source can be represented by either of the symbols shown in Fig.1.9(a) and (b). When the plate structure of the battery is used, the longer/wider plate is placed at the positive terminal.

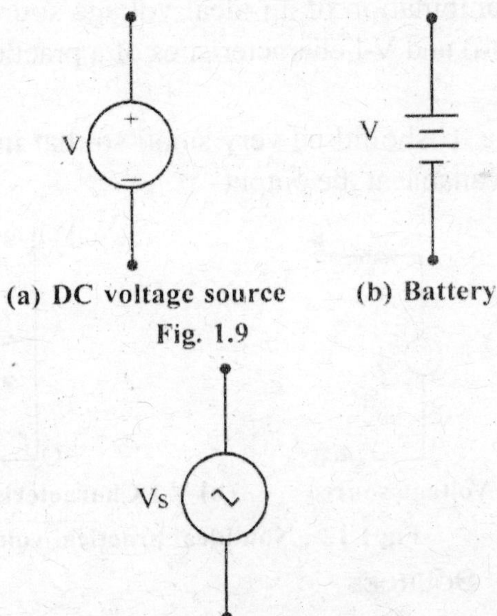

(a) DC voltage source **(b) Battery**

Fig. 1.9

Fig. 1.10 : AC voltage source

An independent AC voltage source is represented by the symbol shown in Fig.1.10 Independent voltage sources can be further classified as :

(i) Ideal voltage sources (ii) Practical Voltage Sources

1.5.1.1.1 Ideal Voltage Source

An ideal voltage source is expected to supply a constant voltage to the circuit irrespective of the amount of current drawn from it. Voltage source shown in Figure 1.11(a) is defined as ideal source as it does not have any internal resistance.

V-I characteristics of an ideal voltage source is shown in Fig 1.11(b).

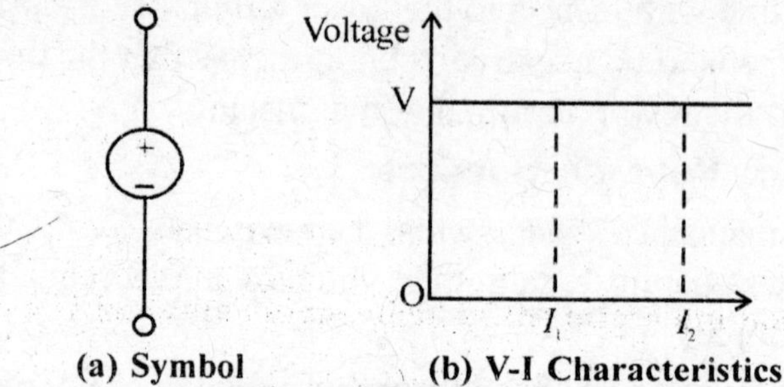

(a) Symbol (b) V-I Characteristics

Fig.1.11 : Ideal Voltage Sources

1.5.1.1.2 Practical Voltage Source

In practice , all physical voltage sources do have some internal resistance due to which their terminal voltage is different from that of the actual source when current is drawn from them. This is due to the internal voltage drop which varies with current . All physical voltage sources may be represented as a combination of an ideal voltage source and some series resistance as shown in Figure 1.12 (a) and V-I characteristics of a practical voltage source is shown in Fig 1.12(b).

Internal resistance 'r' should be very small so that internal voltage drop is small and maximum voltage is available at the output

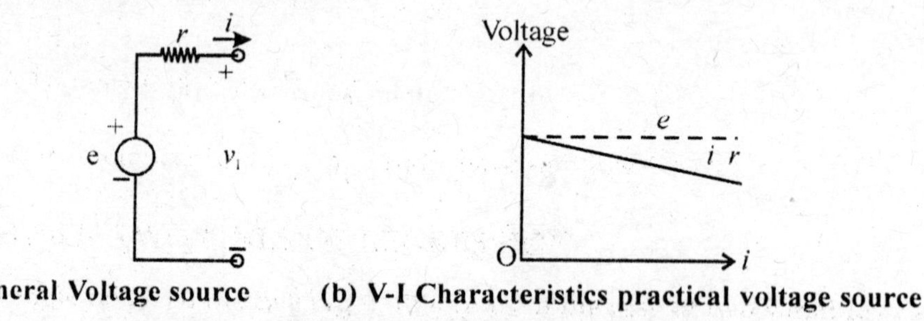

(a) General Voltage source (b) V-I Characteristics practical voltage source

Fig 1.12 : Non ideal practical voltage source

1.5.2 Current Sources

A current source maintains a current of a given magnitude and a prescribed waveform at its terminals independently of the potential difference appearing across its terminals.

It is essential to indicate the reference current direction of a current source by an arrow, as shown in Fig. 1.13.

1.5.2.1 Independent Current Sources

In an independent current source, the current supplied by it is maintained ,fixed or varying with time and is completely independent of any other quantity i.e. all other voltages and currents in the circuit. The symbols for an independent current source are shown in Fig.1.13(a) and (b)

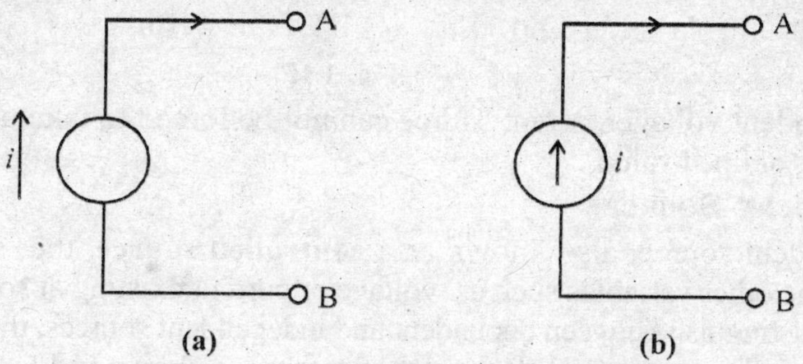

Fig. 1.13 Circuit symbols for Independent current source

Independent current sources can be further classified as :

(i) Ideal current sources (ii) Practical current Sources

1.5.2.1.2 Ideal current sources

In an ideal current source, the current remains constant , irrespective of change in the load. Theoretically it can deliver infinite power from its terminals, because it produces the same finite current for any voltage across it, no matter how large that voltage may be. Current sources shown in Figure 1.14(a) are defined as Ideal current sources as they have infinite shunt resistance. V-I characteristics of an ideal current source is shown in fig 1.14 (b).

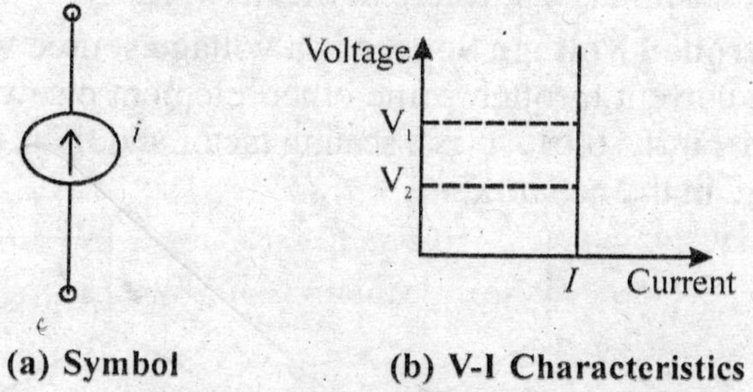

(a) Symbol **(b) V-I Characteristics**

Fig 1.14 : Ideal Current Sources

1.5.2.1.2 Practica Current Sources

A general practical current source is represented as shown in Figure 1.15(a) .It has a finite internal resistance in parallel with it. Internal resistance 'R' of the current source should be very high so that maximum current can be delivered at the output. V-I characteristics of a practical current source is shown in Fig 1.15(b).

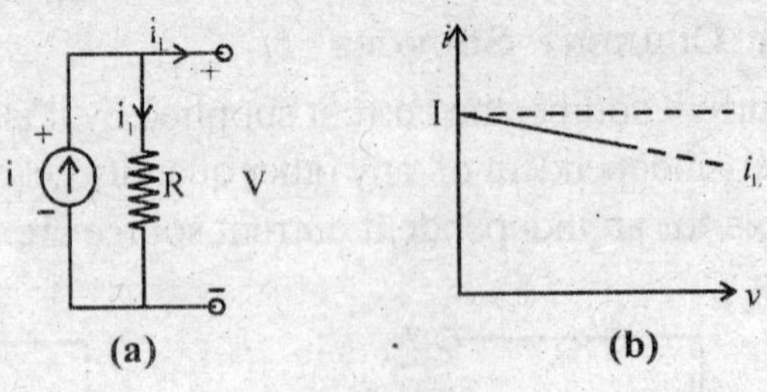

Fig. 1.15

An independent voltage/current source can not be forced to take up a voltage/ current different from its defined value.

1.5.3 Dependent Sources

In a dependent source, also known as a **controlled source**, the source quantity is a function of some other variables such as, voltage or current existing at some other location in the circuit. To distinguish between dependent and independent sources, the **diamond symbols** are used for dependent sources in the circuit diagrams as shown in Fig. 1.16.

The controlling current i_x and the controlling voltage v_x must be defined some where in the circuit. The factor multiplying the controlling current is called the **gain parameter or a scaling factor**.

1.5.3.1 Dependent Voltage Sources

There are two types of dependent voltage sources:

(i)Voltage Controlled Voltage Source

(ii)Current Controlled Voltage Source

(i) Voltage Controlled Voltage Source is a voltage source which has a voltage equal to a constant times the voltage across a pair of terminals elsewhere in the network. It is represented as shown in Fig.1.16(a). 'K' is a dimensionless scaling constant and 'v_x' is the voltage across a pair of terminals elsewhere in the network.

(ii) Current Controlled Voltage Source is a voltage source which has a voltage equal to a constant times the current through some other element elsewhere in the network, It is represented as shown in Fig.1.16(b) . 'r' is a scaling factor and 'i_x' is the current through some other element elsewhere in the network.

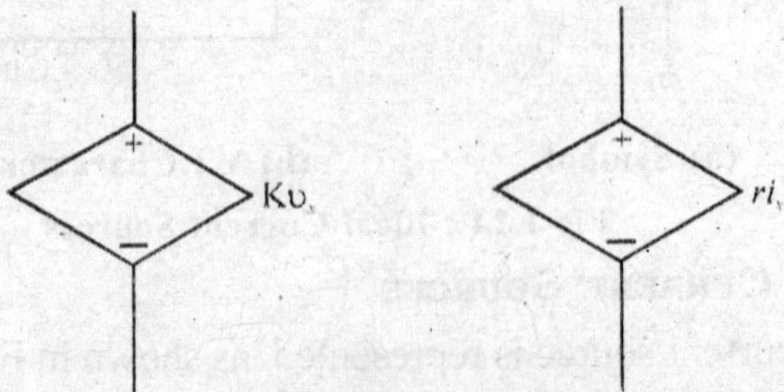

(a) Voltage Controlled Voltage Source (b) Current Controlled Voltage Source

Fig.1.16 : Dependent Voltage Sources

1.5.3.2 Dependent Current Sources

There are two types of dependent current sources:

(i)Voltage Controlled Current Source

(ii) Current Controlled Current Source

(i)Voltage Controlled Current Source is a current source which has a current equal to a constant times the voltage across a pair of terminals elsewhere in the network . It is represented as shown in Fig.1.17(a) .'g' is a scaling factor and v_x is the voltage across a pair of terminals elsewhere in the network

(ii)Current Controlled Current Source is a current source which has a current equal to a constant times the current through some other element elsewhere in the network. It is represented as shown in Fig.1.17(b). 'K' is a dimensionless scaling constant and i_x is the current through some other element elsewhere in the network

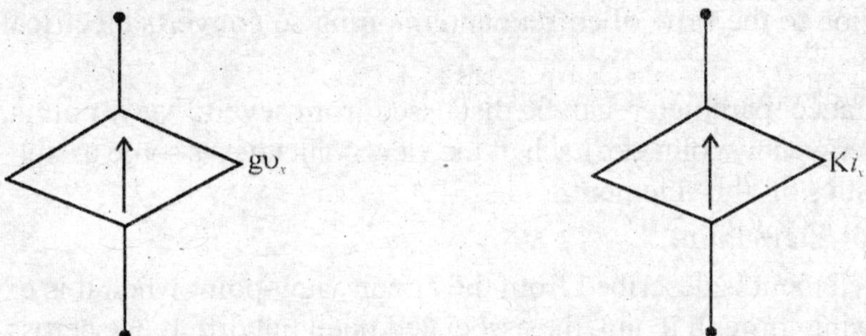

(a) Voltage Controlled Current Source (b) Current Controlled Current Source

Fig.1.17 Dependent Current Sources

Dependent sources are frequently used in Electronics Engineering to model both DC and AC behaviour of transistors etc.

1.6 Network Elements

An electrical network is an interconnection of physical electrical devices such as electrical sources , loads and the conductors. Electrical devices such as light bulbs and generators etc. can be represented by combinations of simple circuit elements. A general circuit element may be shown as a shapeless figure which has two terminals at which connections to other elements may be made (Fig. 1.18)

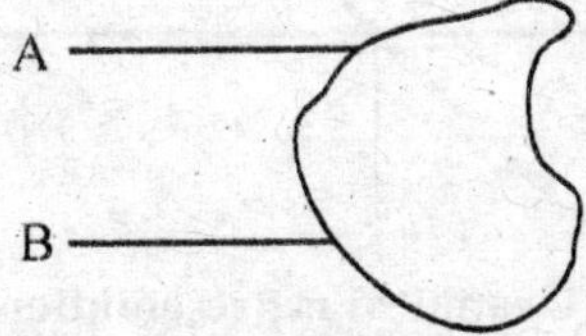

Fig 1.18 A General two terminal circuit element

The purpose of defining the individual elements, connecting them in a network and solving the relevant equations is to analyze the performance of such electric devices as motors, generators, transformers and electronic devices. The solution provides answers about their satisfactory operation and performance.

All the simple circuit elements can be classified according to the relationship of the current through the elements to the voltage across the element. For example, if the voltage across the element is linearly proportional to the current through it, the element is called a **resistor.**

Other types of simple circuit elements have terminal voltages which are proportional to:

(i) the derivative of the current with respect to time (**an inductor**) or

(ii) to the integral of the current with respect to time (**a capacitor**).

1.6.1 The Resistance Parameter

Resistance is one of the three basic parameters of electric circuits and every circuit element exhibits some amount of resistance. Resistance is that property of a circuit element which offers opposition to the flow of current and in doing so converts electrical energy into heat energy.

The resistance parameter can be discussed from several view points, such as : Circuit view point, Energy view point etc. Each of the view point reveals some useful information about the characteristics of this element.

(a) Circuit View-Point

A circuit element is described from the circuit view-point when it is expressed in terms of current flowing through it and the associated potential difference across it.

If 'v' is the voltage across an element and 'i' is the current flowing through it, then by Ohm's law we have :

$v \propto i$ or $v = R\,i$, where R is the proportionality constant.

Or $$\boxed{R = \frac{v}{i}\,\Omega}$$

This equation gives the circuit view point description of resistance and is a linear algebraic expression shown graphically in **Figure 1.19.**

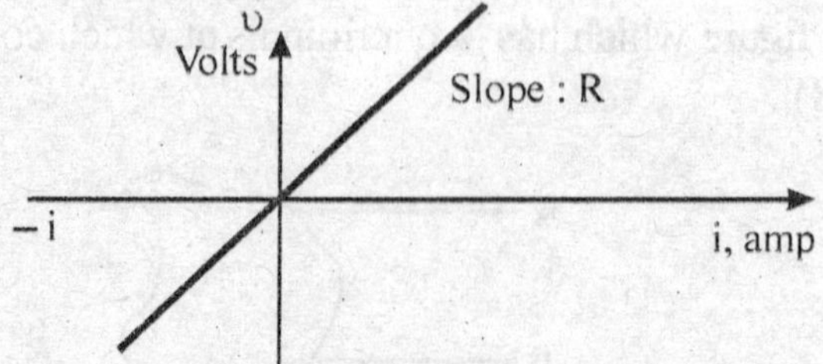

Fig.19 Graphical representation of Ohm's law

It is valid for negative as well positive values of *i*.

(b) Geometrical View-Point

The resistance of a conductor of uniform dimensions depends on its geometry. It is directly proportional to the length, inversely proportional to the cross-sectional area and depends upon conductance properties of the material.

i.e. $$R = \rho \frac{l}{A} \text{ ohm} \qquad (1.4)$$

where ρ = Resistivity of the material in ohm-metres

l = length of the conductor in metres

A = cross-sectional area in metres squared

1.6.1.1 Resistivity/Specific Resistance (ρ)

The resistance per unit length of conductors may not always be available. To allow for all possible combinations of conductor length and cross sectional area, information about **specific resistance or resistivity** of a conductor is useful.

Rewriting equation(1.4), we get:

Resistivity $$\boxed{\rho = \frac{R \times A}{l}}$$

Units of resistivity 'ρ' can be obtained as follows:

Substituting the values of units of different quantities in the above equation, we get:

Units of resistivity $$\rho = \frac{ohm \times m^2}{m} = \text{ohm-m}$$

The specific resistance of a material is the resistance of 1 meter cube of the material.

Table 1.3 gives Resistivity/Specific Resistance of some conductors.

Table1.3 : Resistivity/Specific Resistance of some Conductors

Metal	Resistivity at 20 ^{0} C (Ω - m)
Copper(annealed)	1.69×10^{-8} to 1.74×10^{-8}
Aluminium, hard drawn	2.83×10^{-8}
Tungsten	5.5×10^{-8}
Constantan	49×10^{-8}
Manganin	49×10^{-8}
Silver	1.58×10^{-8}
Carbon	6500×10^{-8}

The best conductor is a material which has lowest value of specific resistance. It can be seen from the table1.4 that Silver is the best conductor of electricity.

Example 1.1 : A wire of length 1m has a resistance of 2 ohms. Obtain its resistance ,if specific resistance is doubled, diameter is doubled and length is increased to three times.

Solution :

We know that:

Area $\propto$ d^2

$\therefore$ new area, $a_2 \propto (2d)^2$ i.e. $4d^2 = 4a_1$

$$\text{New resistance, } R_2 = \frac{\rho_2 l_2}{a_2} = \frac{2\rho_2 \times 3l_1}{4a_1} = 1.5\ R_1 = 1.5\ 2 = 3\ \text{Ohm}$$ **Ans.**

Example 1.2 : The length of a conductor is doubled and its area of cross section is also doubled. What is the effect on resistance of conductor?

Solution :

$$\text{New resistance } R_2 = \frac{P_2 l_2}{a_2} = \frac{P_1 \times 2l_1}{2r} = \frac{P_1 l_1}{a_1} = R_1$$ **Ans.**

Hence, resistance value does not change, it remains the same.

Example 1.3 : Calculate(i) the resistance of 100 m length of wire having a uniform cross sectional area of 0.02 mm^2 and having resistivity of 5 ohm-cm.(ii) If the wire is drawn out four times its original length, Calculate the resistance.

Solution :

(i) $R = \frac{\rho\, l}{a}$ and

$1\ mm^2 = 10^{-6}\ m^2$

resistivity $= 5\mu$ ohm-cm

$= 5\times10^{-8}$ ohm-m,

$a = 0.02\ mm^2 = 2 \times 10^{-8}\ m^2$

Substituting the values of different parameters ,we get:

$$R = \frac{5\times10^{-8}\times100}{2\times10^{-8}} = 250\ \text{Ohm}$$ **Ans.**

(ii) $l_{new} = 4 \times l_{old}$

If length is increased, area of cross section will be reduced, because volume is same.

i.e. volume $= a_{old}\, l_{old} = a_{new}\, l_{new}$
$= a_{new} \times 4\, l_{old}$

or $a_{new} = \dfrac{a_{old}}{4}$

Hence , resistance $R_{new} = \dfrac{\rho \times 4 \times l_{old}}{a_{old}/4} = 16\, R_{old} = 16 \times 250 = 4000$ Ohm **Ans**

Example 1.4 : An electric heater takes a current of 15 amp from a 220 V source. The wire connecting the heater to the electric supply is 50 × 2 = 100 m long. If the voltage drop along the wire is not to exceed 15 volts, find the diameter of the suitable copper wire(annealed)

$$\rho = 1.72 \times 10^{-8}\ \Omega - m$$

Solution :

The maximum allowable resistance $= \dfrac{\textit{Allowable Voltage drop along the wire}}{\textit{current through the wire}}$

$$= \frac{15}{15} = 1\,\text{ohm}$$

Area of cross section of wire $= \dfrac{\rho l}{R} = \dfrac{1.72 \times 10^{-8} \times 100}{1} = 1.72 \times 10^{-6}\ m^2$

Dia of wire $= \sqrt{\dfrac{4a}{\pi}} = \sqrt{\dfrac{4 \times 1.72 \times 10^{-6}}{\pi}} = \sqrt{2.1899 \times 10^{-6}}$

$= 1.479 \times 10^{-3}$ m $= 1.479$ mm **Ans**

1.6.1.2 Effect of Temperature on Resistance of Different Materials

All conductors get heated up when carrying a current. Electrical energy absorbed by the resistance of the conductor is converted in to heat, which leads to temperature rise of the conductor. If the current level exceeds a specified limit, the conductor gets overheated and may burn.

(a) Effect of temperature on resistance of Metals

In resistance calculations using the specific resistance, the temperature is assumed to be constant at 20 °C. However, **the resistance of most metallic conductors increases with increase in temperature**. In a general way, this may be regarded as due to the vibration of atoms or thermal agitation of the ions. These do not occupy fixed positions, but oscillate about a mean position, the amplitude increasing with the increase in temperature and thus become a greater obstruction in the path of electrons.

If the resistance of any metal is plotted against temperature, the graph is a straight line in a temp. range of 0 to 100^0 C as shown in Fig.1.20. Linear variation in resistance for copper occurs in the temp. range of -50^0 C to 200^0 C The variation of resistance of copper with temperature is shown in Fig 1.20.For a certain range of temperature , the rate of increase of resistance is constant. The resistance of copper reduces to zero at -234.5^0 C

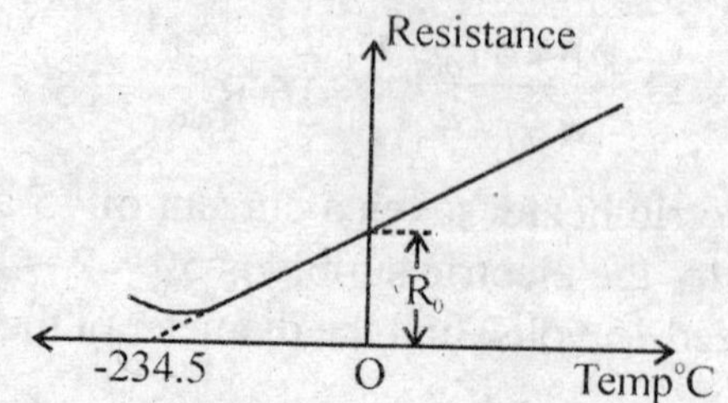

Fig 1.20 Change in resistance of copper due to change in temp.

(b) Effect of temperature on resistance of Alloys

The rate of increase of resistance with temperature is less in alloys than that of metals.

The rate of increase of resistance with temperature is almost zero for some alloys. Certain alloys such as constantan, manganin are used for construction of accurate resistances used in resistance boxes due to this property.

(c) Effect of temperature on resistance of Insulators

Resistance of carbon and insulators **decreases** with increase in temperature. Materials that act as perfect insulators at low temperatures may start conducting at higher temperatures.

(d) Effect of temperature on resistance of Semi Conductors

Resistance of semi conductors decreases with increase in temperature. These have a negative temperature coefficient of resistance/resistivity

Fig 1.21 shows change in resistance due to change in temp.of different materials other than metals.

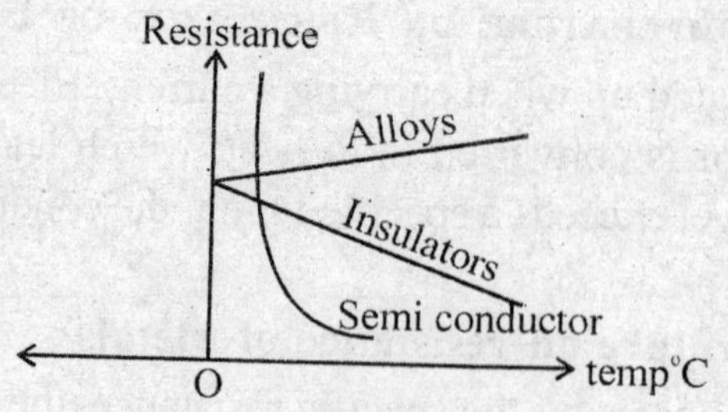

Fig 1.21 Change in resistance due to change in temp.

(e) Super conductors

At very low temperatures, resistance of some materials becomes exactly zero. Such materials are called **super conductors**. The resistance of mercury below 4.2 kelvin becomes zero.

Metals have a **positive temperature coefficient** of resistance

Resistance of some materials such as semi conductors decreases as the temperature increases. These have a **Negative temperature coefficient** of resistance

Example 1.5 : The current in a 2 Ω resistor increases linearly from zero to 5 amp. in 1 ms. At t = 1^+ ms , the current is again zero. and it increases linearly to 5 amp at t = 2 ms. This pattern repeats each 1 ms. Sketch the corresponding voltage drop 'v' in the resistance.

Solution :

Since $v = R\,i$, the maximum voltage is = $(2)\times(5) = 10$ Volts.

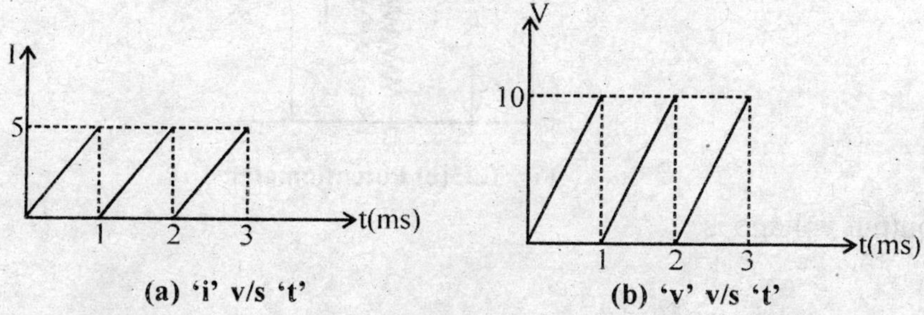

(a) 'i' v/s 't' (b) 'v' v/s 't'

Fig. 1.22

The plots of 'i' and 'v' are shown in Figures 1.22 (a) and 1.22(b) and it is seen that these are of identical nature.

(f) Insulators

Insulators are materials which do not permit flow of current. These are used to prevent the flow of current at places where it is not required.

1.6.1.4 Rheostat and Potentiometers

A wire wound coil with two fixed terminals is called a resistor.

A coil with two fixed terminals and a variable contact terminal which makes contact with the body of the coil is called a rheostat or a variable resistor.

(a) Rheostat

If the moving end ,C is very near the starting terminal, say near terminal A , as shown in Fig. 1.23(a) , the resistance offered by the rheostat will be minimum as .If the moving end is nearer to the farthest end terminal as shown in Fig. 1.23(b), resistance will be maximum.

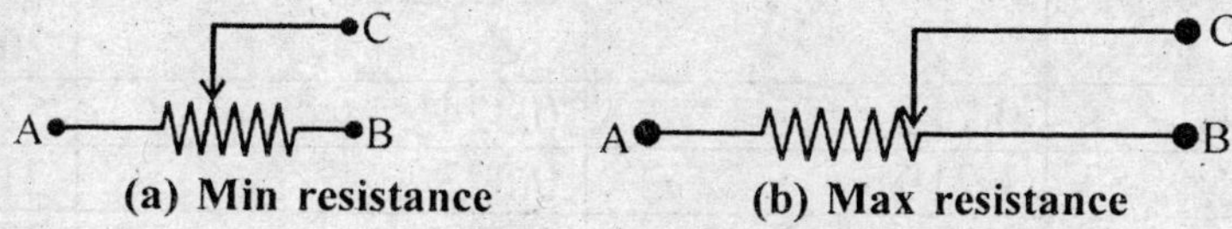

(a) Min resistance (b) Max resistance

Fig.1.23 : Rheostat

(b) Potentiometer

It is a resistor with two fixed end terminals and one moving contact. The two ends of the rheostat are connected across a voltage source which constitutes the input to the potentiometer.

The output is tapped between the moving contact and one of the end terminals as shown in Fig.1.23(c)

A single resistor with end terminals A and B having a resistance R is shown in the figure1.23(c). Terminal 'C' is the moving or adjustable contact.

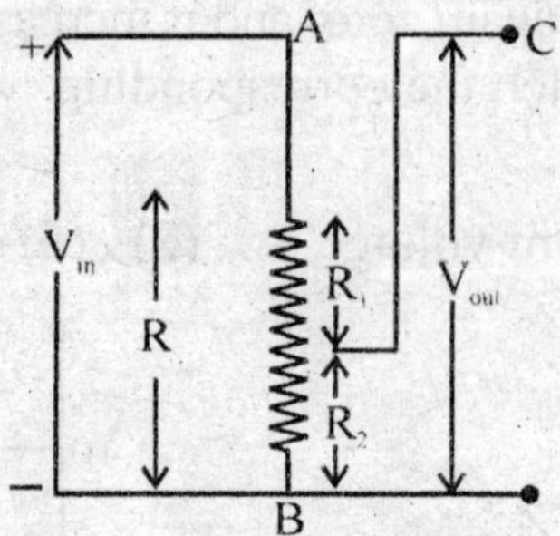

Fig. 1.23(c) Potentiometer

The output voltage is:

$$V_{out} = V_{in}\frac{R_2}{R}$$

where R is the total resistance and R_2 is the resistance of the tapped part. Output voltage is adjustable from zero to maximum input voltage i.e. V_{in}

1.6.1.5 Wire Gauge

Commercial wire sizes are standarised in accordance with the area of cross section of a wire. Each wire size is given a number for identification and each has a maximum allowable current carrying capacity. SWG i.e. Standard or Sterling Wire Gauge is a standard British system of specifying wire size. Smaller SWG corresponds to a larger wire diameter.

Table 1.4 gives some wire gauges and corresponding diameter and resistance of copper wire. Diameter information applies to **solid** wires. Sizes of **stranded** wires are calculated by calculating the equivalent cross sectional area.

Table 1.4 Some Wire Gauges and Resistance of Copper Wire

Conductor size (SWG)	Bare Dia(mm)	Cross sectional area mm^2	Resistance in Ohms per km at 20^0 C
22	0.711	0.324	52.8
18	1.219	0.823	21.0
14	2.032	2.08	8.26
12	2.642	3.31	5.20
6	4.877	13.3	1.29
4	5.893	21.1	0.813

1.6.1.8 Color Code of Resistors

Because resistors used in electronics are physically small, it is inconvenient to print their values on the side. For easy identification of resistance value and tolerance , carbon resistors are color coded. The color bands are painted near one end of the resistor as shown in Fig. 1.24

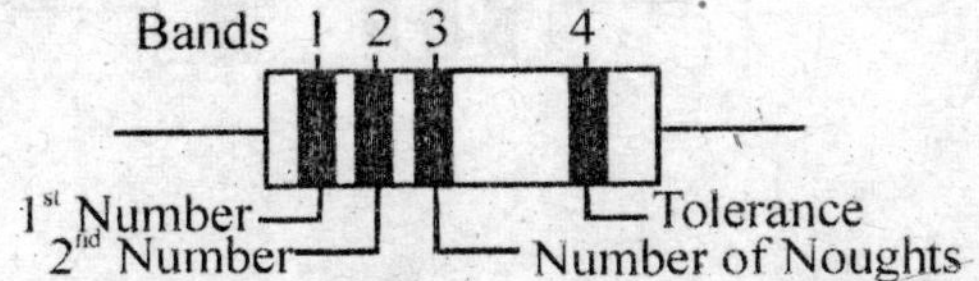

Fig.1.24 Color Coded Bands for Carbon Resistances

The first two color bands indicate the first and second digits of the resistance value and the third band indicates the number of noughts/zeros.

Value of a resistance can be determined by using the relationship, R= (a b)10^c

Table 1.5 COLOR CODE for Carbon Resistances

Color	Value of a,b,c
Black	0
Brown	1
Red	2
Orange	3
Yellow	4
Green	5
Blue	6
Violet	7
Gray	8
White	9

The fourth band indicates tolerance and is always either silver or gold in colour .

Silver indicates $\pm$ 10% tolerance and gold indicates 5% tolerance. If fourth band is black or it not present , default value of tolerance is 20%

Table 1.6 Tolerance Value For Carbon Resistances

Color in the fourth band	Tolerance value (%)
Black or no color	± 20
Silver	± 10
Gold	± 5

Note : The color code can be remembered as: **BBROY** of **G**reat **B**ritain has a **V**ery **G**racious **W**ife.

The procedure of using the colour code is explained in the example given below.

Example 1.6 : First two bands of a color coded resistor are blue and grey. Third band is green and fourth band is **gold**. Find the value of resistance.

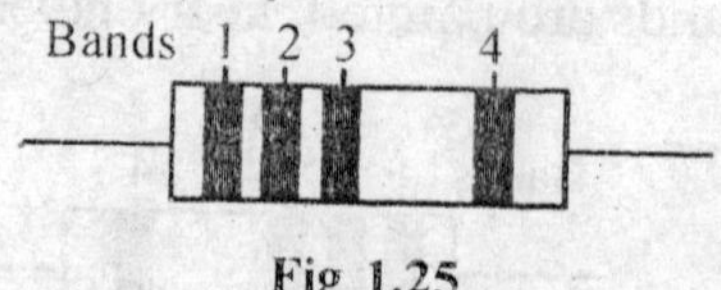

Fig 1.25

Solution :

Referring to Table 1.5, first two colour bands i.e. Blue, Grey give the first two digits of the numerical value of resistance i.e. a = 6 and b = 8 because blue has a value of 6 and grey has a value of 8

Third band colour gives the number of zeros i.e. Green corresponds to five zeros or c = 5, which means 10^5 and the fourth band colour gives tolerance value i.e. Gold means 5 % tolerance

Hence, Resistance value is 6800000 or 68×10^5 ohms i.e. 6.8 Mega ohms with 5 % tolerance.

1.7 Inductance and Capacitance Element

Several types of transducers are based on inductance and capacitance elements. One type of microphone is basically a capacitor in which the capacitance changes with sound pressure.

Some times an electric signal which represents a physical variable is noisy. For example in an electronically controlled suspension of an automobile, the position sensors are affected by road roughness and loading of the vehicle. To obtain an electrical signal representing the displacement of each wheel , the rapid fluctuations due to road roughness must be eliminated. This can be done using inductance and capacitance elements in circuits called filters .

Resistors convert electric energy into heat whereas inductors and capacitors store energy.

1.7.1 The Inductance Parameter

We know that an emf is induced in a conductor / coil if it moves through a magnetic field and the growth of current in a coil can induce an emf in another coil. A voltage can be induced in the coil itself as its current level changes. This effect is called ' self inductance' of a conductor or a coil.

Self inductance of a coil sets up an induced emf which opposes the external applied voltage which drives current through the coil. As this emf is in opposition to the supply voltage, it is called **counter emf or back emf**.

Inductance can be characterized as that property of a circuit element by which energy can be stored in a magnetic flux field. **It makes itself felt in a circuit only when there is a changing current.** A circuit element may have inductance by virtue of its geometrical and magnetic properties, however , its presence in the circuit is exhibited only when there is a time rate of change of current.

An inductor is constructed by coiling a wire around some type of former. The coil former is composed of a magnetic material such as iron or iron oxides which have high permeability and therefore increase the magnetic flux for a given value of current.

1.7.1.1 Circuit View-Point

The current-voltage terminal relationship for the inductance parameter is given by :

$$v_L = \boxed{L\frac{di}{dt}} \tag{1.5}$$

where v_L is the voltage across inductance, L

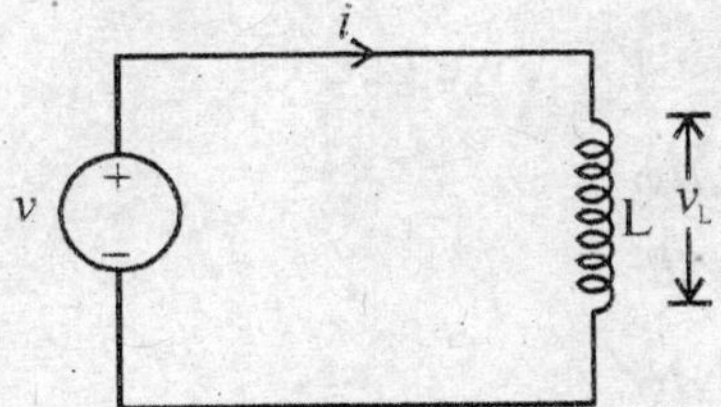

Figure 1.26 Inductance parameter

Any circuit element that exhibits the property of inductance is called an **inductor .**

An Ideal inductor is considered to have no resistance. However, practically it always has some resistance due to the wire of the inductor coil.

Inductance may be expressed as:

$$L = \frac{v_L}{(di/dt)} \text{ volt-sec / amp or henrys} \tag{1.6}$$

Equation (1.6) can be rewritten as :

$$di = \frac{1}{L}\, v_L\, dt$$

Integration of this equation ,gives:

$$\int_{i(0)}^{i(t)} di = \frac{1}{L}\int_0^t v_L\, dt$$

or

$$\boxed{i(t) = \frac{1}{L}\int_0^t v_L\, dt + i(0)} \tag{1.7}$$

where i(0) is the initial current in the coil.

Equation (1.7) shows that the current in an inductor at any time't', depends upon the integral of the voltage across its terminals as well as initial current in the coil.

It also indicates that the current in an inductor cannot change abruptly in zero time i.e. in zero time the contribution from the integral term is zero , so that current is same before and after application of voltage.

1.7.1.2 Geometrical View-Point

The voltage drop across the terminals of an inductor may be expressed using Faraday's law of electromagnetic induction in terms of. rate of change of flux linkages.

Accordingly, we may write :

$$V_L = N\frac{d\phi}{dt}$$

Substituting the value of 'V_L' , we get:

$$L\frac{di}{dt} = N\frac{d\phi}{dt}$$

or

$$L = N\frac{d\phi}{di} \qquad (1.8)$$

If the flux ϕ is directly proportional to current 'i' , equation(1.8), becomes:

$$L = \frac{N\phi}{i} \text{ weber-turns/amp} \qquad (1.9)$$

However, we know that the magnetic flux is given by :

$$\phi = \frac{MMF}{magnetic\, reluctance} = \frac{Ni}{\Re} \qquad (1.10)$$

Where MMF. denotes the magneto-motive-force which produces the flux ϕ in the magnetic circuit having a reluctance .

If an inductor consists of N turns wound around a circular iron core having a mean length of 'l' metres and a cross-sectional area of 'A' metres^2, it is known that the magnetic reluctance is:

$$\Re = \frac{1}{\mu A}$$

Substituting ϕ from equation (1.10) in equation(1.9) , the expression for the inductance parameter becomes

$$L = \frac{N}{i}\frac{Ni}{\Re} = \frac{N^2}{\Re} \qquad (1.11)$$

Substituting the value of reluctance, in to equation(1.11), we get:

$$L = \frac{N^2 \mu A}{l} \qquad (1.12)$$

It is seen that the value of inductance depends upon the physical dimensions and the magnetic properties of the medium. Its value may be increased in any one of following ways :

- Increasing the number of turns.
- Using an iron core of higher permeability.
- Reducing the length of the iron core, and
- Increasing the cross-sectional area of the iron core.

1.7.1.3 Uses of Inductors

Inductors are used in many applications. Some of these are:

(i) Inductors are used in the form of solenoids for operating gate valves to control the fluid flow etc.

(ii) Used in electromagnetic relays for operating the control circuits.

(iii) Used between the transmission lines and the Earth for line compensation.

(iv) Used in Filter Circuits.

(v) Used in Wave Shaping circuits to obtain a desired wave form

(vi) In inductive transducers, as in the case of Linear Variable Differential Transformer (LVDT).

(vii) In Measurement Circuits.

Example 1.7 : An inductance of 4.0 mH has a current i = 10.0 $(1 - e^{-200t})$ amp. Find the corresponding voltage.

Solution

Given that

$$i = 10.0 - 10\, e^{-200t}$$

therefore $$\frac{di}{dt} = -10(-200)\, e^{-200t} = 2000\, e^{-200t}$$

$$v = L\frac{di}{dt} = 4 \times 10^{-3}\,(-10 \times e^{-200t})(-200) = 8.0\, e^{-200t} \text{ volts}$$ **Ans**

In Figure 1.27 (a) and (b), the plots of 'i' and 'v' are given.

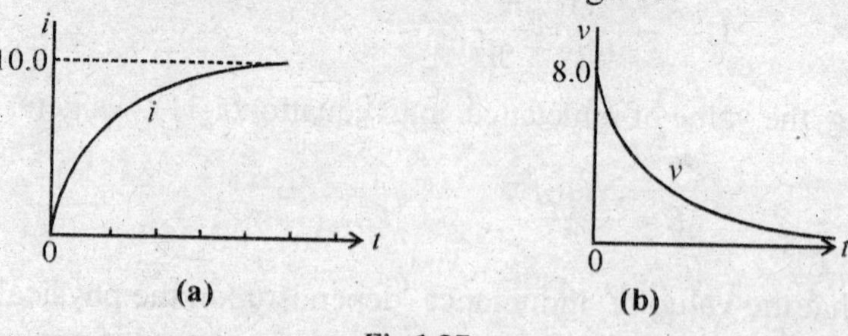

Fig.1.27

1.7.2 The Capacitance Parameter

Two conducting plates having equal and opposite charges, separated by a distance through any insulating material, constitute a capacitor.

Capacitors are constructed by separating two sheets of conductor, which are usually metallic, by a thin layer of insulating material. The insulating material between the plates is called a dielectric, which can be Air, Mylar, Polyester, Polypropylene, Mica etc.

Capacitance can be characterized as that property of a circuit element in which energy is capable of being stored in an electric field. The capacitance value of a capacitor is a measure of the amount of electric charge that can be stored in the element.

Its influence in an electric circuit is manifested only whenever a changing potential difference exists across the terminals of the capacitor.

1.7.2.1 Circuit View-Point

Capacitance is the proportionality factor relating the charge 'q', between two metal surfaces (or conductors), to the corresponding potential difference 'v_C' existing between them.

i.e.

$$\boxed{q = C\, v_c} \tag{1.13}$$

However, $i = \dfrac{dq}{dt} = C\dfrac{dv_C}{dt}$

Any circuit element which shows the property of providing a current which is directly proportional to the rate of change of the voltage across its terminals is called a capacitor.

A capacitor usually consists of large metal surfaces separated by small distances.

Thus equation given above may be written as:

$$C = \frac{i}{(dv_C / dt)} \quad \text{Faradays} \tag{1.14}$$

To express the capacitor voltage as a function of its current, Equation (1.14) may be rearranged as :

$$d\, v_C = \frac{1}{C}\, i\, dt \tag{1.15}$$

Integrating both sides of equation (1.15), we get:

$$\int_0^t d\,v_C = \frac{1}{C}\int_0^t idt$$

i.e. $$v_C(t) = \frac{1}{C}\int_0^t idt + v_C(0) \qquad (1.16)$$

The quantity v_C (0) denotes the initial voltage appearing across the capacitor plates. If there is no initial voltage on the capacitor, Equation (1.16) becomes

$$\boxed{v_C = \frac{1}{C}\int_0^t i\,dt \;\; Volts} \qquad (1.17)$$

Equation (1.17) or (1.16) show an important property of capacitance i.e. the voltage across a capacitor cannot change abruptly.

We know that:

$$i = C\frac{dv_C}{dt}$$

An abrupt change in capacitor voltage v_C in zero time gives a value of infinity for dv_C/dt. i.e. the capacitor current is infinite which is physically not possible.

According to Equation (1.17)

i.e. $$v_C = \frac{1}{C}\int_0^t idt$$

For any finite change in capacitor current, however large, the integral value for t = 0 will always be zero. Hence the capacitor voltage cannot change instantaneously.

1.7.2.2 Geometrical View-Point

When a capacitor is charged , an electrostatic field is established between the conducting plates. The concentration of electric lines of force increases with the presence of dielectric. The ability of a dielectric material to concentrate electric lines of force between plates of the capacitor is called **dielectric constant or relative permittivity of the material.**

Electric permittivity is analogous to magnetic permeability. It specifies the ease with which electric flux can pass through a dielectric material.

The amount of charge which accumulates on the plates of a capacitor is directly proportional to the applied voltage as well as to the capacitance of the capacitor:

i.e. $q = C\ v_C$

The electric flux per unit area at a given cross section in an electric field is called **Electric flux density, D** and is given by :

$$\boxed{D = \varepsilon E} \tag{1.18}$$

Where $D = \dfrac{q}{A}$ coulomb/m²

Multiplying both the sides of equation(1.18), by A, we get:

$$D\,A = \varepsilon\, EA$$

i.e. $$q = \varepsilon\, EA \tag{1.19}$$

where ε is the dielectric constant or permittivity of the material between the plates of the capacitor , A is the area of cross section of the plates

Let us consider a parallel plate capacitor with plates separated by a distance of 'd' metres .The electric field intensity or voltage gradient or electric field strength E , in the space between the plates of a capacitor depends on the applied voltage and the distance between the plates and is given by :

$$E = \frac{v_c}{d} \tag{1.20}$$

Substituting this value of E in (1.19), we get :

$$q = \varepsilon AE = \varepsilon A\frac{v_C}{d} = \left(\frac{\varepsilon A}{d}\right) v_C \tag{1.21}$$

This expression for charge involves the physical dimensions of the capacitor as well as the capacitor voltage.

Comparing the equations (1.21) and (1.13), we get:

$$\boxed{C = \varepsilon\frac{A}{d}} \tag{1.22}$$

This equation defines a capacitor in terms of its geometrical configuration and the physical property of the material which lies between its two metal surfaces.

Equation (1.22) shows that the capacitance parameter is directly proportional to the dielectric constant of the material, inversely proportional to the spacing between the metal surfaces, and directly proportional to the metal surface area.

For vacuum or free space, the dielectric constant is $\varepsilon_0 = 8.84 \times 10^{-12}$ F/m

For other materials , the dielectric constant is

$$\varepsilon = \varepsilon_r\, \varepsilon_0$$

where ε_r is the relative permittivity of the material . Relative Permittivity is also known as dielectric constant.

Hence value of capacitance can be obtained by rewriting equation (1.22) as follows:

$$C = \frac{\varepsilon_r \varepsilon_0 A}{d} \text{ Farads} \tag{1.23}$$

Farad is a very large unit. Hence smaller units like microfarads (μF), nano farads (nF), picofarads (pF) , are frequently used units for capacitance.

Values of relative dielectric constant of some materials are given in Table 1.7

Table 1.7 : Relative Dielectric Constant/Relative Permittivity of Some Materials

Insulating Material	Typical Dielectric Constant / Relative Permittivity
Vacuum	1.0
Air	1.0006
Diamond	5.5
Mica	4.0 - 8.0
Polyester	3.4
Quartz	4.3
Mylar	3
Oxide film	5-25
Teflon	2
Bakelite	4.5-5.5
Rubber	2-3.5
Paper	3-6
Polythene	2-2.5
Insulating oil	3-4
Glass	5-100
Ceramic(high ε_r)	>1000
Ceramic (low loss)	6-20
Porcelain	5
Titanates	100-200

1.7.2.3 Uses of Capacitors

Capacitors are used in many applications. Some of these are:

(i) To tune signals in filter circuits of Radio and TV, etc.

(ii) In operational amplifiers used for integration, differentiation operations, etc.

(iii) In wave shaping circuits to obtain a desired wave form.

(iv) In capacitive transducers, as in a microphone.

(v) For Power Factor improvement

(vi) For compensation of transmission lines to improve the voltage profile

(vii) In Measurements Circuits.

Example 1.8 : A 2 μ F parallel plate capacitor with air as the dielectric medium is charged to a potential of 100 volts. The distance between the two plates is 1 cm. Calculate the electric stress with in the dielectric.

Solution :

Given that

$$d = 1 \text{ cm}, \; C = 2\times10^{-6} \text{ Farad}$$

Therefore

Electric stress with in the dielectric,

$$E = \frac{V}{d} = \frac{100}{0.01} = 10\times10^{3} \text{ volts/m}$$

$$= 10 \text{ kV/m}$$ **Ans**

Example 1. 9 : Calculate the capacitance in a parallel plate capacitor which consists of two metal plates each having an area of cross section 50 cm^2, separated by a dielectric of 2 mm thickness and $\varepsilon_r = 4$, if potential difference of 2000 volts is applied across it.

$$\varepsilon_0 = 8.854\times10^{-12} \text{ F/m}$$

Solution :

Area of each plate =50 cm^2

Thickness of dielectric = 2 mm = 2×10^{-3} m

Pot. Difference across capacitor =2000 volts

Capacitance, $$C = \frac{\varepsilon_0\varepsilon_r A}{d} = \frac{8.854\times10^{-12}\times4\times50\times10^{-4}}{2\times10^{-3}}$$

$$= 8.854\times10^{-16}\times10^{3}\times10^{2}$$

$$C = 8.854\times10^{-11} = 88.54 \text{ µF}$$ **Ans**

Example 1.10 : A capacitance of 50.0 μF has a voltage described as follows :

For $0 < t < 2$ ms,

$v = 20.0 \times 10^{3}\ t$ volts.

Sketch i, and v for the given time interval

Solution :

For $0 < t < 2$ ms,

$$i = C\frac{dv}{dt} = 50 \times 10^{-6} \times \frac{d}{dt}(20.0 \times 10^{-3} t) = 1.0 \text{ amp}$$

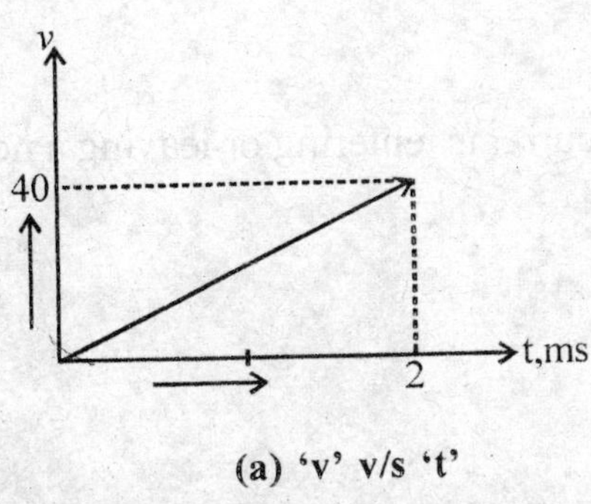

(a) 'v' v/s 't'

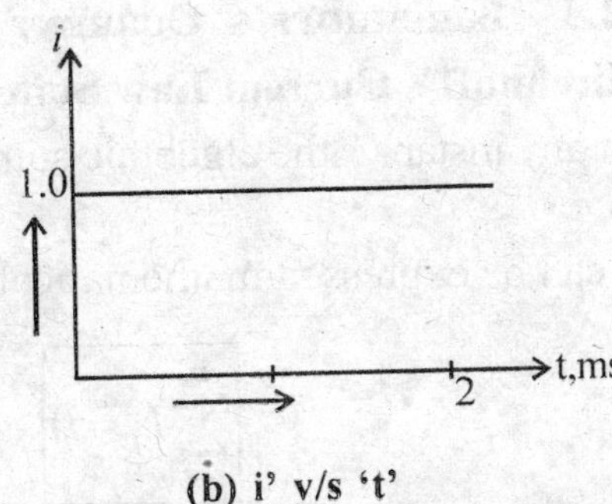

(b) i' v/s 't'

Fig.1.28

1.8 Basic Circuit Laws

Basic concepts of voltage , current and power in electric circuits have been discussed so far. However to determine their values in actual circuits , certain fundamental laws have to be understood. These laws are known as Ohm's law and Kirchhoff's law . In addition to these laws, some techniques which are commonly used for the analysis and design of electrical circuits such as combining resistances is series and parallel, voltage and current division, delta to wye and wye to delta transformation are very important.

1.8.1 OHM'S Law

Resistance represents the current resisting behaviour of a material. Current and Voltage relationship for a resistance is given by Ohm's law.

Ohm's law statement :

Ohm's law states that the potential difference 'V' across the ends of a conductor is directly proportional to the current 'I' flowing through the conductor, temperature being constant.

i.e. $V \propto I$

or $V = R\ I$

where R is constant of proportionality and is defined as resistance of the conductor or $$I = \frac{V}{R}$$

'V' and 'I' may be constant i.e. DC or may be time varying i.e. AC

However this law is not applicable to unilateral circuits consisting of electronic tubes or transistors and non linear elements such as powdered carbon, electric arc etc.

1.8.2 Kirchhoff's Laws

Ohm's law is not sufficient to analyse different electric circuits. However combined with Kirchhoff's Current Law , abbreviated as KCL and Kirchhoff's Voltage Law, abbreviated as KVL , a large variety of circuit problems can be solved.

1.8.2.1 Kirchhoff's Current Law

Kirchhoff's Current Law Statement :

At any instant , the algebraic sum of all the currents entering or leaving a node/junction is zero.

It can be expressed mathematically as :

$$\boxed{\sum_{i=1}^{p} I_j = 0} \qquad (1.24)$$

Where 'p' is the number of circuit elements meeting at the node or connected to the node under consideration and I_j is the current in the jth element.

To apply this law, we need a sign convention.

The sign convention that we will use is :

Currents **leaving** the node will be taken as **positive** and **entering** the node as **negative.** However the opposite convention is also valid .

To illustrate this law , consider the four resistances connected at node 'O' shown in Fig.1.29. Currents i_1 and i_3 are leaving the node 'O' and i_2 and i_4 are entering the node 'O'.

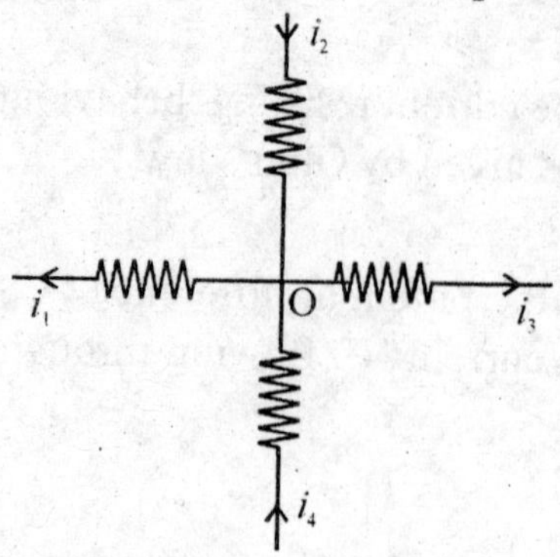

Fig.1.29 Kirchhoff's Current Law

Applying KCL at node 'O' , we get:

$i_1 - i_2 + i_3 - i_4 = 0$

It may be noted that +ve sign has been taken for the outgoing currents i_1 and i_3 and negative sign has been taken for the incoming currents i_2 and i_4

or $i_1 + i_3 = i_2 + i_4$

Kirchhoff's current law can also be stated as :

The sum of currents entering a node is equal to the sum of currents leaving a node.

Example 1.11 : Fig1.30 shows a circuit diagram, consisting of seven resistances. Currents through resistances R_2, R_3 and R_5 are known. Find the magnitude and direction of the currents in other resistances using Kirchhoff's current law.

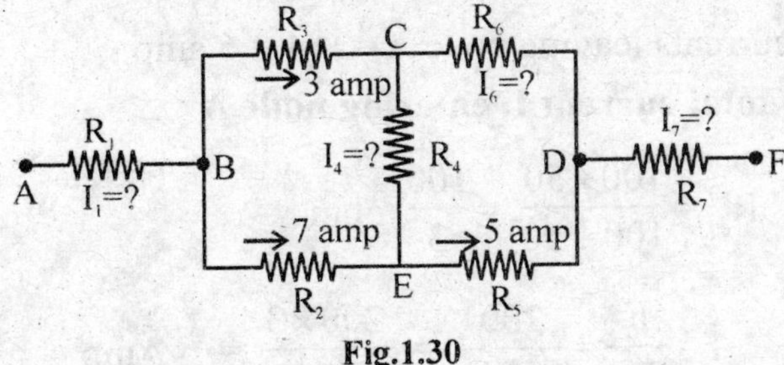

Fig.1.30

Solution :

Applying Kirchhoff's current law at different nodes of Fig. 1.30, we get unknown currents as follows:

At Node B, current I_1 = 7 + 3 =10 amp. **entering** node B (it is equal to the sum of currents leaving node B)

At Node E , current I_4 = 7-5 = 2 amp. **leaving** node E (it is equal to the sum of currents entering node E)

At Node C, current I_6 = 2 + 3 = 5 amp. **leaving** node C (it is equal to the sum of currents entering node C)

At Node D, current I_7 = 5 + 5 =10 amp. **leaving** node D (it is equal to the sum of currents entering node D)

Example 1.12 :

Verify KCL for the circuit shown in Fig.1.31 given below:

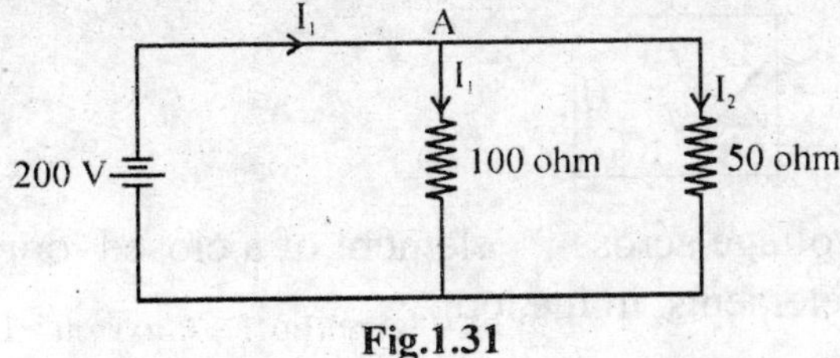

Fig.1.31

Solution :

To verify KCL, we have to prove that:

Sum of the currents entering node A = Sum of the currents + leaving node A

i.e. $I = I_1 + I_2$

Calculation of currents leaving node A

$$I_1 = \frac{V}{R_1} = \frac{200}{100} = 2 \text{ Amp}$$

$$I_2 = \frac{V}{R_2} = \frac{200}{50} = 4 \text{ Amp}$$

Hence sum of currents leaving node A =2+4 = 6 amp.

Calculation of total current I, entering node A

$$R_{eq} = \frac{100 \times 50}{100 + 50} = \frac{100}{3} \text{ ohm}$$

$$\therefore \quad I = \frac{V}{R_{eq}} = \frac{200}{100/3} = \frac{200 \times 3}{100} = 6 \text{ Amp}$$

It is seen that the sum of currents entering node A is equal to the sum of currents leaving node A, which is equal to 6 amp.

Hence KCL is verified.

1.8.2.2 Kirchhoff's Voltage Law

Some of the voltages in the network are due to sources while other voltages are due to the flow of current through passive elements such as resistances etc., known as voltage drops.

Kirchhoff's voltage law statement :

At any instant of time, the algebraic sum of all the voltages around any closed path or loop in a network is zero.

This law can be expressed mathematically as :

$$\boxed{\sum_{i=1}^{r} V_j = 0} \qquad (1.25)$$

Where V_j is the voltage across j^{th} element of a closed loop and 'r' is the number of voltages or number of elements, in the loop.

To implement this law, we need a sign convention. The sign convention that we will follow is:

The terminal where the current enters an element is marked as +ve and the terminal where the current leaves an element is marked as –ve as shown in Fig 1.32(a) and (b)

(a) Current flows from terminal A to B **(b)Current flows from terminal B to A**

Fig 1.32Assignment of voltage polarities

If we move from terminal marked as +ve to terminal marked as –ve, it is called voltage drop. If we move from terminal marked as -ve to terminal marked as +ve, it is called voltage rise.

In the case of voltage sources :

If we move from +ve terminal of a source to a –ve terminal, it is called voltage drop

If we move from -ve terminal of a source to a +ve terminal, it is called voltage rise

A voltage drop is treated as positive value and a voltage rise as a negative value. However the opposite convention is also valid .

For illustration of KVL, consider the circuit /loop shown in Fig.1.33

Traverse this loop in the clockwise direction, starting from below the source voltage V and apply KVL to the circuit .

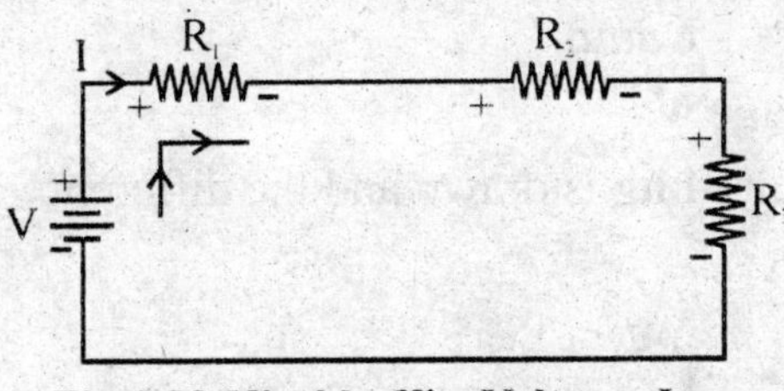

Fig.1.33 Kirchhoff's Voltage Law

We get :

$$-V + I R_1 + I R_2 + I R_3 = 0$$

It may be noted that we have encountered one voltage rise of the voltage sourcen V i.e. negative sign and three voltage drops across three resistances having positive sign , while traversing the loop.

This method for writing KVL equations consists of moving mentally around the closed path in a clockwise direction . The sign of each voltage term corresponds to the polarity of the terminal encountered as per sign convention .

Kirchhoff's voltage law can also be stated as :

Sum of voltage drops in a closed circuit = sum of voltage rises in a closed circuit

Example 1.13 : For the circuit shown in Fig.1.34, determine loop currents I_1 I_2 and voltage V_{ag}

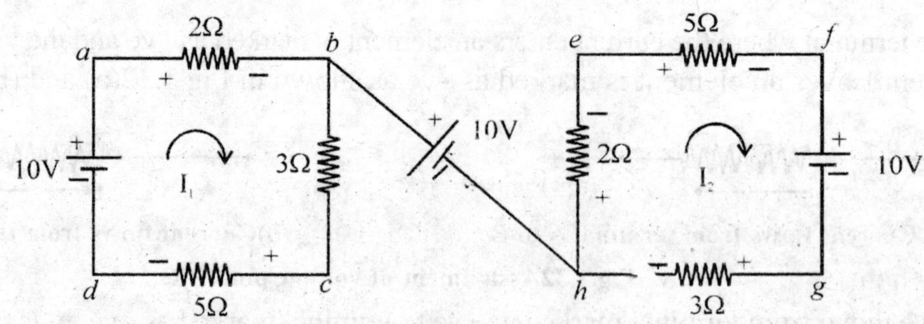

Fig.1.34

Solution :

No current flows in the branch ' bh'.

Applying KVL to the closed loop abcda,we get:

$$(2+3+5)\ I_1 = 10$$

or $$10\ I_1 = 10$$

∴ $$I_1 = 1 \text{ amp.}$$ **Ans**

Applying KVL to the closed loop efghe, we get:

$$(3+2+5)\ I_2 + 10 = 0$$

or $$10\ I_2 + 10 = 0$$

∴ $$I_2 = -1 \text{ amp.}$$ **Ans**

Calculation of Voltage V_{ag}

A portion of the circuit **abhg** is drawn and the different voltages are marked as shown in Fig. given below:

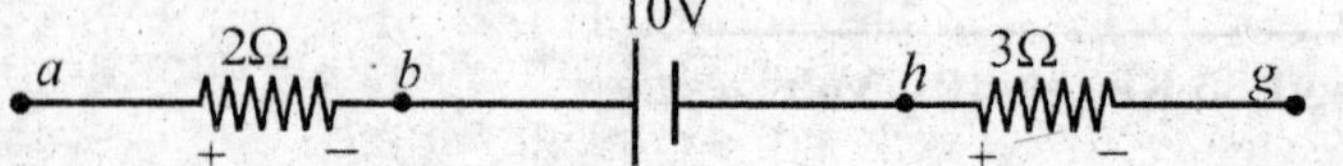

It can be seen that , voltage V_{ag} = 2+10+3 =15 volts **Ans**

Example 1.14 : For the circuit shown in Fig.1.35(a), determine current I , using KVL

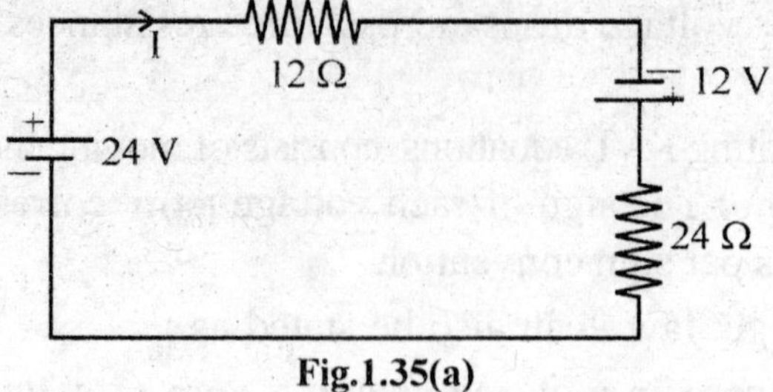

Fig.1.35(a)

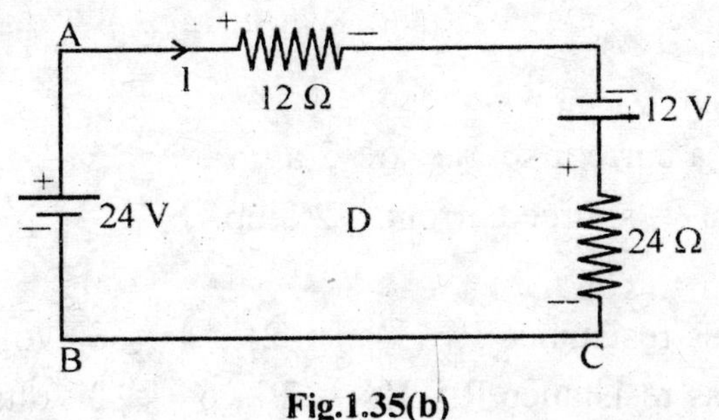

Fig.1.35(b)

Solution :

Step 1

Mark the polarities of voltage drops in different resistances as shown in Fig 1.35(b).

Step2

Trace the circuit ABCDA in the clockwise direction and equate the sums of voltage drops and voltage rises.

Sum of voltage drops $= 12 \times I + 24 \times I$ (i)

Sum of voltage rises $= 12 + 24 = 36$ (ii)

Using KVL i.e. equating equations (i) and (ii), we get:

$$36 \times I = 36$$

or $$I = \frac{36}{36} = 1 \text{ Amp}$$ **Ans**

1.8.2.2.1 Presence of a Current Source in a Loop

If there is an ideal current source in a single loop circuit , the current in the loop is taken to be equal to source current. Voltage across the current source can be obtained by the application of KVL for the loop.

Example 1.15 : For the circuit shown in Fig.1.36, determine(a) loop current (b) voltage drops across all resistances (c) voltage drop across the current source

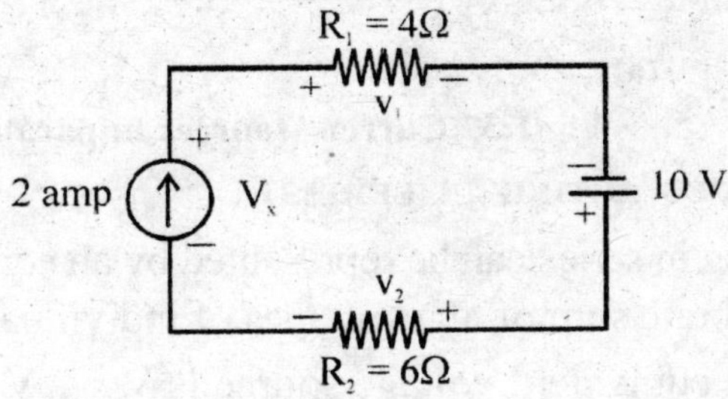

Fig. 1.36

Solution :

(a) Loop current

The loop contains a current source of 2 amp.

Hence loop current = source current = 2 amp. **Ans**

(b)Voltage drops

Voltage drop across resistance R_1, V_1 = 2 × 4 = 8 volts **Ans**

Voltage drop across resistance R_2 V_2 = 2 × 6 =12 volts **Ans**

(c) Voltage across the current source is assumed to be ' and its polarity is assumed as shown in Fig. 1.36

Applying KVL, we get:

$$-V_x + (4\times2) - 10 + (6 \times 2) = 0$$

i.e. $V_x = 10$ volts

Voltage drop across the current source = 10 volts **Ans**

1.9 Circuit Simplification Techniques

Complex circuits can be simplified using some of the techniques described below.

1.9.1 Current Sources acting in parallel

Current sources acting in parallel can be represented by an equivalent current source whose value is equal to the algebraic sum of the currents of individual sources.

Assume the currents in the upward direction as positive and in the downward direction as negative.

For Fig 1.37(a), the equivalent current source, $I = I_1+I_3-I_2$ and is shown in Fig 1.37(b)

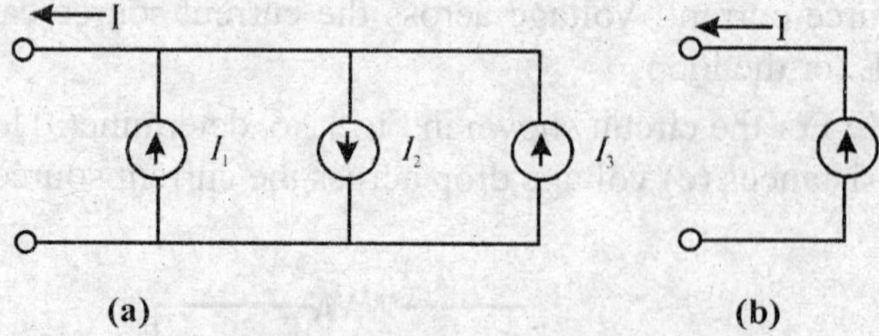

Fig 1.37 Current sources in parallel

1.9.2 Voltage Sources acting in series

Voltage sources acting in series can be represented by an equivalent voltage source whose value is equal to the algebraic sum of the voltages of individual sources.

For Fig 1.38(a) , the equivalent voltage source is $V_{ab} = V_1 + V_2 - V_3$ and is shown in Fig 1.38(b)

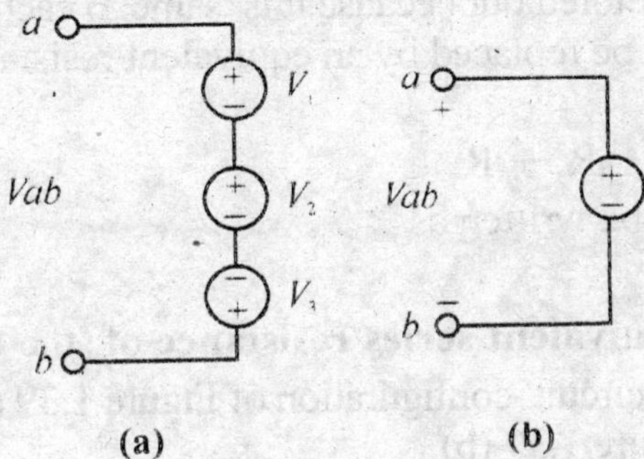

Fig 1.38 Voltage sources in series

This law applies equally well to circuits driven by constant voltage sources and time dependent voltage sources .

Frequently two or more resistances are connected in such a way that these can be combined and represented by a single resistance. However , most commonly encountered configurations of resistances are of the series and parallel type.

1.9.3 Resistances in Series

Circuit elements that carry the same current are said to be in series. In other words , there is only one path for the flow of current.

This is possible if different resistors, R_1, R_2 and R_3 etc. are joined end-on-end basis as shown in Figure 1.39. For a connection to be a series connection, there must not be any branches connected to any of the intermediate nodes.

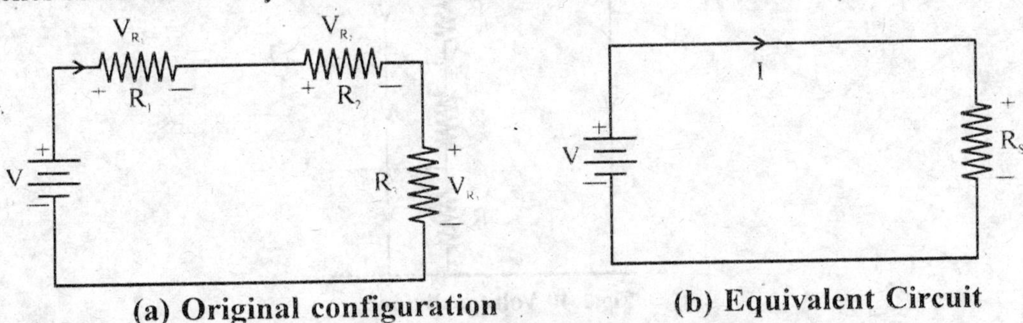

(a) Original configuration **(b) Equivalent Circuit**

Figure 1.39 Resistances in Series

The voltage drop across each element is different as their resistance is different

Applying Kirchhoff's voltage law to the circuit shown in Fig.1.39(a), we have :

$$(V_{R1} + V_{R2} + V_{R3}) - V = 0 \quad (1.26)$$

Substituting the values of V_{R1} , V_{R2} and V_{R3} we get:

or $\quad I\,R_1 + I\,R_2 + I\,R_3 - V = 0$

or $\quad V = I\,(R_1 + R_2 + R_3) \quad (1.27)$

Note that the current I is factored out because it is same in each resistance. Consequently, the quantity in parenthesis may be replaced by an equivalent resistance R_{se} ,which is given by :

$$R_{se} = R_1 + R_2 + R_3 \qquad (1.28)$$

Equation (1.27) may then be written as

$$V = I\,R_{se}$$

Where R_{se} denotes the **equivalent series resistance** of the circuit.

It follows that the original circuit configuration of Figure 1.39 (a) may be replaced by the equivalent circuit shown in Figure 1.39 (b)

In general, if there are '*n*' series-connected resistances in a circuit, the equivalent series resistance, R_{se} is obtained by taking the sum of the individual resistances.

i.e. $$R_{se} = R_1 + R_2 + \ldots\ldots + R_n = \sum_{j=1}^{n} R_j \qquad (1.29)$$

Net resistance of a number of resistances connected in series , is always greater than the largest resistance connected.

1.9.4 Voltage Division

When a voltage is applied to a series combination of resistances , a fraction of the voltage appears across each resistance and a set of series-connected resistors as shown in Figure1.40 is referred to as a voltage divider.

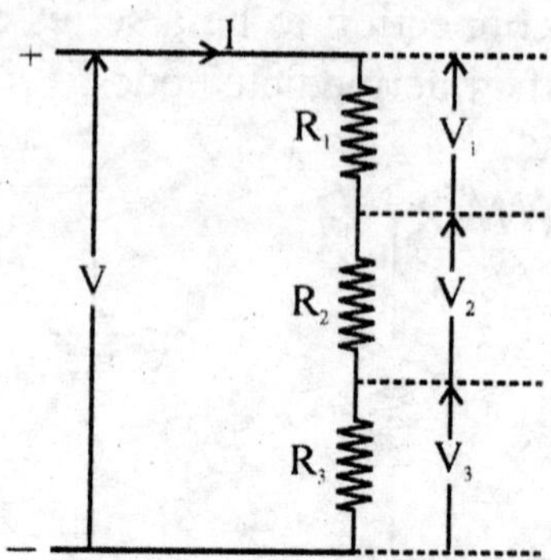

Fig.1.40 Voltage divider.

It can be seen that voltage across resistance R_1 is $V_1 = I\,R_1$ (i)

voltage across resistance R_2 is $V_2 = I\,R_2$ (ii)

voltage across resistance R_3 is $V_3 = I\,R_3$ (iii)

and

Total voltage, $V = I\,(R_1 + R_2 + R_3)$ (iv)

$$\therefore \qquad I = \left(\frac{V}{R_1 + R_2 + R_3}\right) \qquad (v)$$

Substituting equation (v) in to equations(i), (ii)etc., we get:

$$\therefore \quad V_1 = \left(\frac{V}{R_1 + R_2 + R_3}\right) R_1$$

$$V_2 = \left[\frac{V}{R_1 + R_2 + R_3}\right] R_2 \text{ and so on}$$

i.e. Voltage across any resistance is obtained by forming the ratio of that resistance to the total series resistance and multiplying this ratio by the voltage at the input terminals.

This is known as **voltage division principle.**

Example 1.16 : Three resistances R_1, R_2 and R_3 are connected in series to a 220 V supply as shown in the figure given below. Given $R_1 = 60\Omega$, $R_2 = 30\Omega$, current $I_1 = 2$ amp. Calculate R_3.

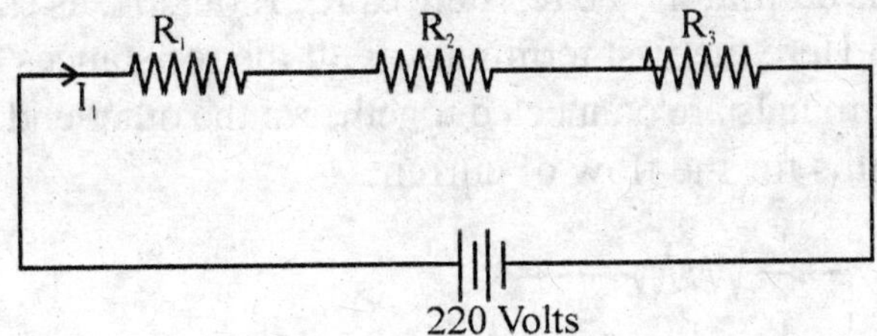

Fig. for Example 1.41

Solution :

$$\text{Equivalent resistance } R_{se} = \frac{V}{I_1} = \frac{220}{2} = 110\ \Omega$$

However, $\quad R_{se} = R_1 + R_2 + R_3$

$$\therefore \quad R_3 = R_{se} - R_1 - R_2$$

$$R_3 = 110-60-30 = 20\ \Omega \quad \text{Ans.}$$

Example 1.17 : A voltage divider circuit of two resistors is designed with a total resistance of the two resistors equal to 50.0 Ω. If the output voltage across one resistance is 10% of the total input voltage, obtain the values of the two resistors in the circuit.

Solution :

We know that:

$$\frac{\textit{output voltage across one resistance}}{\textit{total imput voltage}} = \frac{V_1}{V} = \frac{VR_1}{R_1 + R_2} \times \frac{1}{V} = \frac{R_1}{R_1 + R_2} = 0.10$$

However, $R_1 + R_2 = 50$ ohm

$\therefore$ $\dfrac{R_1}{50.0} = 0.10$

or $R_1 = 50 \times 0.10 = 5\ \Omega$ **Ans**

and $R_2 = 50 - 5 = 45.0\ \Omega.$ **Ans**

1.9.5 Resistances in Parallel

Circuit elements are also found in parallel combination. For example, in a house, all electric light bulbs are connected in parallel with respect to the source voltage. Other electric devices such as the electric iron and the electric heater when used simultaneously are also connected in parallel.

Circuit elements are said to be in parallel when the same potential difference or voltage appears across their terminals.

According to this definition, the resistances R_1, R_2 and R_3 as shown in Figure 1.42 (a)are connected in parallel. Here the first terminals of all the resistances are joined together at one end and the second terminals are connected together at the other end. In a parallel circuit, there are more than one paths for the flow of current.

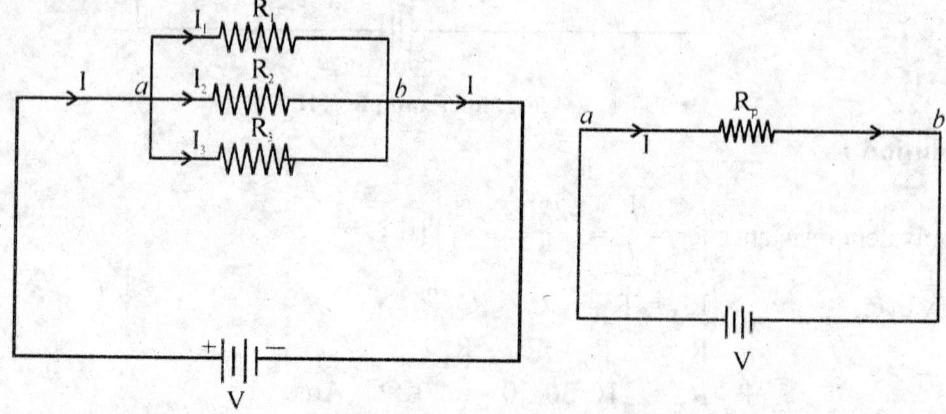

Fig.1.42(a) Resistances in parallel **1.42(b) Equivalent circuit representation**

Applying Kirchhoff's current law , we have the current entering the terminal/node 'a' as equal to the sum of the currents leaving this terminal (or node). Expressed in equation form, we have

$$I = I_1 + I_2 + I_3 \qquad \text{(i)}$$

But $I_1 = \dfrac{V}{R_1}$; $I_2 = \dfrac{V}{R_2}$; $I_3 = \dfrac{V}{R_3}$ (ii)

Substituting set of equations (ii) in to equation(i) , we get:

$$I = \frac{V}{R_1} + \frac{V}{R_2} + \frac{V}{R_3} \qquad \text{(iii)}$$

By factoring out the common variable V, we obtain

$$I = V\left(\frac{1}{R_1} + \frac{1}{R_2} + \frac{1}{R_3}\right) \qquad \text{(iv)}$$

The expression in parenthesis may be replaced by an equivalent value as :

$$\frac{1}{R_1} + \frac{1}{R_2} + \frac{1}{R_3} = \frac{1}{R_p} \qquad \text{(v)}$$

Where R_p denotes the equivalent resistance of the parallel combination of resistances. Upon substituting Equation (v) into Equation (iv), we obtain,

$$I = \frac{V}{R_p} \qquad \text{(vi)}$$

The equivalent circuit representation of Equation (vi) is shown in Figure 1.42 (b).

In general if 'n' resistances are connected in parallel, then their equivalent value can be obtained by the following equation:

$$\frac{1}{R_p} = \frac{1}{R_1} + \frac{1}{R_2} + \ldots\ldots \frac{1}{R_n} = \sum_{j=1}^{n} \frac{1}{R_j} \qquad (1.30)$$

i.e. Reciprocals of resistances add to get the reciprocal of the equivalent resistance.

If n = 2, we can write:

$$\frac{1}{R_p} = \frac{1}{R_1} + \frac{1}{R_2}$$

or

$$\frac{1}{R_p} = \frac{R_1 + R_2}{R_1 R_2}$$

i.e.

$$R_p = \frac{R_1 R_2}{R_1 + R_2}$$

i.e. When two resistances are connected in parallel , their equivalent resistance is equal to the product of the two resistances divided by the sum of the two resistances.

Since $G = \frac{1}{R}$ = Conductance, Hence , Equation (1.30) may also be written in terms of conductances as :

$$G_p = G_1 + G_2 + \ldots\ldots + G_n = \sum_{i=1}^{n} G_j \text{ (mhos)} \qquad (1.31)$$

Where G_p- is the equivalent conductance. It means that conductances in parallel add.

Note : 1.The net resistance of a number of resistances connected in parallel is smaller than the smallest connected resistance.

2.The net resistance of 'N' equal resistances each having a value equal to R, connected in parallel is $\frac{R}{N}$

1.9.6 Current Division

The total current flowing in to a parallel combination of resistances divides and a fraction of the total current flows through each resistance.

A parallel arrangement of resistors as shown in Figure 1.43 leads to division of the total current in different resistors. The ratio of the branch current I_1 to the total current I illustrates the concept of the current division.

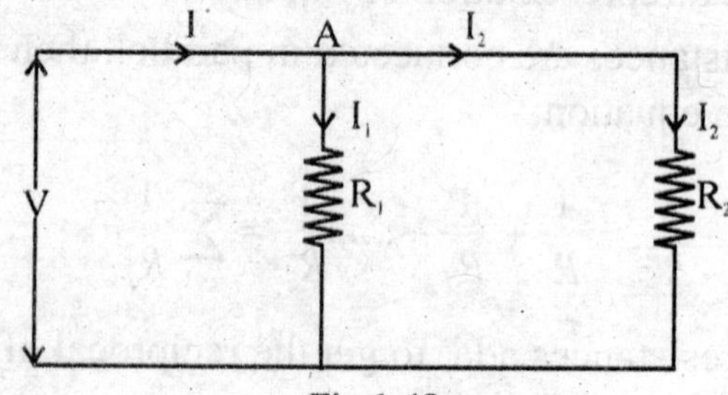

Fig.1.43

In Fig.1.43 , applying KCL at node 'A' , we get:

$$I = I_1 + I_2 \qquad \text{(i)}$$

However, $$I = \frac{V}{R_1} \text{ and } I_2 = \frac{V}{R_2} \qquad \text{(ii)}$$

Substituting the values of I_1 and I_2 , we get:

$$I = \frac{V}{R_1} + \frac{V}{R_2} = V\left(\frac{1}{R_1} + \frac{1}{R_2}\right) \qquad \text{(iii)}$$

Then from equations (ii) and(iii),

$$\frac{I_1}{I} = \frac{1/R_1}{1/R_1 + 1/R_2}$$

Or $$\frac{I_1}{I} = \frac{R_2}{R_1 + R_2}$$

i.e. $$I_1 = \left(\frac{R_2}{R_1 + R_2}\right) I$$

and similarly, $$\frac{I_2}{I} = \frac{R_1}{R_1 + R_2} \text{ i.e. } I_2 = \left(\frac{R_1}{R_1 + R_2}\right) I \qquad (1.32)$$

Hence, the ratio of the current in one branch(of two-branch parallel circuits)to the total current is equal to the ratio of the resistance of the other branch to the sum of the two resistances.

Example 1.18 : In to how many equal pieces , a wire of 100 ohm resistance should be cut so that when they are connected in parallel, resultant resistance is 1 ohm.

Solution :

Let the number of pieces to be cut = n

Hence resistance of each piece = $\frac{100}{n}$ ohm

If these pieces are connected in parallel , their equivalent resistance is given by:

$$\frac{1}{R_T} = \frac{n}{100} + \frac{n}{100} + \frac{n}{100} + \text{-------------} + \text{ up to 'n' terms}$$

or $$\frac{1}{R_T} = n \times \frac{n}{100} = \frac{n^2}{100}$$

However $R_T = 1$ Ohm

Hence $$1 = \frac{n^2}{100}$$

i.e. $n^2 = 100$

or $n = 10$

Hence number of equal pieces of resistances =10 **Ans**

and resistance of each piece $= \frac{100}{n} = \frac{100}{10} = 10$ ohms **Ans**

Example 1.19 : When a resistor is placed across 230 V, DC supply, the current through the resistance is 12 amp. What is the value of resistor that must be connected in parallel to increase the load current to 16 amp.

Solution :

Resistance $R_1 = \frac{V}{I_1} = \frac{230}{12} = 19.167$ ohm

Let R_2 be the resistance which must be connected in parallel to increase the current to 16 amp.

Equivalent resistance of parallel combination $R_{eq} = \frac{V}{I_2} = \frac{230}{16} = 14.375$ ohm

We know that,

$$R_{eq} = \frac{R_1 R_2}{R_1 + R_2}$$

Substituting the values, we get:

$$14.375 = \frac{19.167 R_2}{19.167 + R_2}$$

Simplifying it, we get:

$14.375 \; 19.167 + 14.375 \; R_2 = 19.167 \; R_2$

or $\quad 4.792 \; R_2 = 275.53$

Hence resistance to be connected in parallel,

$$R_2 = \frac{275.53}{4.792} = 57.5 \text{ ohm} \quad \text{Ans}$$

Example 1.20 : Find the effective resistance of two parallel connected resistances $R_1 = 4\Omega$ and $R_2 = 2\Omega$ and also find the current that would flow in the two resistances when these are connected across a voltage V =12 volts as shown in Fig. 1.44.

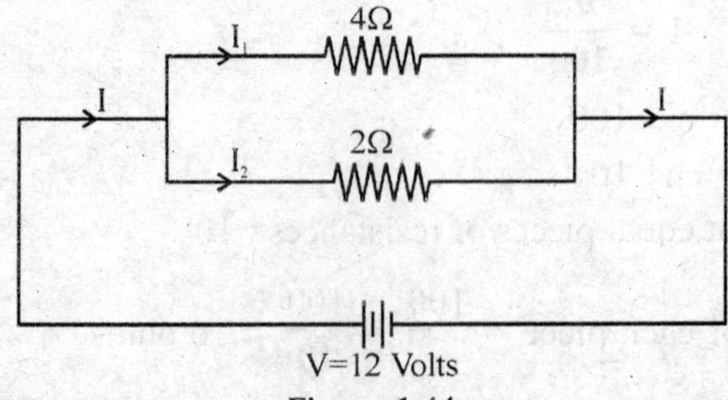

Figure 1.44

Solution :

It is seen that the parallel combination of resistances will have an effective resistance given by

$$R_p = \frac{R_1 R_2}{R_1 + R_2} = \frac{4 \times 2}{4 + 2} = \frac{4}{3} \text{ ohm}$$

Total Current , $I = \frac{12}{4/3} = 9$ amp.

Referring to Figure 1.44 , the currents in the two resistances are given by :

$$I_1 = \left(\frac{R_2}{R_1 + R_2} \right) I = \frac{2}{6} \times 9 = 3 \text{ amp}$$ **Ans**

and

$$I_2 = \left(\frac{R_1}{R_1 + R_2} \right) I = \frac{4}{6} \times 9 = 6 \text{ amp}$$ **Ans**

Example 1.21 : A current of 30.0 amp. is to be divided into two resistances connected in parallel having branch currents of 20.0 amp. and 10.0 amp. by a network which has having an equivalent resistance equal to 10.0 ohms. Obtain the values of two branch resistances.

Solution :

Let R_1 and R_2 be the resistances of the two branches and I_1, I_2 be the two branch currents

Then, $$\frac{I_1}{I} = \frac{20}{30} = \frac{R_2}{R_1 + R_2} = \frac{2}{3} \quad \text{(i)}$$

$$\frac{I_2}{I} = \frac{10}{30} = \frac{R_1}{R_1 + R_2} = \frac{1}{3} \quad \text{(ii)}$$

Equivalent parallel resistance is $R_{eq} = \frac{R_1 R_2}{R_1 + R_2} = 10.0\ \Omega$ (iii)

Substituting equation (i) in equation (iii) gives :

$$\frac{2}{3} R_1 = 10$$

i.e. $R_1 = 15.0\ \Omega$ **Ans**

Substituting equation (ii) in equation (iii) gives :

$$\frac{1}{3}R_2 = 10$$

or $R_2 = 30.0\ \Omega$ **Ans**

1.9.7 Series Parallel Circuits

Practical electrical circuits do not consist of only series or parallel elements. Many of them are combinations of parallel resistors connected in series with other resistors. Such circuits can be analysed by resolving each purely series group into its equivalent resistance and each parallel group of resistances in to its equivalent resistance and repeating the process as many times as required.

Fig 1.45(a) shows a series parallel circuit consisting of resistances R_2 and R_3 in parallel and together they are connected in series with resistance R_1.

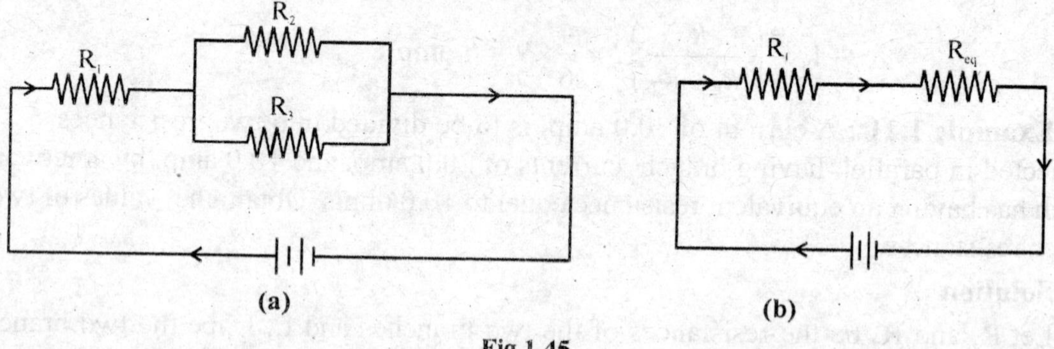

Fig.1.45

To solve this circuit, resistances R_2 and R_3 are first replaced by their equivalent resistance R_{eq} as shown in Fig1.45(b). The circuit gets simplified to a simple series circuit , which can be solved easily.

Example 1.22 : Calculate the current I ,drawn from the supply in the circuit shown in Fig 1.46(a). $R_1 = 24\Omega$, $R_2 = 10\Omega$, $R_3 = 15\Omega$. $V = 60$ Volts

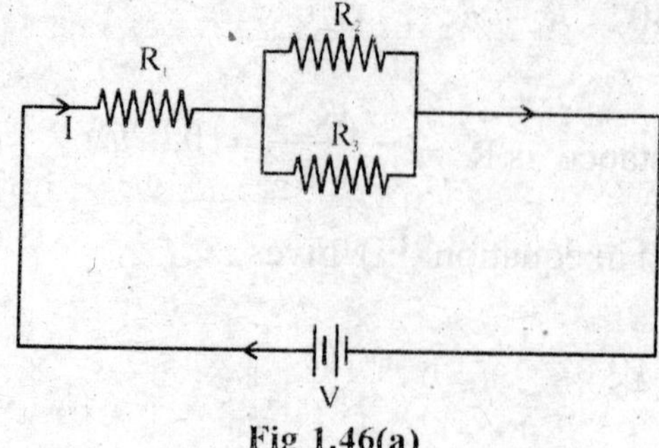

Fig 1.46(a)

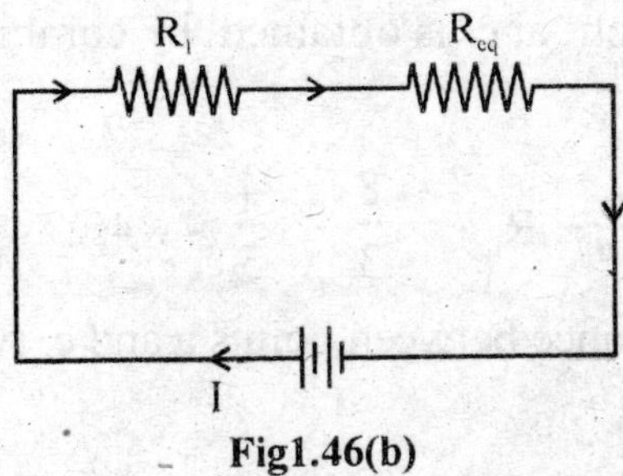

Fig1.46(b)

Solution :

It is seen that ,

$$R_{eq} = R_2 \parallel R_3 = \frac{10 \times 15}{10 + 15} = \frac{150}{25} = 6 \text{ ohms}$$

Draw the simplified equivalent circuit after combining the resistances R_2 and R_3 which are connected in parallel ,as shown in Fig1.46(b).

Now R_1 and R_{eq} are connected in series across a voltage of 60 volts.

Hence. $$I = \frac{V}{R_1 + R_{eq}} = \frac{60}{24 + 6} = 2 \text{ amp.}$$ **Ans**

Example 1.23 : Find the total effective resistance between points 'a' and 'c' for the circuit shown in Figure 1.47. Also find the supply current , I when a potential difference of 80 V is applied between points 'a' and 'c'.

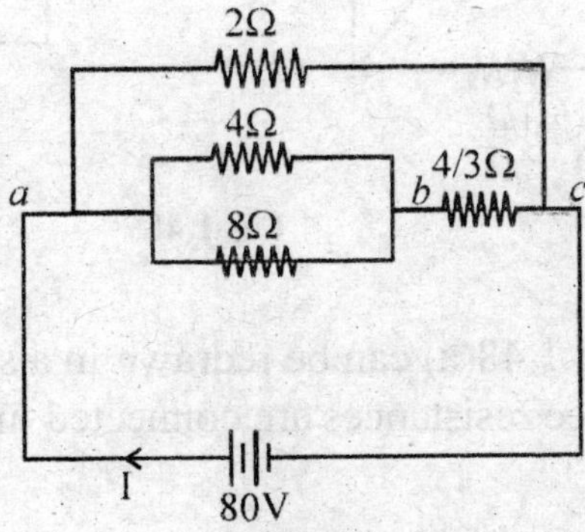

Fig.1.47

4Ω and 8Ω resistances are connected in parallel across terminals 'a' and 'b'. Hence their equivalent resistance between terminals 'a' and 'b' is :

$$R_{ab} = \frac{4 \times 8}{4 \times 8} = \frac{32}{12} = \frac{8}{3}\Omega$$

Total Resistance of branch 'abc' is obtained by combining the resistances R_{ab} and R_{bc} which are in series :

$$\text{i.e.} \qquad R_{abc} = R_{ab} + R_{bc} = \frac{8}{3} + \frac{4}{3} = 4\Omega$$

Hence, Equivalent Resistance between points a and c, (which have 2 and 4 resistances in parallel) is :

$$R_{ac} = \frac{4 \times 2}{4+2} = \frac{8}{6} = \frac{4}{3}\Omega$$

$$\text{Total current, I in the circuit} = \frac{80}{\frac{4}{3}} = 60 \text{ Amp.}$$ **Ans**

Example 1.24 Calculate the current I, drawn from the supply in the circuit shown in Fig 1.48(a)

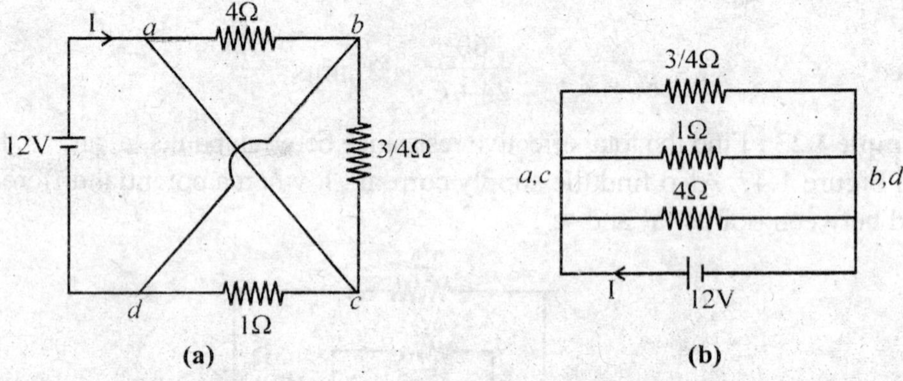

Fig 1.48

Solution :

The circuit shown in Fig 1.48(a) can be redrawn in a simplified form as shown in Fig. 1.48(b). It is seen that the three resistances are connected in parallel across a voltage supply of 12 volts.

It is seen that resistances 4 Ω and $\frac{3}{4}$ Ω are in parallel. Their equivalent resistance R_{eq1} is:

$$R_{eq1} = 4 \| \frac{3}{4} = \frac{4 \times \frac{3}{4}}{4 + \frac{3}{4}} = \frac{3}{19/4} = \frac{12}{19}\Omega$$

Further, this R_{eq1} and 1Ω resistances are in parallel . Their equivalent resistance R_{eq2} is given by:

$$R_{eq2} = \frac{12}{19} \| 1 = \frac{\frac{12}{19} \times 1}{\frac{12}{19} + 1} = \frac{\frac{12}{19}}{\frac{31}{19}} = \frac{12}{31} \Omega$$

Hence, current I drawn from the supply is:

$$I = \frac{V}{R_{eq_2}} = \frac{12}{\frac{12}{31}} = 31 \text{ amp.}$$ **Ans**

1.9.8 Short Circuit and Open Circuit

The two conditions frequently encountered in Electrical Engineering are: Short circuits and Open circuits. These conditions affect the flow of current in a circuit.

1.9.8.1 Short Circuit

A short circuit is a direct connection between any two points. Two points are said to be short circuited when they are connected directly to each other by a conducting wire. A wire made of a good conductor connected across two terminals acts as a short circuit, as shown in Fig 1.49. An element is said to be short circuited, if a short path is connected in parallel with it.

An element / circuit having **zero resistance** is called a **short circuit**

In such cases voltage across the element/circuit is zero.

Because $v = i\,R = i \times 0 = 0$

However the current through a short circuit can have any value. A short circuit increases the current level in a circuit and the power loss is increased in other components

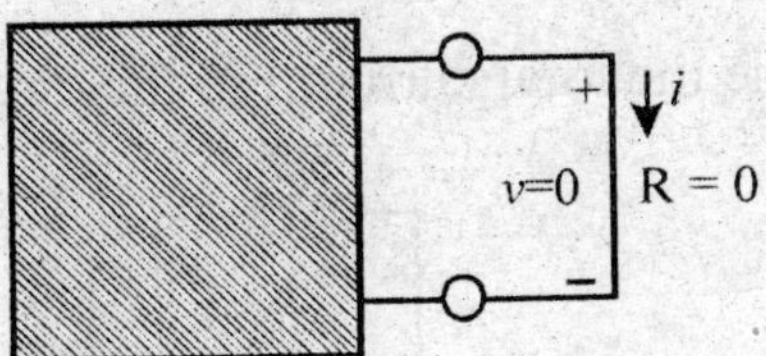

Fig.1.49 Short Circuit

A short circuit in any part of a series circuit increases the current value in those components which are not shorted.

A short circuit in any branch of a parallel circuit , will short circuit the supply source and a very heavy short circuit current will flow, which can damage the source.

1.9.8.2 Open Circuit

An open circuit between any two points means that there is no connection between those two points i.e. any two points are said to be open circuited if there is no direct electrical connection between them.

An open circuit occurs in a series circuit when one of the elements becomes disconnected from an adjacent element . An open circuit in a series circuit , stops flow of current throughout the circuit. There is no voltage drop in any resistance and the supply voltage will appear at the open circuit.

When one of the components in a parallel circuit is open circuited, current does not flow through the branch containing that component of the circuit. The other branches are not affected. An open circuit in one branch of a parallel circuit stops the flow of current in that branch only.

An element/circuit having **infinite resistance** is called an open circuit or an open circuit offers an infinite resistance.

In such cases current through the element/circuit is zero.

$$i = \lim_{R \to \infty} \left(\frac{v}{R} \right) = 0 \qquad (1.33)$$

Fig.1.50 Open Circuit

As terminals are open , no current can flow as shown in Fig 1.50 i.e. i = 0. However the voltage can have any value across an open circuit. An open circuit in the supply line stops the supply current to the whole circuit.

Common operating problems with electrical devices are Short circuits and Open circuits. Looking for Short circuits and Open circuits is a major part of troubleshooting of electrical devices.

Example 1.25 : Determine the total current I in the circuit shown in Fig 1.51(a)

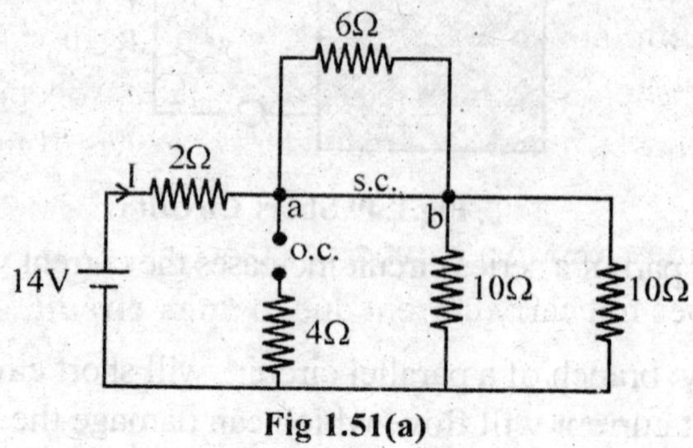

Fig 1.51(a)

Solution :

6 ohm resistance across terminals a and b is shown to be short circuited and 4 ohm resistance branch is open circuited. This circuit can be simplified and redrawn as shown in Fig1.51 (b)

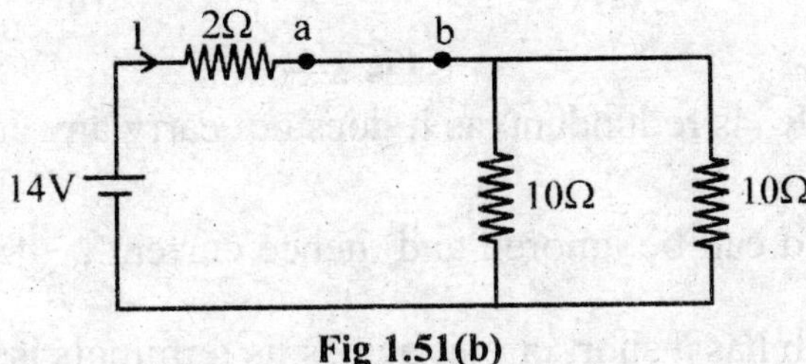

Fig 1.51(b)

Two 10 ohm resistances are connected in parallel as shown in Fig 1.51(b) and their equivalent resistance is :

$$R_{eq} = \frac{10 \times 10}{10 + 10} = 5 \text{ ohm}$$

The simplified circuit can be redrawn as shown in Fig.1.51(c)

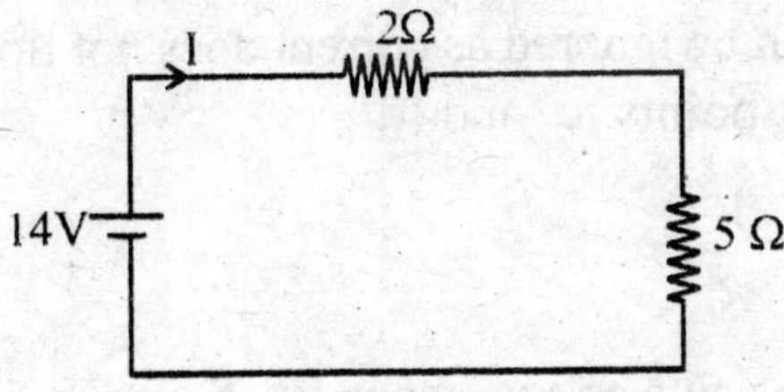

Fig 1.51(c)

It can be seen from Fig 1.51(c) :

Total circuit resistance = 2 + 5 = 7 ohm

$\therefore$ Total current, $I = \frac{14}{7} = 2$ Amp. **Ans**

1.9.9 Removal of Redundant Branches

If a branch which can be removed from the circuit without affecting the circuit , it is called a redundant branch. Sometimes , removal of redundant branches from a circuit leads to circuit simplification.

Following situations may lead to a redundant branch .

1. Any branch that does not carry current due to **open circuit**, as shown in Fig.1.52(a), is redundant.

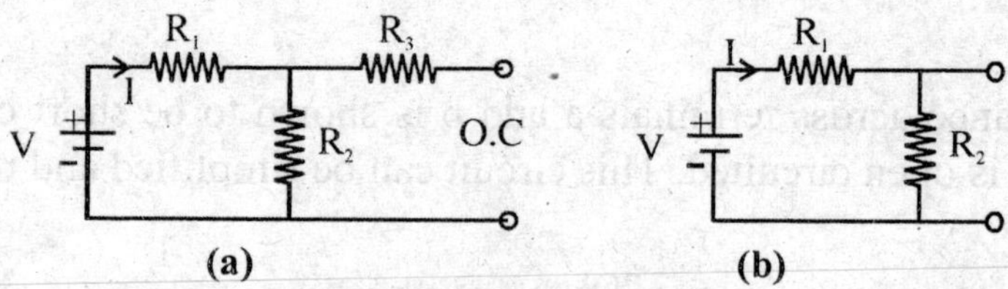

Fig 1.52

Branch containing R_3 is redundant as it does not carry any current due to open circuit at terminals A and Band can be ignored and hence current , . $= \dfrac{V}{R_1 + R_2}$

2. Any branch which has a short circuit across its terminals is redundant , as shown in

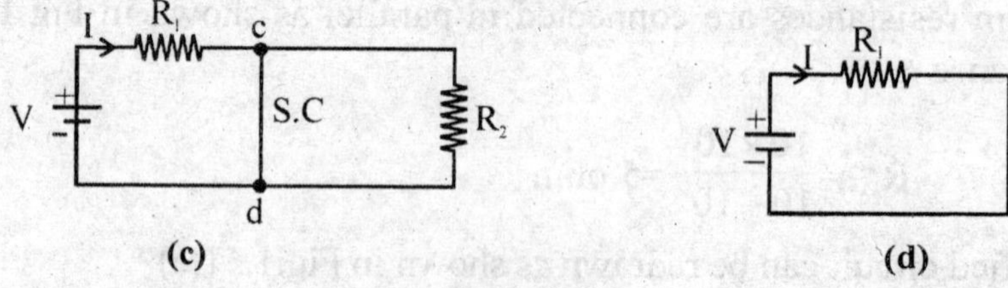

Fig 1.52

Branch containing R_2 can be ignored as current does not flow through the branch because it has been short circuited at points 'c' and 'd'.

Therefore Current , $I = \dfrac{V}{R_1}$

Example 1.26 : Find the value of resistance R , if current in the branch AO in the circuit shown in Fig 1.53(a) is zero.

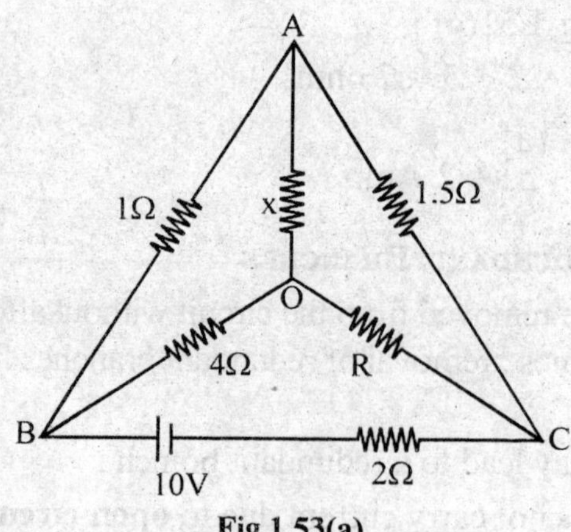

Fig.1.53(a)

Solution :

Since the current in branch 'AO' is zero, hence it is a **redundant branch** and can be removed from the circuit.

The simplified circuit is redrawn as shown in Fig.1.53(b) below:

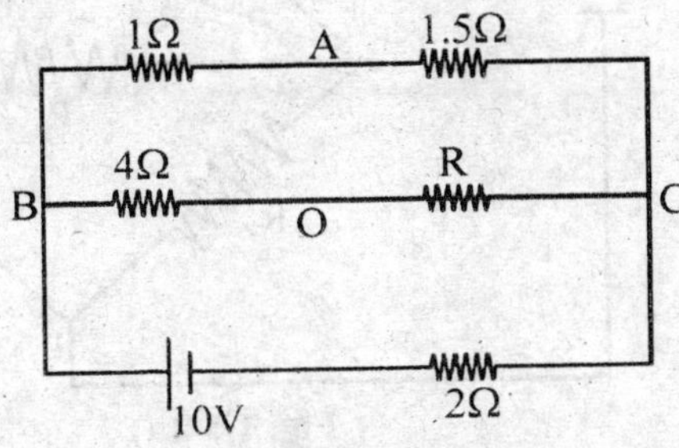

Fig.1.53(b)

As no current flows across points A and O , this means that these points are at the same potential.

i.e. $V_A = V_O$

By applying voltage divider rule to the branch BAC , we get:

$$V_A = \frac{1.5}{1.5+1} V_{BC}$$

By applying voltage divider rule to the branch BOC , we get:

$$V_O = \frac{R}{4+R} V_{BC}$$

Hence, we can write:

$$\frac{1.5}{1.5+1} V_{BC} = \frac{R}{4+R} V_{BC}$$

i.e. $1.5\ (4 + R) = 2.5\ R$

or $6 + 1.5\ R = 2.5\ R$

or $R = 6$ Ohms **Ans**

1.10 Network Reduction by Star Delta Transformation

There are situations when circuits can not be solved by means of simple series or parallel circuit simplification techniques. For example , refer to Fig. 1.54. It can be seen that series / parallel or series-parallel simplification techniques are not applicable in this case, due to the presence of a three terminal circuit , such as **'def'.** However, if this three terminal network can be replaced by an equivalent network, which can remove the cross arm 'ef', then the circuit can be easily solved using series/parallel combination of resistances. The circuit which has this capability is the **wye(Y) or star** network, shown with dashed lines in the figure.

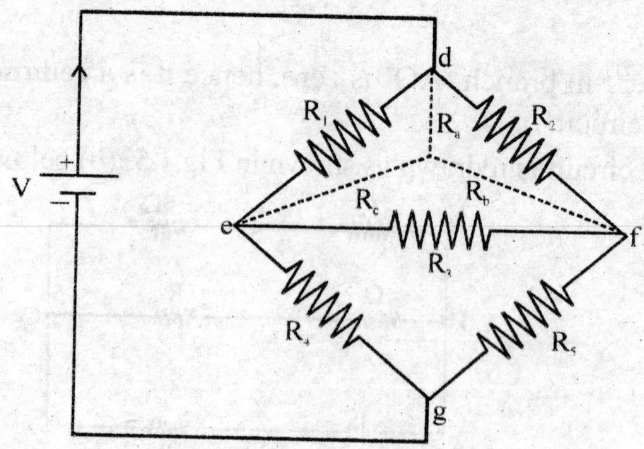

Fig. 1.54(a)

It can be seen that after replacing the three terminal network by an equivalent star circuit, resistance R_c and R_4 are in series and resistances R_b and R_5 are also in series. Hence, further analysis can be carried out easily using series/parallel combinations.

Thus, such circuits can be simplified by using three terminal equivalent circuits, which are: Wye (Y) or Tee (T) networks shown in Fig 1.54 (b)and (c) and the Pi (π) or Delta (Δ) networks shown in Fig 1.54 (d) and (e)

These transformations are applicable to AC as well as DC circuits.

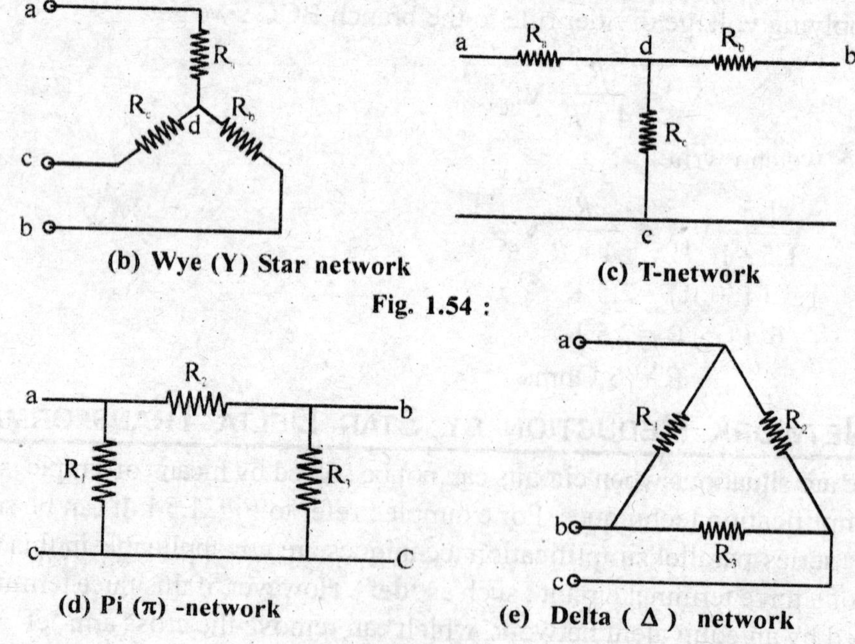

Fig. 1.54

1.10.1 Delta to Star Conversion (D-Y

Consider the two networks of Figure 1.55(a) and (b) . If these networks are identical , then the resistance between any pair of lines should be the same when the third line is opened.

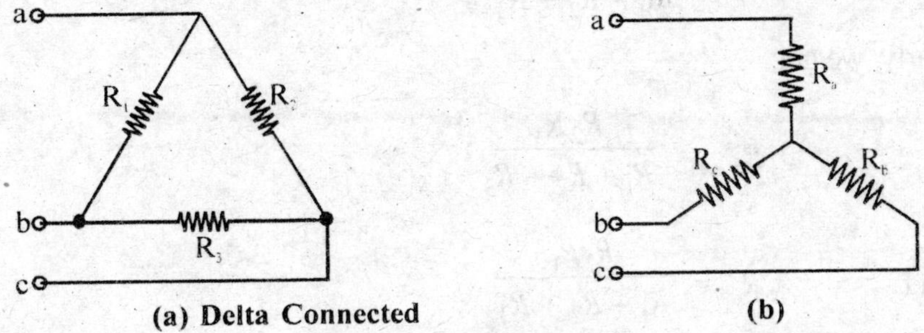

(a) Delta Connected (b)

Fig.1.55 : Star Connection

(i)When line 'a' is open :

In Δ connection, R_1 and R_2 get connected in series and their series combination is in parallel with resistance R_3 (Fig.1.55a)

Hence, resistance between terminals 'b' and 'c', $R_{bc} = \dfrac{(R_1 + R_2)R_3}{R_1 + R_2 + R_3}$ (a)

In Y-connection, R_b and R_c get connected in series, when line 'a' is open.(Fig.1.55b)

Hence,

$$R_{bc} = R_b + R_c \qquad \text{(b)}$$

Equating equations (a) and (b) given above , we get:

i.e. $$R_{bc} = R_b + R_c = \frac{(R_1 + R_2)R_3}{R_1 + R_2 + R_3} \qquad (1.34)$$

Similarly,

(ii)When line'b' is open : $R_{ca} = R_c + R_a = \dfrac{R_1(R_2 + R_3)}{R_1 + R_2 + R_3}$ (1.35)

(iii)When line'c' is open : $R_{ab} = R_a + R_b = \dfrac{(R_1 + R_3)R_2}{R_1 + R_2 + R_3}$ (1.36)

Adding Equations (1.34),(1.35) and (1.36.) and dividing by 2 on both sides, we get:

$$R_a + R_b + R_c = \frac{R_1R_2 + R_2R_3 + R_3R_1}{R_1 + R_2 + R_3} \qquad (1.37)$$

Subtracting Equation (1.34) from Equation (1.37), we get :

$$R_a = \frac{R_1 R_2}{R_1 + R_2 + R_3} \tag{1.38}$$

Similarly, we get:

$$R_b = \frac{R_2 R_3}{R_1 + R_2 + R_3} \tag{1.39}$$

and

$$R_c = \frac{R_3 R_1}{R_1 + R_2 + R_3} \tag{1.40}$$

This gives the conversion of the delta network in to an equivalent star network.

Thus the equivalent star resistance connected to a node is equal to the product of the two delta resistances connected to the same node divided by the sum of all the three delta resistances

1.10.2 Star to Delta Conversion, (Y–Δ) Transformation

For converting a star network into an equivalent delta network (Y–) transformation, we have from Equations (1.38) to (1.40), by multiplying two equations at a time and adding these products:

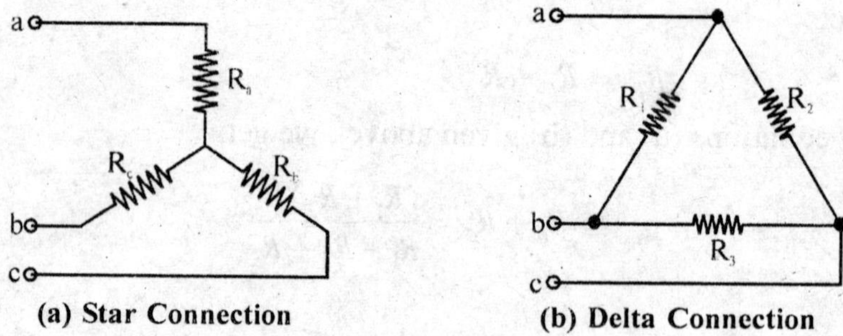

(a) Star Connection (b) Delta Connection

Fig 1.56

$$R_a R_b + R_b R_c + R_c R_a = \frac{R_1 R_2^2 R_3 + R_2 R_3^2 R_1 + R_1^2 R_2 R_3}{(R_1 + R_2 + R_3)^2}$$

$$= \frac{R_1 R_2 R_3 (R_1 + R_2 + R_3)}{(R_1 + R_2 + R_3)^2} = \frac{R_1 R_2 R_3}{(R_1 + R_2 + R_3)} \tag{1.41}$$

Now taking the value of R_a from equation (1.38) and substituting it in equation(1.41),we get:

$$R_aR_b + R_bR_c + R_cR_a = \frac{R_1R_2R_3}{(R_1+R_2+R_3)} = R_aR_3$$

This gives :

$$R_3 = \frac{R_aR_b + R_bR_c + R_cR_a}{R_a} = R_b + R_c + \frac{R_bR_c}{R_a} \qquad (1.42)$$

Similarly , we can obtain the equivalent delta values R_1 and R_2

i.e.

$$R_1 = \frac{R_aR_b + R_bR_c + R_cR_a}{R_b} = R_a + R_c + \frac{R_aR_c}{R_b} \qquad (1.43)$$

and

$$R_2 = \frac{R_aR_b + R_bR_c + R_cR_a}{R_c} = R_b + R_a + \frac{R_aR_b}{R_c}$$

(1.44)

The equivalent delta resistances between two nodes is the sum of two star resistances connected to those nodes plus the product of the same two star resistances divided by the third star resistance .

Example 1.27

Convert the resistances given in delta shown in Fig 1.57(a) to the equivalent star values where $R_1 = 15\Omega$, $R_2 = 10\Omega$ and $R_3 = 25\Omega$

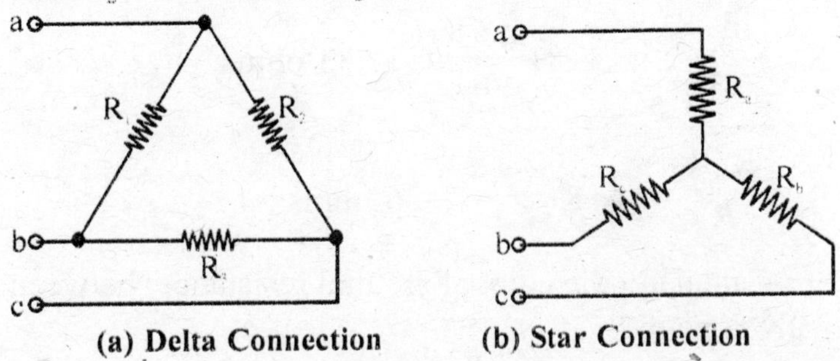

(a) Delta Connection **(b) Star Connection**

Fig.1.57

Solution :

Equivalent star values are calculated as follows:

$$R_a = \frac{R_1R_2}{R_1+R_2+R_3} = \frac{10\times15}{50} = 3 \text{ Ohm}$$ **Ans**

$$R_b = \frac{R_2 R_3}{R_1 + R_2 + R_3} = \frac{25 \times 10}{50} = 5 \text{ Ohm}$$ **Ans**

$$R_C = \frac{R_3 R_1}{R_1 + R_2 + R_3} = \frac{25 \times 15}{50} = 7.5 \text{ Ohm}$$ **Ans**

Example 1.28 : Convert the resistances given in star shown in Fig 1.58(a) to the equivalent delta values such that $R_a = 4\Omega$, $R_b = 6\Omega$ and $R_c = 8\Omega$

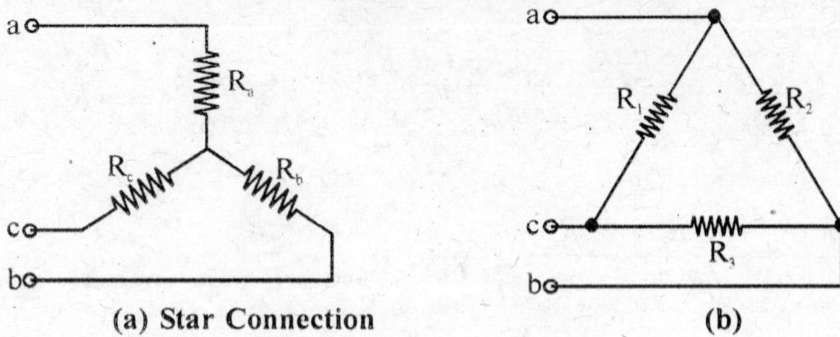

(a) Star Connection **(b)**

Fig 1.58 :Delta Connection

Solution :

Equivalent delta resistances can be obtained as follows:

$$R_1 = R_a + R_c + \frac{R_a R_c}{R_b} = 4+6+\frac{4 \times 6}{8} = 13 \text{ ohms}$$ **Ans**

$$R_2 = R_b + R_a + \frac{R_a R_b}{R_c} = 4+8+\frac{4 \times 8}{6} = 17.33 \text{ ohms}$$ **Ans**

$$R_3 = R_b + R_c + \frac{R_b R_c}{R_a} = 6+8+\frac{6 \times 8}{4} = 26 \text{ ohms}$$ **Ans**

Example 1.29 : Calculate the equivalent total resistance between nodes A and B for the Figure1.59 (a) given below

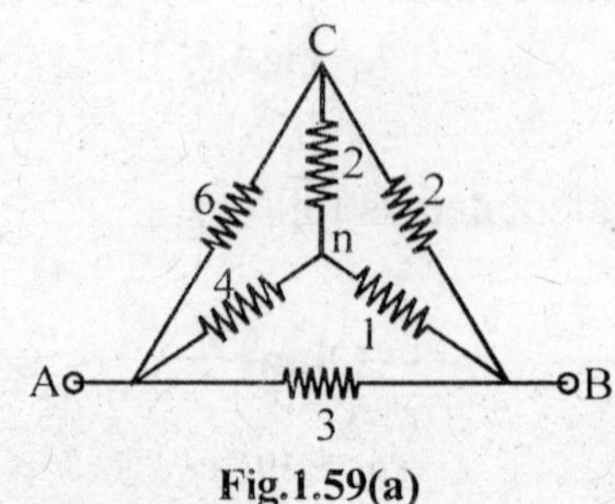

Fig.1.59(a)

Solution

To determine the equivalent resistance between nodes 'A' and 'B', the star connected resistances at terminals A, B and C and node 'n' are converted into an equivalent delta connected resistances as follows and the circuit is redrawn as shown in Fig 1.59(b)

$$R_1 = 4 + 2 + \frac{4 \times 2}{1} = 14\,\Omega$$

$$R_2 = 2 + 1 + \frac{2 \times 1}{4} = \frac{7}{2}\,\Omega$$

$$R_3 = 4 + 1 + \frac{4 \times 1}{2} = 7\,\Omega$$

Fig.1.59(b)

In Fig.1.59(b), resistances connected in parallel are combined to get their equivalent resistances as follows :

$$R_{BC} = \frac{3.5 \times 2}{3.5 + 2} = \frac{14}{11}\,\Omega$$

$$R_{AC} = \frac{14 \times 6}{14 + 6} = \frac{21}{5}\,\Omega$$

$$R_{AB} = \frac{7 \times 3}{7 + 3} = \frac{21}{10}\,\Omega$$

The resultant circuit is shown in Fig 1.59 (c)

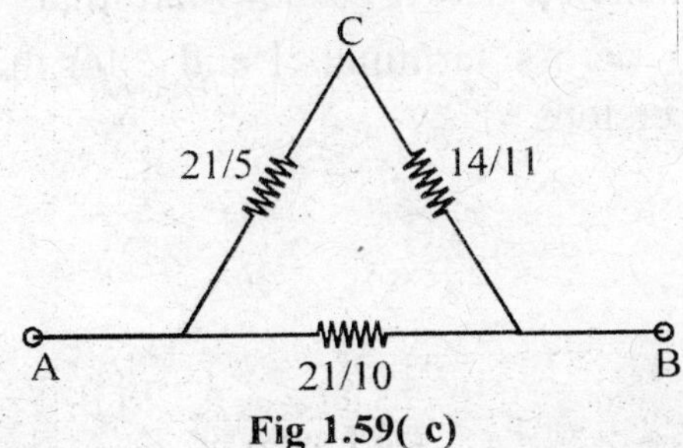

Fig 1.59(c)

In Fig 1.59(c), resistances $\frac{14}{11}$ and $\frac{21}{5}$ are in series. Their equivalent series resistance R_{se}, is:

$$R_{se} = \frac{14}{11} + \frac{21}{5} = \frac{301}{55}\ \Omega$$

The resultant simplified circuit across nodes A and B , is shown in Fig 1.59(d)

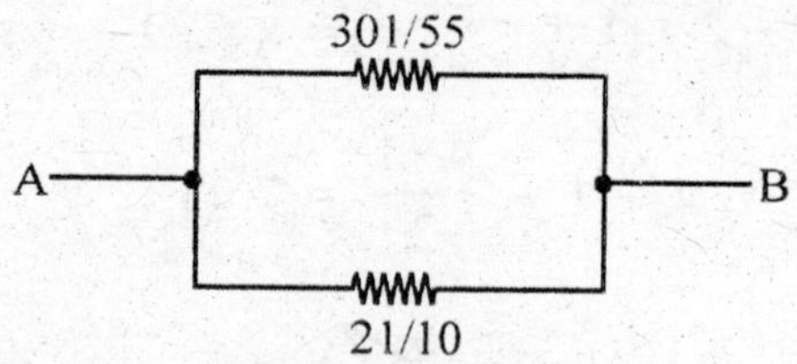

Fig 1.59 (d)

It is seen that resistance $\frac{301}{55}$ ohm is in parallel with resistance $\frac{21}{10}$ Ω **as** shown in Fig 1.59 (d). Their eq. resistance is:

$$= \frac{\frac{310}{55} \times \frac{21}{10}}{\frac{301}{55} + \frac{21}{10}}$$

$$= \frac{301}{55} \times \frac{21}{10} \times \frac{550}{4165}$$

$$= \frac{6321}{4165} = 1.51\Omega$$ **Ans**

Hence the equivalent resistance across A and B terminals is ==1.51Ω **Ans**

Example 1.30 : Find R_{eq} across terminals 1 and 2 for the circuit shown in Fig. 1.60(a), using star/delta conversions.

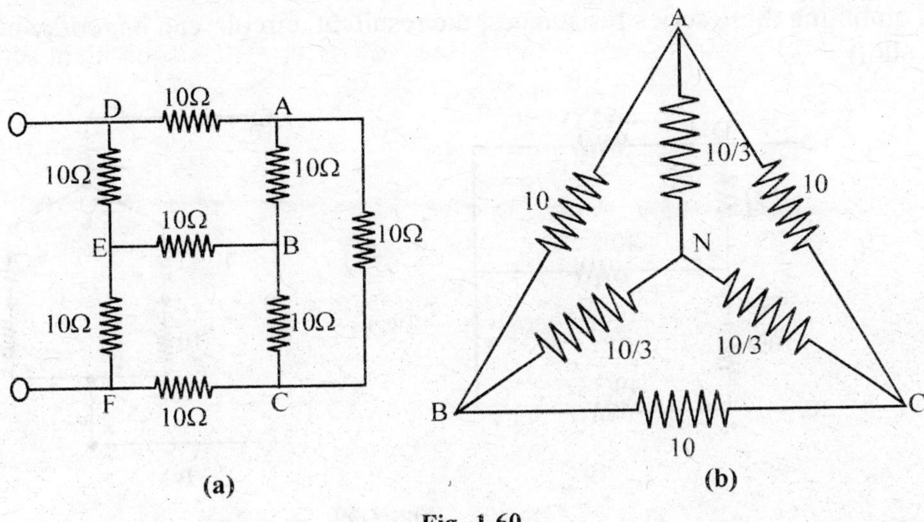

Fig. 1.60

Solution :

(i) Convert delta Δ, between terminals ABC in to an equivalent star Y as shown in Fig 1.60(b).

Equivalent star resistances are :

$$\frac{10 \times 10}{10+10+10} = \frac{100}{30} = \frac{10}{3} \text{ ohm}$$

The resultant circuit of fig. 1.60 (a) is redrawn using these equipment star values and is as shown in Fig 1.60(c)

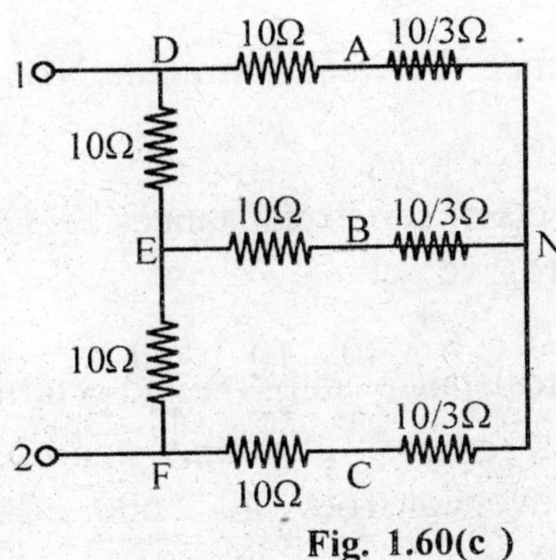

Fig. 1.60(c)

(ii)Individual resistances between terminals DAN , EBN ,and FCN are connected in series in three different branches. Their equivalent resistance in each branch is given by:

$$10 + \frac{10}{3} = \frac{40}{3} \text{ ohm}$$

Combining these series resistances, the resultant circuit can be redrawn as shown in Fig 1.60(d)

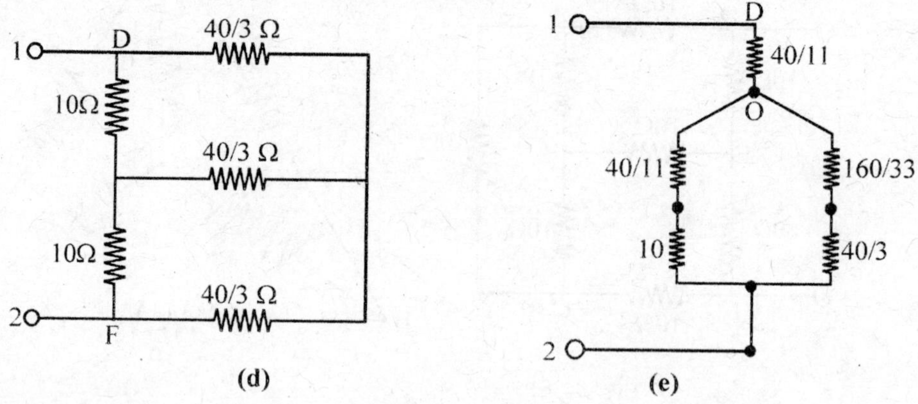

Fig. 1.60

(iii)Convert Delta (Δ)between terminals DNE in to an equivalent star (Y), as shown in Fig 1.60(e), as follows :

$$\text{Resistances between terminals DO and EO} = \frac{\frac{40}{3}\times 10}{\frac{40}{3}+\frac{40}{3}+\frac{10}{1}} = \frac{\frac{400}{3}}{\frac{110}{3}} = \frac{40}{11}\,\Omega$$

$$\text{and resistance between terminals NO} = \frac{\frac{40}{3}\times\frac{40}{3}}{\frac{110}{3}} = \frac{160}{33}\,\Omega$$

(iv)Combining two sets of series resistances i.e. OEF and ONF between terminals 'O' and 'F' , in Fig 1.60(e) we get :

$$R_{OEF} = \frac{40}{11}+\frac{10}{1} = \frac{150}{11}\,\Omega$$

$$R_{ONF} = \frac{160}{33}+\frac{40}{3} = \frac{600}{33} = \frac{200}{11}\,\Omega$$

The resultant simplified circuit will be as shown in Fig 1.60(f)

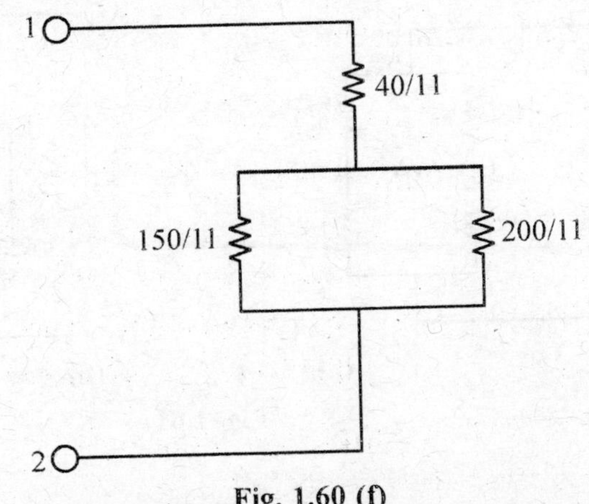

Fig. 1.60 (f)

(v)Combine parallel resistances between terminals 'O' and 'F' and their equivalent resistance R_{OF} is :

$$R_{OF} = \frac{\frac{150}{11} \times \frac{200}{11}}{\frac{150}{11} + \frac{200}{11}} = \frac{600}{77}\ \Omega$$

(vi) Redraw the circuit as shown in Fig. 1.60(g).

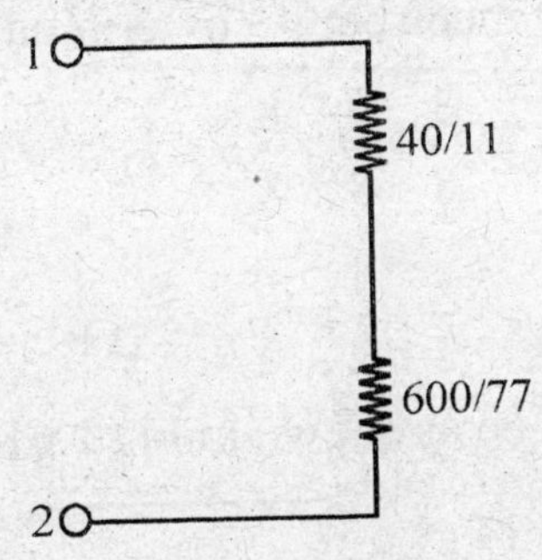

Fig. 1.60 (g)

Combining the resistances in series , resistance R_{eq} across terminals 1 and 2 is :

$$R_{eq} = \frac{40}{11} + \frac{600}{77} = \frac{880}{77} \text{ ohm}$$ **Ans**

Example 1.31 : Find R_{eq} across terminals A and D, for the circuit shown in Fig.1.61(a) using star / delta conversions.

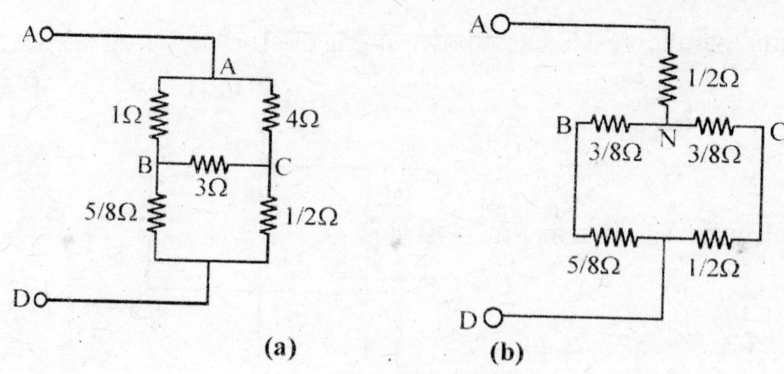

Fig. 1.61

Solution :

(i) Convert Delta(Δ) between terminals ABC in to equivalent (Y) star resistances as shown in Fig1.61(b)

$$R_{AN} = \frac{1\times 4}{8} = \frac{1}{2}\ \Omega$$

$$R_{CN} = \frac{4\times 3}{8} = \frac{3}{2}\ \Omega$$

$$R_{BN} = \frac{1\times 3}{8} = \frac{3}{8}\ \Omega$$

(ii) Combine series resistances between terminals NBD and NCD , we get

$$R_{NBD} = \frac{3}{8}+\frac{5}{8} = \frac{8}{8} = 1\ \Omega$$

$$R_{NCD} = \frac{3}{2}+\frac{1}{2} = \frac{4}{2} = 2\ \Omega$$

The resultant circuit will be as shown in Fig 1.61(c)

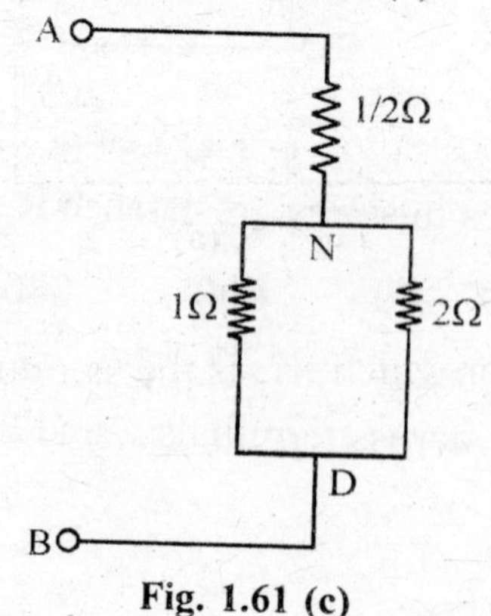

Fig. 1.61 (c)

(iii) Combining parallel resistances shown in Fig.1.61(c) between terminals N and D, we get :

$$R_{ND} = \frac{1 \times 2}{1+2} = \frac{2}{3}\ \Omega$$

Redraw the circuit as shown in Fig. 1.61(d)

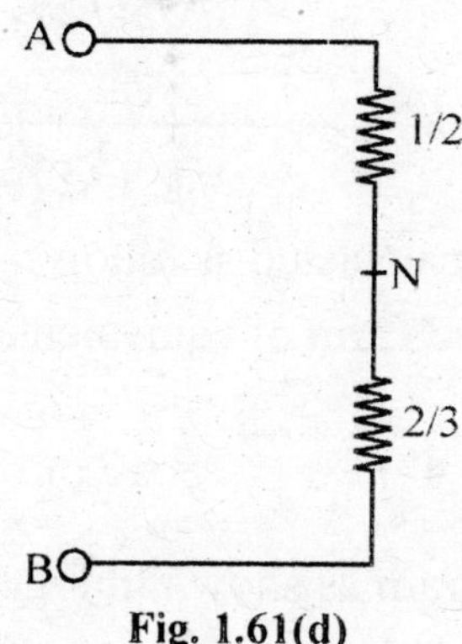

Fig. 1.61(d)

(iv) The equivalent resistance R_{eq} between terminals A and D as shown in Fig 1.61 (d)is:

$$R_{eq} = \frac{1}{2} + \frac{2}{3} = \frac{7}{6} \text{ ohms}$$ **Ans**

Example 1.32 : Find R_{eq} across terminals A and B, for the circuit shown in Fig. 1.62(a) using star / delta conversions.

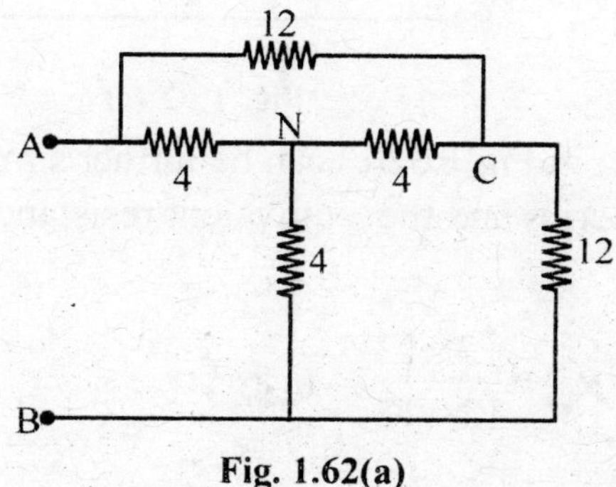

Fig. 1.62(a)

Solution :

(i)Convert the star connection for which 'N' is the star point i.e. three 4 Ω resistances, into an equivalent delta .

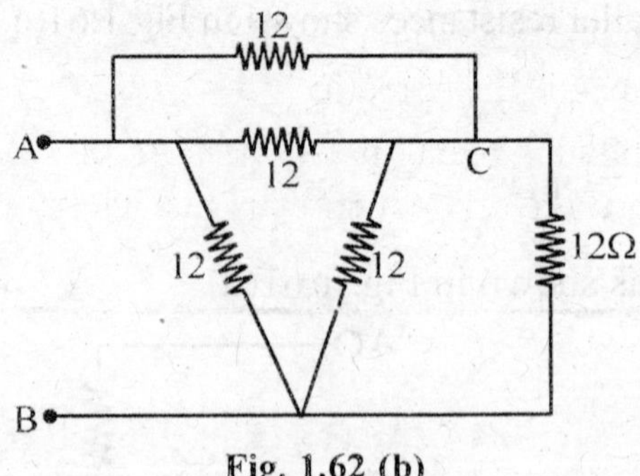

Fig. 1.62 (b)

Equivalent delta values are calculated as follows:

Equivalent delta value of each arm of equivalent. delta connection

$$= 4 + 4 + \frac{4 \times 4}{4} = 12\,\Omega$$

(ii) Redraw the circuit diagram as shown in Fig. 1.62(b)

Simplify parallel connected resistances between terminals (I)A and C (II) C and B .Circuit is reduced to the simplified form as shown in Fig.1.62(c)

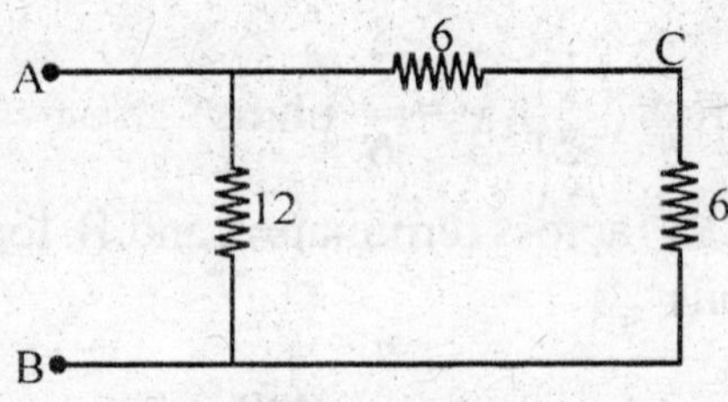

Fig. 1.62 (c)

(iii) The circuit shown in Fig 1.62(c) can be further simplified by combining the resistances connected in series and the equivalent resistance R_{eq} between terminals A and B is:

$$R_{eq} = \frac{12 \times 12}{12 + 12} = 6\ \Omega$$ **Ans.**

1.11 Summary

Basic electrical quantities and their associated units along with the concepts of charge, current, voltage, potential, emf ,dependent and independent voltage and current sources have been explained.

Introduction to the behavior of the circuit elements i.e. resistor , inductance and capacitance has been given.

Ohm's law, Kirchhoff's current law, Kirchhoff's voltage law and concepts of Short circuit and Open circuit have been illustrated

Circuit simplification techniques- series circuit, parallel circuit, series - parallel circuit, redundant branches and network reduction by Star-Delta transformation have been explained.

1.12 HIGHLIGHTS

$$I = \frac{dq}{dt}$$

$$R = \rho \frac{l}{A} \text{ ohm}$$

$$v_L = L\frac{di}{dt}$$

$$i(t) = \frac{1}{L}\int_0^t v_L dt + i(0)$$

$$L = \frac{N^2 \mu A}{l} = \frac{N^2 \mu_0 \mu_r A}{l}$$

$$i = \frac{dq}{dt} = C\frac{dv_C}{dt}$$

$$C = \varepsilon\frac{A}{d} \text{ Farads}$$

$$v_C = \frac{1}{C}\int_0^t idt$$

$$R_{se} = R_1 + R_2 + \ldots\ldots + R_n = \sum_{j=1}^{n} Rj$$

$$\frac{1}{R_p} = \frac{1}{R_1} + \frac{1}{R_2} + \ldots\ldots \frac{1}{R_n} = \sum_{j=1}^{n}\frac{1}{R_j}$$

i.e. Reciprocals of resistances add to get the reciprocal of the equivalent resistance.

$$G = \frac{1}{R} = \text{conductance}$$

$$G_p = G_1 + G_2 + \ldots\ldots. + G_n = \sum_{j=1}^{n} G_j \text{ (mhos)}$$

Where G_p - is the equivalent conductance

Kirchhoff's Current law states that

At any instant , the algebraic sum of the currents at any node/junction is zero

It can be expressed mathematically as :

$\sum_{j=1}^{p} I_j = 0$, Where p denotes the number of elements meeting at the node or connected to the node and I_i is the current in the jth element.

Kirchhoff's Voltage law states that

Around any closed path or loop in a network the algebraic sum of all the voltages is zero.

Some of the voltages are sources while other voltages will result from current in passive elements known as voltage drops.

It can be expressed mathematically as :

$$\sum_{j=1}^{r} V_j = 0$$

Where V_j is the voltage across j th element of a closed loop and ' r' is the number of voltages or number of elements, in the loop.

Delta/Star conversion

The equivalent star resistances connected to a node is equal to the product of the two delta resistances connected to the same node divided by the sum of all the three Delta resistances

$$R_a = \frac{R_1 R_2}{R_1 + R_2 + R_3}$$

$$R_b = \frac{R_2 R_3}{R_1 + R_2 + R_3}$$

$$R_C = \frac{R_3 R_1}{R_1 + R_2 + R_3}$$

Star/Delta conversion

The equivalent delta resistances between two nodes is the sum of two star resistances connected to those nodes plus the product of the same two star resistances divided by the third resistance

$$R_1 = \frac{R_a R_b + R_b R_c + R_c R_a}{R_b} = R_a + R_c + \frac{R_a R_c}{R_b}$$

$$R_2 = \frac{R_a R_b + R_b R_c + R_c R_a}{R_c} = R_b + R_a + \frac{R_a R_b}{R_c}$$

$$R_3 = \frac{R_a R_b + R_b R_c + R_c R_a}{R_a} = R_b + R_c + \frac{R_b R_c}{R_a}$$

1.13 Short Question and Answers

1. Define electric current.

Ans. It is the flow of electrons in a conductor.

2. What is electromotive force?

Ans. The driving force required for the flow of current in a conductor is called electromotive force.

3. What is electrical resistance?

Ans. The property of the conductor by which it opposes the flow of current in a conductor, is called resistance.

4. Define specific resistance.

Ans. The resistance offered by unit cube of the material between its opposite faces is called specific resistance.

5. What is the difference between power and energy?

Ans. Power is defined as the rate of doing work. Energy is defined as the capacity to do the work.

6. State Ohm's law.

Ans. At constant temperature, the current flowing through a conductor is directly proportional to the voltage across the conductor.

7. What is a series circuit?

Ans. It is a circuit in which current passing through all the elements is same.

8. What is a parallel circuit?

Ans. It is a circuit in which voltage across each element connected in parallel in same.

9. How does the resistance of metals and semiconductors vary with temperature?

Ans. The resistance of metals increases with the increase in temperature while the resistance of semiconductors decreases with the increase in temperature.

10 Give reasons why all equipments are connected in parallel across the electric supply.

Ans. All equipments are connected in parallel so that each one may have same voltage across it and can be operated independently.

11. State Kirchhoff's Laws.

Ans. Current Law : The algebraic sum of all the current meeting at a junction is equal to zero i.e., $\sum I = 0$

Voltage Law : In any closed network or mesh, the algebraic sum of product of current and resistances in each of the conductor plus the algebraic sum of electromotive forces in that path is zero, i.e.,

$$\left(\sum IR + \sum e.m.f. = 0\right)$$

Review Questions

1. Calculate the total number of units consumed in operating a 100 watt lamp for 5 hours, 500 W electric iron for 3 hours and 1 kW heater for 4 hours. Find also the cost of energy consumed, if energy is supplied at the rate of Rs. 2.50 paise per unit.

   [Ans.(6 units, Rs.15)]

2. Two resistors of equal values when placed in parallel gives an effective resistance equal to R. What will be the effective resistance if the two resistors are connected in series.

   [Ans. 4R]

3. Two resistances of 4 Ω each connected in parallel are in series with another resistor of 12 Ω. If the current flowing in 12-Ω resistor is 2 A, determine (i) the current

flowing through 4 Ω resistors, (ii) voltage across the whole circuit and (iii) power drawn from the supply.

[Ans.1 amp , 28 V, 56 Watt]

4. Find current 'I' in the circuits (a) and (b) shown below.

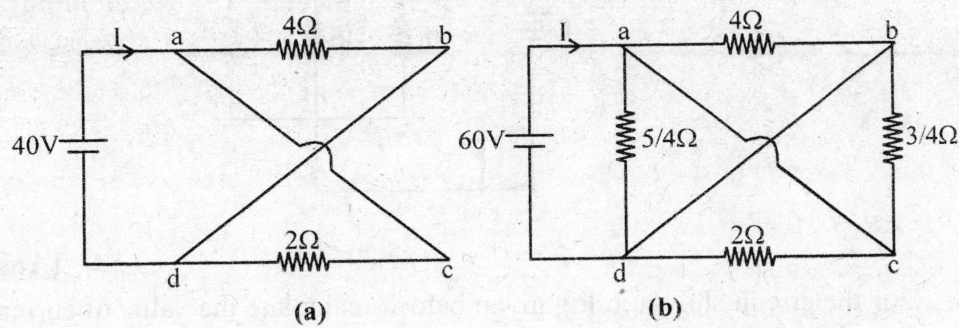

[Ans (a)30 amp, (b)173 amp]

5. Write down KCL equations for nodes a, b and c for the circuit shown in fig. given below and find the currents I_2, I_4 and I_5

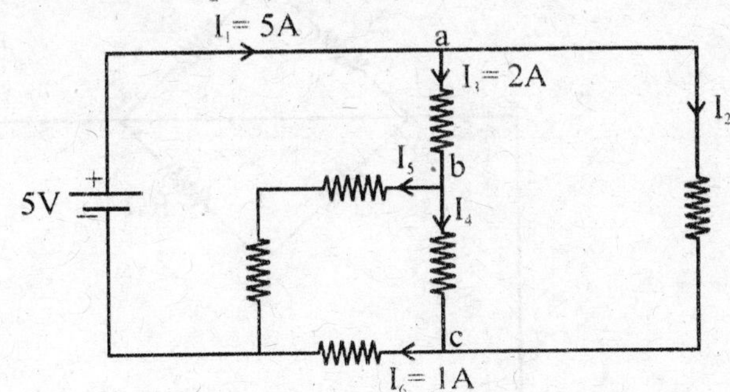

[Ans I_2 =3A , I_4 = -2A , I_5 = 4A]

6. Determine the values of R_1 , R_2, I_1 , I_2 and I_3 for the circuit shown in Fig given below using KCL and KVL

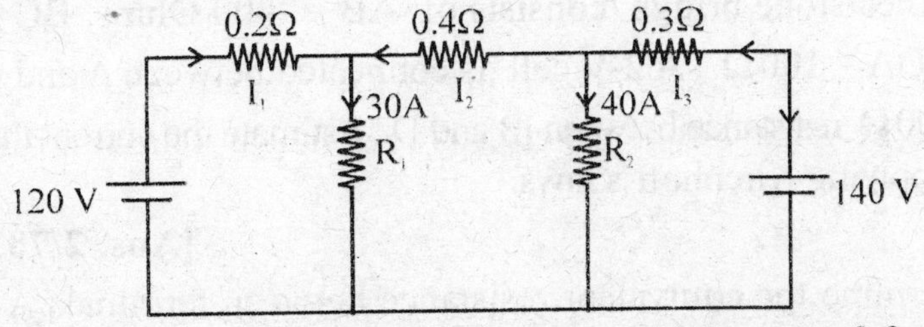

[Ans. 15.5Ω, 14.5Ω, 55.5 amp., 3.9 amp, 3 amp]

7. In the circuit in Figure given below, determine the supply current I.

Given that, $E = 20$ V , $R_1 = \frac{20}{11}$ ohm, $R_2 = 8$ ohm, $R_3 = 12$ ohm, $R_4 = 4$ohm

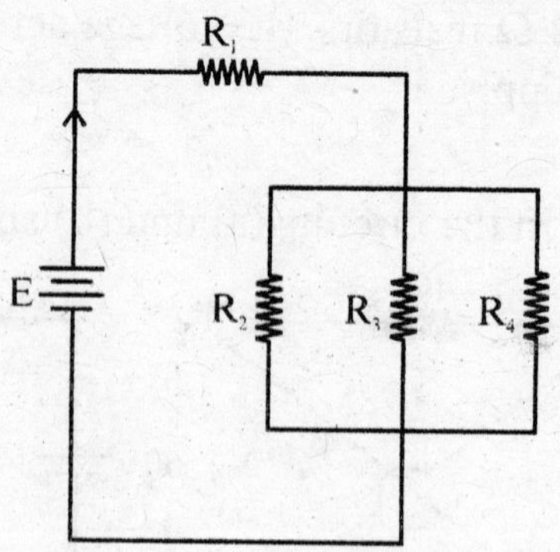

[Ans. 5 amp]

8. For the circuit shown in Fig given below, calculate the value of current in each branch and the value of resistance 'R'. The total current drawn by the circuit is:2.25 amp.

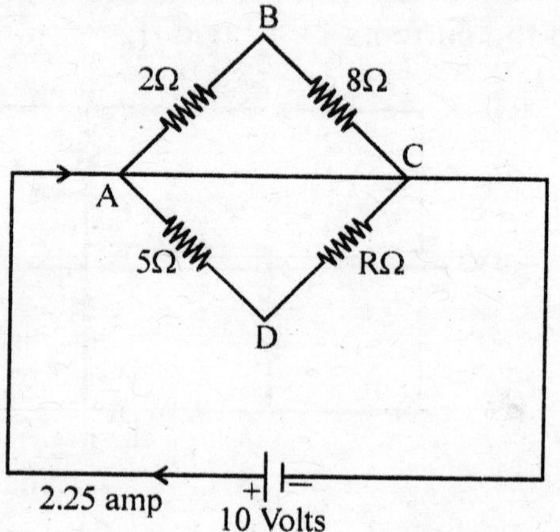

[Ans. 1.0 amp, 1.25 amp, 3 ohm]

9. A Wheatstone bridge consists of AB = 400 Ohms, BC = 220 Ω CD = 225 Ω and DA = 100 Ω . A 2-V cell is connected between A and C and a galvanometer of 200 Ω resistance between B and D . Estimate the current through the galvanometer by applying Kirchhoff's laws.

[Ans. 2/703 amp. from D to B]

10. Determine the equivalent resistance between terminals A and B in the network shown in Fig given below

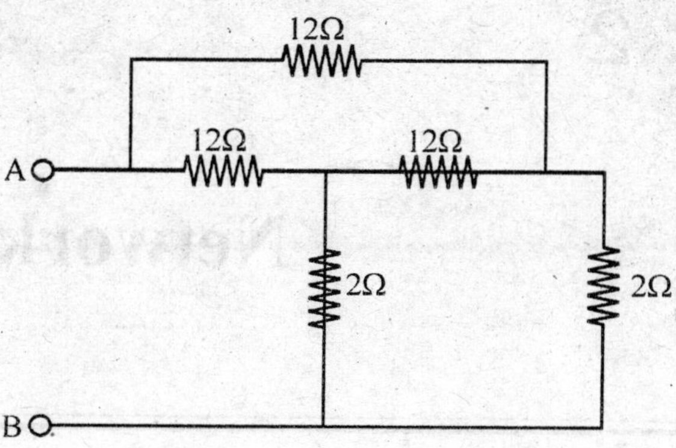

[Ans. 7 ohm]

11. Determine the equivalent resistance between terminals A and B in the network shown in Fig given below

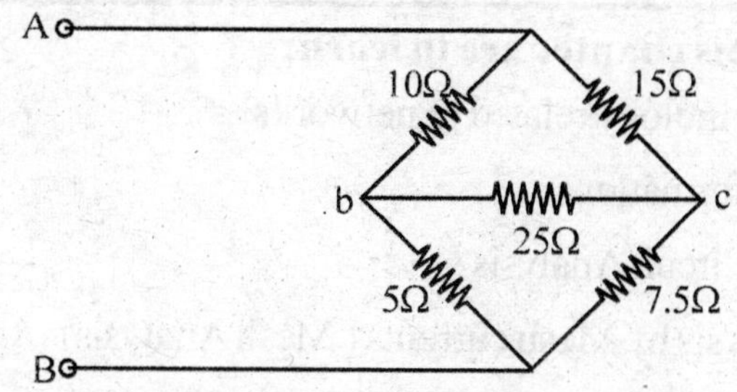

[Ans. 9 ohm]

12. Determine the equivalent resistance between terminals A and B in the network shown in fig given below

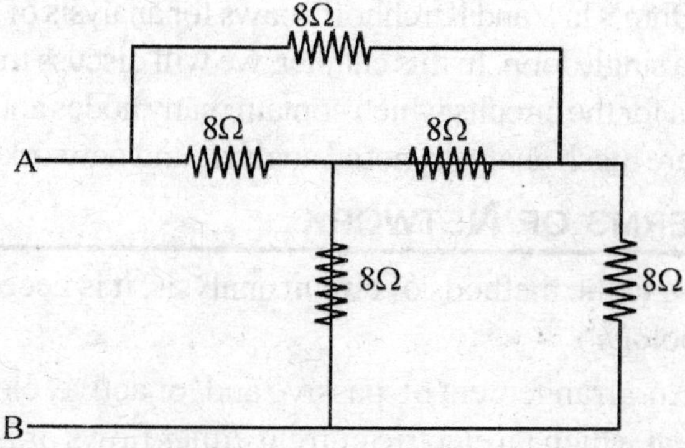

[Ans. 8 ohm]

❑❑❑

Chapter 2

Network Analysis and Theorems

OBJECTIVE

Objectives of this chapter are to learn:

- General terminology related to networks
- Source transformations
- Methods of Circuit Analysis
- Circuit Analysis by Mesh currents (Mesh Analysis)
- Circuit Analysis by Node-pair Voltages (Nodal Method)
- Network Theorems : Superposition Theorem , Thevenin's (or Helmoltz's) theorem

2.1 INTRODUCTION

We have studied Ohm's law and Kirchhoff's laws for analysis of simple circuits which contain either a single node or a single loop. In this chapter, we will discuss methods to calculate the values of voltages and currents for the circuits which contain many nodes and loops. Some of the powerful techniques discussed are mesh analysis, nodal analysis and network theorems.

2.2 GENERAL TERMS OF NETWORK

Before proceeding to the methods of circuit analysis, it is necessary to define the following terms of Network Topology :

(i) **Circuit :** An arrangement of passive and/ or active elements which forms a closed path through which an electric current either flows or is intended to flow is called a **circuit.**

(ii) **Parameter :** The various elements of an electric circuit are called its **parameters** like resistance, inductance and capacitance. These parameters may be lumped or distributed.

(iii) Linear Circuit : A linear circuit is one which obeys laws of superposition and whose parameters are constant i.e. parameters do not change with voltage or current.

(iv) Non-Linear Circuit : It is a circuit whose parameters change their values with change in voltage , current or temperature etc.

(v) Unilateral Circuit : It is that circuit whose characteristics depend on the direction of current flowing through the elements. A diode rectifier is a unilateral circuit, because it cannot perform rectification in both directions. Diode is a unilateral element.

(vi) Bilateral Circuit : It is a circuit whose characteristics do not depend on the direction of current through the various elements. The transmission line is bilateral, because it can perform its function equally well in either direction. Network elements are considered bilateral if voltage and current relations are same irrespective of direction of current e.g. resistance, inductance .

(vii) Passive Network : It is a network which does not contain any energy source i.e. source of e.m.f. or current in it.

(viii) Active Network : It is a network which contains at least one energy source i.e. one or more than one source of e.m.f or current.

(ix) Node of a network is an equi potential surface at which two or more circuit elements are joined or connected together . For example terminals A, B, C and D are nodes in Fig. 2.1.

(x) Junction is that point in a network where three or more circuit elements are joined . In Fig.2.1, B and D are two junctions. The junction is sometimes also referred to as an **independent node.**

(xi) A **branch** is that part of a network which lies between two junction points. In Figure 2.1 ,there are three branches :

1. The branch 'DAB' consists of two circuit elements E_1 and R_1
2. Branch 'BCD' contains circuit elements E_2 and R_3 and
3. Third branch 'BD' containing circuit element R_2.

(xii) A **loop** is any closed path of the network. It is a closed path in a circuit in which no element or node is encountered more than once. In Figure 2.1, we have three loops i.e. : ABDA, DBCD and ABCDA.

(xiii) A **mesh** is the most elementary form of a loop. It cannot be further divided into other loops. It is a loop that contains no other loop within it. It can be called as a **'subset'** of a loop. In Figure 2.1, loops 'ABDA' and 'DBCD' can be called as meshes, but loop 'ABCDA' can not be called a mesh.

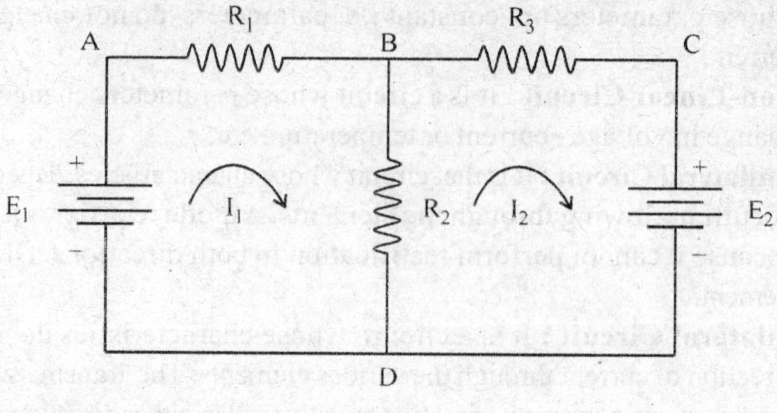

Fig.2.1

2.3 Source Transformations

Some of the networks may contain current sources as well as voltage sources. Such circuits can sometimes be simplified by converting voltage sources in to current sources or vice versa. The source transformations/ conversions are the analysis tools which are used to simplify the circuit analysis by reconfiguring the circuit. There are two types of source transformations :

(a) Converting a Voltage Source to a Current Source

(b) Converting a Current Source to a Voltage Source

In both of these transformations, internal resistance of the source has to be present and these transformations are valid for analyzing the circuits external to the terminals involved.

2.3.1 Conversion of a Voltage Source to a Current Source

This transformation converts a voltage source V in series with a resistance R in to an equivalent current source , I in parallel with the resistance R, as shown in Fig.2.2(a). Both of these configurations are equivalent, and produce the same effect on the external circuit. The value of the equivalent current source is the current that will flow in a short circuit between the points A and B. It is

$$I = \frac{V}{R}$$

Direction of the equivalent current source depends upon the polarity of the voltage source. It should be from negative to positive plate of the voltage source or we can say that it should be leaving the positive terminal. It is shown upwards in the figure2.2 (b) shown.

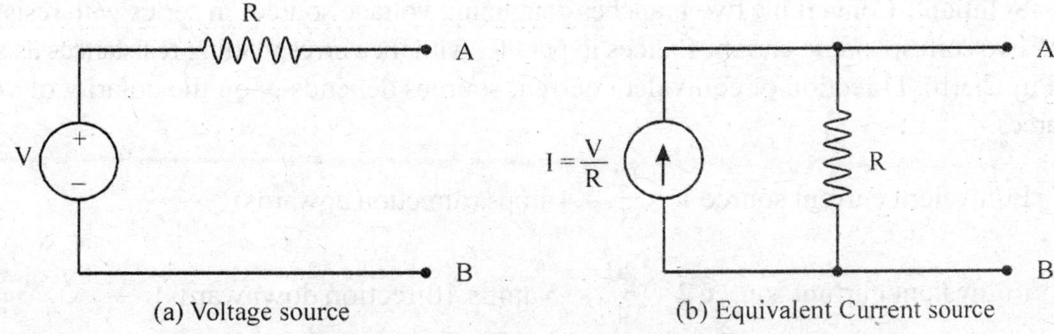

(a) Voltage source (b) Equivalent Current source

Fig. 2.2 : Conversion of a Voltage source to a Current source

2.3.2 Conversion of a Current source to a Voltage source

This transformation converts a current source in parallel with a resistance 'R' Fig 2.2(c) into an equiualent voltage source V, in series with the resistance R as shown in Fig.2.2(d). Both of these configurations are equivalent and produce the same effect on the external circuit. The value of the voltage source is the voltage that would appear across the open circuit terminals A and B. It is: $V = IR$

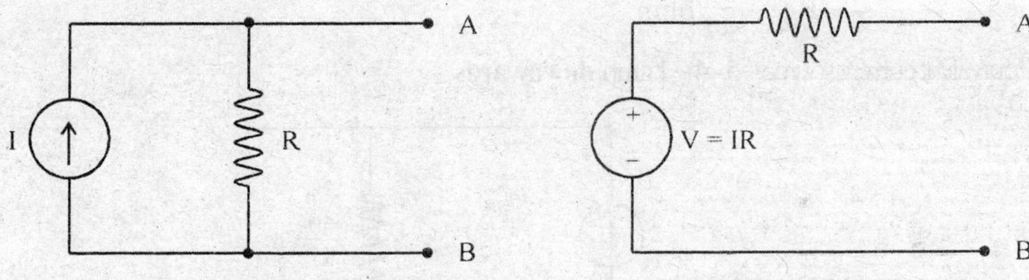

(c) Current source (d) Equivalent Voltage source

Polarity of the voltage source is indicated by placing +ve polarity on the head side of the current source arrow or we can say that positive terminal of the voltage source corresponds to the arrow head side from where current leaves the terminals of the current source.

Example given below illustrates the usefulness of source conversion in circuit simplification.

Example 2.1 : Use source transformation in the network shown in Fig.2.3(a) and find the equivalent network containing only one voltage source in series with a resistance.

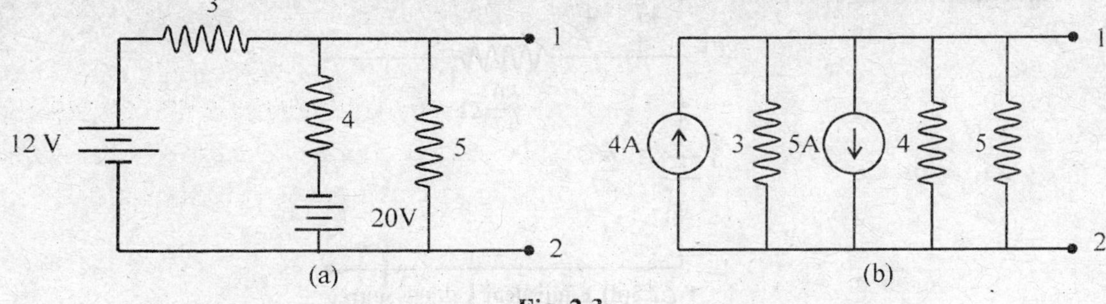

(a) (b)

Fig. 2.3

Solution : Convert the two branches containing voltage sources in series with resistances in to two corresponding current sources in parallel with the corresponding resistances as shown in Fig. 2.3(b). Direction of equivalent current sources depends upon the polarity of voltage sources.

Equivalent current source 1 $= \frac{12}{3} = 4$ amps.(direction upwards)

Equivalent current source 2 $= \frac{20}{4} = 5$ amps. (direction downwards)

It is seen in Fig.2.3(b) that resistances 3Ω , 4Ω and 5Ω are in parallel and the two current sources are in parallel . Combining them , the circuit can be redrawn as shown in Fig.2.3(c)

$$\frac{1}{R_{eq}} = \frac{1}{5} + \frac{1}{4} + \frac{1}{3} = \frac{12+15+20}{60} = \frac{47}{60}$$

i.e. $$R_{eq} = \frac{60}{47} \text{ ohm}$$

Equivalent current source = 5 - 4 = 1 amp. **downwards**

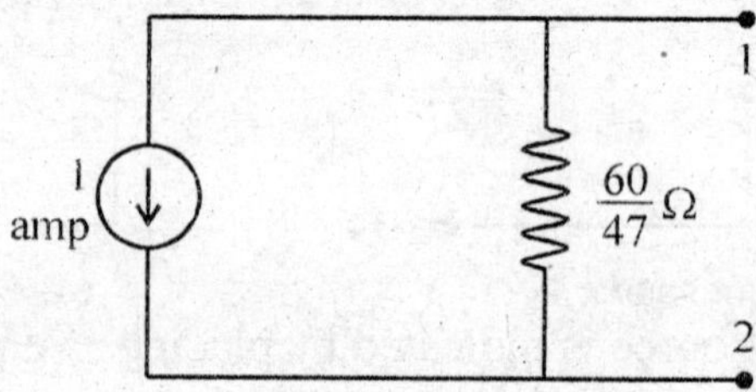

Fig.2.3(c) Equivalent current source

Converting the current source in parallel with the resistance shown in Fig.2.3(**c**) into an equivalent voltage source in series with a resistance, the resultant circuit is shown in Fig. 2.3(d)

Equivalent voltage source $= 1 \times \frac{60}{47} = \frac{60}{47}$ volts.

Fig.2.3(d) Equivalent voltage source

2.4 SERIES AND PARALLEL OPERATION OF VOLTAGE AND CURRENT SOURCES

Voltage and current sources can be operated in series and/or parallel.

(a) Series operation of Voltage Sources

Voltage sources in series can be operated without any difficulty. The equivalent voltage is the sum of individual source voltages. The equivalent resistance is the sum of individual source resistances as shown in Fig 2.4 (a) and(b)

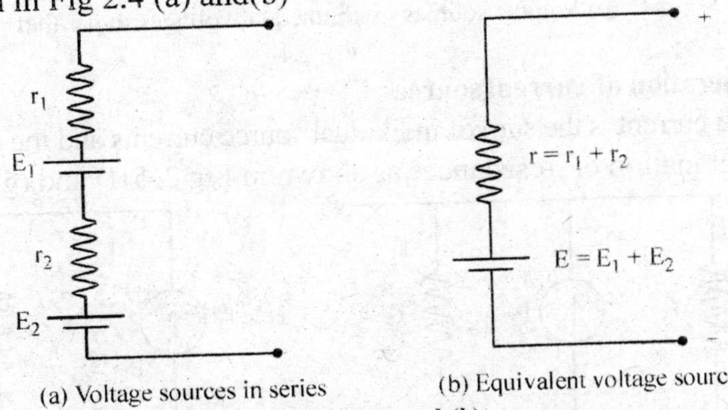

(a) Voltage sources in series (b) Equivalent voltage source

Fig. 2.4 (a) and (b)

The equivalent resistance is the series combination of source resistances and is given as follows.

$$\boxed{r = r_1 + r_2}$$

(b) Parallel operation of voltage sources

Voltage sources can be operated in parallel only if the source voltages are equal, Fig.2.4(c)

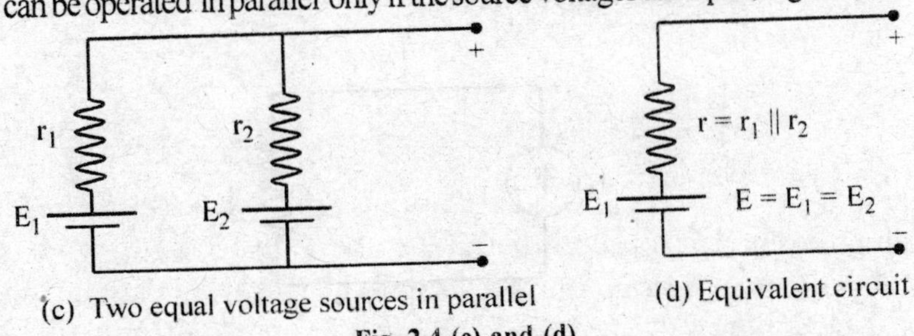

(c) Two equal voltage sources in parallel (d) Equivalent circuit

Fig. 2.4 (c) and (d)

The equivalent voltage is equal to the source voltages. The equivalent resistance is the equivalent parallel combination of source resistances as shown in Fig 2.4 (d)

$$r = \frac{r_1 r_2}{r_1 + r_2}$$

$$E = E_1 = E_2$$

Voltage sources with unequal voltages should not be connected in parallel , because the lower voltage source tends to overload the higher voltage source, since circulating currents will flow as shown in Fig.2.4(e)

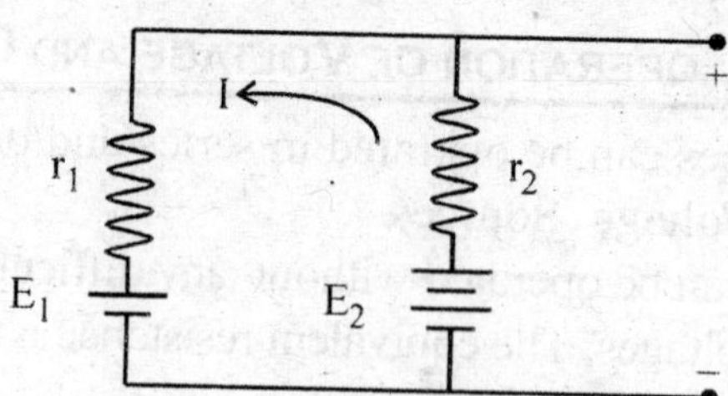

(e) Voltage sources with unequal voltages in parallel

Fig. 2.4

(c) Parallel operation of current sources

The equivalent current is the sum of individual source currents and the equivalent resistance is the parallel combination of resistances as shown in Fig 2.4 (f) and (g)

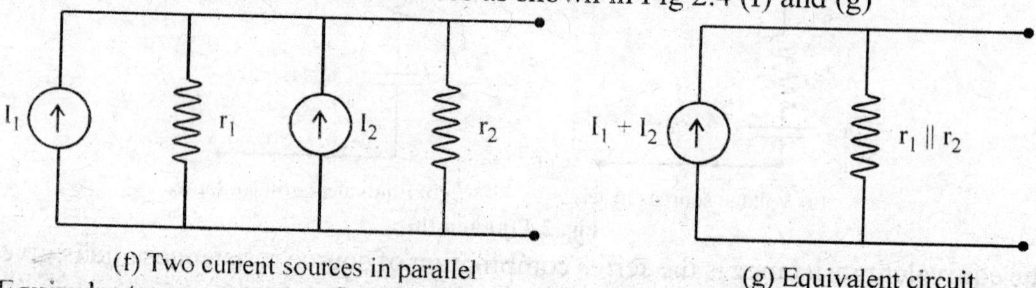

(f) Two current sources in parallel

(g) Equivalent circuit

Equivalent current source $I = I_1 + I_2$

Equivalent resistance, $r = \dfrac{r_1 r_2}{r_1 + r_2}$

(d) Series operation of current sources

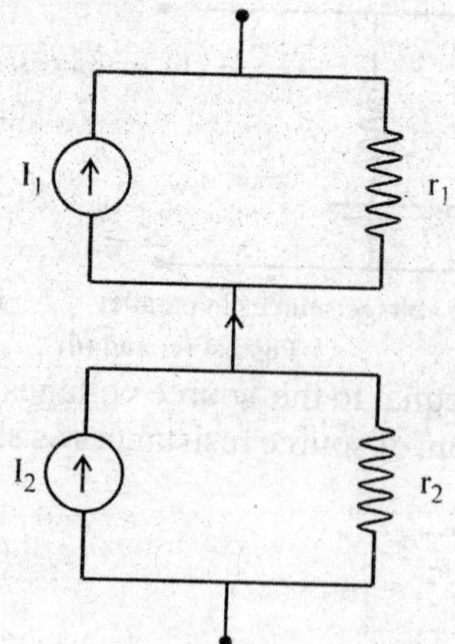

(h) Series operation of current sources

Figure 2.4 (h)

Series operation of current sources having different output currents as shown in Fig 2.4 (h), is not possible.

2.5 METHODS OF NETWORKS ANALYSIS

An electric circuit or network consists of a number of interconnected single circuit elements and it generally consists of at least one voltage or a current source. The different interconnections of elements lead to a different set of constraints between the currents and voltages. These constraints and the current voltage relationships of the individual current elements, provide the solution of the network.

The network analysis methods discussed so far are not applicable to all networks. There are circuits which can not be solved by combining resistances in series or parallel or through star/delta transformation. Voltage division and current division principles may also not be helpful.

In such cases, some other powerful techniques of circuit analysis are used and some of these are:

1. Network analysis using Kirchhoff's laws/The Branch current method
2. Network analysis by Mesh Currents (Mesh Analysis)
3. Circuit analysis by Node-Pair voltages (Nodal Analysis method)

Mesh analysis is based on Kirchhoff's Voltage Law, KVL. Nodal analysis is based on application of Kirchhoff's Current Law, KCL.

Any linear circuit can be solved by using these techniques by writing a set of simultaneous linear algebraic equations and solving them to obtain the required values of currents or voltages in different parts of the circuit.

2.5.1 NETWORK ANALYSIS USING KIRCHHOFF'S LAWS (THE BRANCH CURRENT METHOD)

Complex electrical circuits can be solved using KCL and KVL. To illustrate this method, consider the series parallel circuit shown in Fig. 2.5(a). It is desired to find different currents and voltages in this circuit.

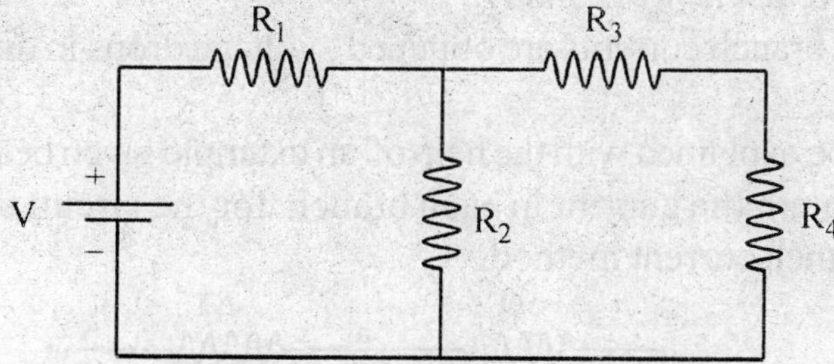

Fig. 2.5(a) Circuit to be analyzed

Following steps are followed in this method:

Step 1 : Identify and mark the nodes as A,B,C,D etc as shown in Fig. 2.5(b)

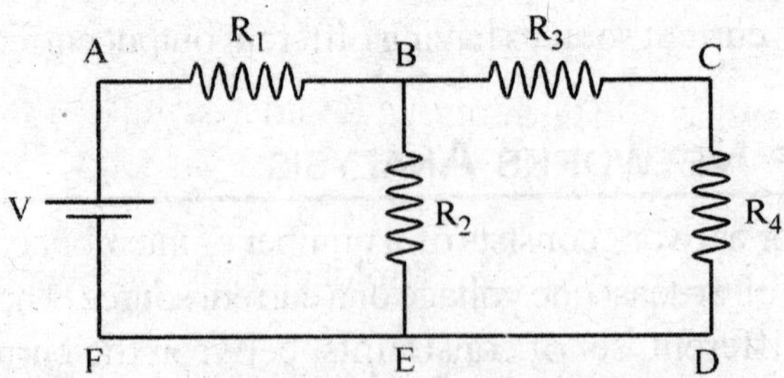

Fig. 2.5(b) Circuit with nodes/junctions identified

Step 2 : Identify the branch current directions and numbe. ʰe branch currents. Label the voltage drops in different resistances and mark their polarities as shown in Fig.2.5(c)

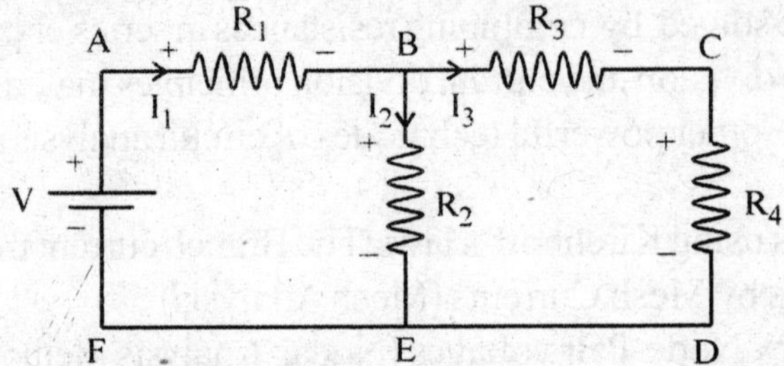

Fig.2.5(c)Branch currents and voltages identified

Step 3 : Write the equations using KVL for the closed paths in the circuit. For the closed path ABEFA, we can write:

$$V = I_1R_1 + I_2R_2 \qquad \text{(i)}$$

For the closed path BCDEB,

$$I_3R_3 + I_3R_4 = I_2R_2 \qquad \text{(ii)}$$

Applying KCL at node B, we get:

$$I_1 = I_2 + I_3 \qquad \text{(iii)}$$

Step 4 : Using the given values of V, R_1, R_2, R_3 and R_4, we can solve these three equations for three unknown currents i.e. I_1, I_2 and I_3

Step 5 : Once the branch currents are obtained, voltage drops in different branches can be found out.

This method can be explained with the help of an example given below:

Example 2.2 : Obtain the current in each branch for the circuit shown in Fig 2.6,using Kirchhoff's laws / Branch current method

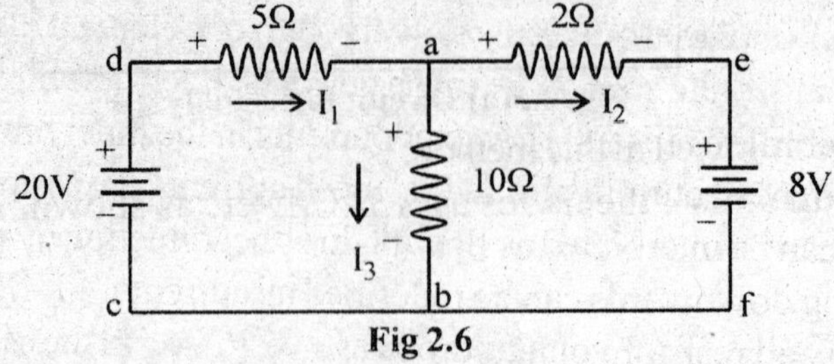

Fig 2.6

Solution

Mark the node'a' and assign currents I_1, I_2, I_3 to the three branches in the directions as shown in Fig 2.6

Applying KCL at node 'a',

i.e. sum of currents entering node'a' = sum of currents leaving node 'a' we get

$$I_1 = I_2 + I_3 \quad (i)$$

Applying KVL in loop 'abcd' ,Voltage at node'a' is = $20 - I_1\times(5) = I_3\times(10)$ (ii)

Applying KVL in loop 'aefb' this voltage is also equal to:

$$20 - I_1(5) = I_2(2) + 8 \quad (iii)$$

Solving equations (i), (ii) and (iii), we get the current in each branch as:

$I_1 = 2$ amp

$I_2 = 1$ amp

$I_3 = 1$ amp

Ans

2.5.2 Network Analysis By Mesh Current (Mesh or Loop Analysis)

In this method, mesh currents are used as circuit variables. Mesh current is that current which flows around the perimeter of a mesh or it is a current that circulates in a closed current path. These currents may not be measurable and are fictitious quantities and are introduced because circuits can be analysed in terms of minimum number of unknowns. Use of mesh currents reduces the number of equations to be solved.

Mesh analysis applies KVL to find unknown currents. This method is applicable to **planar circuits** only. A planar circuit is a circuit which can be drawn in a plane with no branches crossing one another; otherwise it is a non planar circuit. Mesh/Loop method is especially useful for circuits that have many nodes and very few meshes or loops. Mesh analysis is easier to apply when all the sources in a circuit are voltage sources.

Different steps to write mesh equations and to determine mesh currents are as follows:

Step 1 : If the circuit contains current sources, then convert them in to equivalent voltage sources in series with a resistance.(if possible)

Step 2 : Assume that each mesh contains a circulating unknown **mesh** or **loop** current .Assign mesh currents systematically such as $I_1, I_2, I_3, \ldots\ldots\ldots, I_n$ to the 'n' meshes. Although the directions are arbitrary, these may be assumed clockwise for consistency and to simplify the process of writing the relevant equations as shown in Fig.2.7

Step 3 : When several mesh currents flow through one element, the current in that element is considered to be the algebraic sum of the mesh currents. Each current in any outer branch will be equal to the particular mesh current. However currents in the inner branches can be expressed as the difference of two mesh currents, if all the mesh currents are assumed in the same direction. Any branch current can be interpreted as flowing in either direction by redefining it. For example, current $I_1 - I_2$ flowing downwards can be redefined as current $I_2 - I_1$ flowing upward.

Step 4 : Identify all resistor voltage drops as +ve to –ve in the direction of the loop current.

Step 5 : Identify all voltage sources according to their correct polarity.

Step 6 : Apply KVL to each of the 'n' meshes. Use Ohm's law to express voltages in terms of mesh currents. Write the equations for the voltage drops around each loop in turn, by moving in the direction of the particular mesh current and by equating the algebraic sum of the voltage drops to zero. Following this procedure, 'n' simultaneous equations will be obtained.

Step 7 : Solve these 'n' simultaneous algebraic equations by applying **Cramer's rule** or **Matrix Inversion** or any other method of solving the equations to get the values of un known mesh currents.

Step 8 : Current through any branch can then be obtained in terms of these loop currents.

Step 9 : Voltages across the elements and power loss in the element can then be calculated.

The method is explained by using a typical circuit diagram given in Fig. 2.7.

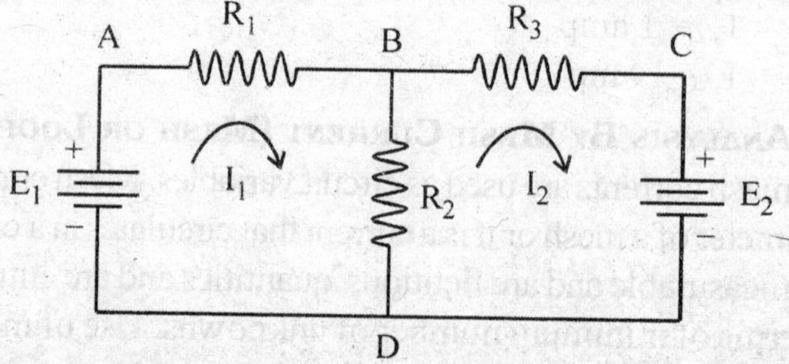

Figure 2.7 Mesh currents in a circuit

There are two independent meshes. Let the mesh currents be I_1 and I_2 in the clockwise direction as shown.

Applying KVL to mesh 1 :

$$R_1 I_1 + R_2 (I_1 - I_2) = E_1$$

Collecting the terms containing I_1 and I_2, we get:

i.e. $$\mathbf{(R_1 + R_2)\, I_1 - R_2 I_2 = E_1} \qquad \text{(i)}$$

Applying KVL to mesh 2:

$$R_3 I_2 + R_2 (I_2 - I_1) + E_2 = 0$$

Collecting the terms containing I_1 and I_2, we get:

i.e. $$\mathbf{-R_2 I_1 + (R_2 + R_3)\, I_2 = -E_2} \qquad \text{(ii)}$$

Thus two simultaneous algebraic equations i.e. (i) and (ii) have been obtained, which can be solved for the mesh currents I_1 and I_2 for given values of the circuit parameters.

2.5.2.1 Solution of Mesh Equations Using Matrix Inversion

Equations (i) and(ii) can be written in compact matrix form as:

$$RI = V \qquad (2.1)$$

Where R is an $n \times n$ resistance matrix, n is the number of meshes

I is a column vector representing independent mesh currents

V is a column vector representing the net voltage source that acts to produce current in the assumed clockwise direction.

From matrix equation (2.1), mesh current vector I can be calculated as follows:

$$I = [R]^{-1}[V] \qquad (2.2)$$

Where R^{-1} is inverse of the resistance matrix R

2.5.2.2 Cramer's Rule For Solving Simultabeous Linear Algebric Equations

Cramer's rule can be used to solve a set of 'n' simultaneous linear algebraic equations to obtain the values of 'n' unknown variables.

Suppose there are 'n' linear simultaneous algebraic equations as given below in which 'n' unknown currents are to be determined.

$$R_{11}I_1 + R_{12}I_2 + R_{13}I_3 + \text{----------} + R_{1n}I_n = E_1$$
$$R_{21}I_1 + R_{22}I_2 + R_{23}I_3 + \text{----------} + R_{2n}I_n = E_2$$
$$R_{31}I_1 + R_{32}I_2 + R_{33}I_3 + \text{----------} + R_{3n}I_n = E_3$$
$$\text{--------------------------------}$$
$$\text{--------------------------------}$$
$$R_{n1}I_1 + R_{n2}I_2 + \text{----------------} + R_{nn}I_n = E_n$$

Where R_{11} to R_{nn} and E_1 to E_{nn} are constants.

Step by step method for the solution

(i) Write the equations given above in matrix form as:

$$\begin{bmatrix} R_{11} & R_{12} & - & - & R_{1n} \\ R_{21} & R_{22} & - & - & R_{2n} \\ R_{31} & R_{32} & - & - & R_{3n} \\ - & - & - & - & - \\ R_{n1} & R_{n2} & - & - & R_{nn} \end{bmatrix} \begin{bmatrix} I_1 \\ I_2 \\ \\ \\ I_n \end{bmatrix} = \begin{bmatrix} E_1 \\ E_2 \\ \\ \\ E_N \end{bmatrix} \qquad (2.3)$$

(ii) Calculate the value of system determinant Δ, by considering only the coefficients of the variables i.e. currents in the above equation and written in matrix form as follows:

$$\text{i.e. System determinant, } \Delta = \begin{vmatrix} R_{11} & R_{12} & - & - & R_{1n} \\ R_{21} & R_{22} & - & - & R_{2n} \\ R_{31} & R_{32} & - & - & R_{3n} \\ - & - & - & - & - \\ R_{n1} & R_{n2} & - & - & R_{nn} \end{vmatrix}$$

(iii)Calculate the values of Sub determinants $\Delta_1, \Delta_2, \text{-----}, \Delta_n$ by replacing the corresponding column of Δ (1st ,2nd , - - - - - etc.) by the column of constants given on the RHS of matrix equation(2.3), containing $E_1, E_2, \text{-----}, E_n$ etc. as illustrated below:

$$\Delta_1 = \begin{vmatrix} E_1 & R_{12} & - & - & R_{1n} \\ E_2 & R_{22} & - & - & R_{2n} \\ E_3 & R_{32} & - & - & R_{3n} \\ - & - & - & - & - \\ E_n & R_{n2} & - & - & R_{nn} \end{vmatrix}$$

Here ,first column of resistance matrix has been replaced by column of constants , containing E_1 , E_2 etc.

$$\Delta_2 = \begin{vmatrix} R_{11} & E_1 & - & - & R_{1n} \\ R_{21} & E_2 & - & - & R_{2n} \\ R_{31} & E_3 & - & - & R_{3n} \\ - & - & - & - & - \\ R_{n1} & E_n & - & - & R_{nn} \end{vmatrix}$$

Here second column of resistance matrix has been replaced by column of constants, containing E_1 , E_2 etc.

(iv)Unknown loop currents can now be calculated as follows, after calculating different values of Δ_i, i =1, 2 , —, n and system determinant, Δ

i.e. $I_1 = \dfrac{\Delta_1}{\Delta}$, $I_2 = \dfrac{\Delta_2}{\Delta}$,-------------------------------------, $I_n = \dfrac{\Delta_n}{\Delta}$

Example 2.3 : Find mesh currents I_1, I_2 in the circuit shown in Figure 2.8

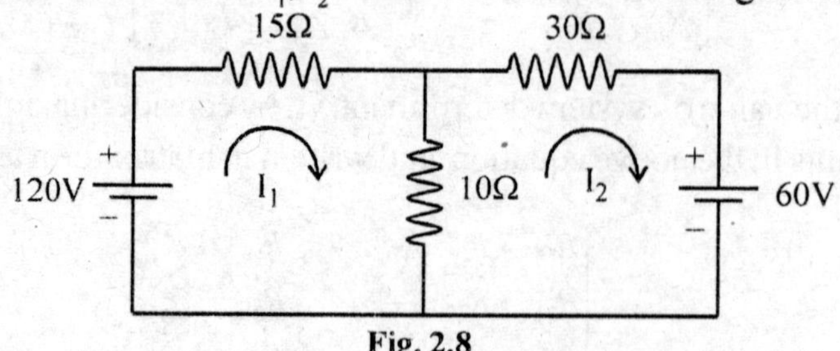

Fig. 2.8

Solution :

Let I_1 and I_2 be the mesh currents in the clockwise direction as shown in the Figure.2.8

Applying KVL to Loop 1

$15I_1 + 10\,(I_1 - I_2) = 120$

Or $(15+10)\,I_1 - 10\,I_2 = 120$

Dividing this equation by 5, we get:

$5I_1 - 2I_{-2} = 24$ (i)

Applying KVL to Loop 2

$$30\,I_2 + 60 + 10(I_2 - I_1) = 0$$

or $$-10\,I_1 + (30+10)\,I_2 + 60 = 0$$

Dividing this equation by 10 and rearranging, we get:

$$-I_1 + 4I_2 = -6 \qquad \text{(ii)}$$

Equations (i) and(ii) can be solved using any one of the following methods.

(i) Using Cramer's rule.

Equations (i) and(ii) can be written in matrix form as follows:

$$\begin{bmatrix} 5 & -2 \\ -1 & 4 \end{bmatrix}\begin{bmatrix} I_1 \\ I_2 \end{bmatrix} = \begin{bmatrix} 24 \\ -6 \end{bmatrix}$$

System determinant, $\Delta = \begin{vmatrix} 5 & -2 \\ -1 & 4 \end{vmatrix} = 20\text{-}2 = 18$

Sub determinant $\Delta_1 = \begin{vmatrix} 24 & -2 \\ -6 & 4 \end{vmatrix} = 96 - 12 = 84$

$\Delta_2 = \begin{vmatrix} 5 & 24 \\ -1 & -6 \end{vmatrix} = -30 + 24 = -6$

Loop/mesh currents can now be obtained as follows:

$$I_1 = \frac{\Delta_1}{\Delta} = \frac{84}{18} = \frac{14}{3} = 4.67 \text{ amp} \qquad \textbf{Ans}$$

$$I_2 = \frac{\Delta_2}{\Delta} = \frac{-6}{18} = \frac{-1}{3} = -0.33 \text{ amp} \qquad \textbf{Ans}$$

(i) Using matrix inversion

Equations (i) and (ii) in matrix form are :

$$\begin{bmatrix} 5 & -2 \\ -1 & 4 \end{bmatrix}\begin{bmatrix} I_1 \\ I_2 \end{bmatrix} = \begin{bmatrix} 24 \\ -6 \end{bmatrix}$$

Solving this matrix equation, we get:

$$\begin{bmatrix} I_1 \\ I_2 \end{bmatrix} = \begin{bmatrix} 5 & -2 \\ -1 & 4 \end{bmatrix}^{-1} \begin{bmatrix} 24 \\ -6 \end{bmatrix} = \frac{1}{20-2}\begin{bmatrix} 4 & 2 \\ 1 & 5 \end{bmatrix}\begin{bmatrix} 24 \\ -6 \end{bmatrix} = \begin{bmatrix} \frac{84}{18} \\ \frac{-6}{18} \end{bmatrix} = \begin{bmatrix} \frac{14}{3} \\ \frac{-1}{3} \end{bmatrix} = \begin{bmatrix} 4.67 \\ -0.33 \end{bmatrix}$$

i.e.Loop/ Mesh currents are : $I_1 = 4.67$ amp and $I_2 = -0.33$ amp **Ans**

2.5.2.3 Mesh Analysis In Circuits Containing Current Sources

Applying mesh analysis to circuits containing current sources may appear complicated, because voltage across its terminals is not known. A common mistake made by the students is to assume the voltage across current source as zero.

However it is easy to handle such cases, because the number of equations to be solved is reduced. Following types of cases may be encountered.

(i) When a current source exists only in one mesh. Value of this current is assigned to that mesh current and KVL equations for other meshes are written in a usual way.

(ii) When a current source exists between two meshes. In such cases, a **supermesh** is created, whose interior is that of the corresponding two meshes, at the periphery of two meshes by excluding the branch containing current source and any elements connected in series with the current source. Two different mesh currents flow in a supermesh.

The Supermesh

Two meshes when they have a common dependent or an independent current source normally form a **supermesh.**

If the current source lies on the perimeter of the circuit, then the single mesh in which it is found is ignored. Kirchhoff's voltage law is applied only to those meshes or supĕr meshes in the reinterpreted network.

Properties of a supermesh

1 The current source in the supermesh provides the constraint equation for the solution of mesh currents

2 A supermesh has no current of its own.

3 A supermesh requires the application of both KVL and KCL

Examples given below illustrate the concept of a 'supermesh'

Example 2.4 Find mesh currents I_1, I_2 and I_3 using mesh analysis in the circuit shown in Fig 2.9

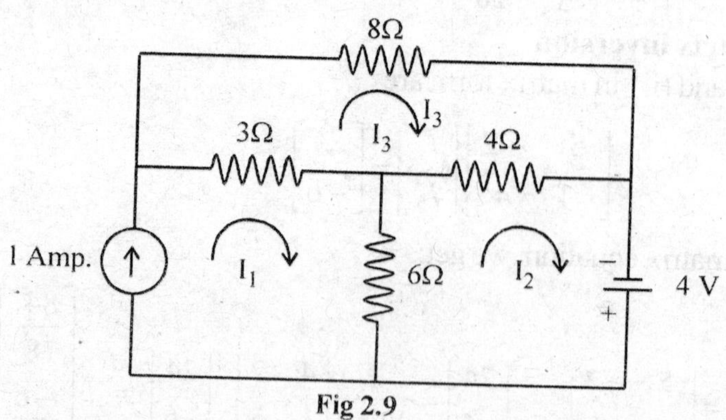

Fig 2.9

Solution

In this circuit, current source lies only in one mesh and lies on its perimeter. Hence value of this current is assigned to that mesh current i.e. I_1 =1.

As $I_1 = 1$ amp is known, KVL for mesh 1 is not required.
Applying KVL to meshes 2 and 3, we get:

Mesh 2

$$6(I_2 - I_1) + 4(I_2 - I_3) = 4 \quad \text{(i)}$$

Mesh 3

$$3(I_3 - I_1) + 8I_3 + 4(I_3 - I_2) = 0 \quad \text{(ii)}$$

Substituting $I_1 = 1$ in equations (i) and (ii), we get:

$$10I_2 - 4I_3 = 10 \quad \text{(iii)}$$

$$-4I_2 + 15I_3 = 3 \quad \text{(iv)}$$

Writing the equations(iii) and (iv) in matrix form, we get:

$$\begin{bmatrix} 10 & -4 \\ -4 & 15 \end{bmatrix} \begin{bmatrix} I_2 \\ I_3 \end{bmatrix} = \begin{bmatrix} 10 \\ 3 \end{bmatrix}$$

i.e.

$$\begin{bmatrix} I_2 \\ I_3 \end{bmatrix} = \begin{bmatrix} 10 & -4 \\ -4 & 15 \end{bmatrix}^{-1} \begin{bmatrix} 10 \\ 3 \end{bmatrix} = \frac{1}{134}\begin{bmatrix} 15 & 4 \\ 4 & 10 \end{bmatrix} \begin{bmatrix} 10 \\ 3 \end{bmatrix} = \begin{bmatrix} \frac{162}{134} \\ \frac{70}{134} \end{bmatrix}$$

Hence, the required loop/mesh currents are:

$$I_2 = \frac{162}{134} = 1.208 \text{ amp.} \quad \textbf{Ans}$$

$$I_3 = \frac{70}{134} = 0.522 \text{ amp} \quad \textbf{Ans}$$

$$I_1 = 1 \text{ amp} \quad \textbf{Ans}$$

Example 2.5 : Obtain the mesh currents for the circuit shown in Fig 2.10(a)

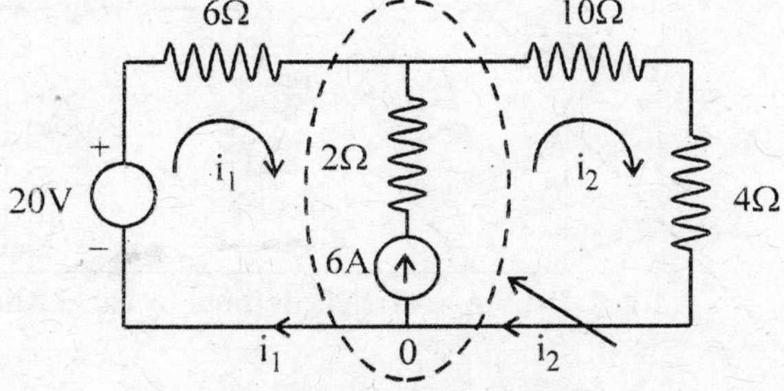

Fig.2.10(a) : Supermesh

Solution :

In this example, a current source of 6 amp. is present between two meshes. Hence, create a supermesh at the periphery of the two meshes excluding the current source and 2 ohm resistance connected in series with it, as shown in Fig.2.10(b) by dashed lines.

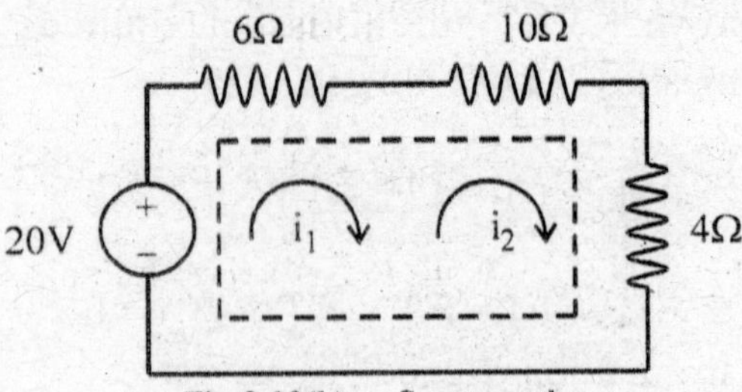

Fig 2.10(b) : Supermesh

It can be seen that current i_1 flows through 6 ohm resistance and current i_2 flows through 10 ohm and 4 ohm resistances i.e. two different currents flow in a supermesh.

A supermesh must satisfy KVL like any other mesh.

(1)Applying KVL to the supermesh in Fig.2.10(b) gives:

$$-20 + 6i_1 + 10i_2 + 4i_2 = 0$$

(voltage drops have been shown with +ve sign)

i.e. $$6i_1 + 14i_2 = 20 \qquad \text{(i)}$$

(2)Apply KCL to a node in the common branch where the two meshes intersect or write the relationship between the source current with the assumed mesh currents.

i.e. Applying KCL to node '0' in Fig. 2.10(a) gives:

$$i_2 = i_1 + 6 \qquad \text{(ii)}$$

Solving Equations (i) and (ii), we get :

$$i_1 = -3.2 \text{ amp}$$

$$i_2 = 2.8 \text{ amp}$$

Ans

Example 2.6 : Use the technique of mesh analysis to evaluate the three mesh currents in Figure 2.11

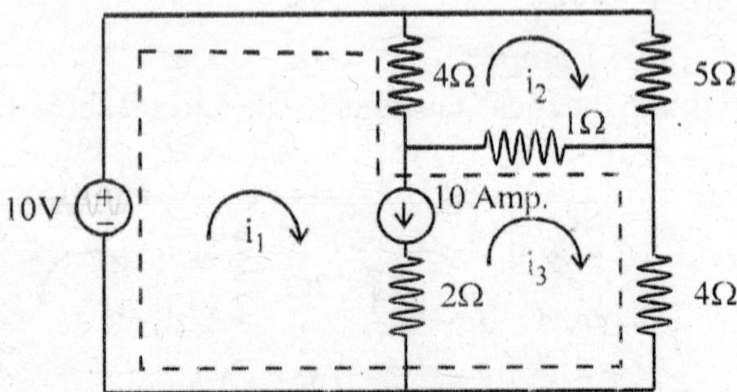

Fig.2.11 : Supermesh is defined by the dashed line.

Solution :

Mesh currents i_1, i_2 and i_3 have already been assigned as shown in Fig2.11. Here a 10 amp independent current source is in the common boundary of two meshes i.e. meshes 1 and 3. This current source requires us to create a **supermesh** whose interior is that of meshes 1 and 3 as shown by dashed lines in Figure 2.11 and the branch containing the current source is excluded from the supermesh.

(i) Applying KVL to the supermesh shown in Figure 2.11, we get:

$$-10 + 4(i_1 - i_2) + 1(i_3 - i_2) + 4i_3 = 0$$

or $$4\, i_1 - 5\, i_2 + 5\, i_3 = 10 \qquad \text{(i)}$$

It may be noted that currents i_1, i_2 and i_3 are assumed to flow, as usual, in the supermesh also.

(ii) Writing KVL around mesh 2, we get:

$$4(i_2 - i_1) + 5\, i_2 + 1\,(i_2 - i_3) = 0$$

or $$-4\, i_1 + 10\, i_2 - i_3 = 0 \qquad \text{(ii)}$$

Adding equations (i) and (ii), we get:

$$5\, i_2 + 4\, i_3 = 10$$

(iii) The independent source current is related to the assumed mesh currents as follows:

$$i_1 - i_3 = 10$$

or $$i_3 = i_1 - 10 \qquad \text{(iii)}$$

Substituting this value of i_3 in equations (i) and (ii) and solving for i_1, i_2 and i_3, we get the loop currents as:

$$i_1 = \frac{110}{13} \text{ amp}$$

$$i_2 = \frac{42}{13} \text{ amp}$$

$$i_3 = \frac{-20}{13} \text{ amp}$$

Example 2.7 : Determine the mesh currents i_1, i_2, i_3 in the circuit of Figure 2.12, using mesh analysis.

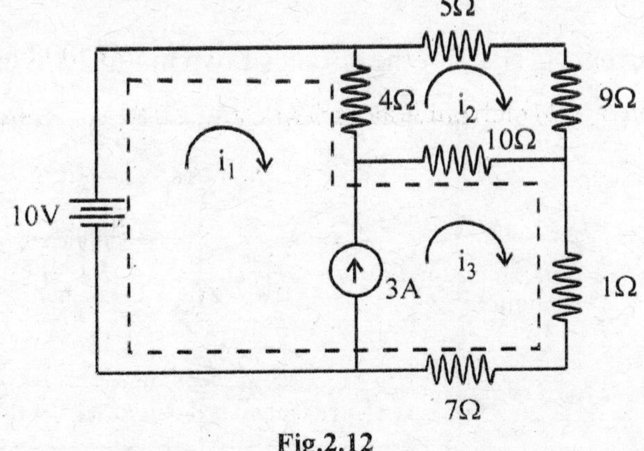

Fig.2.12

Solution

Here a 3 amp. current source is present between meshes 1 and 3. A **supermesh** is created at the periphery of two meshes excluding the current source as shown by dashed lines in Figure 2.12.

(1) Applying KVL to the Super mesh

$$4(i_1 - i_2) + 10(i_3 - i_2) + 1(i_3) + 7\,i_3 = 10$$

i.e. $4i_1 - 14\,i_2 + 18\,i_3 = 10$ (i)

(2) Applying KVL to the Loop 2

$$5\,i_2 + 9\,i_2 + 10(i_2 - i_3) + 4(i_2 - i_1) = 0$$

i.e. $28\,i_2 - 10\,i_3 - 4\,i_1 = 0$ (ii)

(3) Source current of 3 amp. is related to the assumed mesh currents as:

$i_3 - i_1 = 3$ (iii)

Solving equations (i), (ii) and (iii), we get:

$$i_1 = \frac{-29}{15} = -1.93 \text{ amp}$$ **Ans**

$$i_2 = \frac{11}{105} \text{ amp}$$ **Ans**

$$i_3 = \frac{16}{15} \text{ amp}$$ **Ans**

2.5.2.5 Presence of one or more dependent / controlled sources in a circuit

In such cases , at first mesh equations are written exactly as in the case of networks with independent sources. Then controlling variables are expressed in terms of mesh current variables and are substituted in to the mesh equations.

In Example given below , both of the dependent and an independent current sources are included in the network to show how the analysis of such circuits is affected .

Example 2.8

Obtain loop currents i_1, i_2, i_3 for the circuit shown in Fig 2.13, using mesh analysis.

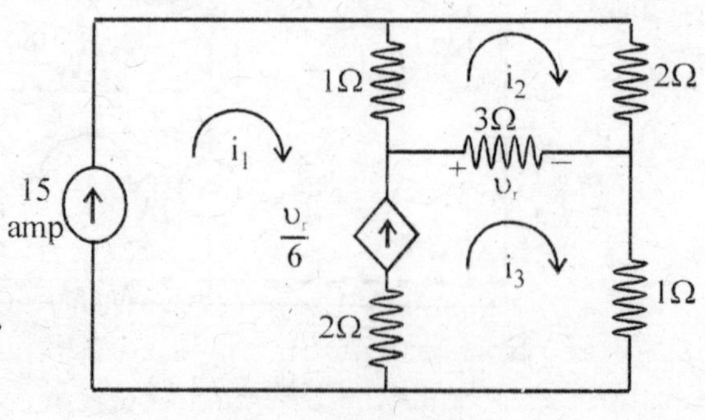

Fig.2.13

Solution

The current sources appear in meshes 1 and 3. Since 15 amp independent current source is located in mesh 1, on the perimeter of the circuit, mesh 1 may be eliminated from consideration

The dependent current source is on the perimeter of mesh 3 . Hence we do not write any equation for mesh 3. Thus only mesh 2 remains for which mesh equation is to be written.

(i) In Mesh 1 , it is seen that current source lies on the periphery of loop 1 , therefore value of this current is assigned to this mesh current , i_1:

i.e. $i_1 = 15$ (i)

(ii) Applying KVL to the Mesh 2

$$1(i_2 - i_1) + 2i_2 + 3(i_2 - i_3) = 0 \quad \text{(ii)}$$

i.e. $6i_2 - i_1 - 3i_3 = 0$

or $6i_2 - 3i_3 = i_1 = 15$

Dividing this equation by 3 , we get:

$$2i_2 - i_3 = 5 \quad \text{(iii)}$$

(iii) Relating the dependent current source with the assumed mesh currents , we get:

$$\frac{v_x}{6} = i_3 - i_1 \quad \text{(iv)}$$

and $v_x = 3(i_3 - i_2)$ (v)

Substituting equation (v) in (iv), we get:

$$\frac{3(i_3 - i_2)}{6} = i_3 - i_1$$

i.e. $i_3 - i_2 = 2i_3 - 2i_1$

or $2i_1 = i_3 + i_2$

or $i_3 + i_2 = 2 \times 15 = 30$ (vi)

Solving equations (iii) and (vi), we get the mesh currents as:

$$i_1 = 15 \text{ amp}$$

$$i_2 = \frac{35}{3} \text{ amp}$$

$$i_3 = \frac{55}{3} \text{ amp}$$

2.5.2.4 Mesh Analysis By Inspection

In general, if the circuit has 'N' meshes, the mesh- current equations can be expressed in matrix form in terms of the resistances as follows:

$$\begin{bmatrix} R_{11} & R_{12} & \cdots & R_{1N} \\ R_{21} & R_{22} & \cdots & R_{2N} \\ \vdots & \vdots & \vdots & \vdots \\ R_{N1} & R_{N2} & \cdots & R_{NN} \end{bmatrix} \begin{bmatrix} I_1 \\ I_2 \\ \vdots \\ I_N \end{bmatrix} = \begin{bmatrix} V_1 \\ V_2 \\ \vdots \\ V_N \end{bmatrix} \quad (2.4)$$

or simply $[R][I]=[V]$

Where,

R_{kk} = Sum of all the resistances in mesh 'k'

$R_{kj} = R_{jk}$ = Negative of the sum of the resistances that are in common with meshes k and j, $k \neq j$

i_k = Unknown mesh current for mesh k in the clockwise direction

v_k = Sum taken clockwise of all independent voltage sources in mesh k, with voltage rise treated as positive and voltage drops as negative.

R is called the Resistance matrix; I is the mesh current vector; and V is the supply voltage vector.

Eq. (2.4) can be solved to obtain the unknown mesh currents.

Consider the circuit shown in Fig. 2.14

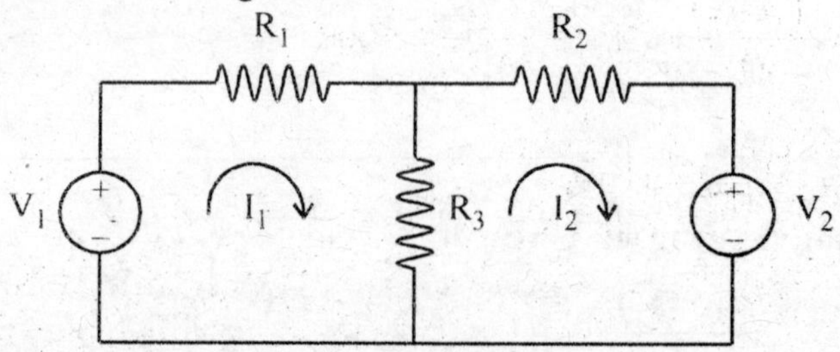

Fig. 2.14 Mesh analysis by Inspection

The circuit has two meshes. Let the mesh currents be i_1 and i_2.

Mesh equations can be written in matrix form by inspection as follows:

$$\begin{bmatrix} R_1+R_3 & -R_3 \\ -R_3 & R_2+R_3 \end{bmatrix}\begin{bmatrix} I_1 \\ I_2 \end{bmatrix}=\begin{bmatrix} V_1 \\ -V_2 \end{bmatrix} \quad (2.5)$$

Where $R_{11} = R_1 + R_3$ and $R_{22} = R_2 + R_3$ i.e. sum of the resistances in meshes 1 and 2 respectively.

and $R_{12} = R_{21} = -R_3$ i.e. negative of the resistance which is common to meshes 1 and 2.

2.5.3 Circuit Analysis By Node-Pair Voltage (Nodal Analysis Methode)

Nodal analysis is a general technique which can be applied to any circuit. Nodal analysis is especially useful for circuits that have many branches and very few nodes.

This method can be applied easily in the following two different cases:

(i) All voltage sources that are present in the circuit are connected to the common ground

(ii) All sources in the circuit are current sources.

2.5.3.1 Selecting the Reference Node

In this method, a common ground point should be selected as a reference node and all voltages should be measured with respect to this common ground. If a ground point is not available, one of the junctions is taken as the **common ground or reference junction.** Solution can be simplified by selecting one end of a voltage source as the reference node.

Then, with reference to this junction ,other junctions are assumed to have node voltages which are designated V_1, V_2, etc. Node equations are then written by applying KCL at different nodes.

Number of independent node pair equations is equal to the number of junctions in the network minus one.

i.e. $n = j-1$

where n = No. of independent node equations

j = No. of junctions

2.5.3.2 Summary of Steps for Solution by Nodal Analysis Method/Node-Pair Voltage Method

For applying the Node-Pair Voltage method , following steps are followed:

Step 1 : Select one of the nodes as a reference node, as explained above.

Step 2 : With respect to this reference node , assume node voltages at all other nodes and assign them the voltages V_1 V_2 etc.

Step 3 : Note the values of voltages at the nodes at which known voltage sources are connected.

Step 4 : Write a KCL for each node. Assume the currents flowing through the resistances as leaving a given node. It is assumed that the node potential is always higher than the other voltages appearing in the equations.

A set of algebraic equations containing unknown voltages will thus be obtained.

Step 5 : Solve these algebraic equations for unknown node voltages.

As an illustration, consider the circuit shown in Figure 2.15.

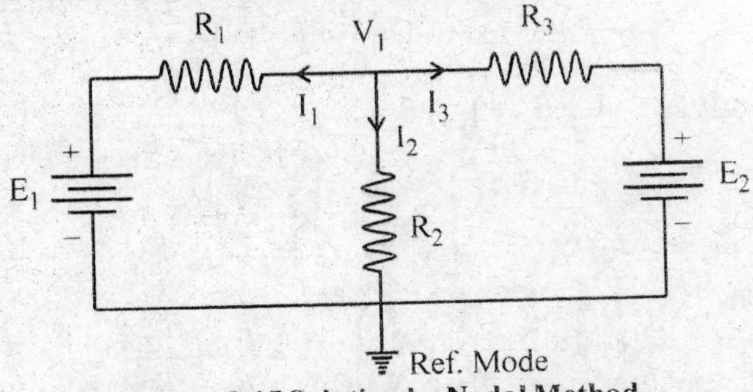

Figure 2.15 Solution by Nodal Method

The only potential which is unknown is that of node1. If this potential, V_1, can be found, then each branch current can be determined because the voltage drop across each resistor will then be known.

Applying Kirchhoff's current law at junction 1, we get:

$$I_1 + I_2 + I_3 = 0 \tag{2.6}$$

However, $I_1 = \frac{V_1 - E_1}{R_1}$, $I_2 = \frac{V_1}{R_2}$ and $I_3 = \frac{V_1 - E_2}{R_3}$

Substituting these values of currents in equation (2.6), we get:

$$\frac{V_1 - E_1}{R_1} + \frac{V_1}{R_2} + \frac{V_1 - E_2}{R_3} = 0$$

This equation is known as node pair equation and can be solved to obtain the value of unknown node voltage V_1, as values of other parameters are known.

Example 2.9 : For the circuit shown in Figure 2.16, write the nodal equations.

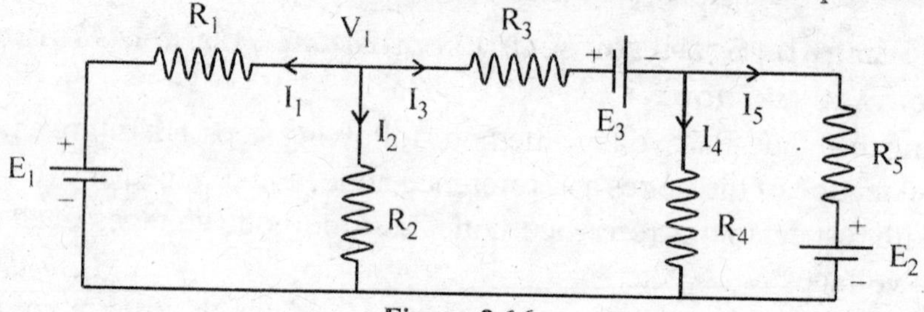

Figure 2.16

Solution :

Mark the reference node, nodes 1 and 2 and currents I_1 I_2, I_3, I_4 and I_5 in the circuit diagram.

KCL at node 1, gives : $I_1 + I_2 + I_3 = 0$

i.e. $$\frac{V_1 - E_1}{R_1} + \frac{V_1}{R_2} + \frac{V_1 + E_3 - V_2}{R_3} = 0$$

or $$V_1\left(\frac{1}{R_1} + \frac{1}{R_2} + \frac{1}{R_3}\right) - \frac{1}{R_3}V_2 = \frac{E_1}{R_1} - \frac{E_3}{R_3} \qquad \text{(i)}$$

KCL at node 2, gives : $-I_3 + I_4 + I_5 = 0$

i.e. $$\left(\frac{(V_2 - E_3) - V_1}{R_3}\right) + \frac{V_2}{R_4} + \frac{V_2 - E_2}{R_5} = 0$$

or $$-\frac{V_1}{R_3} + \left(\frac{1}{R_4} + \frac{1}{R_3} + \frac{1}{R_5}\right)V_2 = \frac{E_3}{R_3} + \frac{E_2}{R_5} \qquad \text{(ii)}$$

Equations (i) and (ii) are the required nodal equations.

Example 2.10 : Find node voltage V_1 and branch currents I_1, I_2 and I_3 for the circuit shown in Fig. 2.17

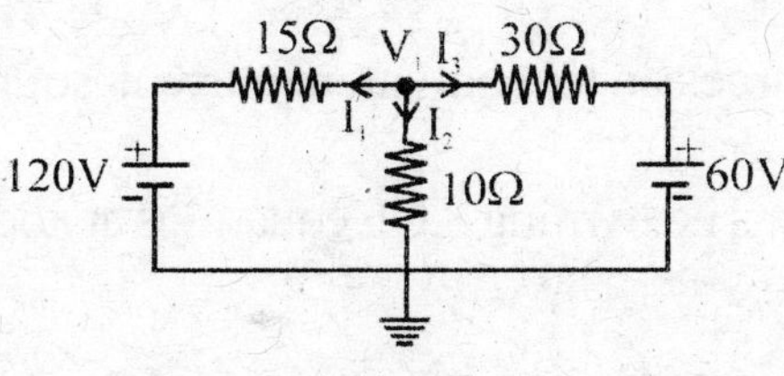

Fig 2.17

Solution :

Let voltage at node'1' be V_1

Nodal equation at node 1 is: $I_1+I_2+I_3=0$

i.e. $$\frac{V_1-120}{15}+\frac{V_1}{10}+\frac{V_1-60}{30}=0$$

i.e. $$\frac{2V_1-240+3V_1+V_1-60}{30}=0$$

or $6\,V_1 = 300$

i.e. $V_1 = \frac{300}{6} = 50$ volts **Ans**

Branch currents I_1, I_2 and I_3 can be calculated as follows:

$$I_1 = \frac{V_1-120}{15} = \frac{-70}{15} = -4.67 \text{ amp}$$ **Ans**

$$I_2 = \frac{V_1}{10} = \frac{50}{10} = 5 \text{ amp}$$

$$I_3 = \frac{V_1-60}{30} = \frac{-10}{30} = -\frac{1}{3} = -0.333 \text{ amp}$$ **Ans**

Example 2.11 : Find V_1 and V_2 in the circuit given in Fig.2.18(a) , using nodal analysis method

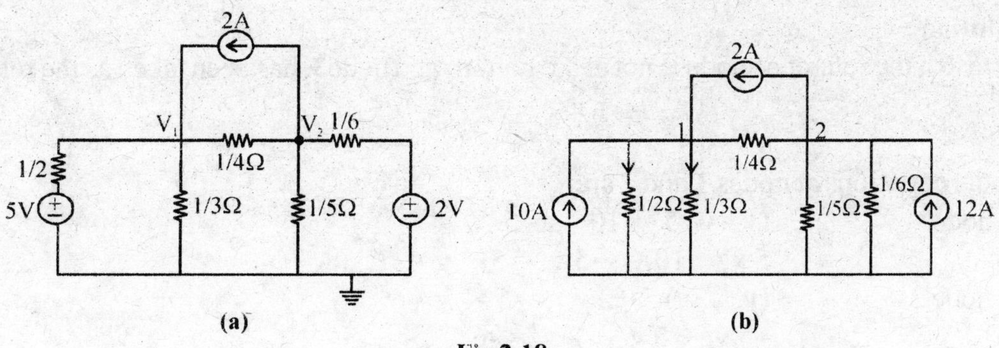

Fig 2.18

Solution :

Convert two voltage sources in to equivalent current sources and redraw the circuit as shown in Fig 2.18(b).

Writing the nodal equations i.e. writing KCL equations at nodes 1 and 2 , in matrix form, we get:

$$\begin{bmatrix} 2+3+4 & -4 \\ -4 & 5+6+4 \end{bmatrix} \begin{bmatrix} V_1 \\ V_2 \end{bmatrix} = \begin{bmatrix} 10+2 \\ 12-2 \end{bmatrix} = \begin{bmatrix} 12 \\ 10 \end{bmatrix}$$

i.e.
$$\begin{bmatrix} V_1 \\ V_2 \end{bmatrix} = \begin{bmatrix} 9 & -4 \\ -4 & 15 \end{bmatrix}^{-1} \begin{bmatrix} 12 \\ 10 \end{bmatrix}$$

i.e.
$$\begin{bmatrix} V_1 \\ V_2 \end{bmatrix} = \frac{1}{119} \begin{bmatrix} 15 & 4 \\ 4 & 9 \end{bmatrix} \begin{bmatrix} 12 \\ 10 \end{bmatrix} = \begin{bmatrix} \frac{220}{119} \\ \frac{138}{119} \end{bmatrix} = \begin{bmatrix} 1.848 \\ 1.159 \end{bmatrix}$$

Hence the required nodal voltages are:

$V_1 = 1.848$ Volts **Ans**

$V_2 = 1.159$ volts **Ans**

Example 2.12 : In the network shown in Fig. 2.19, find the current in the $\frac{1}{10}\Omega$ resistor.

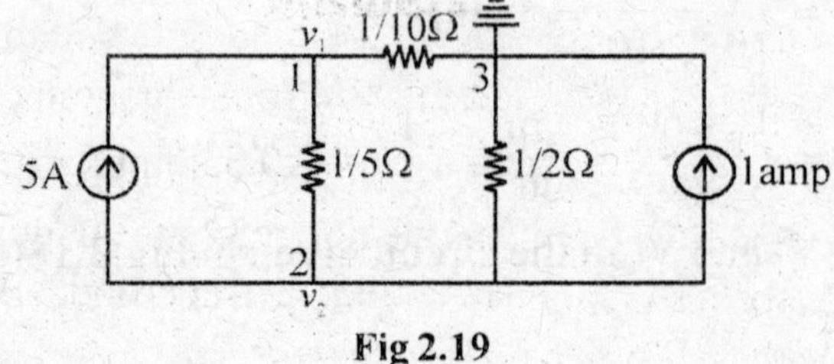

Fig 2.19

Solution :

Note that the reference node is not at the bottom and node3, has been taken as the reference node i.e.

$$v_3 = 0.$$

Nodal equations at nodes 1 and 2 are:

At node 1 : $(v_1 - v_2)5 + 10\, v_1 = 5$

$\Rightarrow$ $5\, v_1 + 10\, v_1 - 5\, v_2 = 5$ (i)

At node 2 : $(v_2 - v_1)\, 5 + 2\, v_2 = -5 - 1 = -6$

$\Rightarrow$ $-5\, v_1 + 5\, v_2 + 2\, v_2 = -6$ (ii)

Writing equations (i) and(ii) in matrix form, we get :

$$\begin{bmatrix} 5+10 & -5 \\ -5 & 5+2 \end{bmatrix} \begin{bmatrix} v_1 \\ v_2 \end{bmatrix} = \begin{bmatrix} 5 \\ -6 \end{bmatrix}.$$

or

$$\begin{bmatrix} v_1 \\ v_2 \end{bmatrix} = \begin{bmatrix} 15 & -5 \\ -5 & 7 \end{bmatrix}^{-1} \begin{bmatrix} 5 \\ -6 \end{bmatrix}$$

i.e.

$$\begin{bmatrix} v_1 \\ v_2 \end{bmatrix} = \frac{1}{80}\begin{bmatrix} 7 & 5 \\ 5 & 15 \end{bmatrix} \begin{bmatrix} 5 \\ -6 \end{bmatrix} = \begin{bmatrix} \frac{5}{80} \\ \frac{-65}{80} \end{bmatrix} = \begin{bmatrix} \frac{1}{16} \\ \frac{-13}{16} \end{bmatrix}$$

Hence the required nodal voltages are:

$$V_1 = \frac{1}{16} \text{ volts}$$ **Ans**

$$V_2 = \frac{-13}{16} \text{ volts}$$ **Ans**

2.5.3.3 Nodal Analysis With Ideal Voltage Sources Present in The Circuit

There are two possibilities:

1. A voltage source is connected between the reference node and a non reference node. In such a case, set the voltage at the non reference node equal to the voltage of the voltage source and the analysis is simplified by knowing the voltage at this node.
2. A voltage source either with parallel resistance or without parallel resistance, dependent or independent, is connected between two non reference nodes. In such cases, the two non reference nodes form a Generalized Node or Supernode. Both KCL and KVL are applied to determine the unknown node voltages.

2.5.3.4 The Supernode

Consider the circuit shown in Figure 2.20 (a), in which a voltage source is connected between two reference nodes.

(1) Assign the node-to-reference voltages v_1, v_2 and v_3.

(2) Apply KCL at each of three non reference nodes.

However, we will run into difficulty at both nodes 2 and 3, because we do not know the current in the branch having the voltage source.

There are two ways to overcome this difficulty.

The more difficult method is to assign an unknown current to the branch which contains the voltage source, apply KCL at different nodes, and then apply KVL once between nodes 2 and 3; the result will be four equations in four unknowns for this example.

The easier method is to treat node 2, node 3, and the voltage source together as a supernode and apply KCL to both nodes at the same time.

The supernode is indicated by the region enclosed by the broken line in Figure 2.20(b). This is possible, because, if the total current leaving node 2 is zero and the total current leaving node 3 is zero, then the total current leaving the combination of the two nodes is zero.

A Supernode is formed by enclosing a dependent or independent voltage source connected between two non reference nodes and any elements connected in parallel with it.

Following properties of a supernode may be noted :

1 The voltage source inside the supernode provides an equation required to solve for the node voltages.

2 A supernode has no voltage of its own.

3. A supernode requires the application of both KCL and KVL.

Example given below illustrates the concept of a supernode.

Example 2.13 : Obtain the nodal voltages v_1, v_2, v_3 for the circuit shown in Fig 2.20(a)

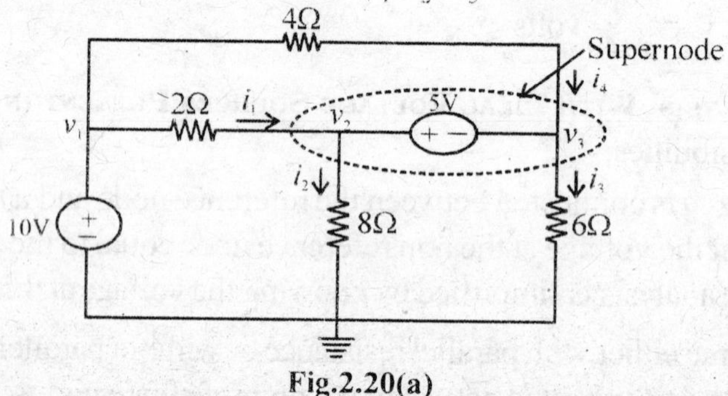

Fig.2.20(a)

Solution :

In Fig.2.20(a) , a voltage source is connected between the non reference nodes nodes 2 and 3 and hence these nodes form a supernode.

(i) However, KCL must be satisfied at a supernode like any other node.

Hence, at the supernode,

Sum of currents entering the supernoode = sum of currents leaving the supernode

i.e. $$i_1 + i_4 = i_2 + i_3 \qquad \text{(i)}$$

or $$\frac{v_1 - v_2}{2} + \frac{v_1 - v_3}{4} = \frac{v_2 - 0}{8} + \frac{v_3 - 0}{6} \qquad \text{(ii)}$$

(ii) However, $v_1 = 10$ volts. Substituting this value of v_1 in equation(ii), we get:

$$\frac{10-v_2}{2}+\frac{10-v_3}{4}=\frac{v_2-0}{8}+\frac{v_3-0}{6}$$

Simplifying, we get:

$$15\,v_2+10\,v_3=180 \qquad \text{(iii)}$$

(iii) To apply Kirchhoff's voltage law to the supernode in Fig.2.20(a), redraw the circuit as shown in Fig. 2.20(b).

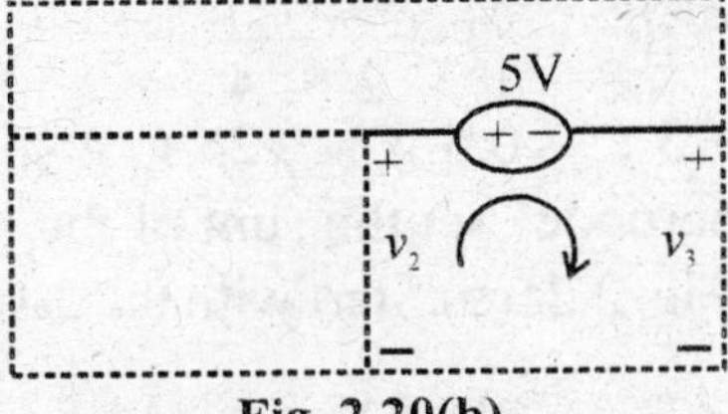

Fig. 2.20(b)

Going around the loop in the clockwise direction gives

$-v_2 + 5 + v_3 = 0$ (assuming voltage rise as negative)

i.e. $v_2 - v_3 = 5$ (iv)

Solving equations (iii) and (iv), we obtain the required node voltages, v_2 and v_3 as:

$$v_2 = \frac{46}{5} \text{ volts} \qquad \textbf{Ans}$$

$$v_3 = \frac{21}{5} \text{ volts} \qquad \textbf{Ans}$$

$$v_1 = 10 \text{ volts} \qquad \textbf{Ans}$$

Example 2.14 : Determine the value of the unknown node voltage v_1 in the circuit of Figure 2.21 (a).

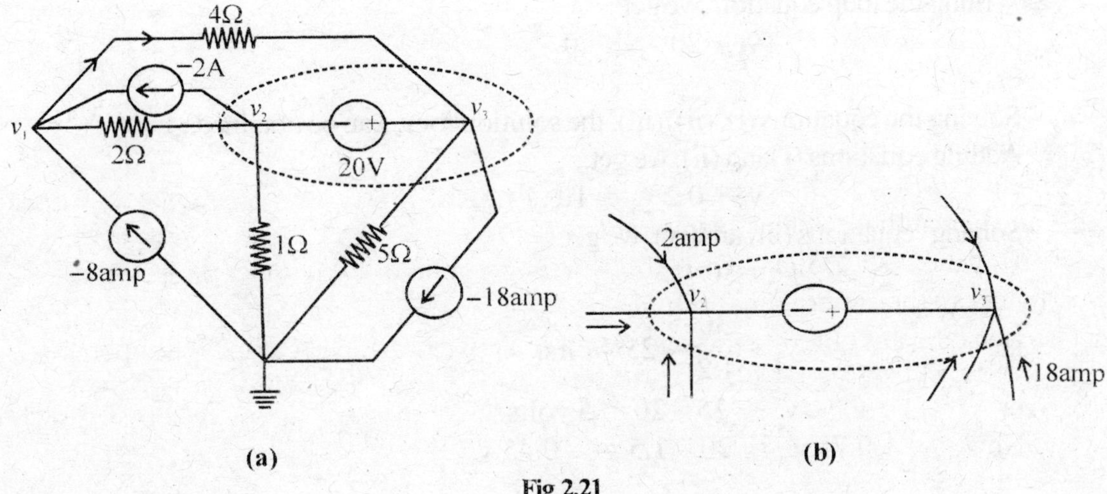

Fig 2.21

Solution :

The KCL equation at **node 1** is :

$$-8-2=\frac{v_1-v_2}{2}+\frac{v_1-v_3}{4}$$

Simplifying the above equation, we get:

$$-10=v_1\left(\frac{1}{2}+\frac{1}{4}\right)-\frac{v_2}{2}-\frac{v_3}{4}$$

or $\quad 0.75\,v_1-0.5\,v_2-0.25\,v_3=-10 \qquad$ (i)

Apply KCL at the supernode .Set the sum of the six currents , which are entering the **supernode** , equal to zero , Fig. 2.21(b).Start with the 2-Ω resistor branch and work clockwise.

$$2+18+\frac{-v_2+v_1}{2}+\frac{-v_3+v_1}{4}-\frac{v_3}{5}-\frac{v_2}{1}=0$$

Simplifying the above equation , we get:

$$-0.75\ v_1+1.5\,v_2+0.45\,v_3=20 \qquad \text{(ii)}$$

We have three unknowns. Hence, we need one more equation. This can be obtained by using the 20 V voltage source between nodes 2 and 3 as shown in Fig 2.21(c)

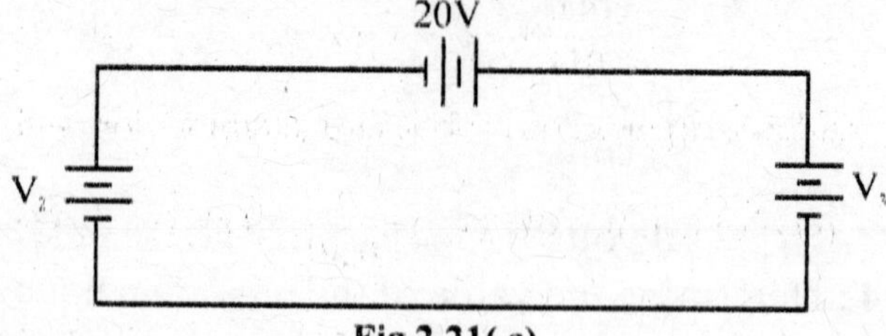

Fig 2.21(c)

Writing the loop equation ,we get:

$$v_2+20-v_3=0$$

or $\quad v_2-v_3=-20 \qquad$ (iii)

Solving the equations (i), (ii), (iii), the solution for v_1 can be obtained :

Adding equations (i)and (ii), we get:

$$v_2+0.2\,v_3=10 \qquad \text{(iv)}$$

Solving equations (iii) and(iv), we get:

$$1.2\,v_3=30$$

i.e. $\quad v_3=\frac{30}{1.2}=25$ Volts.

or $\quad v_2=25-20=5$ volts

Now $\quad -0.75\,v_1=20-1.5\,v_2-0.45\,v_3$

Substituting the values of v_2 and v_3, we get:

$$-0.75\ v_1 = 20 - 7.5 - 11.25 = 1.25$$

or $$v_1 = \frac{1.25}{-0.75} = -1.66 \text{ volts}$$ **Ans**

Example 2.15 : Find the nodal voltages V_1, V_2 and V_3 in Fig 2.22(a) using supernode.

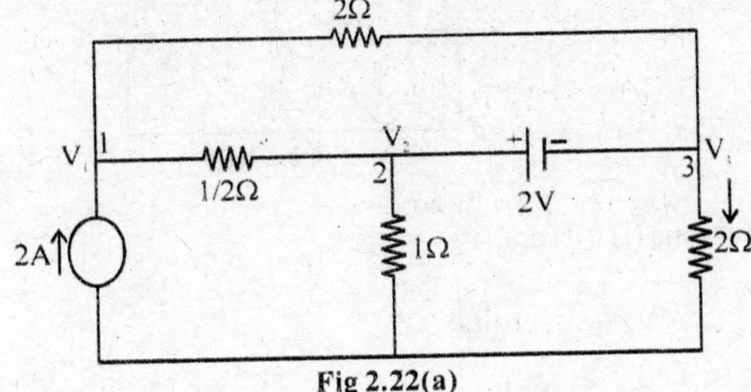

Fig 2.22(a)

Solution :

(i)Applying KCL at node 1 , we get:

$$\frac{2(V_1 - V_2)}{1} + \frac{V_1 - V_3}{2} = 2$$

or $$4V_1 - 4V_2 + V_1 - V_3 = 4$$

or $$5V_1 - 4V_2 - V_3 = 4 \qquad \text{(i)}$$

(ii) Consider the nodes 2 and 3 together and consider the combination as a supernode as shown with dashed lines in Fig.2.22(b)

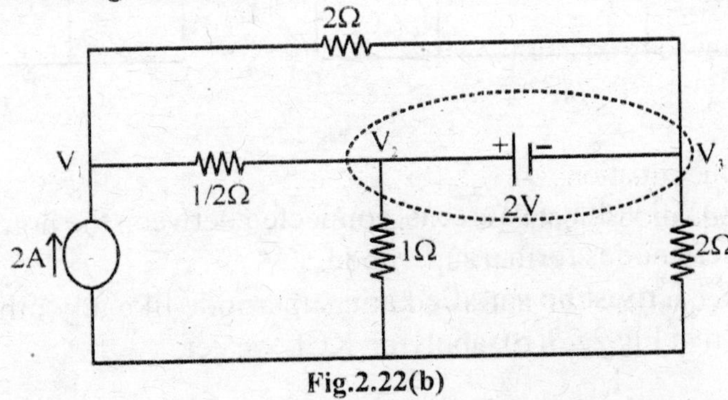

Fig.2.22(b)

Writing KCL at the super node (i.e. sum of currents leaving the supernode is equal to zero.), we get:

$$\frac{2(V_2 - V_1)}{1} + \frac{V_2}{1} + \frac{V_3}{2} + \frac{V_3 - V_1}{2} = 0$$

or $$4V_2 - 4V_1 + 2V_2 + V_3 + V_3 - V_1 = 0$$

or $$-5\,V_1 + 6\,V_2 + 2\,V_3 = 0 \quad \text{(ii)}$$

(iii) Apply KVL to the supernode as shown in Fig 2.22(c)

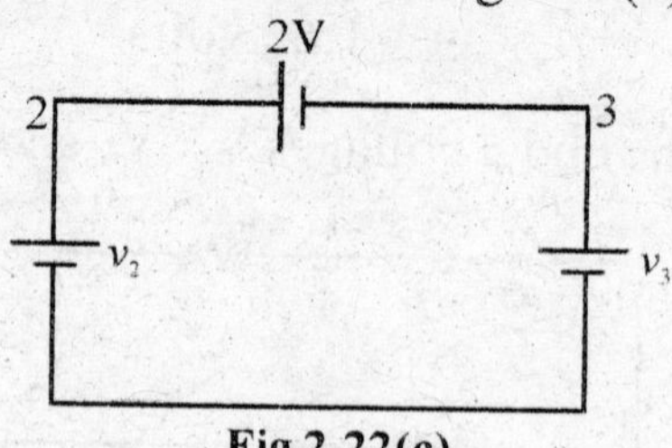

Fig 2.22(c)

i.e. $$V_2 - V_3 = 2 \quad \text{(iii)}$$

Solving equations (i) ,(ii) and(iii), we get:

$$V_1 = \frac{12}{5} \text{ volts}$$ **Ans**

$$V_2 = 2 \text{ volts}$$ **Ans**

$$V_3 = 0 \text{ volts}$$ **Ans**

Example 2.16 : Obtain the nodal voltages V_a and V_b for the circuit shown in Fig 2.23(a)

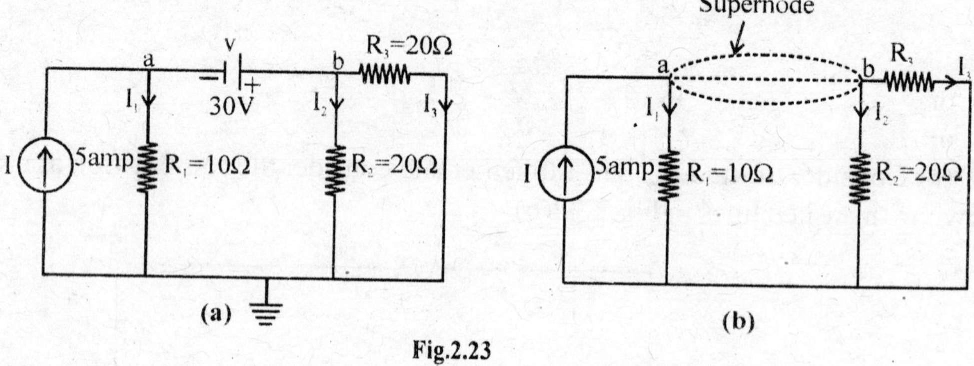

Fig.2.23

Solution :

In Fig.2.23(a) , a voltage source is connected between the non reference nodes ‘a’ and ‘b’ and hence these nodes form a supernode.

However, KCL must be satisfied at a supernode like any other node. Hence, at the supernode shown in Fig 2.23(b), applying KCL we get:

$$I = I_1 + I_2 + I_3$$

$$5 = \frac{V_a}{10} + \frac{V_b}{20} + \frac{V_b}{20}$$

or $$5 = \frac{V_a}{10} + \frac{V_b}{10} \quad \text{(i)}$$

To apply Kirchhoff's voltage law to the supernode in Fig.2.23(a), we redraw the circuit as shown in Fig. 2.23(c).

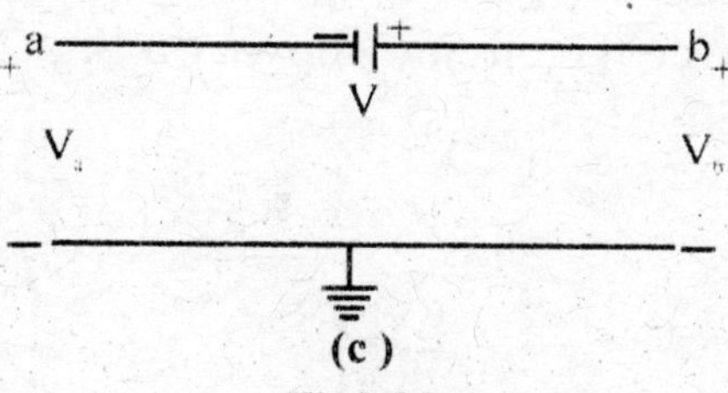

(c)

Fig2.23

Applying KVL at the super node shown in the Fig2.23(c), we get:

$$-V + V_b - V_a = 0$$

i.e. $$V_b - V_a = 30 \qquad \text{(ii)}$$

Solving equations (i) and(ii), we get:

V_a = 10 volts **Ans**

V_b = 40 volts **Ans**

2.5.3.5 Nodal Analysis By Inspection

It is a shortcut approach based on mere inspection of a circuit. When all sources in a circuit are independent current sources, application of KCL to each node to obtain the node-voltage equations is not needed. The equations can be obtained by mere inspection of the circuit.

In general, if a circuit with independent current sources has N non reference nodes, the node- voltage equations can be written in terms of the conductance as :

$$\begin{bmatrix} G_{11} & G_{12} & \cdots & G_{1N} \\ G_{21} & G_{22} & \cdots & G_{2N} \\ \vdots & \vdots & \vdots & \vdots \\ G_{N1} & G_{N2} & \cdots & G_{NN} \end{bmatrix} \begin{bmatrix} v_1 \\ v_2 \\ \vdots \\ v_N \end{bmatrix} = \begin{bmatrix} I_1 \\ I_2 \\ \vdots \\ I_N \end{bmatrix} \qquad (28)$$

i.e. $$[G][v] = [I]$$

where G is called the **conductance matrix**; v is the node voltage vector; and I is the source current vector.

G_{kk} = Sum of the conductances connected to node k.

$G_{kj} = G_{jk}$ = Negative of the sum of the conductances directly connecting nodes k and j, $k \neq j$

v_k = Unknown voltage at node k

I_k = sum of all independent current sources directly connected to node k, with currents entering the node treated as positive

Equation (2.8) can be solved to obtain the unknown node voltages. This is valid for circuits with only independent current sources and linear resistors.

As an example, examine the circuit shown in Fig. 2.24, where G_1, G_2 etc. are the conductances of the respective branches.

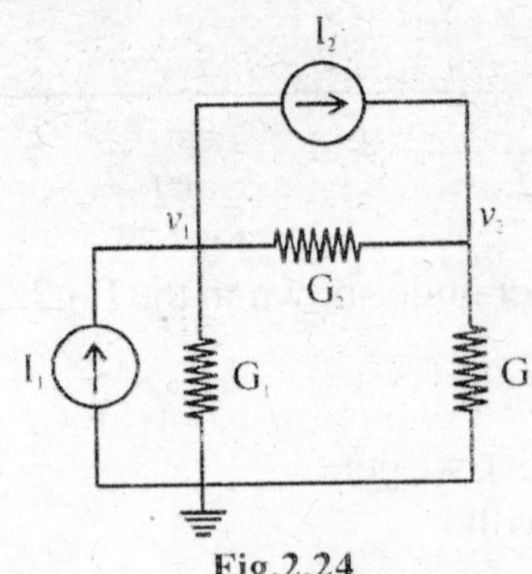

Fig.2.24

The circuit has two non reference nodes i.e. node1 and node2. Let the node voltages be v_1 and v_2

Nodal equations can be written in matrix form as:

$$\begin{bmatrix} G_1+G_2 & -G_2 \\ -G_2 & G_2+G_3 \end{bmatrix} \begin{bmatrix} v_1 \\ v_2 \end{bmatrix} = \begin{bmatrix} I_1-I_2 \\ I_2 \end{bmatrix} \quad (2.9)$$

Each of the diagonal elements of the matrix is the sum of the conductances connected directly to node 1 or 2, while the off- diagonal elements are the negatives of the conductances connected between the nodes.

Also, each term on the right hand side of Equation (2.9) is the algebraic sum of the sources currents entering the node.

2.6 NETWORK THEOREMS

General methods of circuit analysis discussed, tend to become complex and time consuming in a large network. Hence to handle complex circuits, it may be advantageous to use special techniques to obtain the solution of the circuit. Such techniques are termed as network theorems and several theorems have been developed to simplify circuit analysis. These theorems are very useful when the problem involves finding the current in or voltage across a particular branch etc. Following theorems have been discussed in this section:

(i) The Superposition theorem

(ii) Thevenin's (or Helmoltz's) theorem

2.6.1 THE SUPERPOSITION THEOREM

If a circuit has two or more independent sources, the value of voltage or current in any part of the circuit can be determined by finding the contribution of each independent source acting alone and then adding these contributions to obtain the resultant total value. This method is known as **superposition method.**

Superposition Theorem statement :

In a linear bilateral network , which has more than one sources, each branch current (branch voltage) is the algebraic sum of N currents (branch voltages), each of which is determined by considering one source at a time and removing all other sources. In removing the sources, ideal voltage sources are short-circuited (or replaced by resistances equal to their internal resistances for non-ideal sources), while the ideal current sources are open-circuited/(or replaced by their equivalent internal impedance for non-ideal sources)

This theorem is applicable to linear circuits only. To apply this theorem, following points should be noted:

(i) Consider one independent source at a time, turning off the other remaining independent sources.

i.e. Replace voltage sources by zero voltages or a short circuit and replace current sources by zero currents or open circuit. This will result in a simplified circuit.

(ii) Dependent sources are not removed.

The application of the superposition theorem is illustrated with the following example.

Example 2.16 : Determine the current,I flowing in the 10 Ω resistor using superposition theorem in the circuit shown in Figure 2.25(a)

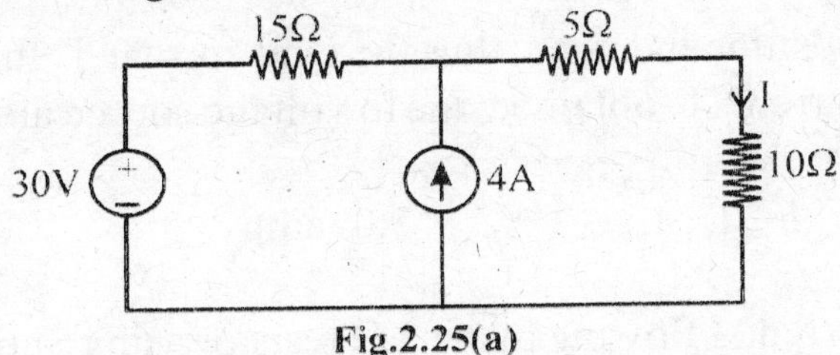

Fig.2.25(a)

Solution :

The circuit has two sources. One of them is a voltage source and the other is a current source

(i) First the current source of 4 amp. is removed i.e. open-circuited (OC) as shown in Figure2.25(b) and the response is obtained due to the voltage source of 30 volts only.

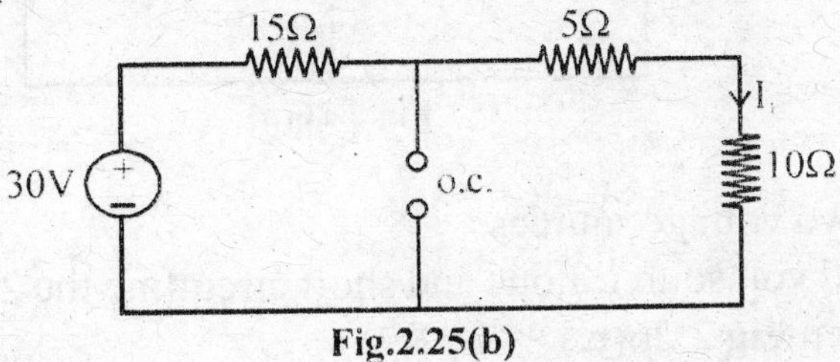

Fig.2.25(b)

The circuit becomes a simple series circuit consisting of a 30 V source in series with three resistances 15Ω, 5Ω, 10Ω in series. The current I_1, through the 10 Ω resistor is :

$$I_1 = \frac{30}{15+5+10} = 1 \text{ amp}$$

(ii) Next the voltage source is removed i.e. it is short-circuited (SC) as shown in Figure2.25(c). and the response is obtained due to 4 amp current source only

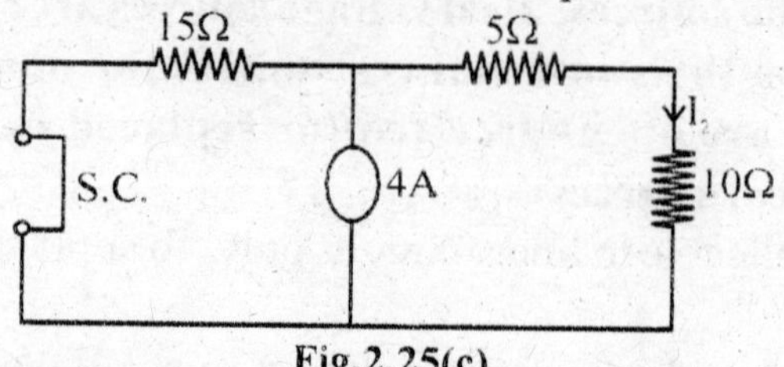

Fig.2.25(c)

The resultant circuit has two parallel branches- the branch at the centre is a 4 Amp current source, with a 15 Ω resistance on the left hand side, and a resistance of (5 + 10)=15Ω on the right hand side of the current source.

The current of 4 amp. is distributed in these two parallel branches.

The current through the 10 Ω resistor is obtained, using the current division formula:

i.e. $$I_2 = 4 \times \frac{15}{30} = 2 \text{ amp}$$

Using the superposition theorem, thus the total current I, through the 10 Ω resistor is the sum of the two currents I_1 obtained due to voltage source alone and I_2 obtained due to current source alone.

i.e. $$I = I_1 + I_2 = 1 + 2 = 3.0 \text{ amp}$$ **Ans.**

Example 2.17

Determine the current, I flowing in the 5 Ω resistor using Superposition theorem in the circuit shown in Fig2.26(a)

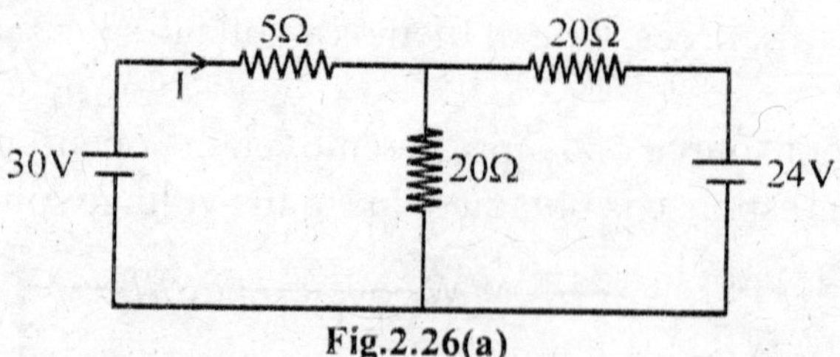

Fig.2.26(a)

Solution

This circuit has two voltage sources.

(i) Considering 30 volt source alone and short circuiting the 24 volt source, circuit can be redrawn as shown in Fig 2.26(b)

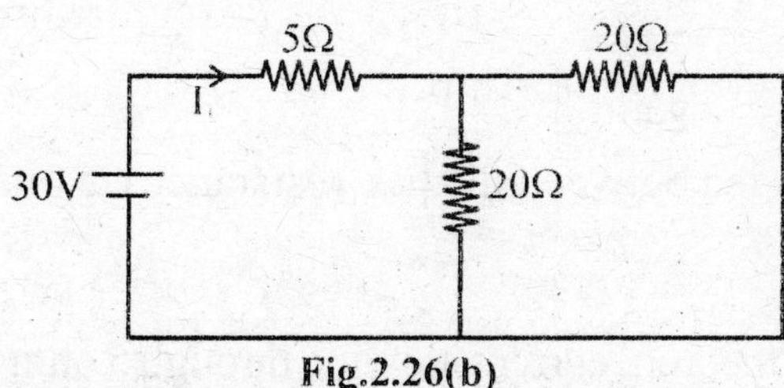

Fig.2.26(b)

Combining the two 20 ohm resistances connected in parallel, circuit of Fig 2.26(b),can be redrawn as shown in Fig.2.26(c)

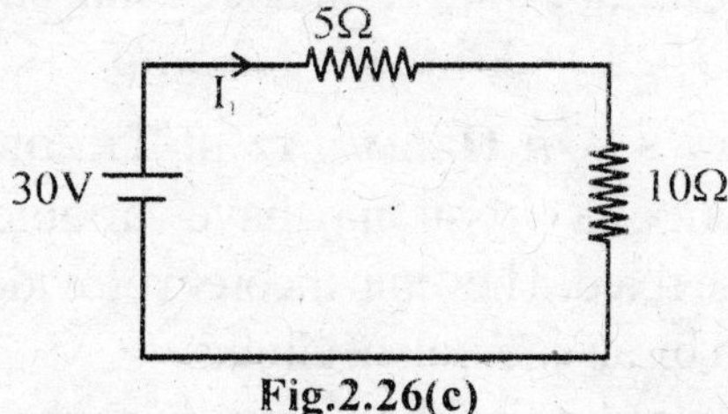

Fig.2.26(c)

Thus Current I_1 flowing through 5 ohm resistance due to 30 volt voltage source $= \frac{30}{15}$

= 2 Amp.

(ii) Considering 24 volt source alone and short circuiting the 30 volt source, circuit of Fig2.26(a) can be redrawn as shown in Fig2.26(d)

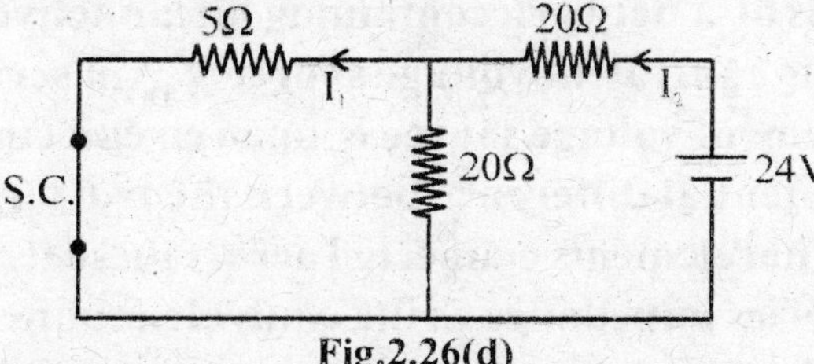

Fig.2.26(d)

Combine the 5 ohm and 20 ohm resistances connected in parallel.

$$R_{eq} = \frac{5\times 20}{5+20} = 4 \text{ ohm}$$

Circuit of Fig2.26(d) can be redrawn as shown in Fig.2.26(e)

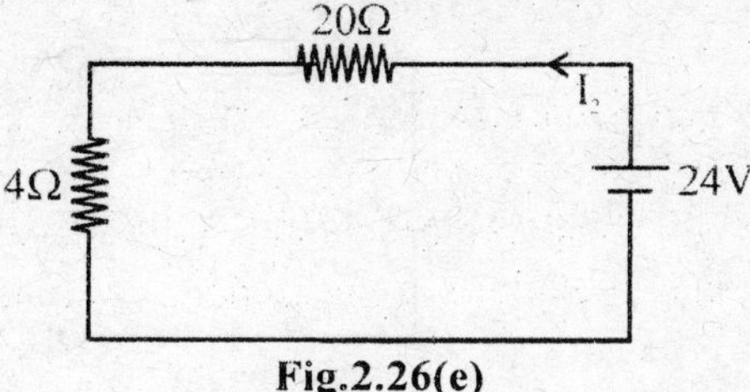

Fig.2.26(e)

Current $I_2 = \frac{24}{24} = 1$ Amp.

This current is divided between the two resistances 5 ohm and 20 ohm connected in parallel.

By current division theorem current flowing through 5 ohm resistance, $I_1 = \frac{1 \times 20}{25} = \frac{4}{5} = 0.8$ amp from right to left.

Net current flowing through 5 ohm resistance using superposition theorem

$$= I_1 - I_1 = 2 - 0.8 = 1.2 \text{ amp}$$ **Ans**

2.6.2 The Thevenin's (or Helmoltz's) Theorem

In practice, one element of a circuit may have variable values while the values of other elements of the circuit may be fixed. Thevenin theorem provides a method by which the fixed part of the circuit can be replaced by an equivalent circuit.

Sometimes it may be desired to find the current through or voltage across a particular branch of a network or it may be desirable to study the variation of current in a particular branch whose resistance is varied without calculating currents or voltages in other branches of the network.

In such cases, it is not necessary to analyse the complete circuit again and again. Such problems can be conveniently handled by using Thevenin's theorem.

Thevenin's Theorem Statement :

Any two terminals of a network containing linear active and passive elements may be replaced by a single equivalent voltage source V_{Th} in series with a single equivalent impedance Z_{Th}. The emf of voltage source is open circuit emf at the network terminals i.e. it is equal to the potential difference between the two terminal points caused by the network with no external elements connected across these terminals (i.e. open circuit the two terminals). The series impedance is the equivalent impedance looking into the two terminal points with all energy sources in the network made inactive and replacing them with their internal impedances.

While finding Thevenin's equivalent, dependent sources are not removed and they remain in the circuit.

Thevenin's equivalent circuit across terminals 'A' and 'B' is shown in Figure 2.27(a).

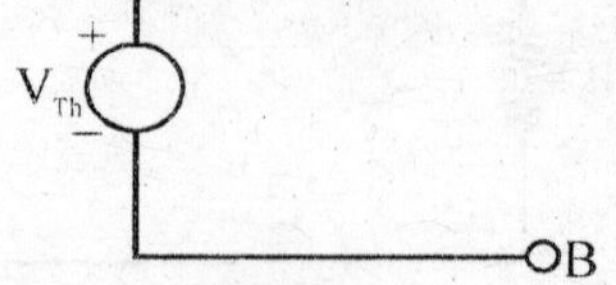

Fig.2.27 (a)

The statement given above is explained as follows:

Consider the circuit shown in Figure 2.27(b).

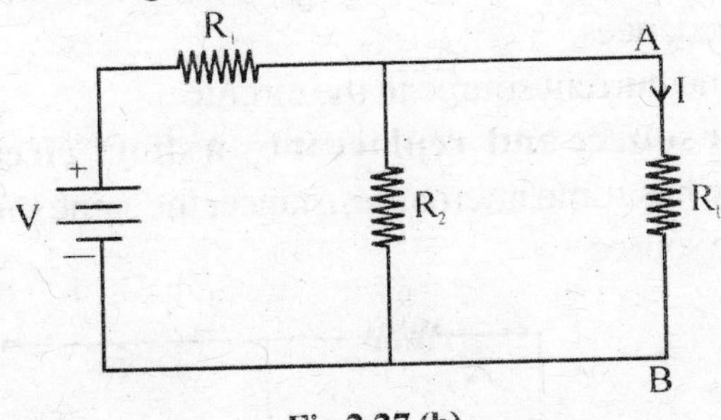

Fig.2.27 (b)

It is desired to find an Equivalent Thevenin's network across terminals 'A' and 'B' and to find the current I through the resistance R_L.

The following steps are followed:

(I) Determination of Open Circuit Voltage, V_{Th}

Step 1 : Remove the external load resistance R_L (i.e. remove the element through which current is to be found out)from the circuit and thus terminals A and B get opened as shown in Fig 2.27(c)

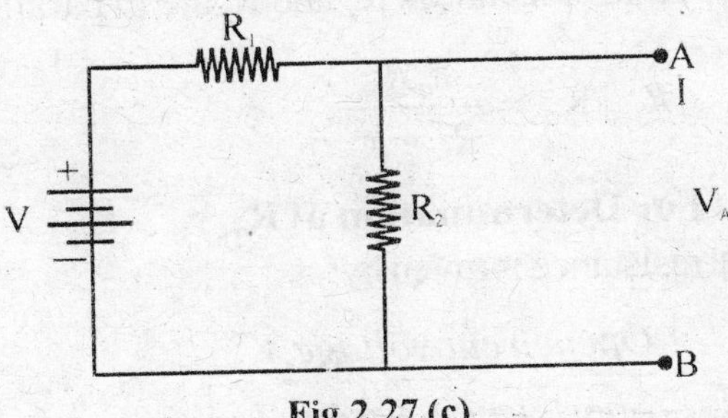

Fig.2.27 (c)

Step 2 : Solve this resultant circuit to find the open circuit voltage, $V_{AB} = V_{oc}$, known as **Thevenin's Voltage , V_{Th}** across the open circuited terminals A and B.

In Fig 2.27(c), it is seen that:

$$I = \frac{V}{R_1 + R_2}$$

and

$$V_{AB} = V_{oc} = I R_2$$

Substituting the value of I, we get:

i.e.
$$V_{oc} = (\frac{V}{R_1 + R_2})R_2 \tag{2.9}$$

Step 3 : Determintation of R_{Th}

To find Thevenin's equivalent resistance R_{Th}, remove all independent sources and replace them by their internal impedances.

In this case, there is no current source in the circuit.

Remove the voltage source and replace it by a short circuit as shown in Fig 2.27(d). In case, the voltage source has some internal resistance, the same must be included in that branch after removing the voltage source

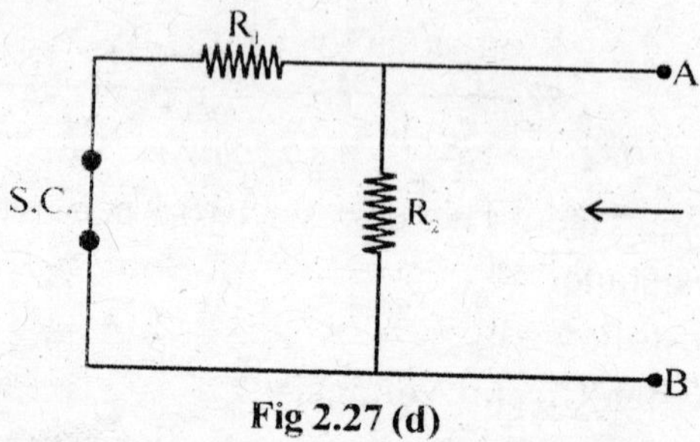

Fig 2.27 (d)

Find the total resistance across the terminals A and B ,looking into the network from terminal-pairs A-B towards the left.

It is seen in Fig 2.27(d) that resistances R_1 and R_2 are in parallel

i.e. $$R_{Th} = R_1 \parallel R_2 = \frac{R_1 R_2}{R_1 + R_2} \qquad (2.10)$$

Alternative Method For Determination of R_{Th}

Thevenin's equivalent resistance is given by:

$$R_{Th} = \frac{Opencircuitvoltage, V_{OC}}{Shortcircuitcurrent, I_{SC}}$$

Assuming that V_{Th} i.e. V_{oc} has been calculated already as explained above, the value of R_{TH} can be calculated as follows :

Calculation of short circuit current

Replace R_L by a short circuit as shown in Fig 2.27(e) and find the short circuit current I_{SC} through the short circuit. Short circuit the terminals A B and redraw the circuit as shown in Fig 2.27(e).

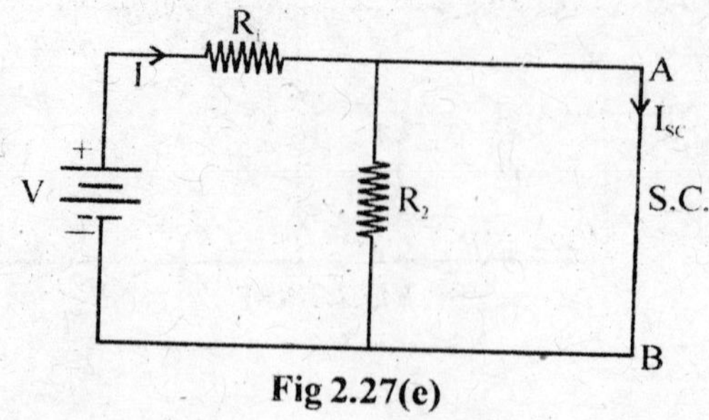

Fig 2.27(e)

It is seen from Fig 2.27(e)that:

$$I_{SC} = \frac{V}{R_1}$$, because resistance R_2 has been short circuited

Thevenin's equivalent resistance is given by: $R_{Th} = \frac{V_{OC}}{I_{SC}}$

Substituting the values of V_{oc} from equation (2.9)and I_{SC} ,we get:

$$R_{Th} = \left(\frac{V}{R_1 + R_2}\right) R_2 \times \frac{R_1}{V} = \frac{R_1 R_2}{R_1 + R_2}$$

It is the same as given in equation (2.10).

The Thevenin's equivalent network obtained across terminals A and B, is depicted in Figure 2.27(f)

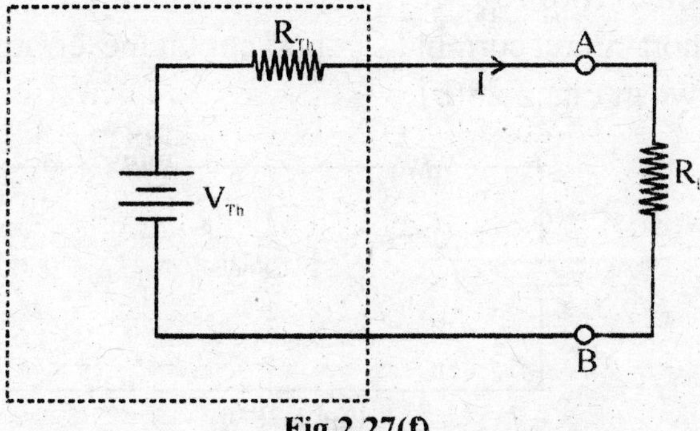

Fig.2.27(f)

Using Thevenin's equivalent network, the current I , flowing through R_L can be obtained as:

$$I = \frac{V_{Th}}{R_{Th} + R_L} \qquad (2.11)$$

Example 2.17 : Find the current , I through R_L=12 ohm in the circuit shown in Fig 2.28(a), using Thevenin's theorem.

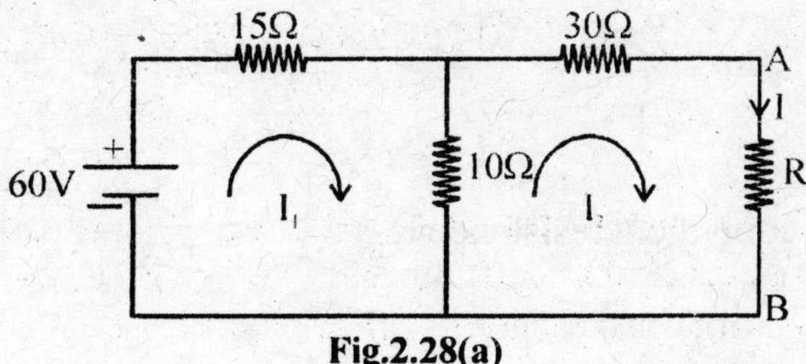

Fig.2.28(a)

Solution :

(a)Calculation of $V_{Th} = V_{OC}$

If we open circuit the terminals AB i.e. remove the resistance R_L , we get the open circuit voltage by applying the voltage divider rule , as follows:

$$V_{OC} = V_{AB} = \frac{60 \times 10}{15+10} = \frac{600}{25} = 24 \text{ Volts} \qquad \textbf{Ans}$$

(ii) Equivalent resistance looking into the network from terminal pair AB towards the left is :30 Ω in series with the parallel combination of resistances 15 Ω and 10 Ω (short circuiting the 60 volt voltage source)

i.e $\quad R_{Th} = 15 \parallel 10 + 30$

or $\quad R_{Th} = \frac{15 \times 10}{15+10} + 30 = 6 + 30 = 36 \text{ ohms} \qquad \textbf{Ans}$

Alternative method to find R_{Th}

To determine the short circuit current I_{SC}, short circuit the terminals AB and assume loop currents I_1 and I_2as shown in Fig. 2.28(b).

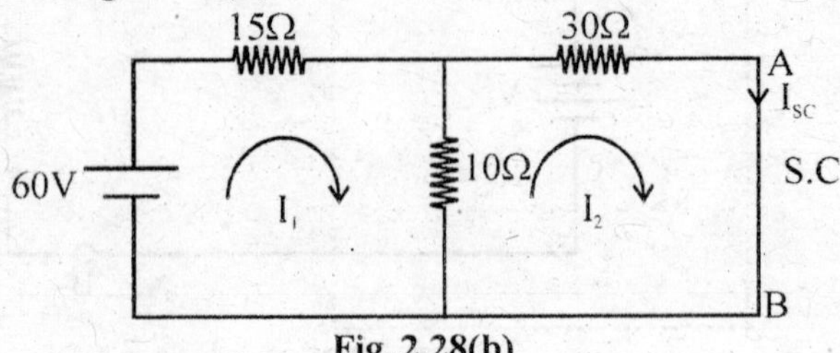

Fig. 2.28(b)

Write the loop equations as follows:

$$15 I_1 + 10(I_1 - I_2) = 60 \qquad \text{(i)}$$
$$30 I_2 + 10(I_2 - I_1) = 0 \qquad \text{(ii)}$$

Solving these equations , we get:

$$I_1 = 8/3 \text{ amp}$$
$$I_2 = 2/3 \text{ amp}$$

It is seen that:

$$I_{SC} = I_2 = \frac{2}{3} \text{ amp}$$

Hence, Thevenin's equivalent resistance $R_{Th} = \frac{V_{Th}}{I_{SC}} = \frac{24}{2/3} = 36 \text{ ohm}$

It is the same value as obtained earlier.

(iii) Hence , current I through R_L is:

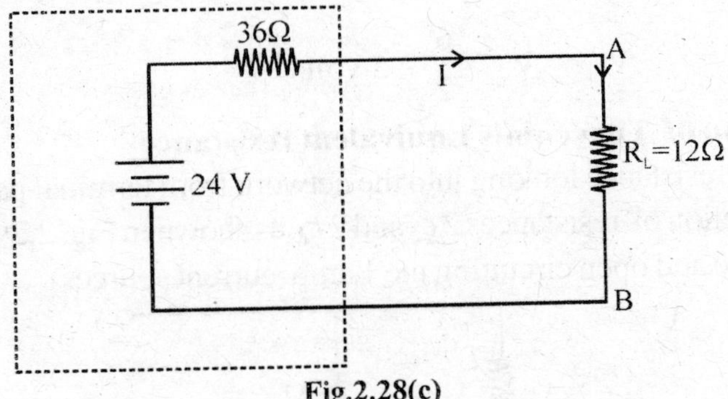

Fig.2.28(c)

$$I = \frac{24}{36+12} = \frac{24}{48} = 0.5 \text{ amp}$$ **Ans.**

Example 2.18 : Find the current through R_L = 5 ohm in the circuit shown in Fig 2.29(a), using Thevenin's Theorem.

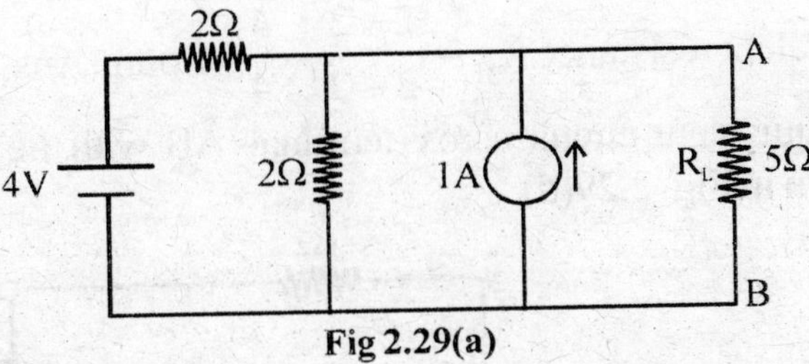

Fig 2.29(a)

Solution :

(a) Calculation of $V_{Th} = V_{OC}$

Open circuit the terminals AB i.e. remove the resistance R_L and redraw the circuit as shown in fig 2.29(b).

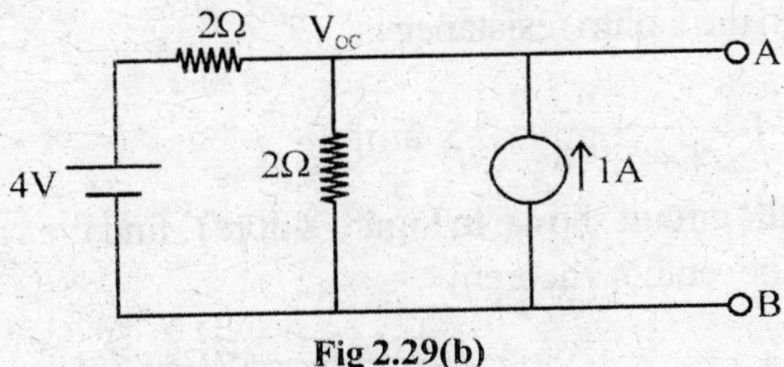

Fig 2.29(b)

We get the open circuit voltage , as follows:

Writing KCL equations at the node, we get:

$$\frac{V_{oc} - 4}{2} + \frac{V_{oc}}{2} = 1$$

i.e. $\quad V_{oc} - 4 + V_{oc} = 2$

or $\quad 2\,V_{oc} = 6$

i.e. $$V_{Th} = V_{oc} = \frac{6}{2} = 3 \text{ volts}$$ **Ans**

(b) Calculation of Thevenin's Equivalent resistance

(ii) Equivalent resistance looking into the network from terminal pair AB towards the left is the parallel combination of resistances 2 Ω and 2 Ω as shown in Fig. 2.29(c) (short circuiting the 4 volt voltage source and open circuiting the 1 amp. current sources)

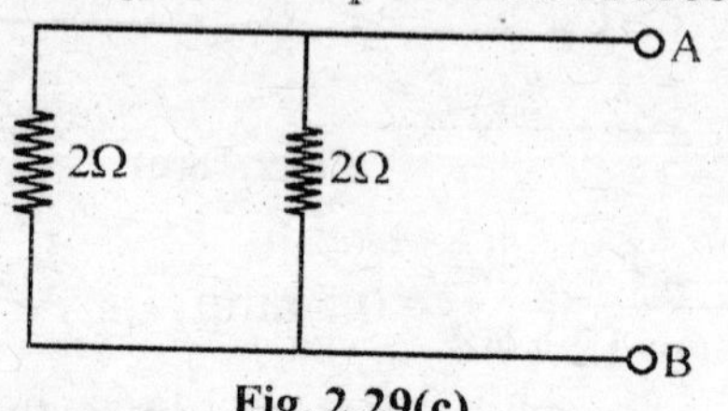

Fig. 2.29(c)

It can be seen from Fig. 2.29 (c) that:

Thevenin's Equivalent resistance $R_{Th} = \frac{2 \times 2}{2+2} = \frac{4}{4} = 1$ ohm Ans

The Thevenin's equivalent circuit across terminals AB with the resistance 5 ohm connected between them, is shown in Fig. 2.29(d).

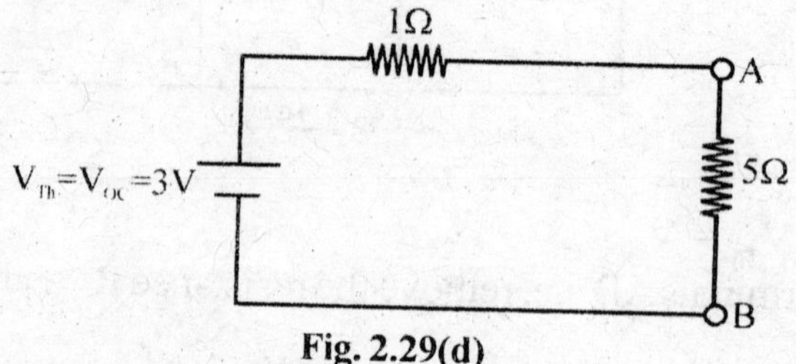

Fig. 2.29(d)

Hence the current in the 5 ohm resistance is:

$$I = \frac{3}{1+5} = \frac{3}{6} = 0.5 \text{ amp}$$ **Ans**

Example 2.19 : In the circuit shown in Figure 2.30(a), find the value of current I through 1.4 ohms resistance ,using Thevenin's Theorem

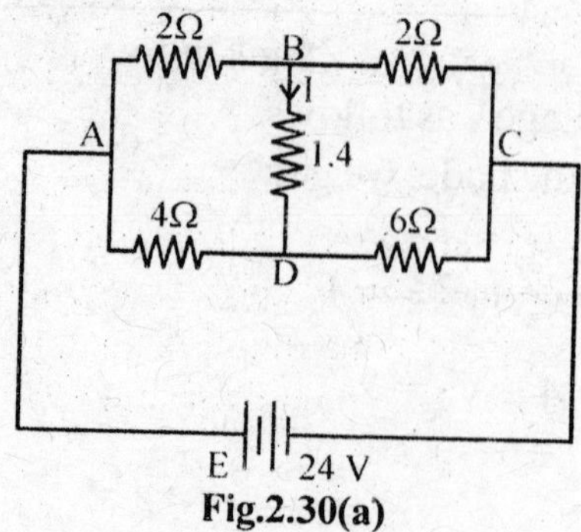

Fig.2.30(a)

Solution :

(i) Calculation of open circuit voltage V_{Th}

Remove 1.4 Ω resistance across the terminals 'B' and 'D' i.e. open circuit the terminals 'B' and 'D' and redraw the circuit as shown in Fig 2.30(b)

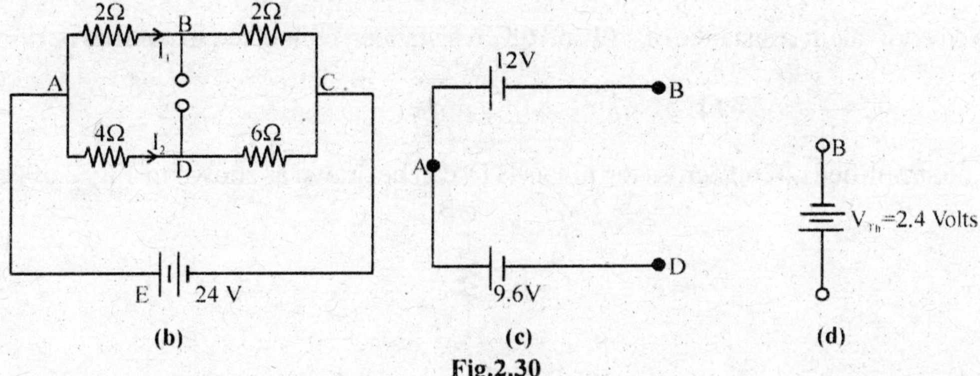

Fig.2.30

It can be seen in Fig 2.30(b) that a voltage of 24 volts is applied across two parallel branches i.e. 'ABC' and 'ADC'.

Hence, Current $$I_1 = \frac{24}{2+2} = \frac{24}{4} = 6 \text{ amp}$$

and Current $$I_2 = \frac{24}{4+6} = \frac{24}{10} = 2.4 \text{ amp}$$

Voltage Drop from A to B = 6 × 2 = 12 Volts

Voltage Drop from A to D = 2.4 × 4 = 9.6 volts

So Potential Difference between points B and D = 12 – 9.6 = 2.4 volts, terminal 'D' is at a higher potential, as shown in FIG 2.30 (c)

So, Open circuit voltage, V_{Th} = 2.4 volts, as shown in Fig 2.30 (d)

(ii) Calculation of R_{Th}

Replace 24 volt, voltage source by a short circuit. The circuit with terminals A and C short circuited, will be as shown in Figure 2.30(e), which can be shown in the simplified form as shown in Fig 2.30(f)

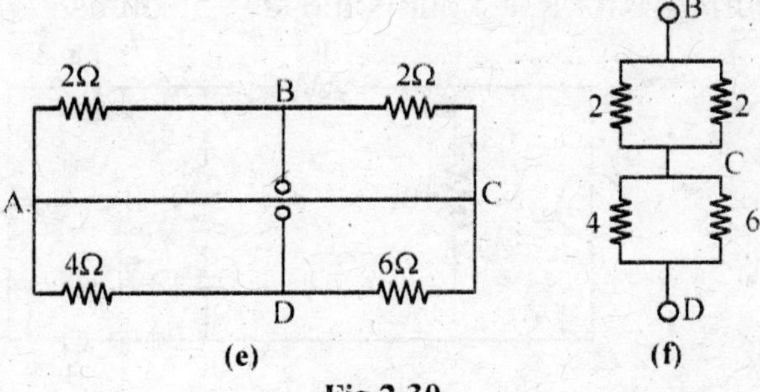

Fig.2.30

Referring to Fig 2.30 (f) ,Equivalent resistance of 2 Ω and 2 Ω resistances connected in parallel is:

$$= \frac{2 \times 2}{2+2} = \frac{4}{4} = 1\Omega$$

And equivalent resistance of 4 Ω and 6 Ω resistances connected in parallel is:

$$= \frac{4 \times 6}{10} = 2.4\,\Omega$$

The simplified circuit across terminals B D can be drawn as shown in Fig. 2.30(g)

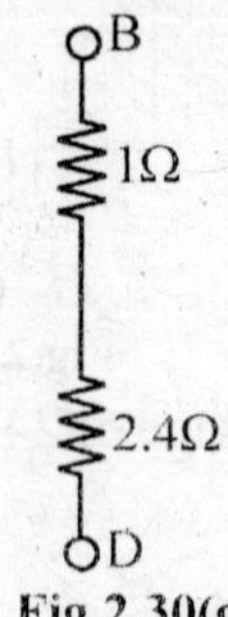

Fig.2.30(g)

$\therefore \quad R_{Th} = 1 + 2.4 = 3.4\,\Omega$ as shown in Figure 2.30(g)

The Thevenin's equivalent network obtained is shown in Figure 2.30(h)

$V_{Th} = 2.4V$ 3.4Ω B 1.4Ω I D

Fig.2.30(h)

Total resistance, $R_T = 1.4 + 3.4 = 4.8\,\Omega$

Hence, Current , $I = \frac{2.4}{4.8} = 0.5$ amp from D to B **Ans**

Example 2.20 : Obtain the Thevenin's. equivalent circuit for the network shown in Fig. 2.31 (a) and calculate the current I for $R_L = 5$ ohms and $R_L = 50$ ohms

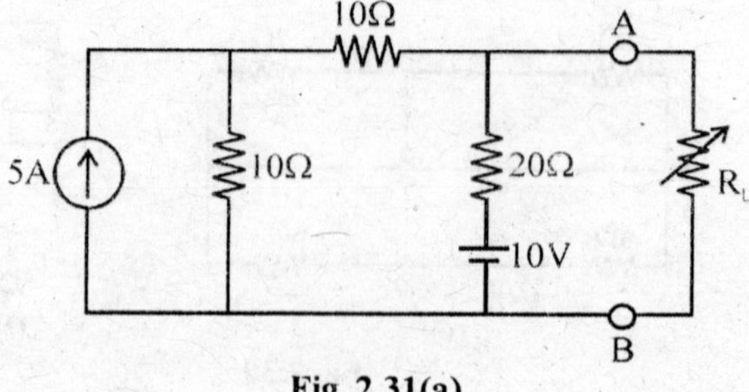

Fig. 2.31(a)

Solution :

Step 1

Convert the 5 amp.current source in parallel with a tresistance in to an equivalent voltage source of 50 volts in series with a resistance and open circuit the terminals A, B and redraw the figure as shown in Fig 2.31(b).

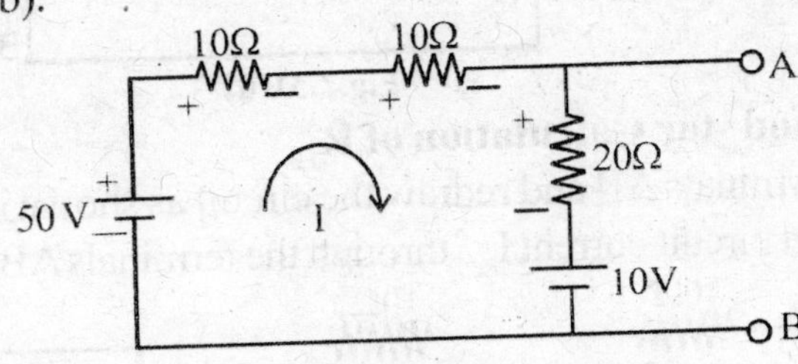

Fig 2.31(b)

The open circuit voltage at the terminals A and B can be calculated as follows:

Assume current I as the loop current.

Writing KVL in Fig. 2.31 (b), we get:

$$40\,I + 10 = 50$$

or
$$I = \frac{40}{40} = 1 \text{ amp.}$$

Hence, Open circuit voltage, $V_{AB} = 10 + 20\,I = 10 + 20 = 30$ Volts **Ans**

Step 2

To obtain the Thevenin's equivalent resistance R_{Th} across terminals A and B. Open circuit the 5 amp current source and short circuit the 10 volt voltage source and redraw the circuit of Fig. 2.31(a) as shown in Fig.2.31(c)

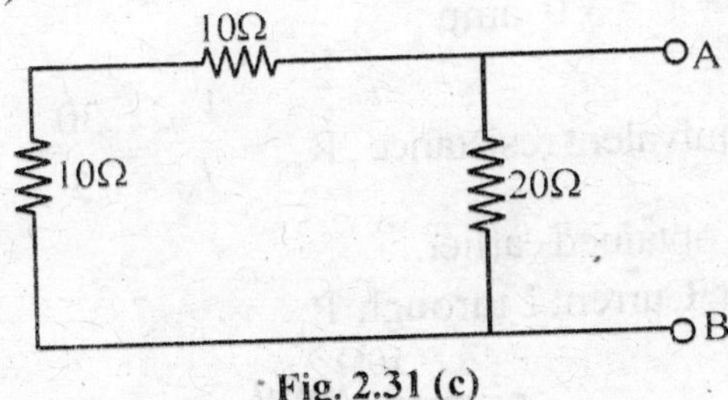

Fig. 2.31 (c)

From Fig 2.31(c) , it is seen that,

$$R_{Th} = 20\ \Omega \parallel (10\Omega + 10\Omega)$$
$$= 20 \parallel 20$$

$$R_{Th} = \frac{20 \times 20}{20 + 20} = 10\Omega$$

Step 3

Thevenin's equivalent circuit is shown in Fig.2.31(d)

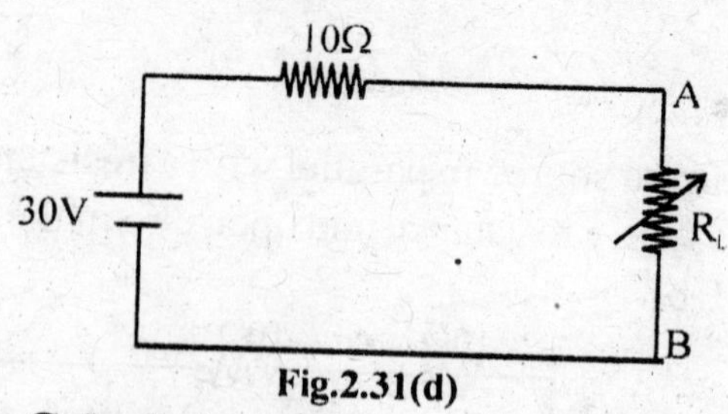

Fig.2.31(d)

Alternative method for Calculation of R_{Th}

Short circuit the terminals AB and redraw the circuit as shown in Fig. 2.31(e). Determine the short circuit current I_{SC} through the terminals AB as follows:

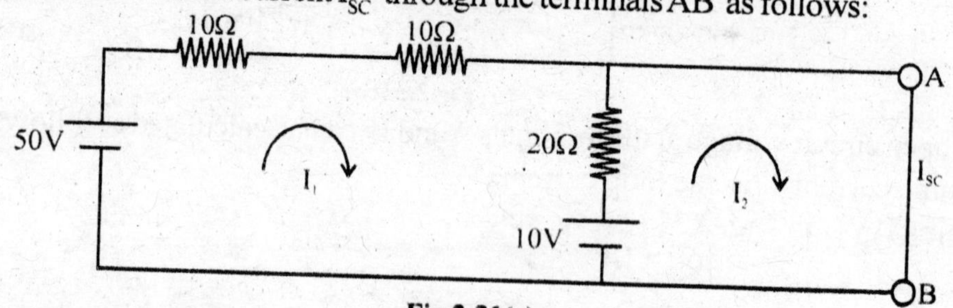

Fig.2.31(e)

Assume loop currents I_1 and I_2 and write the two loop equations as follows:

Loop 1 : $20\,I_1 + 20(I_1 - I_2) + 10 = 50$ (i)

Loop 2 : $20(I_2 - I_1) = 10$ (ii)

Solving these equations (i) and (ii), we get:

$$I_1 = 2.5 \text{ amp}$$

$$I_2 = 3.0 \text{ amp}$$

i.e. $$I_{SC} = I_2 = 3.0 \text{ amp}$$

Ans

Hence, Thevenin's equivalent resistance, $R_{Th} = \dfrac{V_{Th}}{I_{SC}} = \dfrac{30}{3} = 10$ Ohms

It is the same value as obtained earlier.

Step 4 Calculation of Current I through R_L

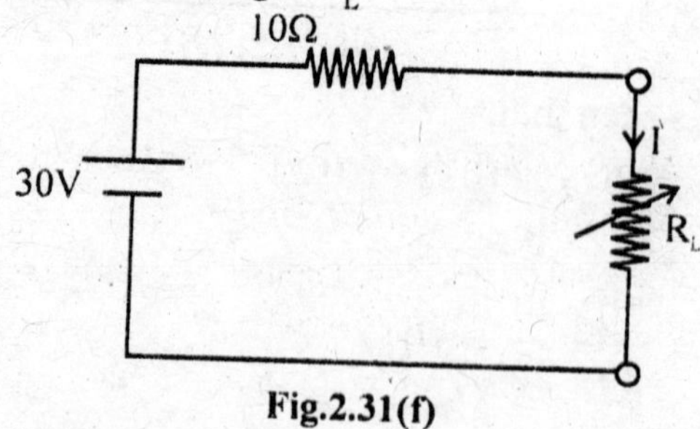

Fig.2.31(f)

It can be seen from Fig.2.31(f), that:

(i) For $R_L = 5$ ohms

Current I through $R_L = \frac{30}{10+5} = 2$ amps. **Ans**

(ii) For R_L=50 ohms

Current I through $R_L = \frac{30}{10+50} = 0.5$ amps **Ans**

2.7 Summary

In this chapter ,basic source transformations have been discussed. Methods of Circuit Analysis i.e. Network Analysis by Mesh currents(Mesh analysis) and by Node-pair Voltages (Nodal method) have been explained.

Use of different Network Theorems i.e. Superposition Theorem , Thevenin's (or Helmoltz's) theorem, for the solution of network problems has been highlighted.

2.8 Short Questions and Answers

1. Write the procedure for network analysis using mesh equations.

Ans. Steps to write mesh equations and to determine mesh currents are as follows:

(i) Assume that each mesh contains a circulating unknown mesh or loop current.Assign mesh currents systematically such as $i_1, i_2, i_3, \ldots\ldots\ldots, i_n$ to the 'n' meshes. Although the directions are arbitrary, these may be assumed clockwise.

(ii) Each current in any outer branch will be equal to the particular mesh current. However currents in the inner branches can be expressed as the difference between two mesh currents if all the mesh currents are assumed in the same direction.

(iii) Apply KVL to each of the 'n' meshes. Use Ohm's law to express voltages in terms of mesh currents.

(iv) Solve the resulting 'n' simultaneous equations by applying Kramer's rule or matrix inversion or any other method of solving the equations to get the values of mesh currents.

(v) Current through any branch can be obtained in terms of loop currents.

2. Write the procedure for Nodal Analysis

Ans. For applying the nodal analysis , following steps are followed:

(i) With respect to reference point, assume node voltages at all other nodes and assign them the voltages V_1, V_2 etc.

(ii) Note the values of voltages at the nodes at which known voltage sources are present.

(iii) Write a KCL for each node. Assume the currents flowing through the resistances as leaving a given node.

(iv) Solve these equations for unknown voltages.

3. State the Superposition Theorem

Ans. In a linear bilateral network which has more than one sources, each branch current (branch voltage) is the algebraic sum of N currents (branch voltages), each of which is determined by considering one source at a time and removing all other sources. In removing the sources, voltage sources are short-circuited (or replaced by resistances equal to their internal resistances for non-ideal sources), while the ideal current sources are open-circuited/ replaced by their equivalent internal impedance.

4. State Thevenin's (or Helmoltz's) Theorem

Ans. Any two terminals of a network containing linear active and passive elements may be replaced by an equivalent voltage source V_{Th} in series with an equivalent impedance Z_{Th}. The voltage source is equal to the potential difference between the two terminal points caused by the active network with no external elements connected to these terminals. The series impedance is the equivalent impedance looking into the two terminal points with all energy sources within the terminal pair made inactive, retaining their internal impedances

Review Questions

1. A constant current source has I = 15 amp and R_S = 5Ω. Determine E and R_S for the equivalent constant voltage source. **[Ans : (75 V, 5 ohm)]**

2. For the circuit given below, find current 'i' using source transformation.

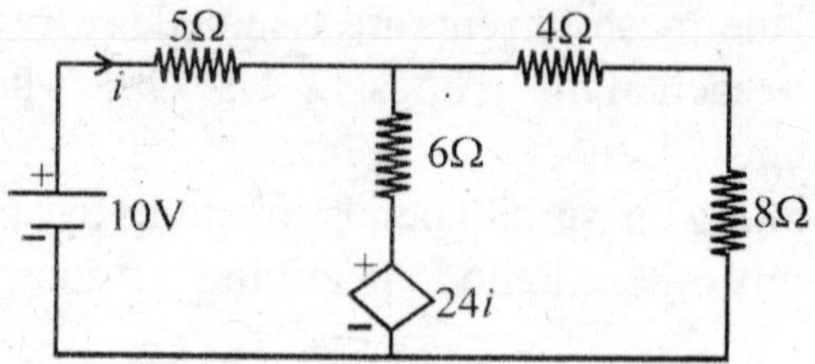

[Ans : *i* = 0.4 amp.]

3. Two batteries A and B are connected is parallel and a load of 12Ω is connected across their terminals. A has an emf of 15V and an internal resistance of 3Ω and 'B' has an emf of 10V and an internal resistance of 1Ω . Use Kirchoff's Law to determine the magnitude of currents and also the directions in each of the batteries. Also determine the potential across external resistance.

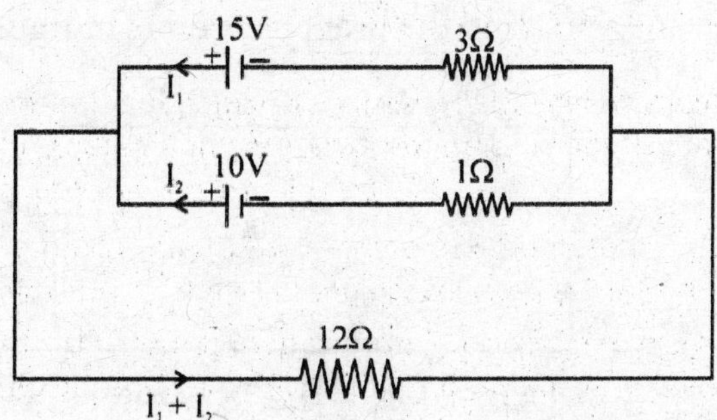

[Ans : 1.47 A, –0.588 A, 10.59 V]

4. Find the value of R and current through each branch if current in branch AO is zero.

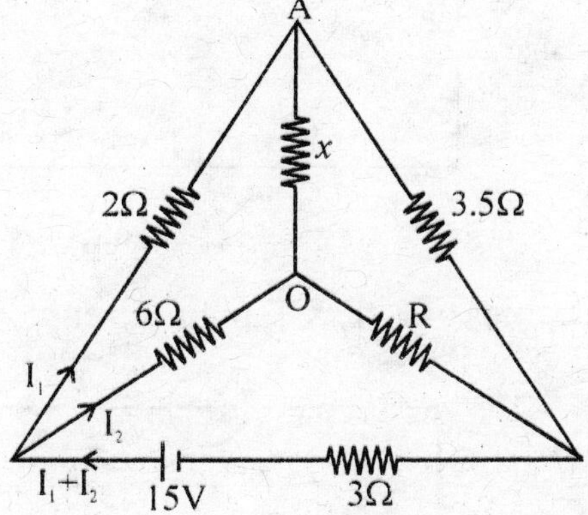

[Ans : 10.5Ω, I_1 = 1.7, I_2 = 0.56 amp]

5. Determine the value and direction of current in 50Ω resistor for the bridge circuit shown in figure given below:

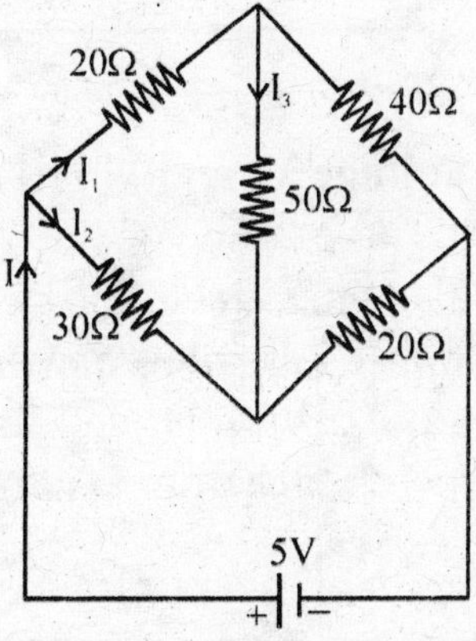

[Ans : 0.0092 amp]

6. Using Mesh analysis, find the branch currents in figure.

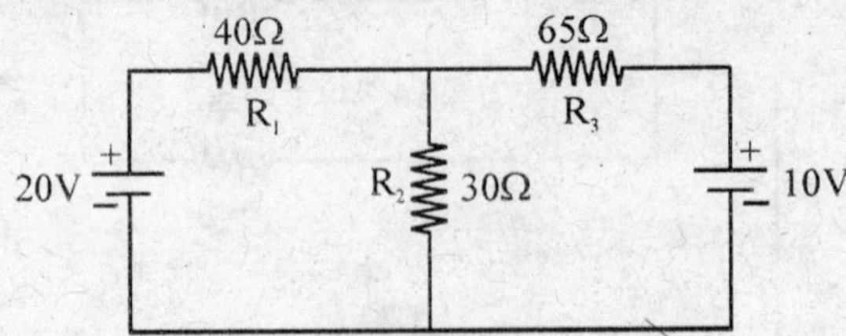

[Ans : IR_1 = 0.278 ampere, IR_2 = 0.295, IR_3 = –0.0174]

7. Calculate the current in each branch of the circuit. Also calculate voltage 'V'.

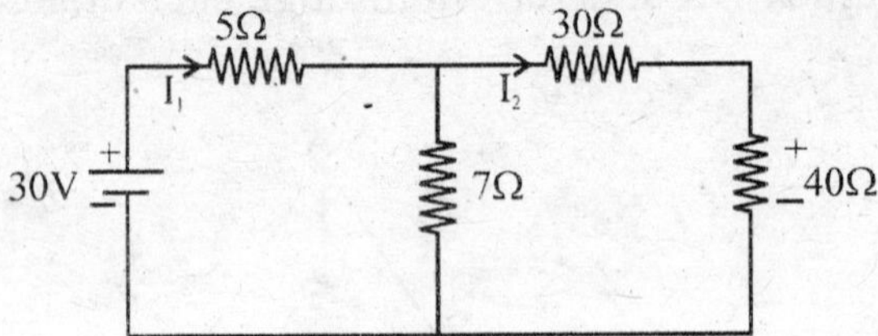

[Ans : I_1 = 2.64, I_2 = 0.24, V = 9.6 V]

8. Determine the values of I_1, I_2 & I_3 in the circuit.

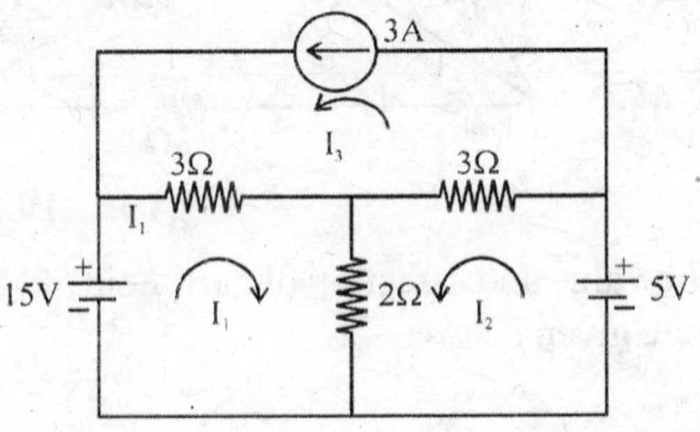

[Ans : 0.12, –2.7,3 amp]

9. Determine (i) node voltage V_b and V_c (ii) currents in various branches for the circuit.

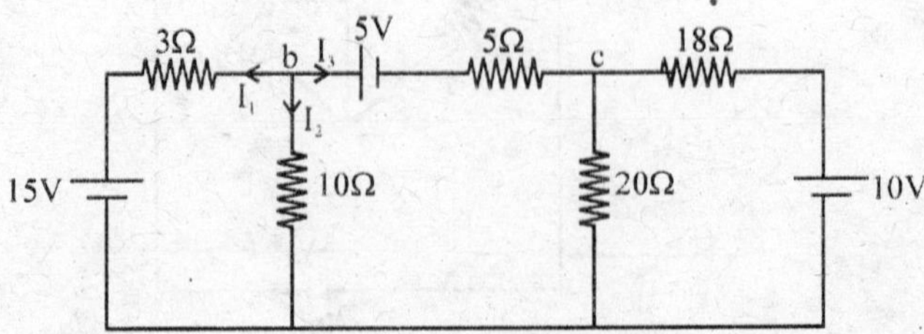

[Ans : 11.3V, 5.9V, 1.23A, 1.13A, –0.08A]

10. Determine voltages at nodes B and C in the network.

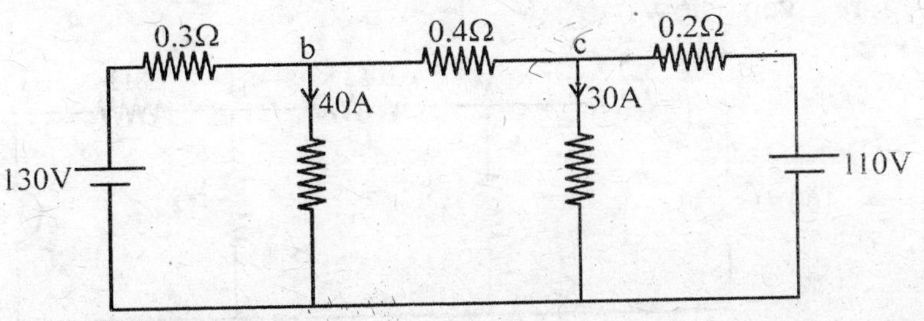

[Ans : 110.4 V, 100 V]

11. Using Nodal analysis, determine voltage 'v' in figure given below.

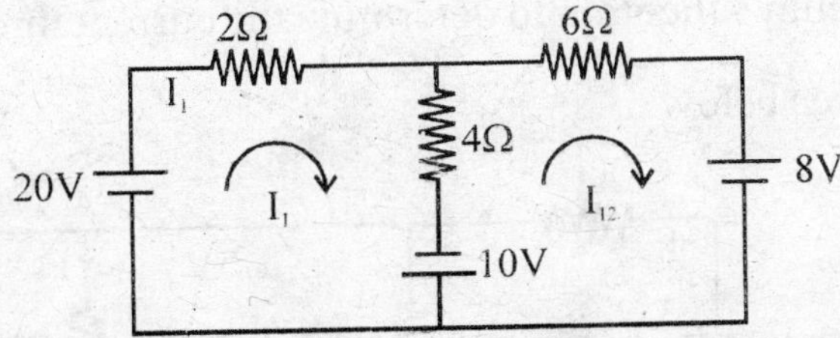

[Ans v = 74/11 volts]

12. Using Nodal analysis, determine I in figure given below.

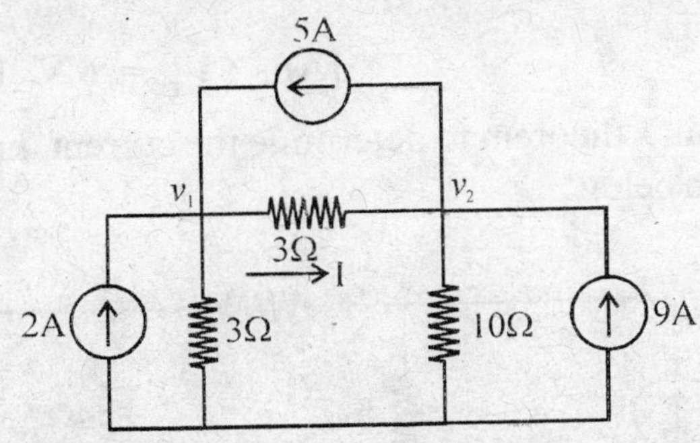

[Ans : 3.2 ampere]

13. Using Nodal analysis, determine voltages V_1 and V_2 in figure given below.

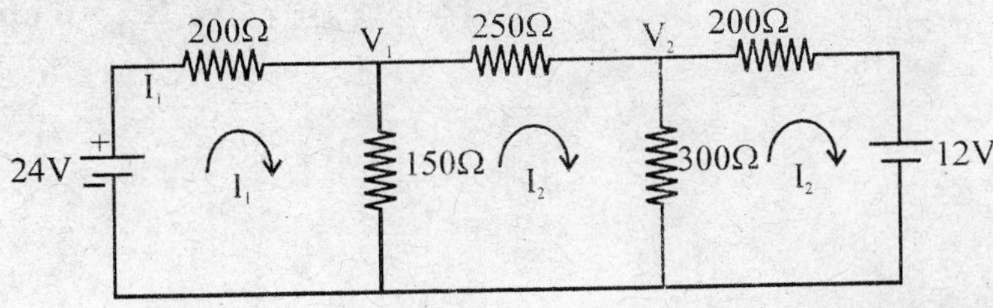

[Ans 9.705,8.013 Volts]

14. Use the Superposition theorem to determine the current through 5 ohm in the Figure given below.

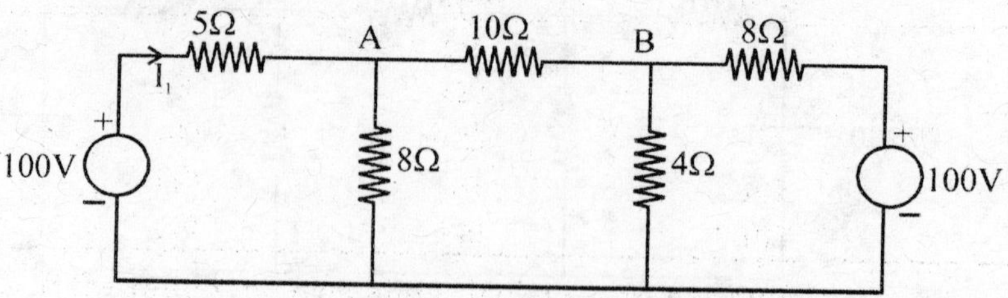

[Ans : 0.162 amp from B to A]

15. Use Thevenin's theorem to determine the current through $\frac{4}{5}$ ohm resistance in Figure given below.

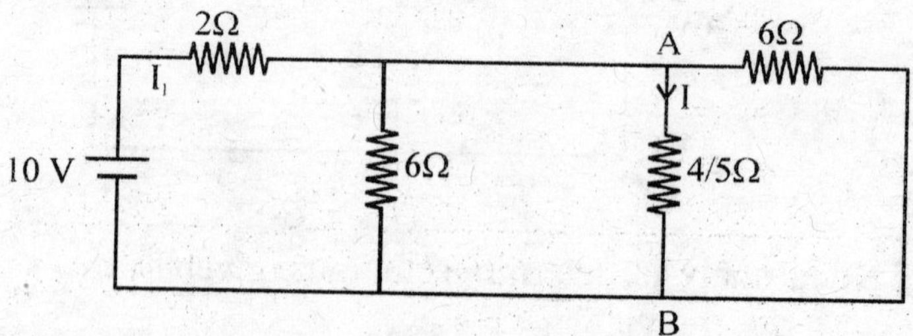

[Ans : V_{Th} = 6 V, R_{Th} = 6/5 ohm, I =3 amp]

16. Use Thevenin's theorem to determine the current through '1' ohm resistance in Figure given below.

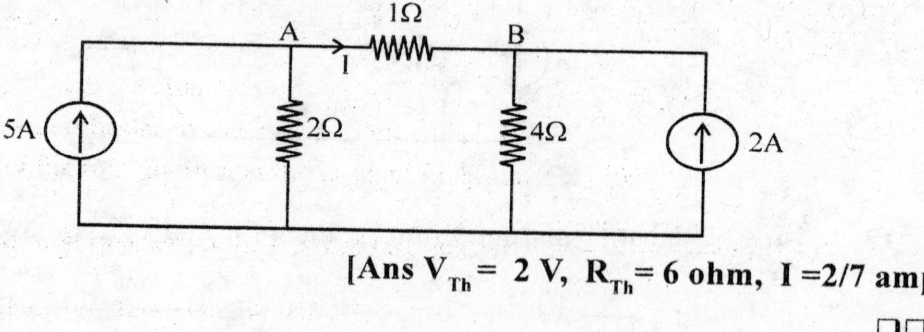

[Ans V_{Th} = 2 V, R_{Th} = 6 ohm, I =2/7 amp.]

❑❑❑

UNIT-2

Chapter 3

Single Phase A.C. Circuits

Chapter 4

Three Phase A.C. Circuits

Chapter 3

Single Phase A.C. Circuit

OBJECTIVE

- To understand the terminology related to AC quantities and phasor representation of alternating quantities
- To analyse single phase AC series and parallel circuits consisting of resistance, inductance and capacitance elements.
- To study resonance in series and parallel AC circuits

3.1 INTRODUCTION

In the earlier chapters, circuits containing only DC sources have been considered. However AC sources and systems are widely used due to their technical feasibility for generation, transmission , distribution and utilization of electrical power.

To understand AC circuits and systems , terminology associated with alternating waveforms has to be understood. In this chapter, the concepts of average and root mean square values are discussed. The analysis of series and parallel circuits consisting of resistance , inductance and capacitance has been explained. The phenomenon of resonance in series and parallel circuits has been discussed. Three phase AC circuits have also been explained.

A waveform is a pictorial representation or a graph of the magnitude of a quantity w.r.t. time. Hence any waveform whose magnitude and direction changes with time is called an alternating waveform The quantity plotted on the x - axis is time and the quantity plotted on y-axis is voltage , current , power etc.

The wave shape of an AC quantity need not always be a sine wave. They can have shapes such as a **triangular wave, square wave or a trapezoidal wave.**

3.2 Why Sinusoidal Wave Form is Preferred?

There are many technical advantages of using sinusoidal voltages and currents.

1. A sine wave can pass through the linear circuit elements such as resistors, capacitors and inductors without distortion. The other waveforms get distorted when passed through these elements.
2. Mathematical handling of sine waves is easy as 'sine' function can be represented simply by an exponential function.The exponential function is the only mathematical function whose original form is preserved even after performing differentiation and integration operations on it.
3. Circuit analysis requires mathematical operations such as differentiation and integration etc. These mathematical operations can be easily performed on sine functions.
4. It is possible to represent any non sinusoidal periodic function in terms of an infinite series of sinusoidal functions.

3.3 Generation of Single Phase AC Alternating Voltage

When a conductor is placed in a varying magnetic field or when a conductor is moved in a magnetic field , voltage is induced in that conductor, which is known as Motional or Rotational emf or dynamically induced emf.

Generation of AC voltage may be explained by considering a rectangular coil, consisting of two series-connected conductors rotating in a magnetic field, as shown in Figure 3.1.

Conductors AB at the bottom and CD at the top form a coil situated within the magnetic field set up between two magnetic poles i.e. North pole and South pole. The ends of the coil i.e. each conductor is connected to two slip rings. These are rigidly fixed to a shaft and are insulated from it and have brushes(X and Y) , for electrically connecting the coil to the external terminals When the coil is rotated, the conductors cut the magnetic flux, established by the magnetic poles, causing an emf to be generated in each conductor.

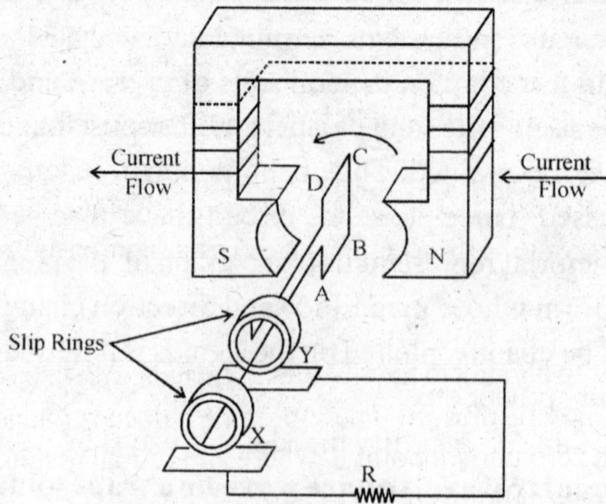

Fig.3.1 : A simplified generator

In the position shown in Fig 3.1, no voltage is being induced because the conductors AB and CD are moving parallel to the magnetic field, and no lines of force are being cut, although flux linking with the conductors is maximum. This position is known as the neutral position. \

Assume this position as the starting position. As the coil rotates counterclockwise at a constant speed, the coil sides begin to cut across the field, slowly at first but at an increasing rate later.

AB moves up and CD moves down. By the left-hand rule, the induced voltage causes electrons to flow from A toward B, in AB, and from C toward D, in CD. Through the outer circuit, current flows from Y to X and back through the brushes to the conductors. As the coil reaches the point where AB is moving down and CD up, the direction of the electron flow reverses in the conductors and in the outer circuit. Since the currents reverse in conductors AB and CD at the same time, and since the voltage induced in AB is in series with the voltage induced in CD, these can be considered as a single conductor

Figure 3.2 shows the end view of a conductor rotating clockwise in a magnetic field whose flux lines are from left to right.

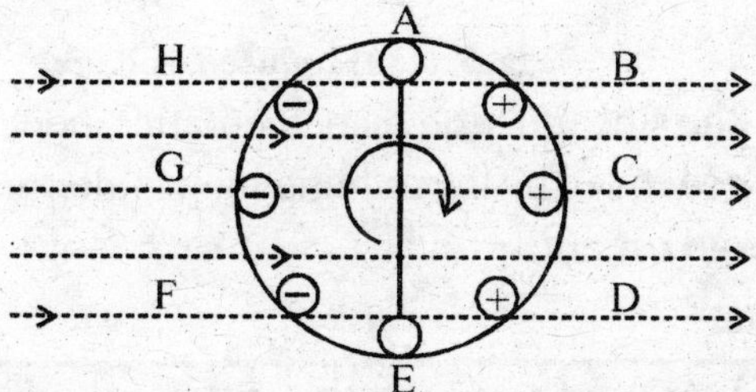

Fig.3.2 : Conductor Rotating Clockwise in a Magnetic Field.

At position A. ,the conductor is travelling parallel to the flux lines; hence no lines of flux are being cut, and the induced voltage is zero.

As the conductor moves from position A to B, the number of flux lines being cut per second is increasing, hence the induced voltage is increasing.

The induced voltage continues to increase until point C . At this point, the conductor is moving perpendicular to the flux lines and cutting a maximum number of flux lines per second, hence maximum voltage is induced.

The number of flux lines being cut per second gradually start decreasing until point D is reached. At point D, the magnitude of voltage induced is the same as the voltage induced at point B.

The induced voltage continues to decrease until point E, where it is zero again. At this point, the conductor is again moving parallel to the flux lines and hence no lines of flux are being cut.

The conductor is now moving upward. Therefore, the induced voltage will be in the opposite direction. It increases from zero until at point F, where, it is equal in magnitude and opposite in polarity to the induced voltage at points B or D.

The induced voltage increases until the point G, where maximum is induced. This maximum voltage is in a direction opposite to the voltage induced at point C.

The voltage starts decreasing until point H, where it is equal to the voltage at point F.

The decrease in induced voltage continues until the conductor returns to its starting position, at point A, where the induced voltage is again zero.

If a graph of the voltage induced in the conductor is plotted at the various points discussed above, a curve as shown in Fig3.3 will be obtained. The voltage induced in one direction is considered as positive and that in the opposite direction as negative.

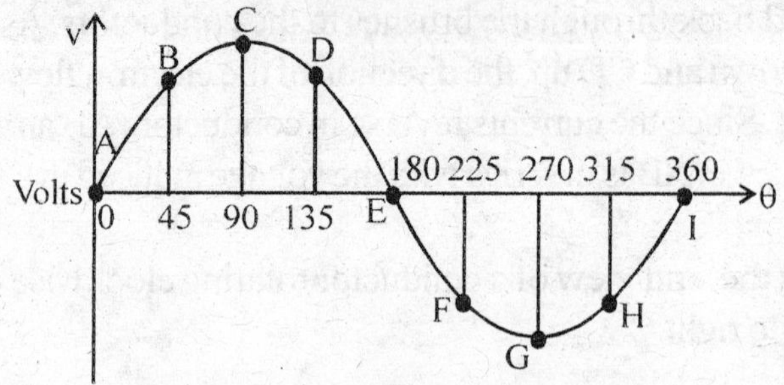

Fig. 3.3 : (a) Induced Voltage.

The curve is similar to the sine function in mathematics, hence this curve is known as a **sine wave**. Because the voltage generated by the rotating loop is alternately positive and negative, it is referred to as an **alternating voltage**.

3.4 EMF Equation

Consider a coil having 'N' turns and rotating anticlockwise in a uniform magnetic field with angular velocity 'ω' as shown in Fig 3.3(b)

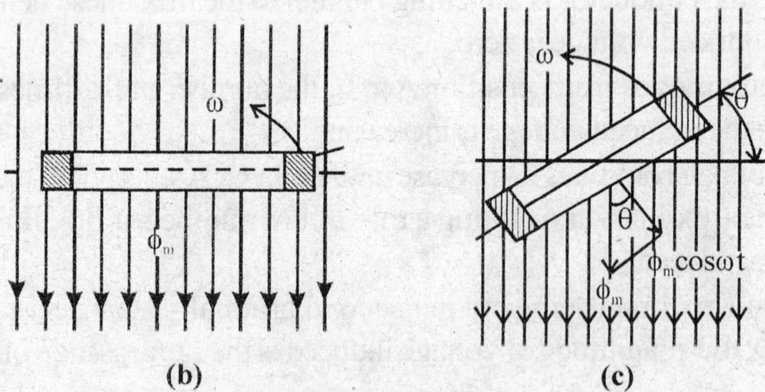

Fig 3.3 : EMF Equation

Let ϕ_m be the maximum flux at t = 0. After 't' seconds, the coil is rotated through an angle, $\theta = \omega$t.

At this position, the component of flux normal to the plane of the coil is ϕ_m cos

$\theta = \phi_m \cos\omega t$, because maximum flux is linked with the coil when the plane of the coil is perpendicular to the magnetic field.

According to Faraday's law of electromagnetic induction, the emf induced in the coil is:

$$e = -\frac{d(N\phi)}{dt}$$

Substituting the value of $\phi = \phi_m \cos\omega t$, we get:

$$e = -N\frac{d(\phi_m \cos\omega t)}{dt}$$

$$= \omega N \phi_m \sin\omega t \qquad (3.1)$$

The emf 'e' is maximum when $\sin\omega t = 1$ i.e. $\omega t = 90^0$

This maximum emf is denoted by E_m.

i.e. $E_m = \omega N \phi_m$

Hence, equation (3.1), can be rewritten as :

$$e = E_m \sin\omega t \qquad (3.2)$$

This is the required emf equation.

3.5 Terminology Related to AC Quantities

To study various aspects of alternating quantities, it is necessary to define and understand the commonly used terminology associated with it. Some important AC terms are defined in the following sub-sections, which will be frequently used in this chapter.

1. Alternating Quantity : Any quantity whose magnitude changes continuously with respect to time is called an **Alternating Quantity.**

An alternating voltage or current is one, which changes continuously in magnitude and alternates in direction at regular intervals of time as shown in Figure 3.4(a)

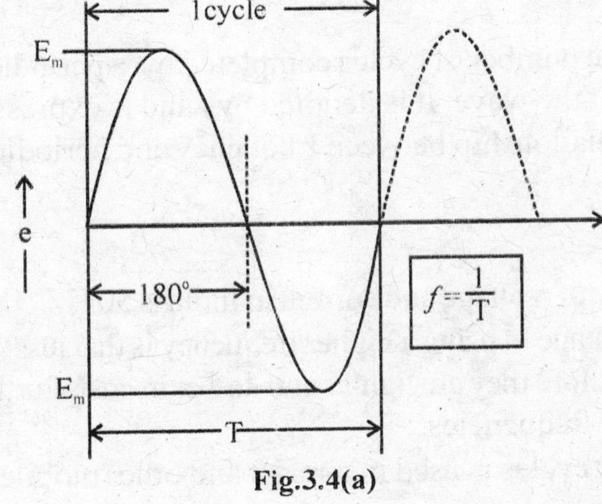

Fig.3.4(a)

2. Periodic Function : A function f (t) is said to be a periodic function in time 't' with a period of 'T' seconds, if the same values of the function are repeated after every 'T' seconds.

i.e. If a function satisfies the condition:

$f(t) = f(t+T)$, then it is a periodic function with a period of 'T' seconds

3. Instanteous Value : The magnitude of alternating voltage or current at any particular instant of time is called its **instantaneous value** and is represented by a lower case letter i.e., 'e' for emf, '*v*' for voltage and 'i' for current etc. The instantaneous value varies from one instant to another.

4. Wavefrom : The graph of the variation of magnitude of the voltage or current (along Y-axis)when plotted against time or angle (along X-axis) is called a **waveform** or wave shape. It provides an instant to instant value of the magnitude of a signal. The waveform of alternating emf varying sinusoidally is shown in Fig.3.4(a)

5. Cycle : An AC wave periodically passes through a definite cycle of changes, during one of which the voltage acts in one direction and during the other, in the opposite direction. Hence, a complete set of positive and negative values of an alternating voltage or current is termed as a **cycle** of AC wave or A positive and a negative alternation make up one cycle of a sine wave.

If we have angle in place of time on the x-axis ,then a cycle is specified in terms of angular measure spread over 360° or 2π radians.

6. Periodic Time/Time Period (T) : The time taken by an alternating voltage or current to complete one full cycle is termed its **periodic time**. It is denoted by 'T' and is expressed in seconds. After every T seconds , the cycle repeats itself. The period can be measured from the positive or negative peak in one cycle to the corresponding positive or negative peak in the next cycle.The relationship between frequency and periodic time is given by :

$$\text{Periodic time(secs)}, T = \frac{1}{f}$$

7. Frequency : The number of cycles completed by a periodic function in one second is called the **frequency** of the wave. It is denoted by *f* and is expressed in hertz (Hz) or cycles per second (c/s). The relationship between frequency and periodic time is given by :

$$\text{Frequency}, f = \frac{1}{T}$$

Standard frequency of voltage and current in India is 50 Hz.

The principal advantage in using a higher frequency is that less iron and copper is required in transformers and therefore they are lighter and lower in cost. Further flicker in incandescent lamps is lower at higher frequencies.

A frequency of 400 cycles is used in aircraft and other mobile equipment .

8. Angular Velocity (ω) : As a phasor rotates, the speed at which it rotates about the origin will depend on the frequency (f). In one second the phasor will rotate through f revolutions, or through $f \times 360°$ (one revolution = 360°).

It is common to use the angular unit RADIAN , where 360° = 2π rads. The phasor can therefore be said to rotate through $2\pi f$ radians per second. This is called the ANGULAR VELOCITY of the phasor, and is represented by the symbol ω (omega).

The angular distance covered per second is defined an **angular velocity** or it is defined as the rate of change of angle' θ' with respect to time.

i.e. $$\omega = \frac{d\theta}{dt}$$

The relationship between angular velocity and frequency can be obtained as follows:

$$\omega = \frac{\text{Angle traversed in one cycle}}{\textit{Time taken to complete one cycle}}$$

i.e. $$\omega = \frac{2\pi}{T} = 2\pi f$$

9. Amplitude/Peak Value : The maximum value or peak value of the positive or negative half of an AC wave is called its **amplitude / peak value**. The amplitude of an alternating emf, voltage and current is expressed by E_m, V_m and I_m, respectively

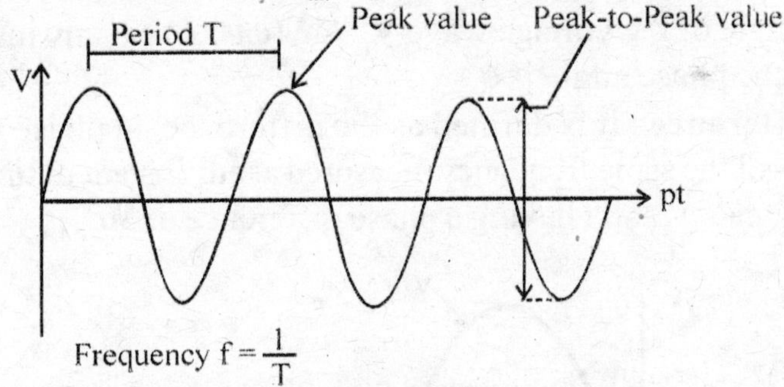

Fig.3.4(b)

10. Peak to Peak value : It is the voltage or current from positive to the negative peak. It is twice the peak value. If the amplitude of an alternating emf, voltage and current is expressed by E_m, V_m and I_m, respectively then the corresponding peak to peak values are $2E_m$, $2V_m$ and $2I_m$.

11. Phase Angle : The **phase angle** of a point on a periodic wave is the distance between the point and a specified reference point, expressed using an angular measure.

The phase angle of a vector may be written as $M \angle$ "θ, where M is the magnitude of the vector and θ is the phase angle relative to the specified reference point.

The phase angle of a waveform can be used to calculate the instantaneous value.

For one complete revolution, the conductors rotate through 360^0. As T is the time to complete one revolution, the angle through which the conductors move in one second is:

$$\text{Angular velocity} = \frac{1}{T} \text{ rev/sec}$$

$$= \frac{360}{T} \text{ deg/sec}$$

$$\text{Or Angular velocity}, \omega = \frac{2\pi}{T} \text{ rad/sec}$$

Phase angle at any instant t, measured from t =0 is:

$$\theta = \frac{2\pi}{T} \times t \text{ radians}$$

or

$$\theta = \frac{360}{T} \times t \text{ degrees}$$

Phase angle equation can also be written as:

$$\theta = 2\pi f t \text{ radians or } \theta = \omega t \text{ radians}$$

The phase angle is the angle of a wave at time, t = 0

The phase angle of the voltage wave $V_m \sin(\omega t + \phi)$, at any instant of time is $(\omega t + \phi)$ but at time t =0, the phase angle is ϕ

12 Phase difference : It is defined as the difference in phase angle between the two alternating signals of the same frequency measured at any instant of time. Figure.3.4(c) shows two alternating signals V and I having a phase difference of 30^0

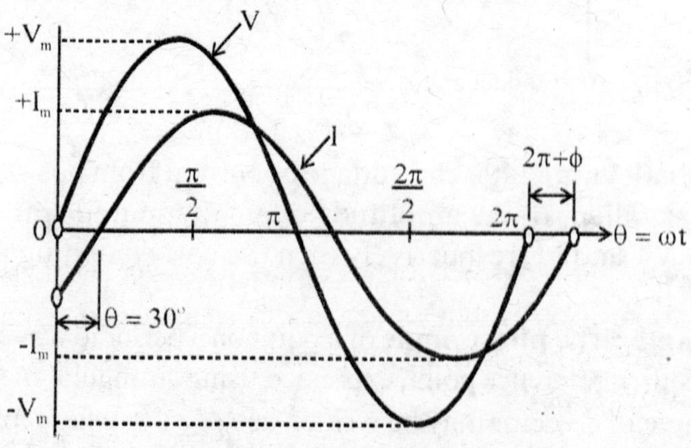

Fig.3.4(c) Phase difference of 30^0

To determine the phase difference between two sine waves, locate the points on the time axis where the two waves cross the time axis traveling in the same direction. The number of degrees between the crossing points is the **phase difference**. The wave that crosses the axis at the later time (to the right on the time axis) is said to **lag** the other wave.

Sine wave in Phase

It is seen in Fig. 3.4(d) that the sine wave of voltage and the resulting sine wave of current are superimposed on the same time axis and as the voltage increases in a positive direction, the current increases along with it, and that when the voltage reverses direction, the current also reverses direction. When two sine waves, are in step with one another, they are said to be **IN PHASE**. To be in phase, the two sine waves must go through their maximum and minimum points at the same time and in the same direction.

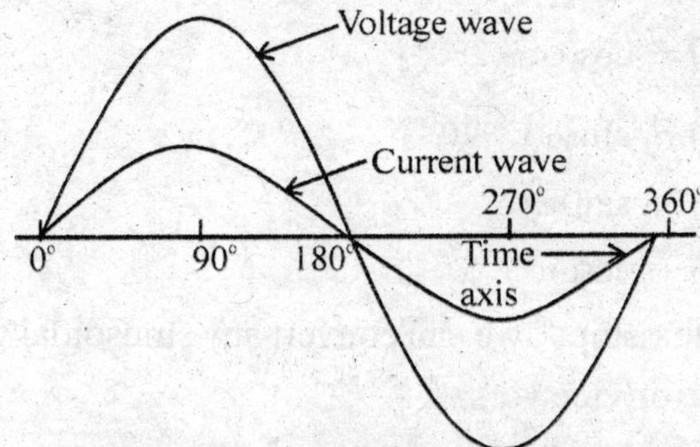

Fig. 3.4(d) Voltage and current waves in phase.

Phase Difference between a Sine wave and a Cosine wave

Figure 3.4(e) shows a sine wave and a cosine waves 90° out of phase.

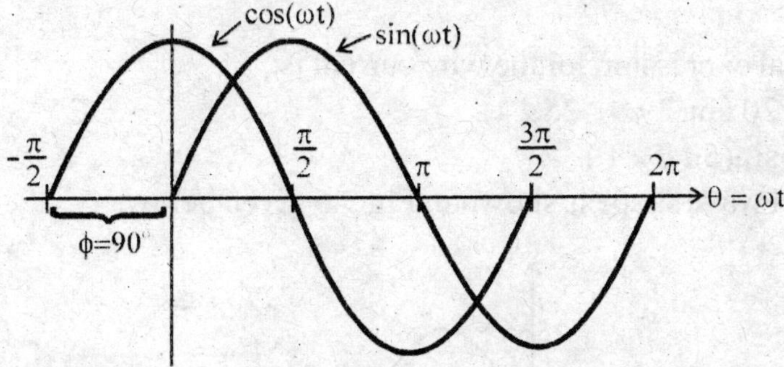

Fig.3.4(e) - Sine wave and a Cosine waves 90° out of phase.

Alternatively, we can also say that a sine wave is a cosine wave that has been shifted in the other direction by -90°. When dealing with sine waves or cosine waves with an angle the following rules will always apply:

Sine and Cosine Wave Relationships

$$\cos(\omega t + \phi) = \sin(\omega t + \phi + 90^o)$$

$$\sin(\omega t + \phi) = \cos(\omega t + \phi + 90^o)$$

When comparing two sinusoidal waveforms it is common to express their relationship as either a sine or cosine function by using the following trigonometric identities :

$$-\sin(\omega t) = \sin(\omega t + 180^o)$$

$$-\cos(\omega t) = \cos(\omega t + 180^o)$$

$$-\cos(\omega t) = \sin(\omega t + 270^o)$$

$$\pm\sin(\omega t) = \cos(\omega t + 270^o)$$

$$\pm\cos(\omega t) = \sin(\omega t + 90^o)$$

$$-\sin(\omega t) = \sin(-\omega t)$$

$$\cos(\omega t) = \cos(-\omega t)$$

By using these relationships , we can convert any sinusoidal waveform from either a sine wave into a cosine wave or vice versa.

EXAMPLE 3.1

A wire carries both AC and DC current with maximum value of AC 20 amp. and 5 amp. of DC. The frequency of resultant current is 25 Hz. Write mathematical expression for the wire current as a function of time and draw the wave shape.

Solution

Mathematical expression for the wire current is:

$$i(t) = 5 + 20\sin(2\pi \times 25 \times t)$$

$$= 5 + 20\sin(50\pi \times t)$$ **Ans**

The required wave shape is shown in Fig.3.5 given below:

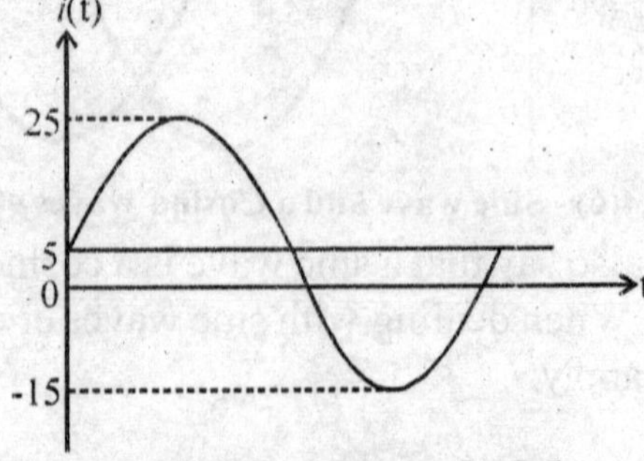

Fig 3.5 : for example 3.1

3.6 Average and Effective Values of Periodic Functions

An alternating quantity can be completely defined by its amplitude , frequency and phase. In addition to these parameters, it is useful to know their effective or rms value, average value and the form factor.

3.6.1 Average Value of a Function

In general, the **average value** of a function 'f' on an interval [a,b] is :

$$\bar{f} = f_{avg} = \frac{1}{b-a}\int_a^b f(x)dx$$

We know that the average of 'n' numbers, y1 , y2 ,-----,yn is :

$$\bar{y} = \frac{1}{n}\sum_{k=1}^{n} yk$$

For the average value of the function f, suppose we take its values at a few equally spaced points in the interval [a,b], $x_1, x_2, ... x_n$. Then the spacing of the points is $\Delta x = (b-a)/n$ and

$$\bar{f} = \frac{1}{n}\sum_{k=1}^{n} f(xk)$$

From the spacing between the values of x, we notice that

$$\frac{1}{n} = \frac{\Delta x}{b-a}$$

As the number of sample points gets larger ($n \to \infty$), this sum approaches the definite integral, so that, in this limit,

$$\bar{f} = \frac{1}{b-a}\int_a^b f(x)dx$$

Average value of any function f(t) is defined as equal to the average of all the instantaneous values considered over a specified period.

The average value of any function, f(t) over the specified time interval between time 't_1' and 't_2' is expressed mathematically as :

$$F_{av} \cong \frac{1}{t_2 - t_1}\int_{t_1}^{t_2} f(t)dt$$

When f(t) is a periodic function having a period of 'T' seconds, (3.3)

$$F_{av} \cong \frac{1}{T}\int_0^T f(t)dt$$

If the time function is expressed in terms of radians ($\alpha = \omega t$), then an alternate form for the average value of the function can be expressed as :

$$F_{av} = \frac{1}{2\pi}\int_0^{2\pi} f(\omega t)d(\omega t) = \frac{1}{2\pi}\int_0^{2\pi} f(\alpha)d\alpha$$

Where T = 2π

For symmetrical ac waves, the average value for a complete cycle is zero because both positive and negative half cycles are exactly identical. Hence, their average value is defined for a half cycle.

The average value of ac waves can be obtained by the following methods:
(i) Graphical method or Mid ordinate method (ii) Analytical method

3.6.2 Average Value of a Sinusoidal wave

(i) Graphical method or Mid ordinate method

Consider 'n' equal intervals of the half cycle as shown in Fig.3.6(a). Let $i_1, i_2, \text{-----}, i_n$ be the mid ordinate magnitude of currents in different time intervals.

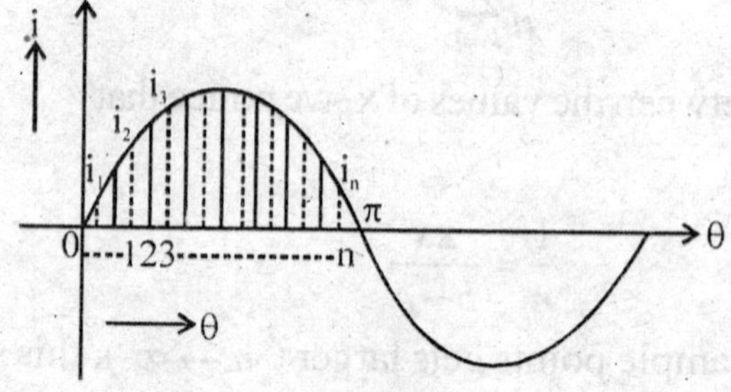

Fig.3.6(a)

Average value of current over a half cycle = mean of mid ordinate currents

i.e. $$I_{av} = \frac{i_1 + i_2 + \text{-----} i_n}{n}$$

(ii) Analytical method

The average value of the sinusoidal current variation i.e. $i(t) = I_m \sin \omega t$, which is symmetrical and is shown in Figure 3.6(b) can be obtained as follows:

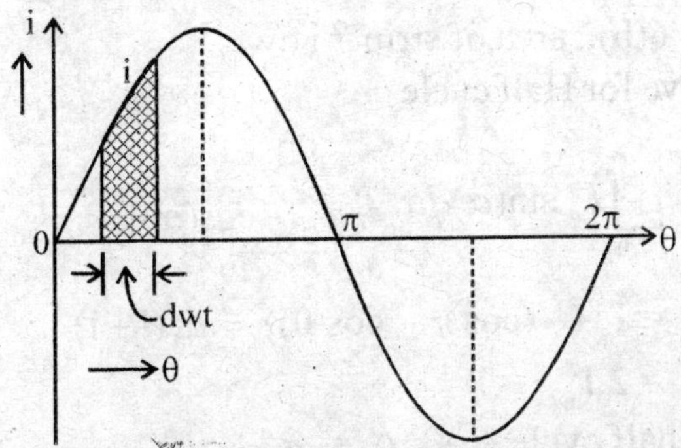

Fig 3.6(b) Wave form of i(t)= I_m sin ωt

Consider a small strip of width'dwt' and let ' i' be its mid ordinate i.e. average instantaneous value of current.

Then area of this strip = i dwt .

Area under the curve for full cycle

$$= \int_0^{2\pi} i\, d(\omega t) = 0$$

Length of Base over full cycle = 2 π − 0 = 2 π

Hence, Average value of full cycle

$$= \frac{Area \text{ under the curve for full cycle}}{length\ of\ Base} = \frac{0}{2\pi} = 0$$

Alternatively,

$$I_{av} = \frac{1}{2\pi}\int_0^{2\pi} I_m \sin(\omega t)\, d(\omega t) = \frac{1}{2\pi}\int_0^{2\pi} I_m \sin(\alpha)\, d\alpha \qquad (3.4)$$

Where α = ωt

Integrating equation (3.4) and inserting limits, we get ,

$$I_{av} = \frac{I_m}{2\pi}[-\cos(\alpha)]_{\alpha=0}^{2\pi} = \frac{I_m}{2\pi}[-\cos(2\pi)+\cos(0)] = 0$$

This equation shows that the **average value of a sinusoid over one complete cycle is equal to zero**.

This is obvious because the area under positive half cycle is equal to the area under negative half cycle.

3.6.3 Average value of a Sinusoidal wave over half cycle

For a sinusoidal wave , the average value of the half cycle is taken to be the average value (because average value of a sinusoid over one complete cycle is identically equal to zero as shown above) and is denoted as I_{av}.

As shown in Fig. 3.6(b), area of strip = i dwt .
Area under the curve for Half cycle

$$= \int_0^{\pi} I_m \sin(\alpha) d\alpha = I_m[-\cos(\alpha)]_0^{\pi}$$

$$= I_m (-(\cos\pi - \cos 0)) = I_m(1+1)$$

$$= 2\ I_m$$

Length of Base for half cycle = $\pi - 0 = \pi$

Hence, Average value over half cycle

$$= \frac{\textit{Area} \text{ under the curve for half cycle}}{\textit{Base over half cycle}} = \frac{2\,I_m}{\pi} = 0.637\ I_m$$

Alternatively,

$$[I_{av}]_{half\ cycle} = \frac{1}{\pi}\int_0^{\pi} I_m \sin(\alpha)\, d\alpha \qquad (3.5)$$

In Equation 3.5, the lower and upper limits of integration have been selected to include the positive area i.e. positive half cycle only.

Integrating equation (3.5) and substituting the limits , we get :

$$I_{av} = \frac{I_m}{\pi}[-\cos(\alpha)]_0^{\pi} = I_m\,(-(\cos\pi\ -\cos 0))$$

$$= \frac{I_m}{\pi}[1+1] = \frac{2I_m}{\pi}$$

Or $\qquad I_{av} = 0.637\,I_m \qquad (3.6)$

i.e. Average value of a sinusoidal signal is 0.637 times its maximum value

3.6.4 The Effective (or RMS) value of Alternating Quantity

For symmetrical, periodic functions, we have a more suitable criterion to measure the effectiveness of a periodic function. Such criterion is based on the energy or power capability associated with the periodic function and is discussed below :

The effective or RMS value of an AC current is equal to the DC current , which is required to produce the same amount of heat in a resistance as produced by the AC current provided that the resistance and time for which these currents flow are identical.

Consider that the sinusoidal current, $i(t) = I_m \sin(\omega t)$ flows through a resistance R ohms.

Then by Joule's law , the instantaneous power absorbed by the resistor and converted to heat is:

$P = i^2R$ watts

The average power absorbed by the resistor over one full cycle is given by :

$$P_{av} = \frac{1}{T}\int_0^T i^2 R dt = \left[\frac{1}{T}\int_0^T i^2 dt\right]R \qquad (3.7)$$

The quantity in brackets is the average or mean value of the function $f(t) = i^2(t)$.

The term inside the bracketed quantity is current squared and can be interpreted as the square of an effective current, which when multiplied by R, yields the average power.

Mathematically, we can write the square of an effective current as:

$$I_{eff}^2 = \frac{1}{T}\int_0^T i^2 dt$$

or

$$I_{eff} = \sqrt{\frac{1}{T}\int_0^T i^2(t)dt} = I_{rms} \qquad (3.8)$$

Hence it follows that the effective current is the **Root mean square value of the current.** The effective (or r.m.s.) quantity is usually written without a subscript.

i.e. $\quad I_{eff} = I$

An analogy with the direct current case can be established.

A direct current I_{dc} flowing through a certain resistance R for a certain time, causes an average power dissipation of I_{dc}^2 R and a certain amount of heat will be produced.

The flow of a steady alternating current through the same resistor for the same time leads to an average power dissipation of I_{eff}^2 R or exactly the same amount of heat is produced for the given time as in the case of direct current.

Thus this value of steady direct current is taken as equivalent to the alternating current.

Comparing this result with that of the direct current case, another interpretation of effective current follows i.e. the effective current is that current which produces the same heating effect as the steady direct current.

3.6.5 Effective or RMS value of sinusoidal wave

Effective or RMS value of a sinusoidal function $i(t) = I_m \sin \omega t = I_m \sin \theta$ can be obtained by any one of the following methods:

(i) Graphical method or Mid ordinate method(ii) Analytical method

(i) Graphical method or Mid ordinate method

Consider 'n' equal intervals of the half cycle as shown in Fig.3.6(c). Let $i_1, i_2, - - - -, i_n$ be the mid ordinate magnitudes of currents in different time intervals.

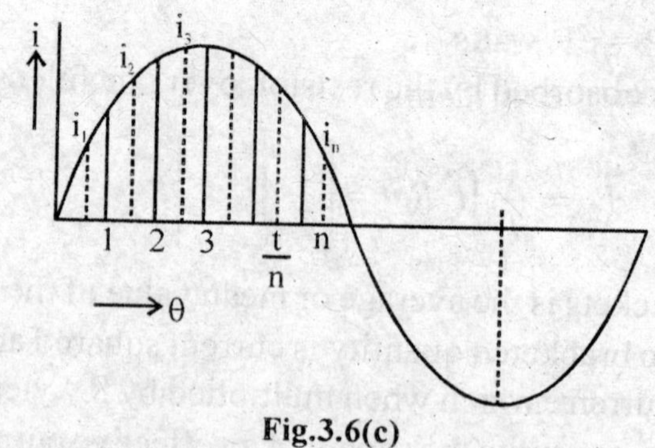

Fig.3.6(c)

Then ,I_{RMS} = Square root of the average value of i^2

$$= \sqrt{\frac{i_1^2 + i_2^2 + - - - - + i_n^2}{n}}$$

(ii) Analytical method

The average value of the sinusoidal current variation i.e. $i(t) = I_m \sin \omega t$, which is symmetrical and is shown in Figure 3.6(d) is obtained as follows:

Step 1 Obtain the square of the given current i(t)

i.e. $i^2(t) = I_m^2 \sin^2 \theta$.

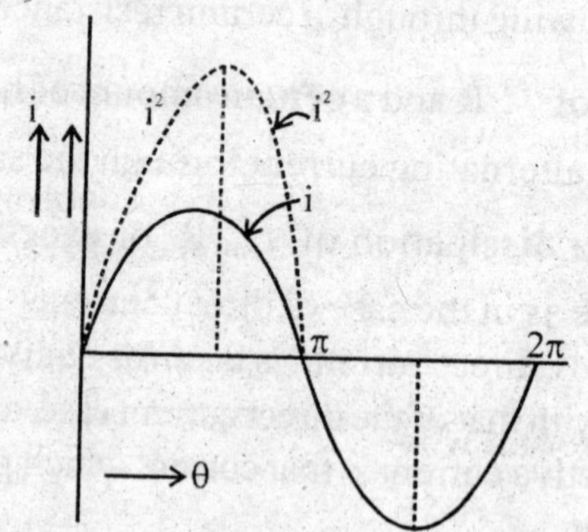

Fig 3.6(d) Wave form of $i^2(t) = I_m^2 \sin^2 \omega t$

Its waveform is shown in Fig 3.6(d)

Step 2 Obtain average value of squared current over a half cycle

i.e. Avg value of squared current over a half cycle

$$= \frac{1}{\pi} \int_0^{\pi} I_m^2 \sin^2 \theta \; d\theta \tag{3.9}$$

Substituting the trigonometric identity, $\sin^2\theta = \frac{1-\cos 2\theta}{2}$ in equation (3.9), we get:

Avg value of squared current

$$= \frac{I_m^2}{2\pi}\int_0^{\pi}(1-\cos 2\theta)d\theta \qquad (3.10)$$

Integrating equation (3.10), we get :
Avg value of squared current

$$= \frac{I_m^2}{2\pi}(\theta + \frac{\sin 2\theta}{2})\Big|_0^{\pi}$$

Substituting the limits , we get:

Avg value of squared current

$$= \frac{I_m^2}{2\pi} \times \pi = \frac{I_m^2}{2} \qquad (3.11)$$

Step 3 Obtain the square root of the average value of i^2
I_{RMS} = Square root of the average value of i^2

$$\text{or} \qquad I_{RMS} = \sqrt{\frac{I_m^2}{2}} = \frac{I_m}{\sqrt{2}} = 0.707 I_m \qquad (3.12)$$

i.e. **RMS value of a sinusoidal signal is 0.707 times its maximum value.**
This important result is used often in the study of AC circuits.

3.6.6 Form Factor and Peak Factor of Sinusoidal Wave

RMS value , average value and peak values are related to each other and their relationship is given by the two factors i.e. Form factor and Peak factor

(i) Form factor : The ratio of effective value or RMS value to the average value is defined as the **form factor** of the wave form.

i.e. **Form factor of sinusoidal wave**

$$= \frac{I_{rms}}{I_{av}} = \frac{0.707 I_{max}}{0.637 I_{max}} = 1.11 \qquad (3.13)$$

(ii) Peak factor

Peak factor of sinusoidal wave (also called the Crest or amplitude factor)

$$= \frac{\textit{Maximum}\ \text{value}}{\textit{RMS}\ \text{value}}$$

$$= \frac{I_{max}}{.707 I_{max}} = 1.414 \qquad (3.14)$$

Example 3.2(a) Write the mathematical expression for a 50 Hz sinusoidal voltage supplied for domestic purpose at 230 volts.

Solution :

Declared Voltage values and those given by measuring instruments are RMS values.

i.e. $V_{RMS} = 230$ volts

Corresponding value of

$$V_{max} = \sqrt{2}\ (230) = 325.27 \text{ volts}$$

The required mathematical expression is:

$$v(t) = 325.27 \sin(wt)$$

i.e. $v(t) = 325.27 \sin(2\pi\ ft)$

Substituting the value of 'f' we get:

$$v(t) = 325.27 \sin(2 \times \pi \times 50 \times t)$$

i.e. The required mathematical expression is ,$v(t) = 325.27 \sin(314.16t)$ **Ans**

Example 3.2(b) An alternating voltage is expressed as $v = 10 \sin 100\,t$. Determine(a) Average value(b) RMS value (c) Frequency (d) Instantaneous voltage when $t = 2 \times 10^{-3}$ seconds

Solution :

(a)Average value, $V_{av} = 0.636\ V_m = 0.636 \times 10 = 6.36$ volts **Ans**

(b) RMS value, $V_{RMS} = 0.707 V_m = 0.707 \times 10 = 7.07$ volts **Ans**

(c) $2\pi f = 100$

Hence freq, $f = \dfrac{100\pi}{2\pi} = 50$ Hertz **Ans**

(d) Instantaneous voltage is $10 \sin 100\pi t$

$= 10 \sin 100\pi 2\times10^{-3} = 10 \sin 0.2\pi = 10 \sin 36^0$

$= 10\times0.59 = 5.9$ volts Ans

Example 3.3

Find the average and RMS values of the current:

$$i\,(t) = 10 + 10 \sin 314t$$

Solution :

$$\text{Average value of current} = \frac{1}{2\pi}\int_0^{2\pi}(10 + 10\sin 314t)\,dwt$$

$$= \frac{1}{2\pi}\int_0^{2\pi}(10)\,dwt + \frac{1}{2\pi}\int_0^{2\pi}(10\sin 314t)\,dwt$$

$= 10 + 0 = 10$ amp **Ans**

$$\text{RMS value of current} = \left\{\frac{1}{2\pi}\int_0^{2\pi}(10+10\sin 314t)^2\,dwt\right\}^{0.5}$$

$$= \left\{\frac{1}{2\pi}\int_0^{2\pi}(100)dwt + \frac{1}{2\pi}\int_0^{2\pi}(100\sin^2 314t)dwt + \frac{1}{2\pi}\int_0^{2\pi}(200\sin 314t)dwt\right\}^{0.5}$$

$$= \left\{100 + \frac{1}{2\pi}\int_0^{2\pi}\frac{100(1-\cos 628t)}{2}dwt + 0\right\}^{0.5}$$

$$= \left\{100 + \frac{1}{2\pi}\int_0^{2\pi}(50)dwt - \frac{1}{2\pi}\int_0^{2\pi}(50\;\cos 628t)dwt\right\}^{0.5}$$

$= (100+50-0)^{0.5} = \sqrt{150} = 12.25$ amp **Ans**

Example 3.4

Find for the full wave rectified sine wave shown in Fig.3.7(i) Average value (ii) Effective or RMS value (iii)Form Factor (iv) Peak factor

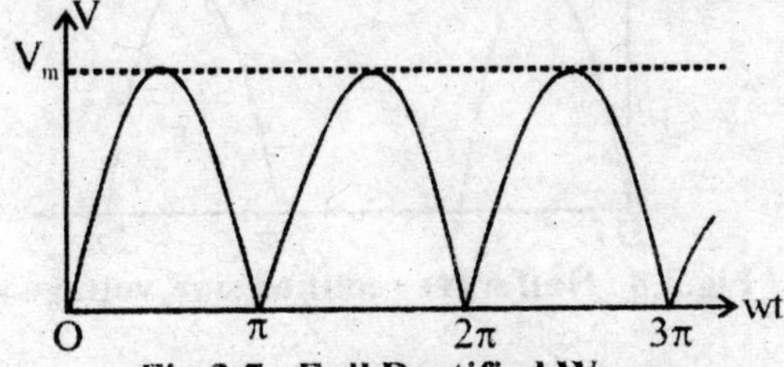

Fig.3.7 : Full Rectified Wave

Solution:

$$\text{(i) Average value} = \frac{1}{\pi}\int_0^{\pi}V_m \sin\omega t\; d(\omega t)$$

$$= \frac{V_m}{\pi}[-\cos\omega t]_0^{\pi} = \frac{V_m}{\pi}[-\cos\pi + \cos 0] = \frac{V_m}{\pi}[1+1] = \frac{V_m}{\pi}[2]$$

$= 0.637V_m$ ans

(ii) RMS value

$$V_{RMS} = \left[\frac{1}{\pi}\int_0^{\pi}V_m^2\sin^2\omega t\; d(wt)\right]^{0.5}$$

$$= \frac{V_m^2}{2\pi}\left[\int_0^{\pi}(1-\cos 2\omega t\; d(wt)\right]^{0.5}$$

$$(V_{rms})^2 = \frac{V_m^2}{2\pi}\left[wt - \frac{\sin(2wt)}{2}\right]_0^\pi$$

$$(V_{rms})^2 = \frac{V_m^2}{2}$$

$$V_{RMS} = 0.707\ V_m$$

(iii)Form factor $= \frac{V_{RMS}}{V_{AV}} = \frac{0.707\ V_m}{0.637\ V_m} = 1.11$

(iv)Peak factor $= \frac{V_{max}}{V_{RMS}} = \frac{V_m}{0.707\ V_m} = 1.414$

Example 3.5

Find the form factor and peak factor of half wave rectified sine voltage wave shown in **Fig.3.8**

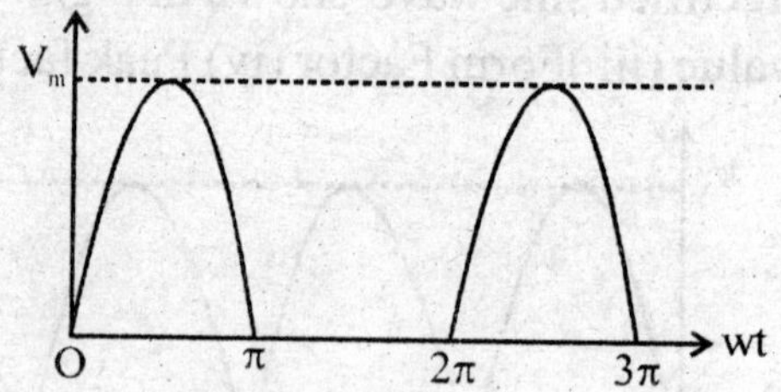

Fig.3.8 : Half wave rectified sine voltage wave

Solution :

Average value

$$= \frac{1}{2\pi}\left[\int_0^\pi V_m \sin\omega t\, d(\omega t) + \int_\pi^{2\pi} o\ d(\omega t)\right]$$

$$= \frac{1}{2\pi}\int_0^\pi V_m \sin\omega t\, d(\omega t) = \frac{V_m}{2\pi}[-\cos\omega t]_0^\pi = \frac{V_m}{\pi} = 0.318 V_m$$ **Ans**

RMS value $= \left[\frac{1}{2\pi}\int_0^\pi (V_m \sin\omega t)^2\, d(\omega t)\right]^{0.5} = 0.5\ V_m$

Form factor $= \frac{RMS\ \text{Value}}{Avg\ \text{Value}} = \frac{0.5 V_m}{0.318\ V_m} = 1.57$ **Ans**

Peak factor $= \frac{Peak\ \text{Value}}{RMS\ \text{Value}} = \frac{V_m}{0.5\ V_m} = 2$ **Ans**

Example 3.6

Find the RMS and Average values of the voltage wave shown in Fig.3.9

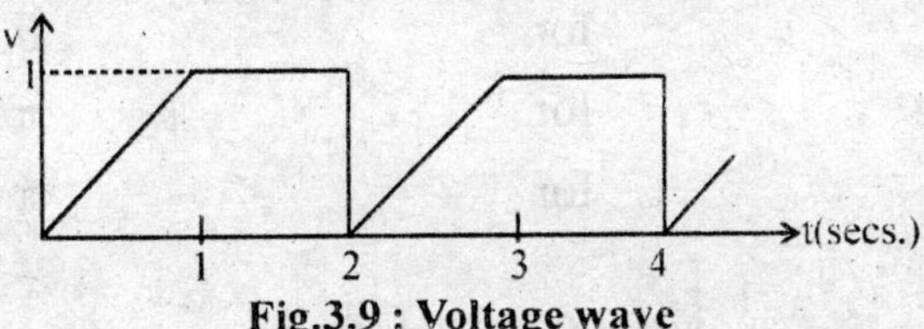

Fig.3.9 : Voltage wave

Solution :

One cycle of the wave from t = 0 to 2 seconds, can be expressed as :

v = t for 0 < t < 1

v = 1 for 1 < t < 2

$$\text{Average value} = \frac{1}{2}\left(\int_0^1 t\,dt + \int_1^2 dt\right) = \left\{\frac{1}{2}\left[\frac{t^2}{2}\right]_0^1 + \frac{1}{2}[t]_1^2\right\}$$

$$= \frac{1}{2}\left(\frac{1}{2} - 0\right) + \frac{1}{2}(2-1) = \frac{1}{4} + \frac{1}{2} = \frac{3}{4} \text{ Volts}$$ **Ans**

$$\text{RMS value} = \left[\frac{1}{2}\left(\int_0^1 t^2\,dt + \int_1^2 dt\right)\right]^{0.5} = \left\{\frac{1}{2}\left[\frac{t^3}{3}\right]_0^1 + \frac{1}{2}[t]_1^2\right\}^{0.5} = \left(\frac{1}{6} + \frac{1}{2}\right)^{0.5}$$

$$= \left(\frac{4}{6}\right)^{0.5} = \sqrt{0.67} = 0.8165 \text{ volts}$$ **Ans**

Example 3.7

A delayed half wave rectified wave is shown in Figure. 3.10(a). (i)Find its Average and RMS values.

(ii)Determine the average value of the triangular wave shown in the Fig.3.10(b)

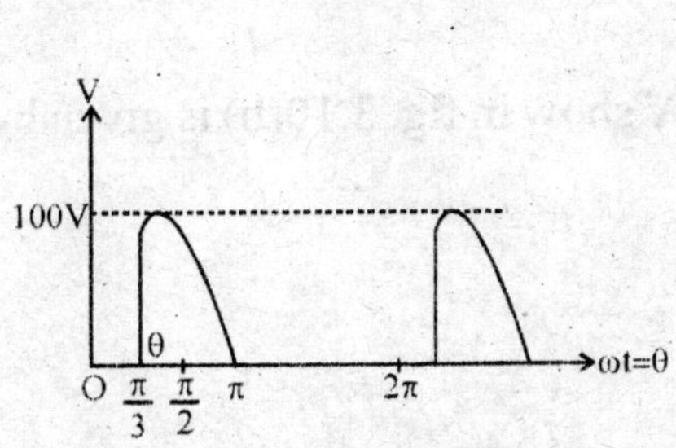

(a) Delayed half wave rectified wave

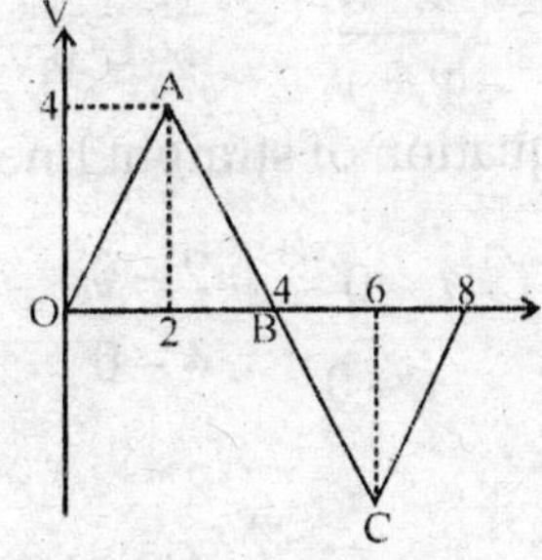

(b) Triangular wave

Fig.3.10

Solution :

(i) The wave can be mathematically expressed as :

$v = 0$	for	$0 < \theta < \pi/3$
$v = 100 \sin \theta$	for	$\pi/3 < \theta < \pi$
$v = 0$	for	$\pi < \theta < 2\pi$

$$\text{Average value} = \frac{1}{2\pi}\left[\int_{\pi/3}^{\pi} 100 \sin \theta \, d(\theta)\right]$$

$$= \frac{100}{2\pi}\left[\cos(\theta)\right]_{\pi/3}^{\pi} = \frac{50}{\pi}\left[-1-0.5\right] = 15.91(1.5) = 23.87 \text{ Volts}$$

$$[\text{RMS value}]^2 = \frac{1}{2\pi}\left[\int_{\pi/3}^{\pi} (100 \sin \theta)^2 \, d\theta\right]$$

$$= \frac{10000}{2\pi}\left[\int_{\pi/3}^{\pi}\left(\frac{1-\cos 2\theta}{2}\right)d\theta\right]$$

$$= \frac{10000}{4\pi}\left[\theta - \frac{\sin 2\theta}{2}\right]_{\pi/3}^{\pi}$$

$$= \frac{2500}{\pi}\left[(\pi - 0) - (\frac{\pi}{3} - \frac{\sin 120}{2})\right]$$

$$= 795.77(2.527) = 2010.9$$

$$\text{RMS value} = (2010.9)^{0.5} = 44.83 \text{ Volts}$$ **Ans.**

(ii) We know that equation of a straight line passing through two points (x_1, y_1) and (x_2, y_2) is given by:

$$\frac{x - x_1}{y - y_1} = \frac{x_2 - x_1}{y_2 - y_1}$$

Hence , equation of straight line 'OA' show in fig. 3.10(b) is given by:

$$\frac{t - 0}{v - 0} = \frac{2 - 0}{4 - 0}$$

$$\frac{t}{v} = \frac{2}{4}$$

or $v = 0 < t < 2$ **(i)**

Equation of straight line 'AB' fig. 3.10(b), is given by:

$$\frac{t-2}{v-4} = \frac{4-2}{0-4} = \frac{2}{-4}$$

Or v = -2 t +8 ⇒ for 2< t < 4 **(ii)**

Average value ,

$$V_{av} = \frac{1}{4}[\int_0^2 2tdt + \int_0^4 (-2t+8)dt]$$

$$= \frac{1}{4}[2(\frac{t^2}{2})_0^2 + [-2\frac{t^2}{2} + 8t]_2^4]$$

$$= \frac{1}{4}[(4-0) + [(-16+32) - (-4+16)]$$

$$= \frac{1}{4}[4 + [16-12]]$$

$$= \frac{1}{4}[4+4] = \frac{8}{4} = 2$$ **Ans**

Example 3.8 :

Find the average value (over the full cycle) of voltage wave shown in Fig 3.11

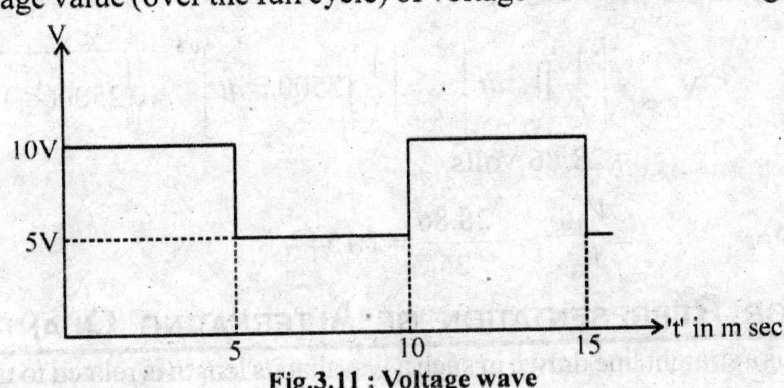

Fig.3.11 : Voltage wave

Solution :

The area under the curve from 0 to 5 ms = 10 (5 × 10⁻³) = 50 × 10⁻³

The area under the curve from 5 ms to 10 ms

= 5 (5 × 10⁻³) = 25 × 10⁻³

Total area over the full cycle =50 × 10⁻³ +25 × 10⁻³ = 75 × 10⁻³

The time period is 10 ms.=10× 10⁻³

$$\text{Average value} = \frac{\textit{Area} \text{ under the curve over full cycle}}{\textit{Time period}} = \frac{75\times10^{-3}}{10\times10^{-3}} = 7.5 \text{ Volts Ans}$$

Example 3.9 :

Find the average value, effective value and form factor of saw tooth voltage wave shown in Fig 3.12

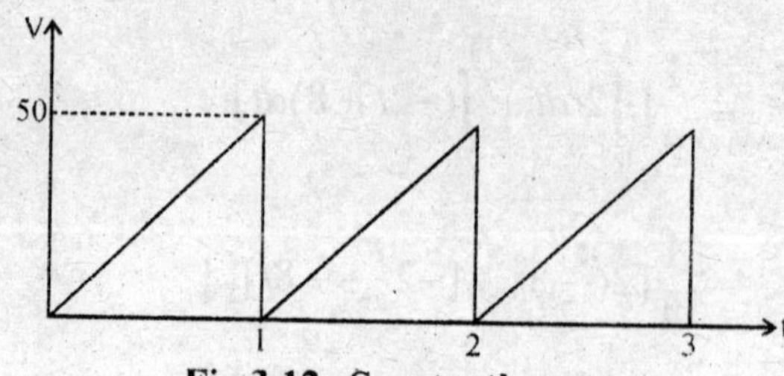

Fig.3.12 : Saw tooth wave

Solution :

Area of rectangle=50 × 1= 50

Area under a triangle is half the area of the rectangle,

$$V_{av} = \frac{50}{2} = 25 \text{ V}$$

$$V_{av} = \frac{\textit{Area of triangle}}{\textit{Base of triangle}} = \frac{25}{1} = 25 \text{ Volts} \qquad \textbf{Ans}$$

For interval $0 < t < 1$, $V = 50\,t$

$$V_{RMS} = \left[\frac{1}{T}\int_0^t V^2 dt\right]^{0.5} = \left[\frac{1}{1}\int_0^1 2500\,t^2 dt\right]^{0.5} = \sqrt{2500\left(\frac{t^3}{3}\right)_0^1} = \frac{50}{1.732}$$

$$= 28.86 \text{ Volts} \qquad \textbf{Ans}$$

$$\text{Form factor} = \frac{V_{RMS}}{V_{AV}} = \frac{28.86}{25} = 1.154 \qquad \textbf{Ans}$$

3.7 Phasor Representation of Alternating Quantities

A phasor is a straight line drawn in such a way that its length is related to the amplitude of the sine wave represented, and its angular position relative to other phasors indicates the phase difference between the quantities.

3.7.1 Basic concepts of Phasor Representation

One method of carrying out the mathematical operations on sinusoidally varying quantities is to plot each sinusoidal quantity and then perform point-by-point mathematical operations to get the result graphically. However, such a method is very time consuming.

The analytical method using trigonometric identities is an alternative to the graphical method of solution. Although the analytical procedure requires less time compared to the graphical method, it is laborious when more than two such quantities are involved.

A simple and direct method of dealing with sinusoidal quantities is to use the P**hasor representation.** A phasor is used to represent a time varying quantity in terms of both magnitude and direction.

A sinusoidal current or voltage at a given frequency is characterized by only two parameters i.e. an amplitude V_m and a phase angle'ϕ' and can be written as:

$V_m \sin(\omega t + \phi)$

This voltage can be represented in exponential form as: $V_m e^{j(\omega t + \phi)}$

A real sinusoidal voltage or current is transformed into a phasor as explained below :

A real sinusoidal voltage ,v (t) = $V_m \sin(\omega t + \phi)$, is expressed as the imaginary part of a complex quantity by invoking Euler's identity :

$$\text{i.e. } e^{\pm j\omega t} = \cos\omega t \pm j\sin\omega t \qquad (3.15)$$

$$\text{e.g. } v(t) = \text{im}\{V_m e^{j(\omega t + \phi)}\}$$

where 'im' means 'imaginary part of'

Further simplification is achieved by dropping the factor $e^{j\omega t}$ and ignoring the designation 'im'. The factor $e^{j\omega t}$ can be removed because frequency is assumed to be constant and phasor value is being calculated at a particular given instant of time.

i.e. sinusoidal voltage at a particular instant can be written as :$\mathbf{V} = V_m e^{j\phi}$

In polar form , it can be written as :

$$\mathbf{V} = V_m \angle\phi$$

This abbreviated complex representation of a sinusoidal quantity is called its **phasor representation** .

A phasor is a complex number that contains the amplitude and phase angle information of a sinusoidal function.

The concept of phasor representation is explained graphically in Fig 3.13 (a) and(b), where OB represents the maximum value of the voltage phasor V_m i.e. OB = V_m.

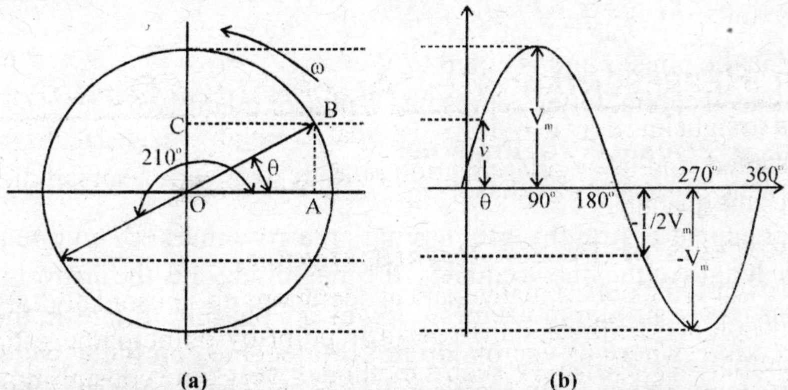

Fig. 3.13 Phasor representation of an Alternating quantity

Suppose the line OB rotates in the anti-clockwise direction about origin 'O' of a circle at a uniform angular velocity 'ω' rad/sec. An arrow head is marked at the end of the phasor to indicate the length of the phasor and to indicate the phasor end which moves.

Figure 3.13 (a) shows the phasor 'OB' when it has moved through an angle $\theta = \omega t$. from the position it occupied when the voltage was passing through its zero value.

If BA and BC are drawn perpendicular to horizontal and vertical axes respectively

Then, $OC = AB = OB \sin\theta = OB \sin\omega t$

$= V_m \sin\omega t$

Hence the projection of V_m along the vertical axis represents the instantaneous value of the voltage.

When $\omega t = 0°$, OB is horizontal. The vertical component of OB is zero at this instant, and hence the instantaneous values of the voltage $v = 0$.

i.e. when $\omega t = 0°$, $v = 0$

Similarly, when $\omega t = 90°$, $v = V_m \sin\omega t = V_m \sin 90^0 = V_m$;

and when $\omega t = 180°$, $v = V_m \sin\omega t = V_m \sin 180^0 = 0$;

and so on.

Thus OB rotates through one revolution or 2π radians or 360^0 corresponding to one cycle of sinusoidal variation of the voltage wave.

If 'f' is the frequency in cycles per second, the phasor revolves through $2\pi f$ radians in 1 sec. Then the angular displacement in t seconds is $\theta = 2\pi f t$ radians.

The instantaneous value of voltage is given by :

$$v = V_m \sin\theta$$
$$= V_m \sin 2\pi f t \text{ volts}$$

As measuring instruments are calibrated to measure RMS values, the length of the phasor may be made equal to the RMS value instead of the maximum value. Hence the sinusoidal voltage may be written as :

$$V_m \sin(\omega t + \phi) = \sqrt{2}\, V e^{j\phi} = \sqrt{2}\, \mathbf{V}$$

Where **V** is the phasor and is equal to $V e^{j\phi} = |V| \angle \phi$

The boldfaced capital letter represents a phasor quantity.

3.7.2 Rules Applicable to Phasors

The following rules apply to phasors.

1. A phasor is a straight line having an arrow marked on one side.
2. The length of the line is equal to the magnitude and the arrow gives the direction
3. Phasors are similar to vectors. However, phasors rotate in the anti clock wise direction, where as vectors do not rotate. One complete cycle of sine wave is represented by one complete rotation of the phasor through 360^0

4. Two phasors are in phase when they are parallel to each other and point in the same direction (Figure 3.14).

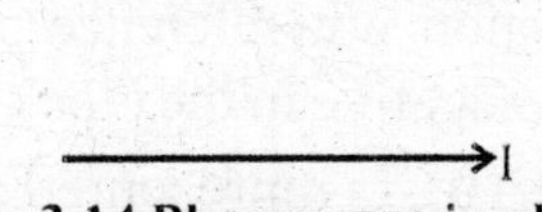

Fig 3.14 Phasors are in phase.

5. Two phasors are out of phase when they point in different directions. The phase angle between them is the angle through which one of the phasors has to be rotated to make it point in the same direction as the other (Figure3.15). Phasor 'I' has to be rotated counterclockwise by an angle 'θ' to make it point in the same direction as phasor V.

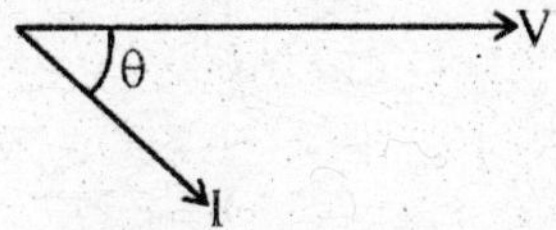

Fig 3.15 Phasors are out of phase.

6. If a phasor V has to be rotated clockwise to make it point in the same direction as phasor I, then phasor V is said to **lead** phasor I. In Figure 3.15, phasor V leads phasor I by θ degrees or I **lags** behind V by θ degrees.
7. The **smaller angle** of the two phase angles between the two phasors is selected to designate the lag or lead situation.
8. Phasors need not have a common origin and may be entirely separate from each other.
9. The angle 'θ' between two phasors is a measure of the time by which the peak positive values of the two phasors are separate from each other.

3.7.3 Representation of Phasors in Complex Form

Phasors in the complex form can be represented as follows:

(i)Polar Representation (ii) Rectangular or Cartesian Representation (iii)Trigonometric Representation (iv) Exponential Representation

3.7.3.1 Polar Representation

A phasor **OA**=$|OA| \angle \theta_1$ can be located by the magnitude |OA| and its angleθ_1 from the refern horizontal axis as shown in Fig.3.16

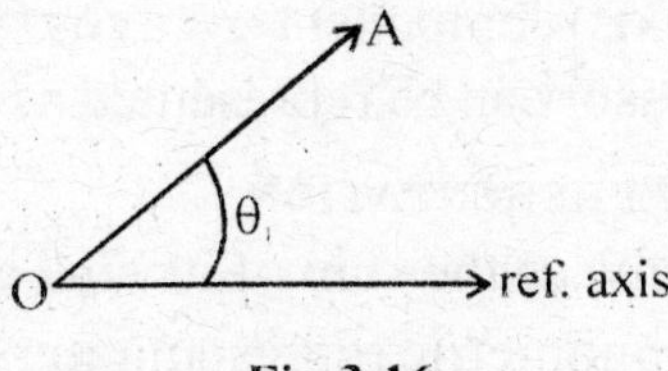

Fig.3.16

General polar form representation of a complex quantity is : $r\angle \pm \phi$

Here r = peak value / magnitude of the phasor and $\pm\phi$ is the phase angle of the phasor measured in the anticlockwise direction w.r.t reference positive x-axis. Phase angle is positive in the anticlockwise direction and negative in the clockwise direction .

If , $i = I_m \sin(wt) = I_m \sin(wt + 0)$, is represented as $I_m \angle 0$

Then a phasor ,$v = V_m \sin(wt - \phi)$ will be represented as $V_m \angle -\phi$ assuming phasor 'i'as a reference phasor.

Its polar representation is shown graphically in Figure.3.17, i.e. phasor 'v' lags reference phasor i, by an angle ϕ.

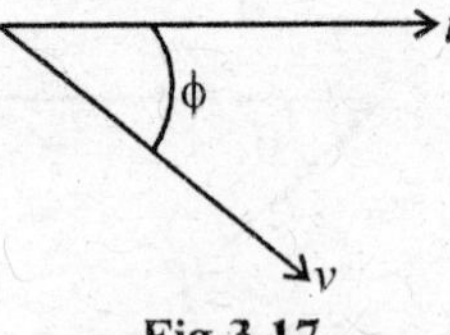

Fig.3.17

3.7.3.2 Rectangular or Cartesian Representation

A complex phasor can be represented by a complex number.

A phasor 'z' can be resolved into real and imaginary components 'x' and 'y' respectively and can be represented in rectangular form as :

$$z = x \pm jy$$

Where x = real component i.e. projection of the phasor 'z' on the x-axis i.e. z cos ϕ.

and y = imaginary component of the phasor 'z' on the y-axis i.e, z sin ϕ.

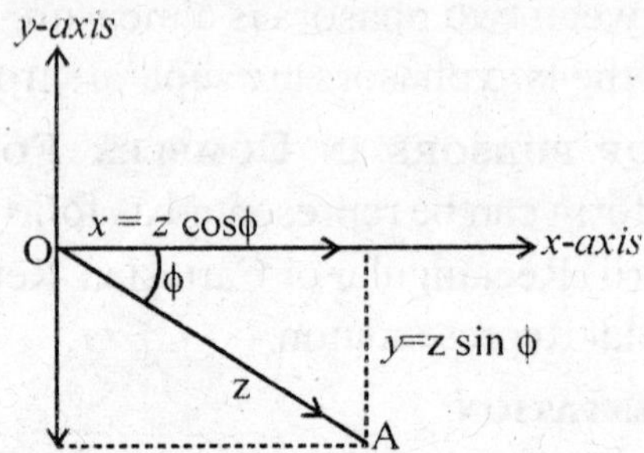

Fig.3.18 (a)Rectangular or Cartesian representation

Thus a phasor ,**z** = OA shown in Fig.3.18(a), can be resolved in to:

a horizontal (real axis or x-axis) component x = z cos f and a vertical (imaginary or j-axis) component y = z sin f and the phasor can be represented as : **z** = x - jy = z (cos f - j sinf)

3.7.3.3 Trigonometric Representation

A phasor can be representante as the sum of its sine and cosine projections

Fig. 3.18(b) shows the Trigonometric representation of a phasor.

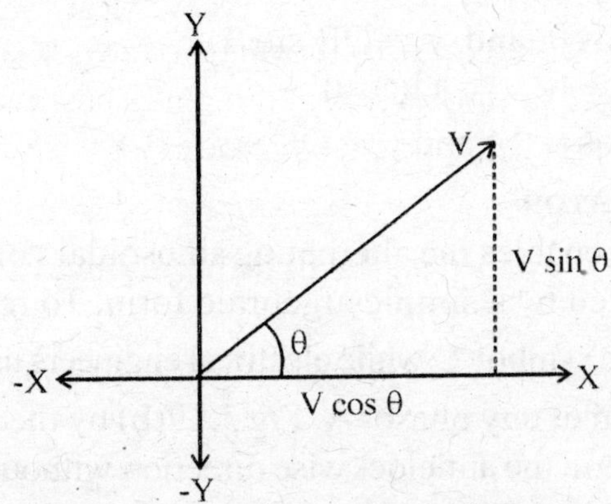

Fig.3.18(b) Trigonometric Representation

Phasor **V** has x-component = |V| cosθ and y- component = |V| sinθ .

Hence, **V** can be written in Trigonometric form as:

V =|V|(cosθ + j sinθ)

3.7.3.4 Exponential Representation

We know that an exponential function e ±jθ = cosθ ± jsinθ

Hence we can write the phasor **V** as :

V = V e ±jθ

This representation is known as Exponential Representation

3.7.4 Resolution of Phasors

The various phasors shown in Figure3.19 can be resolved in to their rectangular components and written in their resolved form as shown below :

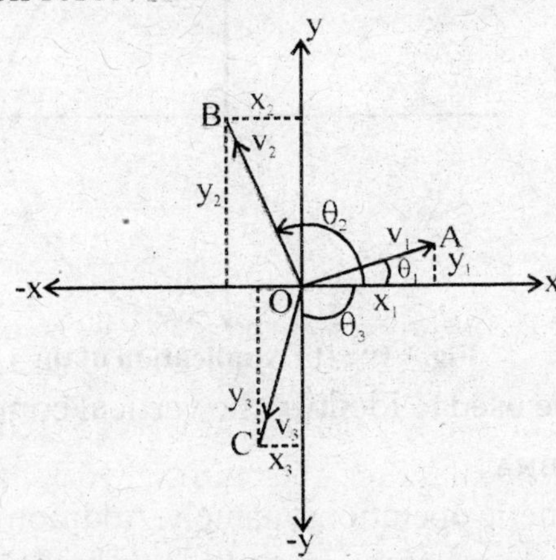

Fig.3.19 : (a) Resolution of Phasors

Phasor **OA** = $x_1 + jy_1 = |OA| \angle \theta_1$

Where, $x_1 = OA \cos \theta_1$ and $y_1 = OA \sin \theta_1$

Phasor **OB** = $-x_2 + jy_2 =| OB| \theta_2$

Where x_2 = OB cos θ_2 and y_2 = OB sin θ_2,

Phasor **OC** = $-x_3 - jy_3 = |OC| \angle -\theta_3$

Where x_3 = OC cos $(-\theta_3)$ and y_3 = OC sin $(-\theta_3)$

3.7.5 The j Operator

Complex algebra enables the alternating sinusoidal voltages, currents, and their phase angles to be represented by a simple algebraic form. To represent an imaginary number, mathematicians use the symbol 'i' while electrical engineers use the symbol 'j', where $j = \sqrt{-1}$.

The multiplication of any phasor A, Fig 3.19(b) by the operator 'j' rotates that phasor through an angle of 90° in the anticlockwise direction without changing its magnitude. When phasor jA is multiplied by the operator 'j', it means a rotation of 180° from the original position.

i.e. $j(jA) = j^2A = -A$

Hence,

$$j^2 = -1 \text{ or } j = \sqrt{-1}$$

Similarly,

$$j(j^2A) = j^3A = -j\,A \text{ and } j(j^3A) = j^4A = A$$

Figure 3.19(b) shows the effect of the operator j when applied to phasor A.

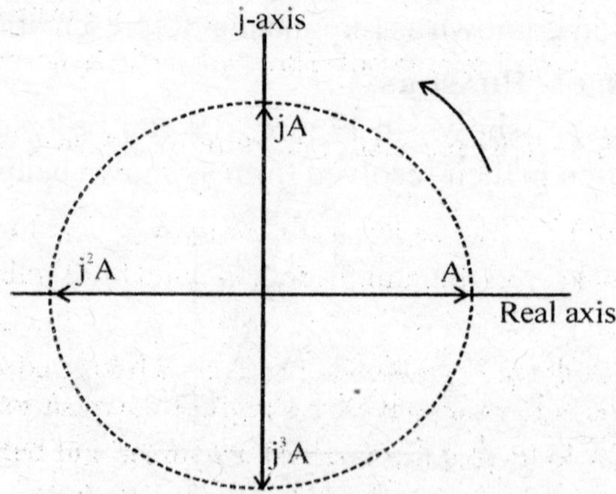

Fig.3.19 : (b) Application of the j operator

The j operator can be used to identify the vertical component of a phasor.

3.7.6 Phasor Algebra

The four basic arithmetic operations, namely, Addition, Subtraction, Multiplication, and Division on different phasors, can be carried out as explained below:

Addition and Subtraction of phasors has to be carried out frequently to draw the phasor diagrams of AC circuits. These operations on phasors are carried out to find out the resultant voltage or current.

3.7.6.1 Addition of Phasors

There are two methods for the addition of two phasors :

(i) Graphical Method

(ii) Analytical Method

(i) Graphical Method :

Following steps are followed in the graphical method:

1. Draw the two phasorsto be added to a suitable scale corresponding to the alternating signals of the same frequency as per their magnitudes and phase angles.
2. Complete the parallelogrm and use the parallelogram method to add these phasors.
3. Diagonal of the parallelogram represents the resultant signal i.e. sum of the two phasors.

Let the two phasors to be added be given by :

$$\mathbf{OA} = A\angle\phi_1 \text{ and } \mathbf{OB} = B\angle\phi_2$$

Draw these phasors **OA** and **AB** as shown in Fig 3.20, at angles ϕ_1 and ϕ_2 w.r.t. reference axis.

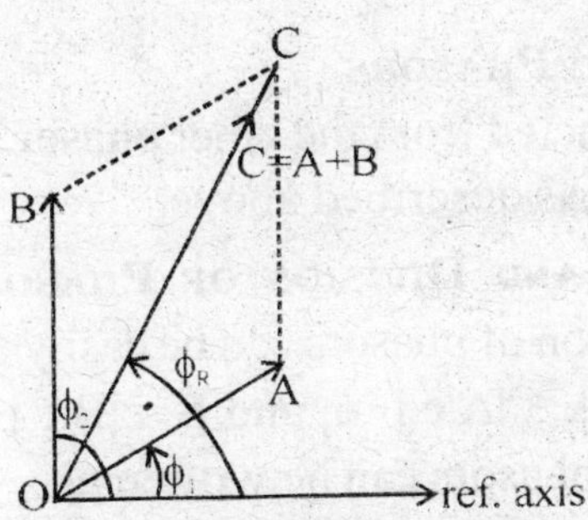

Fig.3.20 : Addition of phasors

Draw BC || to OA and AC parallel to OB and complete the parallelogram OBCA

Their phasor sum A+B is given by the phasor **OC**, the diagonal of the parallelogram OACB and it makes an angle ϕ_R with the reference axis.

(ii) Addition of Phasors by Analytical Method

The graphical methods are time consuming. However, analytical method is less time consuming and is a systematic method. Following steps are followed in this method :

1. Resolve the two phasors into their x-components i.e. real components (components along the horizontal axis) and y- components i.e. imaginary components (components along the vertical axis)
2. Add the x and y components i.e. real and imaginary components of all the individual phasors to obtain the 'x' and 'y' components of the resultant phasor.
3. Calculate the resultant phasor from these 'x' and 'y' components of the resultant phasor.

Let the 'x' and 'y' components of the two phasors P and Q be given by:

$$P = x_1 + jy_1$$
$$Q = x_2 - jy_2$$

Then the resultant phasor R which is the sum of phasors P and Q can be obtained by adding the 'x' and 'y' components as follows:

$$R = P+Q$$
$$= x_1 + jy_1 + x_2 - jy_2$$

i.e. $$R = (x_1 + x_2) + j(y_1 - y_2)$$

In Polar form, this can be written as, $|R| \angle\theta$:

where $$|R| = \sqrt{(x_1 + x_2)^2 + (y_1 - y_2)^2}$$

and $$\tan\theta = \frac{sum\ of\ imag.components}{sum\ of\ real\ components} = \frac{y_1 - y_2}{x_1 + x_2}$$

or $$\theta = \tan^{-1}\frac{y_1 - y_2}{x_1 + x_2}$$

3.7.6.2 Subtraction of Phasors

One phasor can be subtracted from the other phasor by reversing the proper phasor and then adding the two phasors as described above.

3.7.6.3 Multiplication and Division of Phasors

Multiplication and division of phasors can be easily done using polar co ordinates.

Let the two phasors be **A** = $|A| e\, j{+}\phi_1$ and **B** = $|B| e\, j{+}\phi_2$

The **product** of the two phasors can be written as

A×**B** = $|A| e\, j{+}\phi_1 \times |B| e\, j{+}\phi_2 = |A||B|\ e\, j(\phi_1 + \phi_2)$

or $|A|\angle\phi_1\ |B|\angle\phi_2 = |A||B|\angle(\phi_1 + \phi_2)$

Similarly, the **division** of the two phasors can be written as :

$$\frac{A}{B} = \frac{|A|e^{j\phi_1}}{|B|e^{j\phi_2}} = \frac{|A|}{|B|}e^{j(\phi_1 + \phi_2)}$$

or $$\frac{A}{B} = \frac{|A|\angle\phi_1}{|B|\angle\phi_2} = \frac{|A|}{|B|}\angle(\phi_1 - \phi_2)$$

Hence it is clear that to add and subtract phasors , it is convenient to represent complex quantities in the **Rectangular form**, while for multiplication and division of phasors , it is convenient to represent complex quantities in the **Polar form.**

3.7.8 PHASOR DIAGRAM

A phasor diagram is used to show the phase relationships between two or more sine waves having the same frequency. In use, phasor diagrams are static. Imagine that the phasors are rotating in an anticlockwise (counter clockwise) direction.

Every phasor in the diagram will have the same angular velocity because they represent sine waves of identical frequency. The length of the each phasor arm is directly related to the amplitude of the wave it represents, and the angle between the phasors is the same as the angle of phase difference between the sine waves.

It is a diagram in which RMS values of different A.C. quantities of the **same frequency** are represented by phasors and it shows their correct phase relationship with respect to each other.

Some salient points of phasor diagrams are:

1. One phasor is chosen as a reference phasor and is usually drawn coinciding with the positive x-axis i.e. horizontal direction.
2. Phasors are assumed to be rotating in the anti clock wise direction and hence a phasor which is ahead in the **anticlockwise** direction from the reference phasor, is said to **lead** the reference phasor.
3. A phasor which is behind in the **anticlockwise** direction from the reference phasor, is said to **lag** the reference phasor, Fig.3.21(a), (b) and(c)

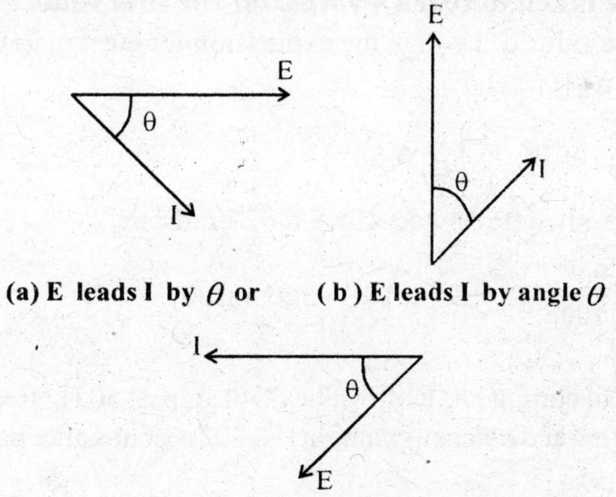

(c) E leads I by an agle θ I lags E by angle θ

Fig.3.21 : Lagging/Leading phasors

4. The angle between two phasors represents the phase difference between the two sinusoidal quantities.

Example 3.10

An alternating current of frequency 50 Hz has a maximum value of 5 amp.

(i) Write the expression for its instantaneous value.

(ii) Find the value of current after $t = \frac{1}{400}$ seconds.

(iii) Find time taken to reach 4 amps. for the first time.

Assume the time is zero when current is zero.

Solution :

Part(i) The expression for the instantaneous value of :urrent is :

$$i(t) = I_m \sin(2f\pi t)$$

Substituting the value of I_m and 'f', we get the required expression :

$i(t) = 5 \sin(100\pi t)$ **Ans**

Part(ii) Value of current after $t = \frac{1}{400}$ seconds

Substituting the value of 't' in the expression for current i(t), we get :

Current, $i = 5 \sin\left(100\pi \times \frac{1}{400}\right) = 5 \sin\left(\frac{\pi}{4}\right) = \frac{5}{\sqrt{2}} = 3.535$ amps **Ans**

Part(iii) Time taken to reach 4 amps. for the first time.

Substituting the value of i = 4, in the expression for current i(t), we get

$$4 = 5 \sin(100\pi t)$$

$$\text{or } \sin(100\pi t) = \frac{4}{5} = 0.8$$

$$\text{or } 100\pi t = \sin^{-1}(0.8) = 53.13^\circ = 0.927 \text{ radians}$$

i.e. $t = \frac{0.927}{100\pi} = 2.95 \times 10^{-3} = 2.95$ msec **Ans**

Example 3.11

An AC sinusoidal current has RMS value of 40 amp. at 50 Hz frequency. Write expression of instantaneous current and obtain its value at t=0.002 seconds after passing through maximum positive value.

Solution :

$I_{max} = \sqrt{2} \times I_{RMS} = \sqrt{2} = (40) = 56.57$ amp

Expression for instantaneous current is:

$i(t) = 56.57 \sin(2\pi\, 50\, t)$

$= 56.57 \sin(314.16\, t)$ **Ans**

Value of current at t_2, 0.002 secs after positive peak

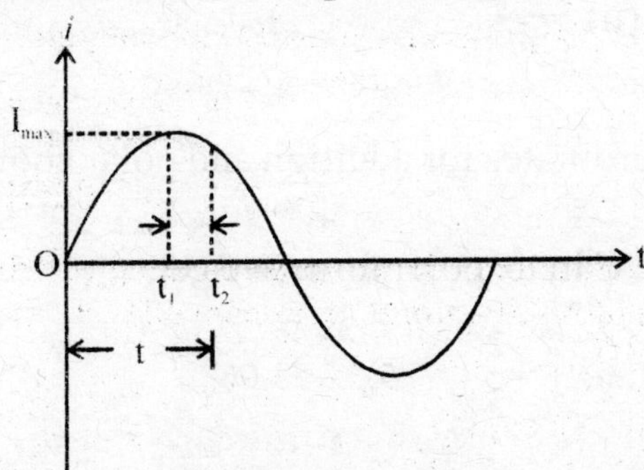

Fig.for Example 3.11

At first , find t_1 ,where the maximum value of current occurs

We know that maximum value of current occurs when, $\sin(314.16\,t_1)=1$

or $\quad 314.16\,t_1 = 90^0 = \dfrac{\pi}{2}$ radians

or $\quad t_1 = \dfrac{\pi}{2\times 314.16} = 5\times 10^{-3}$ seconds

However, we have to find the value of 'i' at $t = t_1 + t_2 = 0.005+0.002 = 0.007$sec.=7 msec.

Hence value of current , i at time, t = 7 ms is $=56.57\sin(314.16\times7\times10^{-3})$

$= 56.57\sin(2.199)$

Converting the angle in radians into degrees , we get:

i.e. 2.199 radians $= \dfrac{2.199\times360}{2\pi} = 126^0$

Hence current i $= 56.57\sin(126^0) = 56.57\,(0.809) = 45.77$ amp **Ans**

Example 3.12

Calculate the following complex quantities :

(i) p = (8+j6)(10-j7.5)

(ii) q = (8+j6) (-10-j7.5)

Solution

(i) p = 80 - j60+ j60+45.0

=125.0 **Ans**

(ii) q =(8+ j6) (-10-j7.5)

= - 80 - j60 - j60 + 45.0

= -35-j 120 **Ans**

Example 3.13

If A =3+j2 , B=-9+j6 and C=3-j4

Calculate (i) $\frac{A+B}{C}$ (ii) $\frac{A\times B}{C}$

Solution :

Use rectangular coordinates for addition and polar coordinates for multiplication and division.

Converting A,B and C in to polar form ,we get:

$$C=3-j4 = \sqrt{9+16}\ \tan^{-1}\left(-\frac{2}{3}\right) = 5\angle-53.06^{o}$$

$$A = \sqrt{9+4}\ \tan^{-1}(\frac{2}{3}) = \sqrt{13}\ \tan^{-1} 0.67 = 3.6\angle 33.7^{\circ}$$

$$B = \sqrt{81+36}\tan^{-1}(\frac{6}{-9}) = \sqrt{117}\ \tan^{-1}-0.67 = 10.81\angle-33.7^{\circ}$$

Adding A and B we get:

A+B=(3+j2)+(–9+j6) = (–6+j8) In polar form , it can be written as:

$$A+B = \sqrt{36+64}\ t^{-1}\left(\frac{-8}{6}\right) = \sqrt{100}\ \tan^{-1}-1.33$$

Or A+B =10-$\angle$53.06⁰

$$\text{(i)}\ \frac{A+B}{C} = \frac{10\angle-53.06}{5\angle-53.06} = 2\angle 0 \qquad \textbf{Ans}$$

$$\text{(ii)}\ \frac{A\times B}{C} = \frac{(3.6\angle 33.7)(10.81\angle-33.7)}{5\angle-53.06} = 7.78\angle 53.06^{0} \qquad \textbf{Ans}$$

Example 3.14

The instantaneous values of the currents i_1 and i_2 are given by the expressions : $i_1 = 2\sin 100t$ and $i_2 = 3\sin(100t + 90^{o})$.

Obtain the expressions for instantaneous values of : (a) $i = i_1 + i_2$ (b) $i = i_1 - i_2$

(c) $i = i_1 \times i_2$ (d) $i = \frac{i_1}{i_2}$ Use Analytical Method

Solution :-

Currents i_1 and i_2 can be written in polar form as:

$$i_1 = 2\angle 0 \quad \text{and } i_2 = 3\angle 90^{o}$$

Part (a)Calculation of $i = i_1 + i_2$

Convert i_1 and i_2 into rectangular coordinates

$$i_1 = 2(\cos 0 + j\sin 0) = 2(1 + j0) = 2 + j0$$

$i_2 = 3(\cos 90 + j \sin 90^o) = 3(0 + j1) = 0 + 3j$

Adding the real and imaginary components of i_1 and i_2 we get:

$i_1 + i_1 = 2 + j3$

Converting these rectangular components into polar co ordinates , we get:

$$i_1 + i_2 = \sqrt{2^2 + 3^2} \angle \tan^{-1}\frac{3}{2} = 3.60\angle \tan^{-1} 1.5 = 3.60\angle 56.3^o$$

Hence the required expression is : $i = 3.6 \sin(100t + 56.3^o)$ Ans

Part (b) Calculation of $i = i_1 - i_2$

$i_1 = 2 + j0$

Reversing the current phasor i_2 , we get:

$i_2 = 0 - j3$

Adding the real and imaginary components of i_1 and $-i_2$, we get:

$i = i_1 - i_2 = 2 - j3$

Converting the rectangular components into polar co ordinates , we get:

$$i_1 - i_2 = \sqrt{2^2 + 3^2} \tan^{-1} \angle\frac{-3}{2} = 3.6 \angle -56.3^o$$ **Ans**

Hence the required expression is : $i = 3.6 \sin(100t - 56.3^o)$ Ans

Part (c) Calculation of $i = i_1 \times i_2$

$= 2\angle 0 \times 3\angle 90 = 6\angle 90$

Hence the required expression is : $i = \mathbf{6} \sin(100t + 90^o)$ **Ans**

Part (d) Calculation of $i = \frac{i_1}{i_2}$

$$\frac{2\angle 0}{2\angle 90^o} = 0.667\angle -90^o$$

Hence the required expression is : $i = 0.667 \sin(100t - 90^o)$ Ans

3.8 Response of Single Network Elements to AC Voltage Input

Three basic AC circuits having single network element are :

Purely Resistive Circuits , Purely Inductive Circuits , Purely Capacitive Circuits.

All these circuits are discussed below:

3.8.1 Purely Resistive Circuits

A purely resistive circuit is a circuit in which inductance can be neglected as compared to its resistance.

3.8.1.1 Voltage and Current Relations

Consider an AC circuit containing a non – inductive resistance R ohms connected across a sinusoidal voltage source represented by $v(t) = V_m \sin \omega t$, as shown in Fig.3.22(a)

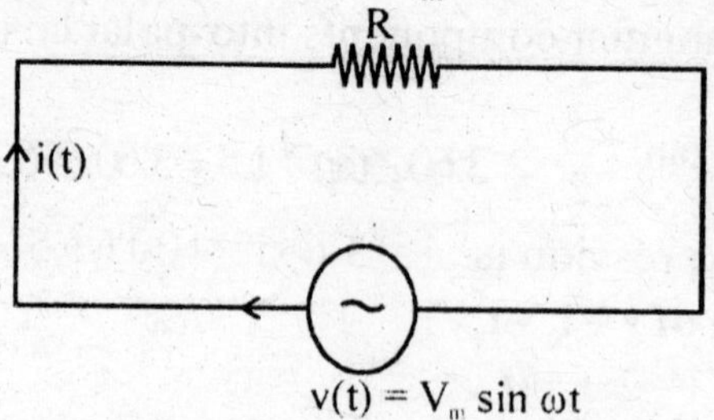

Fig 3.22 : (a) Purely resistive Circuit

Applying ohm's law to the circuit, we get:

i.e. $i(t)\,R = v(t)$

or $$i(t) = \frac{v(t)}{R} = \frac{V_m}{R} \sin \omega t \qquad (3.16)$$

Let $$\frac{V_m}{R} = I_m$$

∴ Instantaneous value of current may be expressed as , $i(t) = I_m \sin \omega t$

Hence it can be seen that in a pure resistive circuit, the applied voltage and current are sinusoidal and are in phase with each other as shown in fig. 3.22(b)

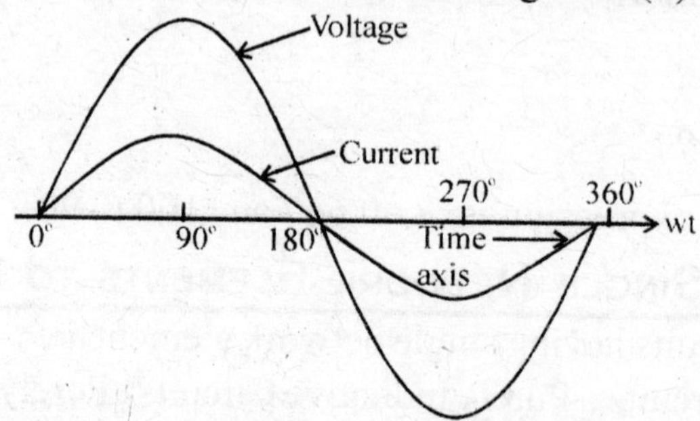

Fig 3.22 : (b) Voltage and current waveform

3.8.1.2 Phasor diagram

Phasor diagram is shown in Fig. 3.22(c). It can be seen that Voltage and current are in phase i.e. phase difference between the voltage and current phasors in a purely resistive circuit is **zero.**

$\xrightarrow{\quad i \quad}\!\!\longrightarrow \upsilon$

Fig 3.22 (c) Phasor diagram

3.8.1.3 Power in Purely Resistive Circuits

The instantaneous power delivered to the circuit is the product of the instantaneous values of applied voltage and current, Fig 3.22(d).

i.e. Instantaneous power, $p = v\,i = (V_m \sin \omega t)(I_m \sin \omega t)$

$$= V_m I_m \sin^2 \omega t$$

or $$p = \frac{V_m I_m}{2}(1 - \cos 2\omega t), \text{ because } \sin^2 \omega t = \frac{1 - \cos 2wt}{2}$$

Hence $$p = \frac{V_m I_m}{2} - \frac{V_m I_m}{2}\cos 2wt$$

$$\text{Average Power, } P = \text{Average of } \frac{V_m I_m}{2} - Avg.of \frac{V_m I_m}{2}\cos 2wt \frac{V_m I_m}{2}$$

We know that, Average of $\frac{V_m I_m}{2} \cos 2wt$ over a complete cycle is zero and Average of constant.

$$\therefore \text{ Average power, } P = \frac{V_m I_m}{2} = \frac{V_m}{\sqrt{2}} \cdot \frac{I_m}{\sqrt{2}} \qquad \ldots(3.17)$$

Where V_{rms} and I_{rms} are the RMS values of applied voltage and current respectively.

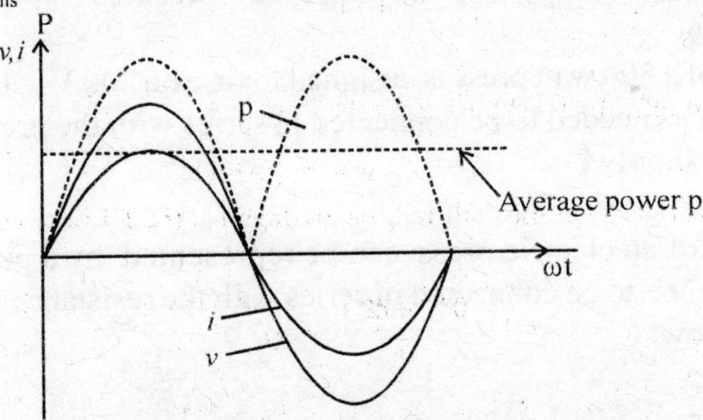

Fig 3.22 (d) Power in Purely Resistive Circuits

Thus for purely resistive circuits, power is equal to the product of the rms values of voltage and current and it is always positive, because the instantaneous values of voltage and current are always either both positive or both negative and, therefore the product is always positive. This means that the voltage source constantly delivers power to the purely resistive circuit and the circuit consumes it.

Example 3.15

A 230 V, 50 Hz voltage is applied across a circuit consisting of a pure resistance of 10 Ω. Determine (i) the current flowing through the circuit and (ii) power absorbed by the circuit. (iii) Write the expressions for the voltage and current.

Solution : (i) Current flowing through the circuit,

$$I_{rms} = \frac{V}{R} = \frac{230}{10} = 23 \text{ Amp Ans.}$$

(ii) Power absorbed by the circuit,

$$P = VI = 230 \times 23 = 5290 \text{ watts Ans}$$

(iii) Peak value of applied voltage,

$$V_{max} = \sqrt{2}\,V = \sqrt{2} \times 230 = 325.27 \text{ Volts}$$

Peak value of current,

$$I_{max} = \sqrt{2}\,I_{rms} = \sqrt{2} \times 23 = 32.527 \text{ Amp}$$

Expression for voltage, $v = V_{max} \sin 2\pi ft$

Assuming that at $t = 0$, $v = 0$ and is increasing in positive direction, we can write expression for voltage as:

$v = 325.27 \sin 2\pi \times 50\,t = 325.27 \sin 314\,t$ **Ans.**

Similarly expression for current is :

$i = I_{max} \sin 2\pi ft = 32.527 \sin 314\,t$ **Ans.**

Example 3.16

The element of a 500 watt press is designed for use on 200 V, 50 Hz supply. What value of extra resistance is needed to be connected in series with the press; so that press can be operated at 230 V supply.?

Solution

The element of an electric press can be represented by a pure resistance R and an external resistance 'r' to be connected in series with the resistance of the press is shown in the figure given below:

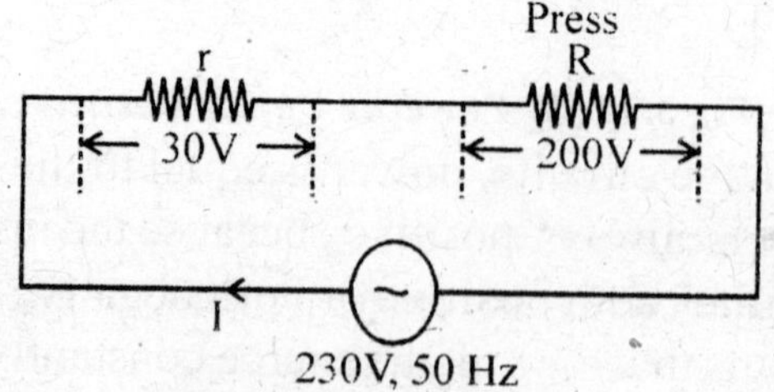

Fig : for Example 3.16

Permissible voltage drop across electric press element =200 volts

Power consumed by the element =500 watts

Hence, resistance of press element,

$$R = \frac{V^2}{P} = \frac{200^2}{500} = 80 \text{ Ohm}$$

Since the press has to be operated on 230 volts instead of 200 volts, hence voltage to be dropped across the external resistance, v=230-200=30 volts

Current, $$I = \frac{200}{80} = 2.5 \text{ Amp}$$

External resistance, $$r = \frac{V}{I} = \frac{30}{2.5} = 12.5 \text{ ohms}$$ **Ans**

3.8.2 Purely Inductive Circuits.

A purely inductive circuit is a coil which has negligible resistance.

When an alternating voltage is applied to a purely inductive coil, an emf, known as self induced emf, is induced in the coil which opposes the applied voltage. Since coil has no resistance, therefore, at every instant , applied voltage has to overcome this self – induced emf only. This induced emf is also known as **counter emf or back emf.**

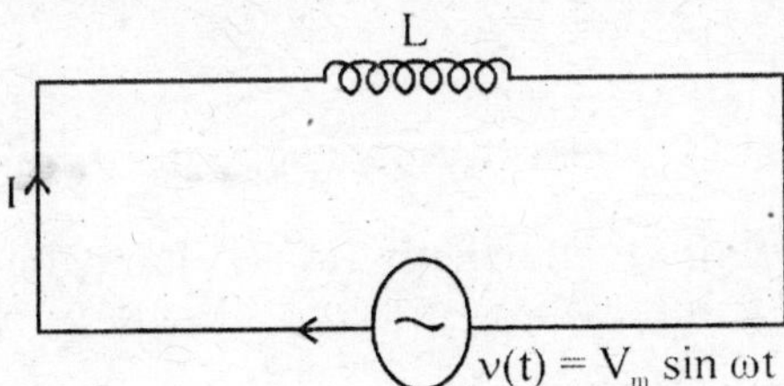

Fig 3.23 : (a) Purely Inductive Circuit

3.8.2.1 Voltage and Current Relations

Consider an AC circuit containing a pure inductance 'L' henry connected across a sinusoidal voltage source represented by $v(t) = V_{max} \sin \omega t$ as shown in Fig. 3.23(a).

Applied voltage $v(t) = V_{max} \sin wt$

Self inductance of coil = L Henry

Self induced emf in the coil, $e_L = L\frac{di}{dt}$

Since applied voltage at every instant is equal and opposite to the self induced emf i.e. $v = e_L$, we can write:

$$V_{max} \sin wt = L\frac{di}{dt} \qquad \text{(i)}$$

Equation(i) can be rewritten as:

$$di = \frac{V_{max}}{L} \sin wt\, dt \qquad \text{(ii)}$$

Integrating both sides of equation (ii), we get:

$$i = \frac{V_{max}}{L} \int \sin wt\, dt$$

or

$$i = \frac{V_{max}}{L} (-\cos wt) + k$$

Where 'k' is a constant of integration and is assumed to be zero.

i.e. $i = -\frac{V_{max}}{wL} \cos wt$ (iii)

Using the Trigonometric Identity, $-\cos wt = \sin\left(wt - \frac{\pi}{2}\right)$, equation (iii) can be rewritten as:

or, $i = \frac{V_{max}}{wL} \sin\left(wt - \frac{\pi}{2}\right)$ (iv)

(iv)

Let $\frac{V_{max}}{wL} = I_{max}$

Then instantaneous current given by equation (iv) may be expressed as :

$$\mathbf{i = I_{max} \sin\left(wt - \frac{\pi}{2}\right)} \qquad (3.18)$$

Hence by comparing the phase angles of applied voltage and current, it is observed from equation (3.18) that in a purely inductive circuit ,the current has a sinusoidal waveform and it lags behind the applied voltage by an angle of radians or 90^0.

3.8.2.2 Voltage and Current Waveforms

Waveforms of voltage and current and induced emf in a purely inductive circuit are shown in Fig.3.23(b) .

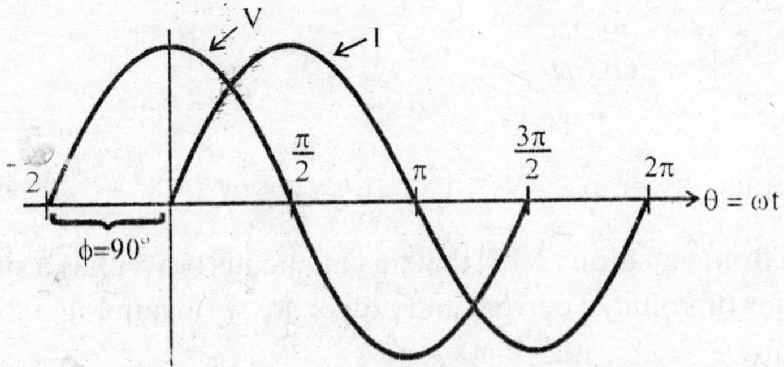

Fig 3.23 (b) Waveforms of voltage and current

Phasor diagram

The phasor diagram is shown in Fig 3.23(c)

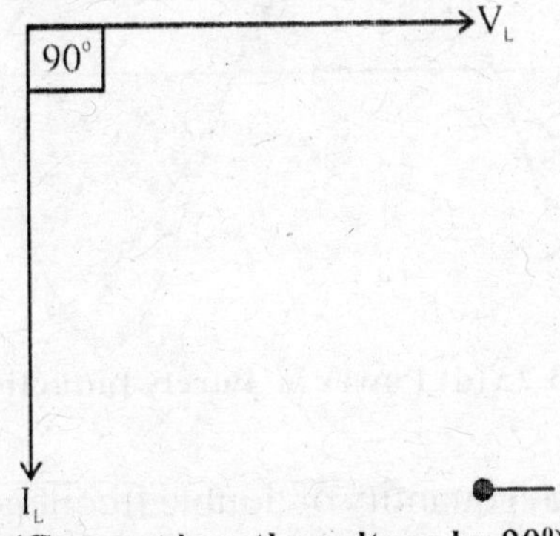

(Current lags the voltage by 90°)

Fig 3.23(c)Phasor diagram of a purely inductive circuit

3.8.2.3 Inductive Reactance

Inductive reactance can be defined as the opposition offered by an inductance to the flow of an alternating current.

From equation(iv) in section 3.8.2.1 , we see that:

$$I_{max} = \frac{V_{max}}{\omega L},$$

The term, ωL is known as **Inductive Reactance** and is denoted by X_L

i.e. $X_L = \omega L = 2\pi fL$

If 'L' is in Henry and 'ω' is in radians per second , then X_L will be in ohms.

Inductive reactance increases linearly with frequency and its value is zero for direct current.

3.8.2.4 Power in Purely Inductive Circuit

Instantaneous power, $p = v \times i = (V_{max} \sin \omega t) \times I_{max} \sin\left(\omega t - \frac{\pi}{2}\right)$

But $\sin\left(\omega t - \frac{\pi}{2}\right) = -\cos\omega t$

$\therefore$ Instantaneous power, $p = -V_{max} I_{max} \sin\omega t \cos\omega t = -\frac{V_{max} I_{max}}{2} \sin 2\omega t$ (3.19)

It can be seen from equation (3.19) that instantaneous power has a sinusoidal waveform. The waveforms of voltage, current and power are shown in Fig 3.23(d)

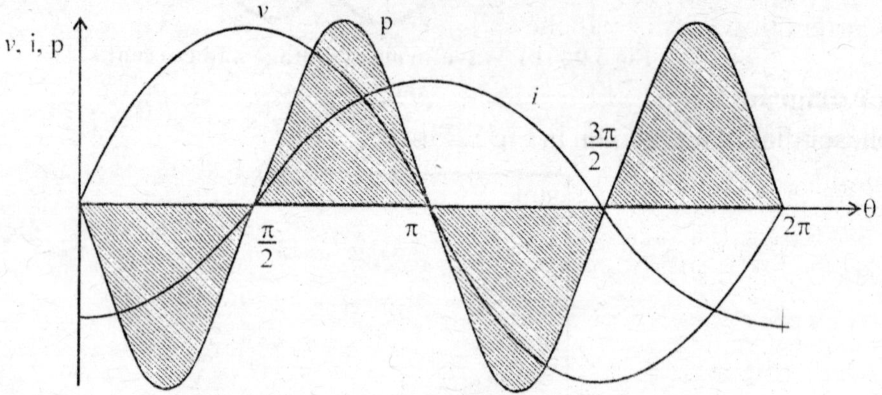

Fig 3.23 (d) Power in Purely Inductive Circuits

Average power

Since average of a sinusoidal quantity of double frequency over a complete cycle is zero, therefore in a pure inductive circuit, power absorbed is **zero.**

Example 3.17

(a)A choke coil having negligible resistance takes a current of 5 amp. ,when connected to 250 volts, 50 Hz supply.(i) Determine the value of inductance(ii) Determine the current, if frequency is reduced to 40 Hz.

(b)A 0.01 H coil with negligible resistance is connected to a 200 V, 50 Hz supply. Determine (i) inductive reactance of the coil (ii) current flowing through the coil (iii) power consumed by the coil . Write down the expressions for voltage and current.

Solution

(a) Inductive reactance $X_L = \frac{V}{I} = \frac{250}{5} = 50$ ohm

However, Inductive reactance $X_L = 2\pi f L$

substituting the values, we get:

$50 = 2 \times \pi \times 50 \times L$

Or $L = \frac{50}{100\pi} = \frac{1}{2\pi} = 0.159$ henry **Ans**

(ii) Inductive reactance $X_L = 2\pi f L = 2 \times \pi \times 40 \times 0.159 = 40.21\Omega$

Hence , current $I = \frac{V}{X_L} = \frac{250}{40.21} = 6.21$ amp **Ans**

(b)Applied Voltage, $V_{rms} = 200$ V, Supply frequency, f = 50 Hz

(i) Inductive reactance of coil, $X_L = 2\pi f L = 2\pi \times 50 \; 0.01 = 3.14\Omega$ **Ans**

(ii) Current flowing through the coil, $I = \frac{V}{X_L} = \frac{200}{3.14} = 63.69$ amp. **Ans**

(iii) Power consumed by the coil, P = VI cos ϕ = VI cos $90^0 = 0$

Expression for the voltage, $v(t) = V_{max} \sin 2 f t$

Assuming that at time t = 0, v = 0 and is increasing in positive direction

Expression for the voltage, $v(t) = 200\sqrt{2} \sin 2\pi \times 50t = 282 \sin 314\,t$ **Ans**

Expression for the current,

$$i(t) = I_{max} \sin\left(2\pi ft - \frac{\pi}{2}\right) = 63.69\sqrt{2} \sin\left(314t - \frac{\pi}{2}\right) = 90.07 \sin\left(314t - \frac{\pi}{2}\right)$$ **Ans**

3.8.3 Purely Capacitive Circuits

When a capacitor is connected to an alternating voltage, the capacitor is charged when the applied voltage is increasing and it is discharged when applied voltage is decreasing during alternate cycles.

3.8.3.1 Voltage and Current Relations

Let an alternating voltage represented by $v(t) = V_{max} \sin \omega t$ be applied across a capacitance 'C' as shown in Fig 3.24 (a)

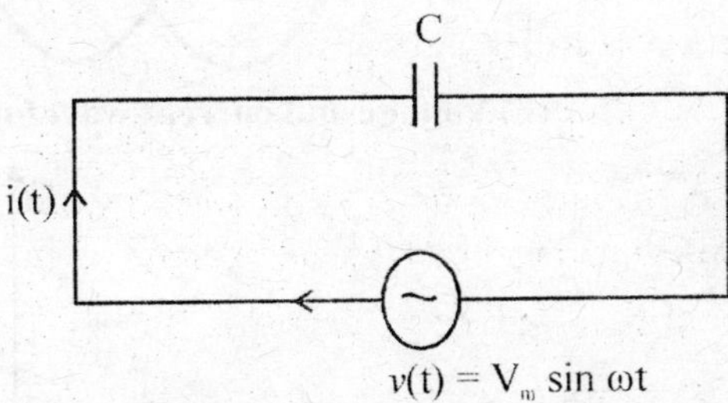

Fig.3.24 (a) Purely Capacitive Circuit

The expression for instantaneous charge is :

$$q = C\,v; \qquad \text{(i)}$$

Substituting the value of 'v' in equation(i), we get:

$$q = C\,V_{max} \sin \omega t \qquad \text{(ii)}$$

However, Current , $i(t) = \frac{dq}{dt} = C\frac{dv}{dt} = C\frac{d(V_{max} \sin \omega t)}{dt}$

$$= \omega C V_{max} \cos wt \qquad \text{(iii)}$$

Using the Trigonometric Identity, $\cos wt = \sin\left(wt + \frac{\pi}{2}\right)$ and rewriting equation (iii), we get:

$$i(t) = \frac{V_{max}}{1/wc} \sin\left(wt + \frac{\pi}{2}\right)$$

$$\text{i.e. } i(t) = I_{max} \sin\left(wt + \frac{\pi}{2}\right) \qquad (3.20)$$

where $\frac{V_{max}}{1/we} = I_{max}$

It is observed from equation (3.20) that the **current' i(t) ' leads the applied voltage'v' by an angle of $\frac{\pi}{2}$ radians or 90^0.**

3.8.3.2 Voltage and CurrentWaveforms and Phasor diagram

Waveforms of voltage and current in a purely capacitive circuit are shown in Fig. 3.24(b).

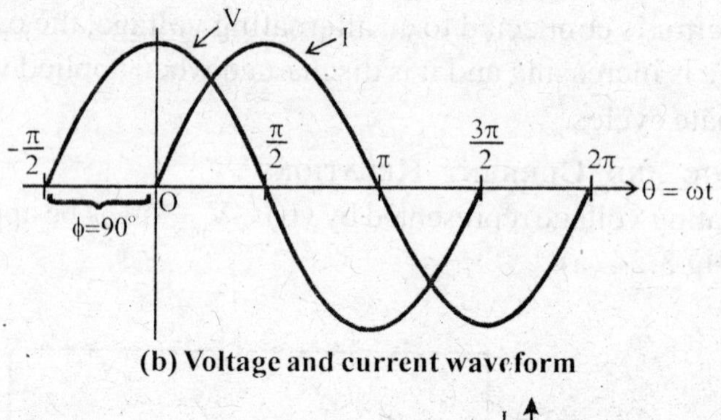

(b) Voltage and current wavcform

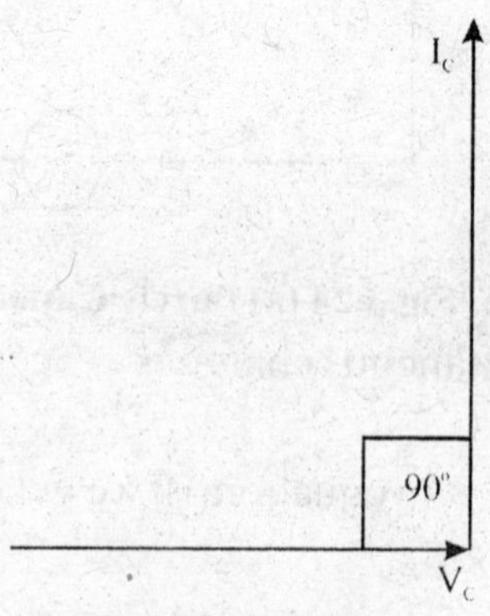

(c) Phasor diagram

Fig.3.24 : Current leads voltage by90^0

It can be seen from Fig 3.24(b), that the current leads the voltage by 90^0 in a purely capacitive circuit.

The phasor diagram is shown in Fig 3.24(c)

3.8.3.3 Capacitive Reactance

It is defined as the opposition offered to the alternating current by pure capacitor. In the above expression, $\frac{1}{\omega C}$ is known as **Capacitive Reactance** and is denoted by X_C.

i.e. $X_C = \frac{1}{\omega C}$ or $X_C = \frac{1}{2\pi f C}$

Thus capacitive reactance is inversely proportional to the frequency and it decreases with increase in frequency.

The reactance offered by a capacitor to direct current is '∞'

If 'C' is in farads and 'ω' is in radians/sec, then X_C will be in ohms.

3.8.3.4 Power in Purely Capacitive Circuit

Instantaneous power, $p = v\,i = (V_{max} \sin \omega t).I_{max} \sin\left(\omega t + \frac{\pi}{2}\right)$

Substituting, $\sin\left(\omega t + \frac{\pi}{2}\right) = \cos \omega t$ in the above equation, we get:

$$p = V_{max} I_{max} \sin \omega t \cos wt = \frac{V_{max} I_{max}}{2} \sin 2wt \tag{3.21}$$

It is seen from equation (3.21), that instantaneous power has a sinusoidal waveform of double frequency.

The waveforms of voltage, current and power in a purely capacitive circuit, are shown in Fig 3.24(d)

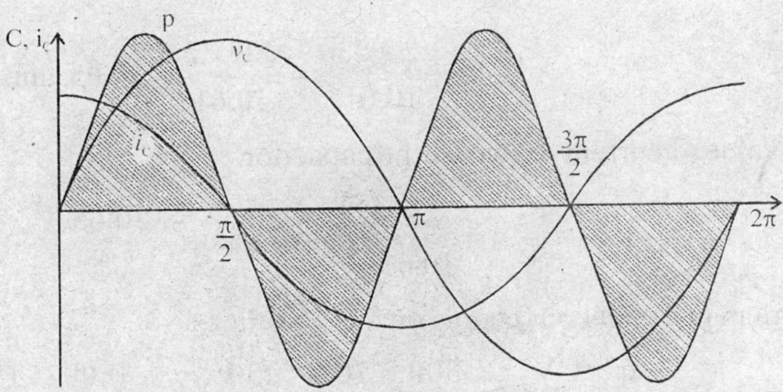

Fig 3.24 (d)Waveforms of voltage, current and power in a purely Capacitive Circuits Average power

Average power over a complete cycle, $P_{av} = \frac{V_{max} I_{max}}{2} \times$ average of $\sin 2\omega t$ = zero.

Hence power absorbed in a pure capacitive circuit is **zero.**

The energy taken from the supply circuit is stored in the capacitor during the **first quarter cycle** and returned during the next.

Example 3.18

(a)An AC circuit with pure capacitance has a reactance of 314 ohm at 50 Hz. The circuit is given a supply of 200 volts at 50 Hz. Find the value of capacitance and current.

(b)A 300 μF capacitor is connected to 230 V, 50 Hz supply. Determine (i) the capacitive reactance offered by the capacitor (ii) maximum current (iii) rms value of the current drawn by the capacitor (iv) maximum instantaneous charge on the capacitor (v) Write the time equations for the voltage and current.

Solution

(a) We know that:

$$C = \frac{1}{2\pi f X_C} = \frac{1}{2\pi \times 50 \times 314} = 1\times 10^{-5} \text{ Farad Ans}$$

$$I = \frac{V}{X_C} = \frac{200}{314} = 0.636 \text{ amp Ans}$$

(b) Applied voltage, $V_{rms} = 230$ V , Supply frequency, = 50 Hz

$$\omega = 2\pi f = 2\pi \times 50 = 314 \text{ radians per second}$$

(i) Capacitive reactance, $X_C = \frac{1}{\omega C} = \frac{1}{314\times 300\times 10^{-6}} = 10.61\,\Omega$ **Ans.**

(ii) Maximum current drawn by the capacitor

$$I_{max} = \frac{V_{max}}{X_C} = \frac{\sqrt{2}\times 230}{10.61} = \frac{325.22}{10.61} = 30.65 \text{ amp}$$ **Ans.**

(iii) RMS value of current drawn by the capacitor,

$$I_{rms} = \frac{V_{rms}}{X_C} = \frac{230}{10.61} = 21.677 \text{ amp} = \frac{30.65}{\sqrt{2}}$$ **Ans.**

(iv) Maximum instantaneous charge on the capacitor

$$q = CV_{max} = 300 \times 10^{-6} \times 230\sqrt{2} = 0.0975 \text{ Coulombs}$$ **Ans.**

(v) Expression for the voltage, $v = V_{max} \sin \omega t$ assuming that at time t = 0, v = 0 and is increasing in positive direction i.e. $v = 230\sqrt{2} \sin 314\,t$.

Or $v = 325.22 \sin 314\, t$ **Ans.**

Expression for current, $i = I_{max} \sin\left(wt + \frac{\pi}{2}\right) = 30.65 \sin\left(314t + \frac{\pi}{2}\right)$ **Ans.**

3.9 SINGLE PHASE AC SERIES CIRCUITS

Most of the useful electrical circuits consist of two or more of the passive elements i.e. R ,L,C, connected in Series, Parallel or Series - Parallel . In this section Series RL, RC and RLC circuits are discussed.

3.9.1 R-L SERIES CIRCUIT

This circuit is generally met in practice .

Consider an AC circuit consisting of a resistance, R ohms and inductance . L Henrys connected in series across a voltage $v = V_{max} \sin wt$, as shown in **Fig.3.25 (a)**

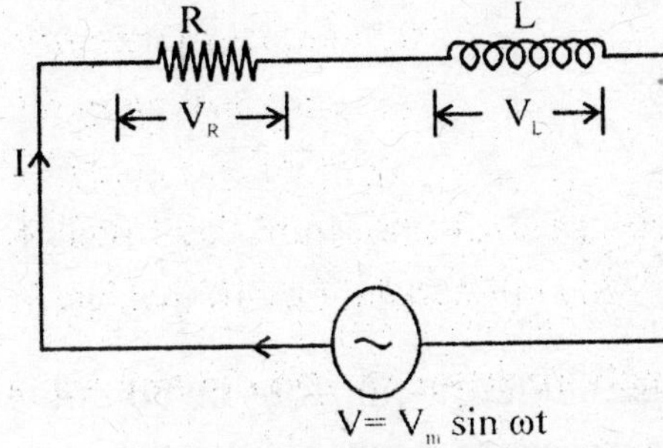

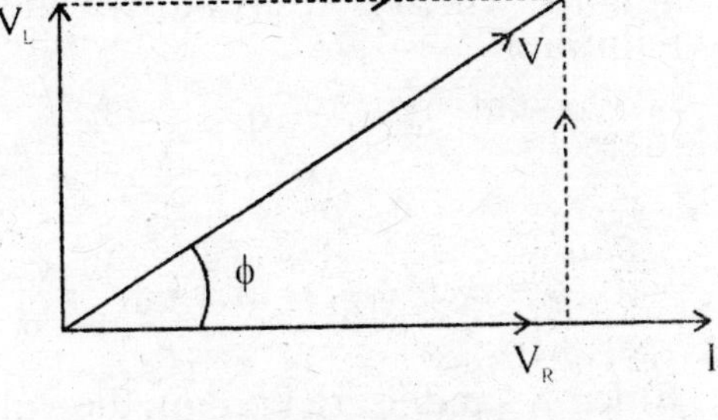

Fig.3.25 : (a) R-L Series Circuit **Fig.3.25 : (b) Phasor diagram**

Let the supply frequency be 'f' and current flowing through the circuit be 'I' amperes. RMS value of current I is taken as a reference phasor for drawing the phasor diafgram , because current is same through both of the circuit elements R and L.

Voltage drop across resistance R, $V_R = I\,R \angle 0$ and is in phase with the current.

Voltage drop across inductance L, $V_L = j\,I\,X_L = jI\,\omega L$, leading the current I by 90^0 or $\frac{\pi}{2}$ radians

3.9.1.1 IMPEDANCE

Applying KVL to the series R-L circuit shown in **Fig.3.25(a)** , we get:

$V = V_R + V_L$ (phasor addition) , Where V is the applied voltage.

$$= I\,R + j\,I\,X_L = I(R + j\,X_L)$$

$$= I\,Z \angle \Phi$$

Where $Z = (R + j\,X_L)$ is the **impedance** and is expressed in ohms.

and $|Z| = \sqrt{R^2 + X_L^2}$ is known as magnitude of complex **impedance , Z**

3.9.1.2 Phasor diagram

Phasor diagram shown in Fig 3.25(b) is drawn with current as a reference phasor ,because current is the common quantity in a series circuit.

Following steps are followed in drawing the phasor diagram.

(i) Draw a horizontal line representing current phasor , I .

(ii) Draw V_R in phase with I , i.e. a horizontal line alongside I, to represent voltage drop across resistance, V_R

(iii) Draw V_L leading I by 90^0 i.e. vertically at an angle of 90^0 with respect to phasor 'I'.

(iv) Complete the rectangle. Diagonal of this rectangle gives the resultant supply voltage, V

From this phasor diagram, it is evident that the current lags behind the applied voltage V by angle Φ, which is less than 90^0 and is given by :

$$\tan \Phi = \frac{V_L}{V_R} = \frac{I X_L}{I R} = \frac{X_L}{R}$$

or

$$\Phi = \tan^{-1} \frac{X_L}{R}$$

Since X_L and R are known, the value of phase angle Φ can thus be computed.

If the applied voltage $V = V_{max} \sin \omega t$, then expression for the current in a series RL circuit, will be:

$$i = I_{max} \sin(\omega t - \Phi) \qquad (3.22)$$

Where $I_{max} = \frac{V_{max}}{Z}$

3.9.1.3 Impedance Triangle

Z, R, can be represented on a vector diagram. These are not phasor quantities because they have fixed values. To distinguish it from a phasor diagram, it is drawn in triangular form and is called **impedance triangle**.

By dividing each side of the **voltage triangle** Fig 3.25(a), by I , we obtain **impedance triangle** AOB shown in Fig 3.25(b), where

$$AO = \frac{V_R}{I} = R, \quad OB = \frac{V_L}{I} = X_L \text{ and } AB = \frac{V}{I} = Z$$

The angle ϕ between Z and R sides of the impedance triangle is known as **phase angle** and **cosine** of this angle is known as **power factor** of the circuit.

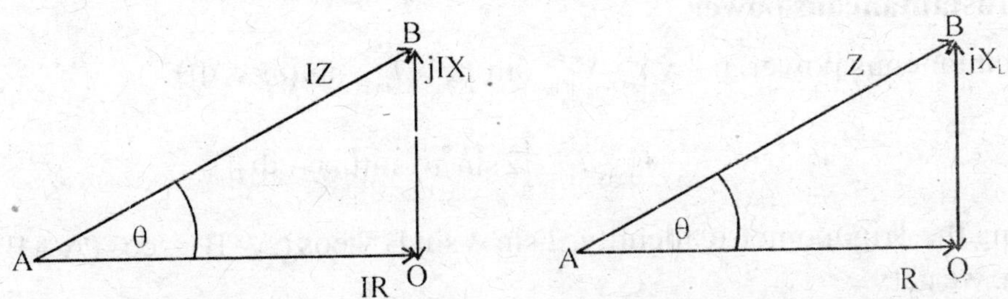

(a) Voltage triangle (b) Impedance triangle

Fig.3.25 : Voltage triangle and Impedance triangle for R-Lseries circuit

It can be seen from Fig. 3.25(b), that

$$\cos\phi = \frac{R}{Z}$$

Complex impedance Z, can be expressed in polar form as:

$\mathbf{Z} = |Z| \angle \phi$

In cartesian co ordinates ,

Impedance, $\mathbf{Z} = |Z|(\cos\phi + j\sin\phi)$

i.e. $R = |Z| \cos\phi$

and $X_L = |Z| \sin\phi$

3.9.1.4 Power in R–L Series Circuit

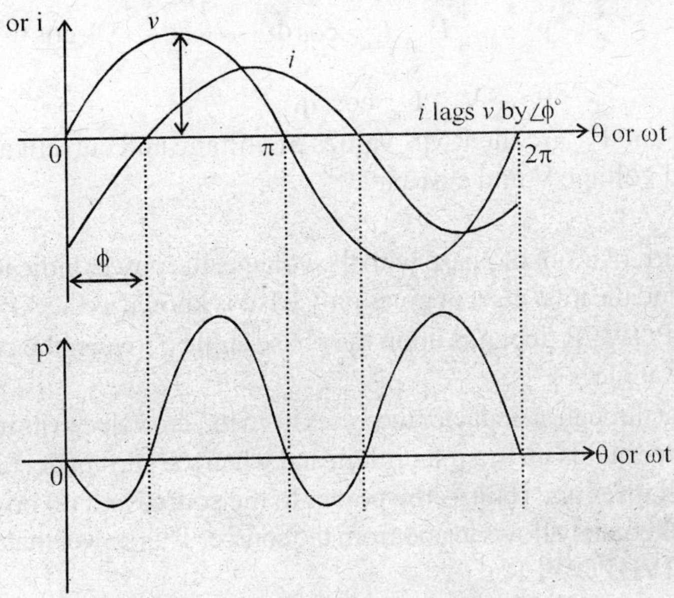

Fig.3.25 : (c) Power in R–L Series Circuit

(a) Instantaneous power

Instantaneous power, $p = v\,i = V_{max} \sin \omega t \times I_{max} \sin(\omega t - \Phi)$

$$= \frac{1}{2} V_{max} I_{max} [2 \sin \omega t . \sin(\omega t - \Phi)]$$

Using the Trigonometric Identity, 2 sin A sin B = cos (A - B) – cos (A + B), in the above equation, we get:

$$p = \frac{1}{2} V_{max} I_{max} [\cos \Phi - \cos(2\omega t - \Phi)]$$

i.e.

$$p = \frac{1}{2} V_{max} I_{max} \cos \Phi - \frac{1}{2} V_{max} I_{max} \cos(2\omega t - \Phi)$$

It is seen that for a fixed value of Φ, the instantaneous power consists of two components a constant part and a time varying part, which has a frequency twice that of voltage and current waveforms.

(b) Average power

Average value of time varying component, $\frac{1}{2} V_{max} I_{max} \cos(2\omega t - \Phi)$ over a complete cycle is zero, therefore, average power of the circuit :

$$P_{av} = \frac{1}{2} V_{max} I_{max} \cos \Phi = \frac{V_{max}}{\sqrt{2}} . \frac{I_{max}}{\sqrt{2}} \cos \Phi$$

or

$$P_{av} = V_{rms} I_{rms} \cos \Phi \qquad (3.23)$$

Where V_{rms} and I_{rms} are the RMS values of voltage and current and ϕ is the phase angle between applied voltage V and current I.

When the current is out of phase with the voltage, the power indicated by the product of the applied voltage and the total current gives only what is known as the APPARENT POWER.

The TRUE POWER depends upon the phase angle between the current and voltage. The symbol for phase angle is ϕ.

As the current through an inductor increases from its zero value to its maximum value ,the field around the inductor builds up to a maximum, and when the current decreases from maximum to zero the field collapses and returns the power to the source and no power is used up in either case.The power alternately flows to and from the source. This power that is returned to the source is called REACTIVE POWER.

In a purely resistive circuit all of the power is consumed and none is returned to the source; in a purely reactive circuit no power is consumed and all of the power is returned to the source.

The part of the power curve which is shown below the horizontal reference line is the result of multiplying a positive instantaneous value of current by a negative instantaneous value of the voltage, or vice versa. Therefore the power at that instant is considered as negative power. In other words, during this time the reactance was returning power to the source.

3.9.1.5 Power triangle

Power triangle is obtained from the voltage triangle by multiplying each side of this triangle by 'I'

i.e. $\mathbf{V\,I = V_R\,I + V_L\,I}$ **(phasor addition)**

But $V_R = V\cos\phi$ and $V_L = V\sin\phi$

$V\,I = I\,V\cos\phi + I\,V\sin\phi$

These terms represent the three sides of the power triangle.

$S = V\,I$ is the Apparent power

$P = I\,V\cos\phi$

is the Real Power

$Q = I\,V\sin\phi$

is the Reactive Power.

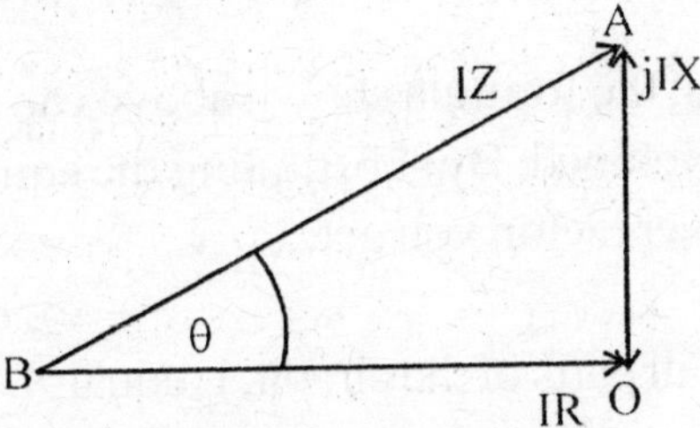

Fig.3.25 : (d)Voltage triangle

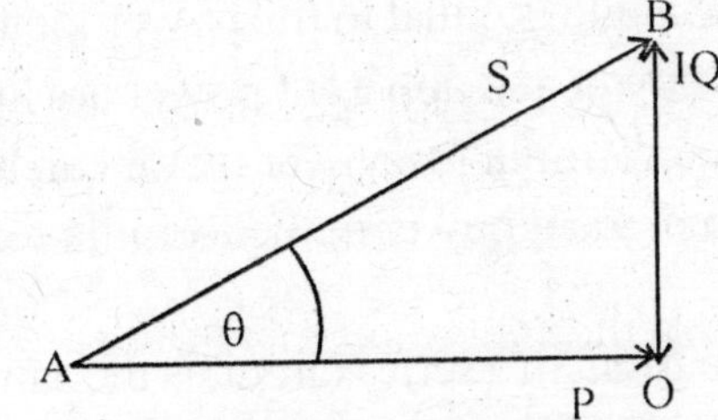

Fig3.25 : (e) Power triangle

By referring to the power triangle shown in Fig.3.25(e), different types of power can be defined as follows:

(i) Apparent Power

Apparent power is the power that appears to the source because of the circuit impedance. Apparent power is the combination of true power and reactive power.

The product of rms value of current and rms value of voltage, V I is called the **Apparent Power**. It is denoted by S and is measured in volt-amperes or kilo-volt amperes (KVA).

i.e. Apparent power, $S = V\,I$ **volt amperes**

$$\text{or} \quad S = \frac{V\,I}{1{,}000}\ \textbf{KVA}$$

(ii) Real Power or True Power or Active power or in-phase power.

The true power of a circuit is the power actually used in the circuit. This power, measured in watts, is the power associated with the total resistance in the circuit. To calculate true power, the voltage and current associated with the resistance must be used.

It is defined as the average power taken by or consumed by the series R-L circuit. The true power in an AC circuit is obtained by multiplying the apparent power by the power factor and is expressed in watts or kilo-watts (Kw)

True power , $P = V I \cos \phi$ watts (3.24)

Where ϕ is the angle between V and I

(iii) Reactive Power / Wattless power/imaginary power/quadrature power.

The product of apparent power, V I and the sine of the angle between voltage and current phasors i.e. sin , is called the **Reactive Power.**

This is also known as **wattless power** and is expressed in reactive volt- amperes or kilo-volt amperes reactive (KVAR).

Reactive power, $Q = V I \sin \phi$ **VAR** (3.25)

(iv) Power Factor

The power factor is a number (represented as a decimal or a percentage) that represents the portion of the apparent power dissipated in a circuit.

The easiest way to find the power factor is to find the cosine of the phase angle è. The cosine of the phase angle is equal to the power factor.

If true power and apparent power are known, the formula shown above can be used.

Another formula for power factor can be developed. By substituting the equations for true power and apparent power in the formula for power factor, you get:

Since current in a series circuit is the same in all parts of the circuit, I_R equals I_Z. Therefore, in a series circuit,

Power factor may be defined by any of the following relations :

(i) cosine of the phase angle between voltage and current. i.e. $\cos \phi$

(ii) The ratio of the resistance to impedance i.e. $\cos \phi = \dfrac{R}{Z}$ (3.26)

(iii) The ratio of true power to apparent power i.e. $\cos = \dfrac{VI \cos\phi}{VI}$

The power factor can never be greater than unity.

The power factor is expressed either as fraction or as a percentage.

Uusually , the word **'lagging' or 'leading'** is attached with the numerical value of power factor to signify whether the current lags behind the voltage or leads it.

(v) Active/ Wattful/In-phase Component of Current

The current component which is in phase with circuit voltage (i.e I cos) and contributes to active or true power of the circuit is called the **active component (wattful or in-phase)**of current

(vi) Reactive/Wattless Component of Current

The current component which is in quadrature (or 90°out of phase) to circuit voltage i.e. I sin and contributes to reactive power of the circuit, is called the **reactive component (or wattless)** of current

3.9.1.6 Summary of Calculation Procedure (RL Series Circuit)

1. Calculate the inductive reactance $X_L = = 2\pi f L$
2. If there is more than one resistance, calculate the total resistance, $R = R_1 + R_2 + ...$
3. Calculate the impedance $Z = \sqrt{R^2 + X_L^2}$
4. Calculate the phase angle, $\phi = \tan^{-1}\frac{X_L}{R}$
5. Calculate the current, $I = \frac{V}{2}$
6. Determine the voltage drop across resistance, $V_R = IR = V\cos\phi$
7. Determine the voltage drop across inductance, $V_L = I X_L = V\sin\phi$

Check

These calculations can be checked by calculating the value of supply voltage V,as follows:

Supply voltage, $V = \sqrt{V_R^2 + V_L^2}$

Compare it with the given value. The two values should be equal.

Example 3.19

A voltage v(t)=340 sin(377t+20) is applied to a RL series circuit. It causes a steady state current to flow which is described by i(t)= 28.28 sin(377t - 10). Determine the pf and the average power delivered to the circuit.

Solution

A comparison of voltage and current expressions indicates that current lags the voltage by an angle ϕ of $(20-(-10)) = 30^0$

Hence pf = cos = $\cos 30^0 = 0.866$ **Ans**

$$V_{RMS} = \frac{V_{max}}{\sqrt{2}} = \frac{340}{\sqrt{2}} = 240 \text{ volts}$$

$$I_{RMS} = \frac{I_{max}}{\sqrt{2}} = \frac{28.28}{\sqrt{2}} = 20 \text{ amp}$$

∴ Average power $P_{av} = V I \cos\phi = (240)(20)(0.866) = 4160$ Watts **Ans**

Example 3.20

A series circuit consists of R = 10 ohms and L = 10 mH and an AC supply of 200 V with f = 50 Hz. Calculate the current, the voltage across R, the voltage across L and the phase angle of current with respect to the supply voltage

Solution

$X_L = 2\pi f L = 2\times50\times10\times10^{-3} = 3.14$ ohm

$Z = t = \sqrt{R^2 \times X_L^2} = \sqrt{(10)^2 + (3.14)^2} = 10.488\Omega$

Phase angle, $\phi = ta^{-1}\dfrac{X_L}{R} = ta^{-1}\dfrac{3.14}{10}$ $\tan^{-1}0.314 = 17.43^0$

Current, $I = \dfrac{V}{Z}$ or $I = \dfrac{200}{10.488} = 19.07$ amp. **Ans**

$\cos\phi = \cos 17.43 = 0.954$

$\sin 17.43 = 0.299$

Voltage across R, $V_R = V\cos\phi = 200\times0.954 =$ 190.8 volts

Also $V_R = IR = 19.07\times10 = 190.7$ Volts

Voltage across L, $V_L = I_{X_L} = V\sin\phi = 200\times0.299 = 59.90$ volts

Also $V_L = IX_L = 19.07\times3.14 = 59.9$ volts

Choosing proper scales for current and voltage phasors, phasor diagram can be drawn as shown below

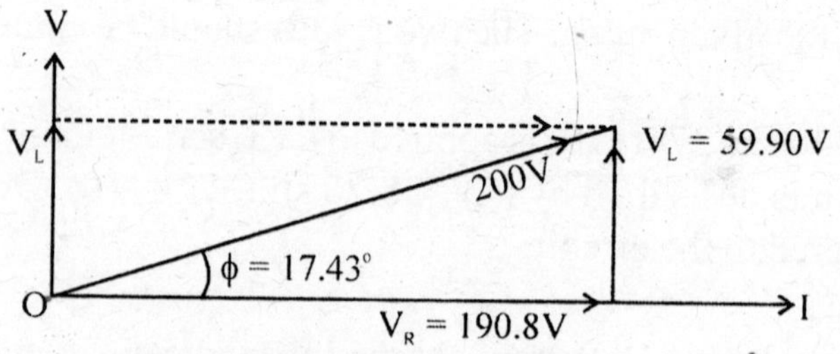

Phasor diaram for Example 3.20

Check

$V = \sqrt{V_R^2 + V_L^2} = \sqrt{(190.8)^2 + (59.9)^2}$

$= \sqrt{36328.36 + 3582.02} = \sqrt{39910.38} =$ Approx.200 V

Which is equal to the supply voltage.

Example 3.21

A choke coil, (consisting of an unknown resistance 'r' in series with an unknown inductance, L), connected in series with a resistance R = 15 ohms is connected across a supply of 230V, 50 Hz as shown in Fig(a) given below. Voltage drop across the resistance R is 75 volts, Voltage drop across the choke coil is 200 volts.

Calculate (a) pf of the circuit(b)power consumed in the resistance(c)power consumed in the coil

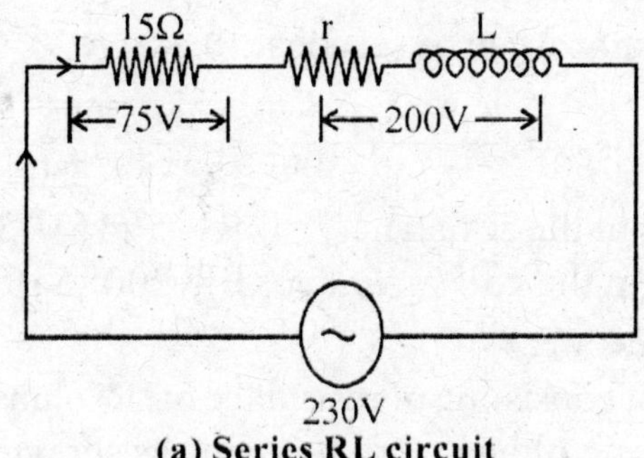

(a) Series RL circuit

for Example 3.21

Solution

Current, $I = \frac{75}{15} = 5$ amp

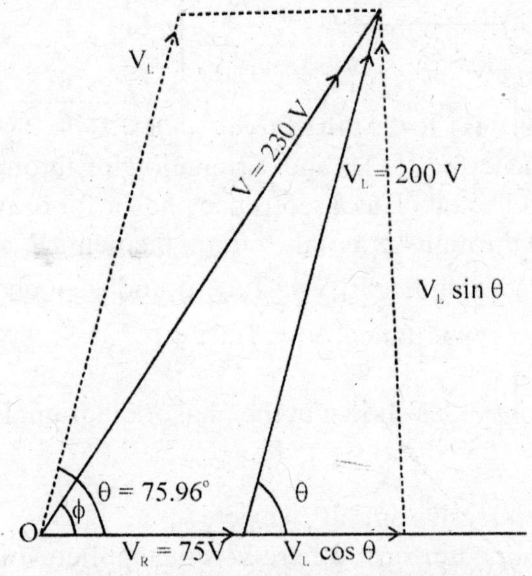

(b) Phasor diagram for example3.21

It can be seen from the phasor diagram given above

$(V_R + V_L \cos\theta)^2 + (V_L \sin\theta)^2 = V^2$

i.e. $V_R^2 + V_L^2 + 2 V_R V_L \cos\theta = V^2$

or $\cos\theta = \frac{V^2 - VR^2 - V_L^2}{2V_RV_L} = \frac{230^2 - 75^2 - 200^2}{2(75)(200)} = 0.2425$

i.e. $\theta = \cos^{-1}(0.2425) = 75.96^0$

or $\sin\theta = 0.97$

(a) Power factor of the circuit, cosϕ

From the phasor diagram , it can be seen that:

$$\cos\phi = \frac{V_R + V_L + \cos\theta}{V} = \frac{75 + 200 \times 0.2425}{230} = \frac{75 + 48.5}{230} = 0.537 \textbf{ Ans}$$

(b) Power consumed in the resistance = $I^2 R = 5^2(15) = 375$ watts Ans

(c)Power consumed in the coil $= V_L I \cos\theta = 200 \times 5 \times 0.2425 = 242.5$ Watts **Ans**

3.9.2 R-C Series Circuits

Consider an AC circuit consisting of resistance of 'R' ohms and capacitance of 'C' farads, connected in series across a voltage $v(t) = V_{max}$ sin , as shown in **Fig. 3.26(a)**

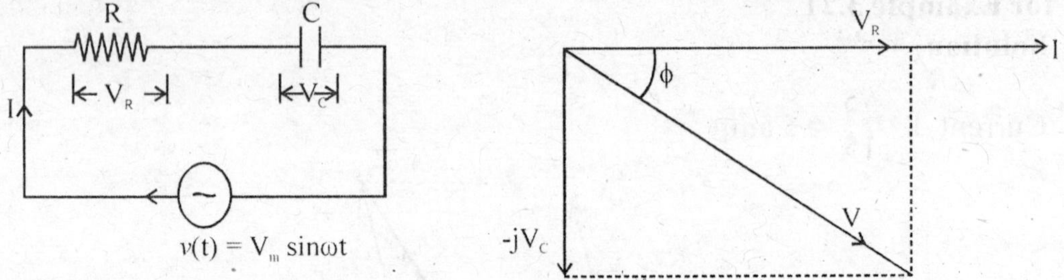

Fig.3.26(a) R-C Series Circuit Fig.3.26(b) Phasor diagram

Let the supply frequency be 'f' Hz and current flowing through the circuit be'I' amperes. . RMS value of current I is taken as a reference phasor for drawing the phasor diafgram , because current is same through both of the circuit elements R and C.

Voltage drop across resistance R, $V_R = I R \angle 0$ and is in phase with the current.

Voltage drop across capacitance, $V_C = I X_C \angle -90 = -j I X_C$, lagging behind current phasor I by $\frac{\pi}{2}$ radians or 90°, as shown in the phasor diagram, Fig.3.26(b).

3.9.2.1 Impedance

Applying KVL to the series circuit , we get:

$V = V_R + V_C$, (phasor addition) ,Where V is the applied voltage.

$$= I R - j I X_C = I(R - j X_C)$$

$$= I Z \angle -\phi$$

Where $Z = (R - j X_C)$ is the **impedance** and is expressed in ohms.

$|Z| = \sqrt{R^2 + X_C^2}$, is known as magnitude of **impedance**

In **polar form, impedance** $\mathbf{Z} = |Z| \angle\phi$

Where $\mathbf{Z} = \sqrt{R^2 + X_c^2}$ and $\angle\phi = \tan^{-1}\frac{-X_c}{R}$

3.9.2.2 Phasor diagram

Draw V_R in phase with phasor I and draw V_C lagging phasor I by 90^0, as shown in Fig. 3.26(b)

Complete the rectangle.Diagonal of this rectangle gives the resultant supply voltage, V

If the applied voltage is $v(t) = V_{max} \sin \omega t$, then expression for the current is series RC circuit will be :

$i = I_{max} \sin(\omega t + \phi\chi)$

3.9.2.3 Impedance Triangle

By dividing each side of the voltage triangle **Fig 3.26 (c)** by I, we obtain an **impedance triangle, Fig 3.26(d)**

The sides of the impedance triangle are : R, X_C and Z

The applied voltage lags behind the current by an angle ϕ

$$\text{Where} \quad \tan\phi = \frac{V_C}{VR_1} = \frac{IX_C}{IR} = \frac{X_C}{R} = \frac{1}{wCR}\phi$$

$$\text{or} \quad \phi = \tan^{-1}\frac{1}{RwC}$$

$$\text{Power factor, } \cos\phi = \frac{R}{Z}$$

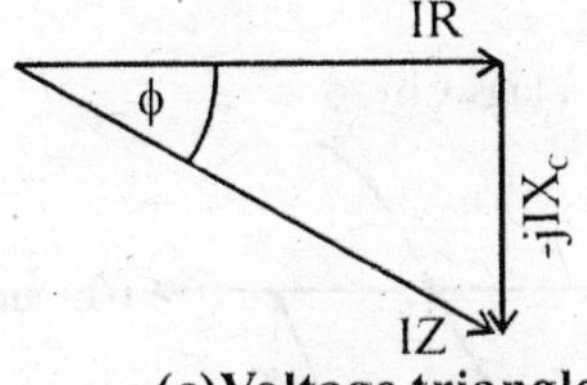

(c)Voltage triangle

R
ϕ
-jXc
Z

(d) Impedance triangle

Fig 3.26

Hence complex impedance can be expressed in polar form as:

$\mathbf{Z} = Z \angle -\Phi$

In cartesian co- ordinates ,

$\mathbf{Z} = Z(\cos\phi - j\sin\phi)$

or $R = Z\cos\phi$ and $X_C = Z\sin\phi$

3.9.2.4 Power in Series RC circuit

(a)Instantaneous Power

Instantaneous power, $p = v\,i$

$$= V_{max} \sin wt \times I_{max} \sin(wt + \phi)$$

$$= \frac{1}{2} I_{max} V_{max} [2 \sin wt \times \sin(wt + \phi)]$$

Using the Trigonometric Identity, 2 sin A sin B = cos (A - B) − cos (A + B), in the above equation, we get:

$$p = \frac{1}{2} V_{max} I_{max} [\cos\Phi - \cos(2\omega t + \Phi)]$$

i.e. Instantaneous power

$$p = \frac{1}{2} V_{max} I_{max} \cos\Phi - \frac{1}{2} V_{max} I_{max} \cos(2\omega t + \Phi)$$

Waveform of this Instantaneous power is shown in Fig.3.26(e)

(b)Average power

Average value of pulsating component $\frac{1}{2} V_{max} I_{max} \cos(2\omega t + \Phi)$, over a complete cycle is zero, therefore, average power of the series RC circuit is :

$$P_{av} = \frac{1}{2} V_{max} I_{max} \cos\Phi = \frac{V_{max}}{\sqrt{2}} \cdot \frac{I_{max}}{\sqrt{2}} \cos\Phi$$

or

$$P_{av} = V_{rms} I_{rms} \cos\phi$$

Where V_{rms} and I_{rms} are the RMS values of voltage and current and ϕ is the phase angle between applied voltage V and current I.

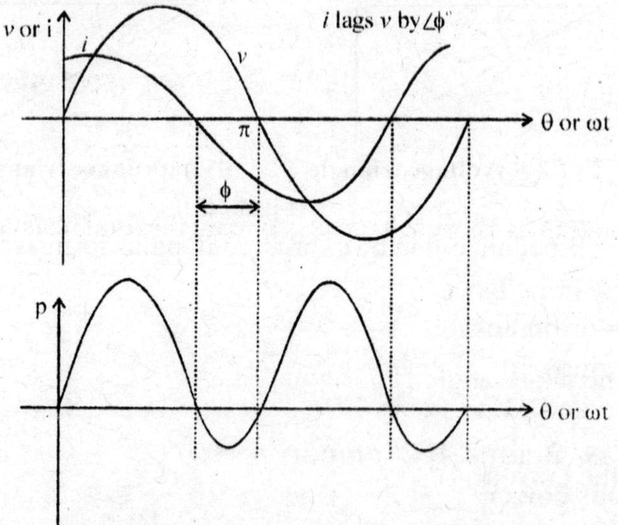

Fig.3.26 : (e) Power in series R-C circuit

When an alternating voltage is impressed across a capacitor, power is taken from the source and stored in the capacitor as the voltage increases from zero to its maximum value. Then, as the impressed voltage decreases from its maximum value to zero, the capacitor discharges and returns the power to the source

3.9.2.5 Power Triangle of Series RC Circuit

The power triangle can be obtained by multiplying each side of voltage triangle by current 'I' as shown in Fig 3.26(f)

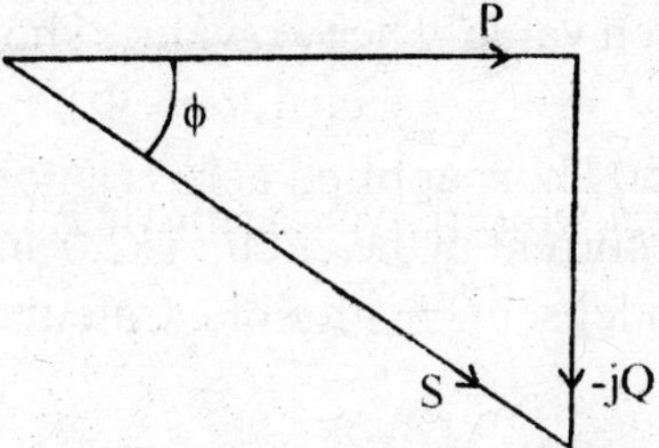

Fig.3.26 : (f) (Power triangle)

Three sides of the Power triangle are:

(i)V_R I i.e. V I cos Φ , (ii) V_CI i.e. V I sin ϕ (is negative) and (iii) V I

Apparent power = V I

True or real power = V_R I or V I cos ϕ

Reactive power = V_CI = V I sin ϕ

Power factor

$$\text{p.f.} = \frac{True\ \text{Power}}{Apparent\ \text{power}} = \frac{VI\cos\Phi}{VI} = \cos\phi$$

3.9.2.6 Summary of Calculation Procedure (RC Series Circuit)

1. Calculate the capacitive reactance $X_C = \frac{1}{\omega C} = \frac{1}{2\pi f C}$
2. If there is more than one resistance, calculate the total resistance. $R = R_1 + R_2 + ...$
3. Calculate the impedance $Z = \sqrt{R^2 + X_C^{\ 2}}$
4. Calculate the phase angle, $\Phi = \tan^{-1}\frac{X_C}{R}$
5. Calculate the current, $I = \frac{V}{Z}$
6. Determine the voltage drop in resistance, $V_R = IR = V\cos\phi$
7. Determine the voltage drop across capacitance, $V_C = IX_C = V\sin\phi$

Check

These calculations can be checked by calculating the value of supply voltage V, as follows:

Supply voltage, V= $\sqrt{V_R^2 + V_C^2}$

Compare it with the given value. The two values should be equal.

Example 3.22.

A voltage of 150 V at 50 Hz is applied across a non- inductive resistor connected in series with a condenser. The current in the circuit is 2.0 amp., the power loss in the resistor is 100 watts and that in the condenser is negligible. Calculate the resistance and capacitance.

Solution :

Applied voltage, V = 150 volts,

Circuit current, I = 2.0 amperes

$$\text{Impedance, } Z = \frac{V}{I} = \frac{150}{2} = 75\Omega$$

Power loss in the resistor = I^2 R = 100 watts

$$\text{Resistance of the circuit, } R = \frac{100}{(2)^2} = 25\Omega$$ **Ans.**

Since power loss in the condenser is negligible, therefore, resistance of the capacitor is zero.

$$\text{Capacitive reactance of the circuit, } X_C = \sqrt{Z^2 - R^2} = \sqrt{(75)^2 - (25)^2}$$

$$= \sqrt{5625 - 625} = \sqrt{5000} = 70.71 \text{ Ohms}$$

$$\text{Capacitance , } C = \frac{1}{2\pi\ f\ X_C} = \frac{1}{2\pi \times 50 \times 70.71} = \frac{1}{314 \times 70.71} = 45.03\mu F$$ **Ans.**

Example 3.23.

A resistor of 200Ω is connected in series with a 50μFcapacitor to a supply at 230 V, 50 Hz. Find (i) the impedance (ii) the current (iii) the phase angle (iv) the pf (v) the voltage across the resistor and voltage across the capacitor. Draw the phasor diagram.

Solution :

Resistance of the circuit, R = 200Ω

$$\text{Capacitive reactance of the circuit, } X_C = \frac{1}{2\pi fC} = \frac{1}{2\pi \times 50 \times 50 \times 10^{-6}} = 63.69\Omega$$

(i) Impedance, $Z = \sqrt{R^2 + X_C^2} = \sqrt{(200)^2 + (63.69)^2} = 209.89\Omega$ **Ans.**

(ii) Current, $I = \frac{V}{Z} = \frac{230}{209.89} = 1.09$ amp. **Ans.**

(iii) Phase angle, $\Phi = \tan^{-1}\frac{X_C}{R} = \tan^{-1}\frac{63.69}{200} = \tan^{-1} 0.318 = 17.66^{o}$ **Ans.**

(iv) Power factor, $\cos\Phi = \cos 17.66° = 0.952$ (lead) **Ans.**

(v) Voltage across resistor, $V_R = IR = 1.09 \times 200 = 218$ V **Ans.**

Voltage across capacitor, $V_C = I X_C = 1.09 \times 63.69 = 69.42$V Ans.

Choosing proper scales for voltage and current phasors, phasor diagram can be drawn as shown below.

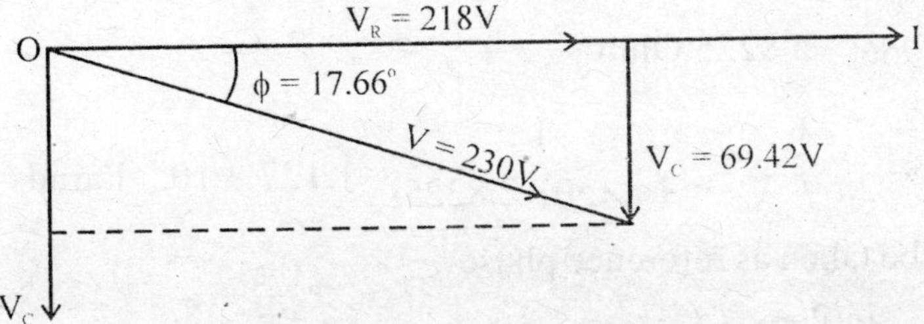

Phasor diagram for Example 3.23

Example 3.24.

A series RC circuit takes a power of 7000 watts when connected to 200V ,50Hz supply. The voltage across the resistor is 130 volts. Calculate (i) the resistance ,current ,pf , impedance and capacitance (ii) Write expressions for v(t) and i(t).

Solution

The circuit is shown in Fig 3.27 given below:

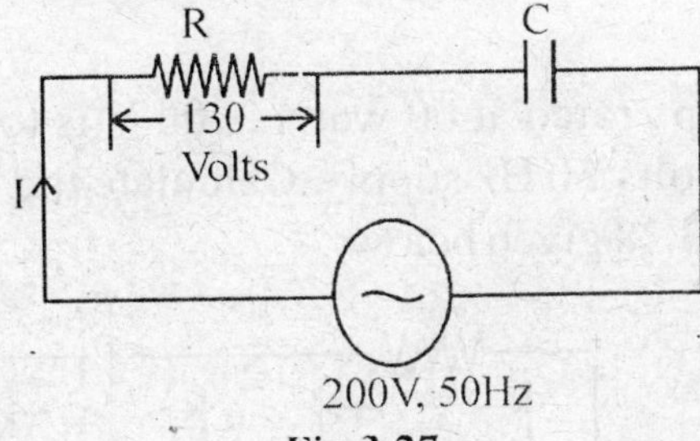

Fig.3.27

(i) Power, $P = \frac{V_R^2}{R}$

Or Resistance, $R = \frac{V_R^2}{P} = \frac{(130)^2}{7000} = 2.414$ Ohm **Ans**

Current ,I $= \frac{V_R}{R} = \frac{130}{2.414} = 53.85$ Amp Ans

Power ,P = V I cos ϕ

$\therefore$ Pf ,cos $\phi = \frac{P}{VI} = \frac{7000}{200 \times 53.85} = 0.65$ **Ans**

Hence, $\phi = \cos^{-1} 0.65 = 49.46^0$

Impedance,Z $= \frac{V}{I} = \frac{200}{53.85} = 3.714$ Ohms **Ans**

However, $X_C^2 = Z^2 - R^2 = (3.714)^2 - (2.414)^2 = 7.966$

Or X_C $\sqrt{7.966} = 2.8225$ Ohm

Hence,C= $= \frac{1}{2\pi f X_C} = \frac{1}{2\pi \times 50 \times 2.8225} = 1.127 \times 10^{-3}$ Farads **Ans**

(ii) Let i(t) be taken as reference phasor.

i.e. $i(t) = I_{max} \sin(2\pi \times f t)$

$I_{max} = \sqrt{2 \times 53.85} = 76.16$ amp.

Hence, Expression for $i(t) = 76.16 \sin(2\pi \times 50\, t)$ **Ans**

Since voltage lags the current phasor by an angle ϕ ,

Hence, Expression for $v(t) = V_m \sin(2\pi f t - \phi)$)

$= 200\sqrt{2} \sin(100\pi t - 49.46^0)$ volts **Ans**

Example 3.25.

A metal filament lamp , rated at 60 watts , 200 V is to be connected in series with a capacitance across a 250 volts,50 Hz supply. Calculate the value of capacitance required. The circuit is shown in Fig 3.28 given below:

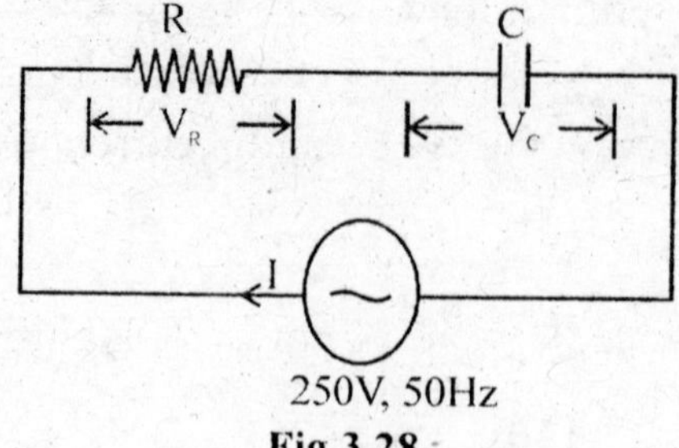

Fig.3.28

Solution

Lamp power , P = VI

$$I = \frac{P}{V} = \frac{60}{200} = \frac{3}{10} = 0.3 \text{ amp}$$

Voltage across capacitance, $V_C = \sqrt{V^2 - V_R^2} = \sqrt{(250)^2 - (200)^2} = \sqrt{62500 - 40000}$

$$= \sqrt{22500} = 150 \text{ Volts}$$

Capacitive reactance, $X_C = \frac{V_C}{I} = \frac{150}{0.3} = 500$ Ohm

$$\text{Or } C = \frac{1}{2\pi f X_C} = \frac{1}{2\pi \times 50 \times 500} = \frac{1}{\pi \times 100 \times 500} = \frac{1}{\pi \times 5 \times 10^4}$$

Hence capacitance required = $0.063 \times 10^{-4} = 6.3\ \mu$ Farads **Ans.**

3.9.3 R-L-C Series Circuit

Consider an AC circuit containing resistance R ohms, inductance L Henries and capacitance C farads connected in series, across a voltage $v = V_{max} \sin \omega t$, as shown in Fig.3.27.

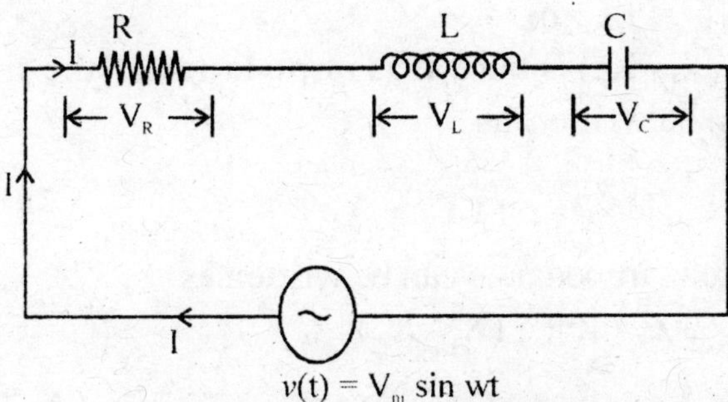

$v(t) = V_m \sin wt$

Fig.3.27 : R-L-C Series Circuit

Let the current flowing through the circuit be 'I' amperes and supply frequency be 'f' Hz. RMS value of current I is taken as a reference phasor for drawing the phasor diafgram , because current is same through both of the circuit elements R , L and C.

Voltage drop across resistance, $V_R = IR$, in phase with current phasor, I.

Voltage drop across inductance, $V_L = jIX_L$, leading current phasor, I by $\pi/2$ radians or 90°.

Voltage drop across capacitance, $V_C = -jIX_C$ or $\frac{-jI}{\omega C}$ lagging behind current phasor, I by $\pi/2$ radians or 90°

V_L and V_C are 180° out of phase with each other i.e. they oppose each other.

This series circuit can either be effectively inductive or capacitive depending upon which voltage drop (V_L or V_C) is dominant.

3.9.3.1 Impedance, Phasor Diagrams and Voltage Triangle

Depending upon the values of X_L and X_C, three different situations may arise as described below:

1. $X_L > X_C$
2. $X_L < X_C$
3. $X_L = X_C$

Case 1. $X_L > X_C$

In this case, Voltage drop across L i.e. V_L is greater than Voltage drop across C, i.e. V_C.

i.e $\quad V_L > V_C$

Therefore the circuit is **inductive**, if $X_L > X_C$

Applying KVL to the series circuit, we get:

$$V = V_R + V_L + V_C$$

i.e. The applied voltage V is equal to the phasor sum of voltages V_R, V_L and V_C

i.e. $\quad V = I R + j I X_L - j I X_C$

$$= I(R + j(X_L - X_C))$$

$$= IZ \angle \Phi$$

where $Z = R + j(X_L - X_C)$ is known as **impedance** of the circuit

Its magnitude can be written as :

$$|Z| = \sqrt{R^2 + (X_L - X_C)^2}$$

In complex notation, impedance can be written as:

$$Z = R + j(X_L - X_C) \tag{3.28}$$

Or

$$Z = R + jX_{eq}$$

Where $\quad X_{eq} = (X_L - X_C)$

As this circuit is inductive, current will lag behind the supply voltage by an angle Φ, which is given by :

$$\Phi = \tan^{-1}\frac{V_L - V_C}{V_R} = \tan^{-1}\frac{I X_L - I X_C}{I R}$$

$$= \tan^{-1}\frac{X_L - X_C}{R} = \tan^{-1}\frac{X_{eq}}{R} \tag{3.29}$$

Sign of angle, depends on the sign of X_{eq} i.e. $X_L - X_C$

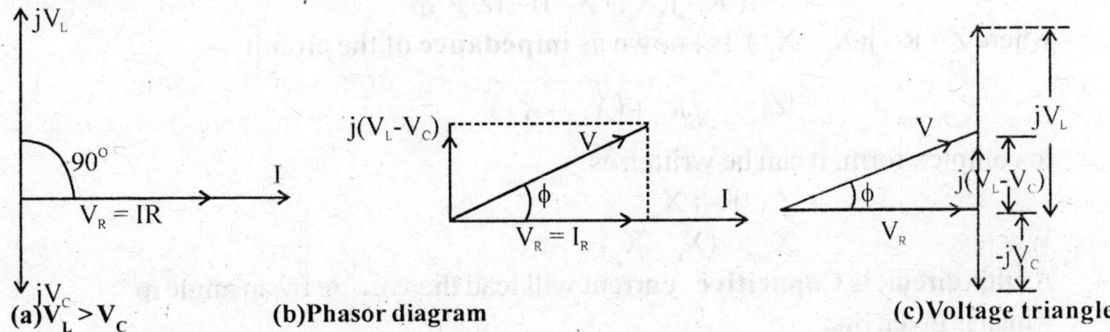

(a)$V_L > V_C$ (b)Phasor diagram (c)Voltage triangle

Fig 3.28 Phasor diagram and voltage triangle for series RLC circuit ($X_L > X_C$)

3.9.3.2 Impedance triangle for RLC series circuit ($X_L > X_C$)

The Impedance triangle is obtained from the voltage triangle by dividing its each side by current, I. and is as shown in Fig.3.28 (d)

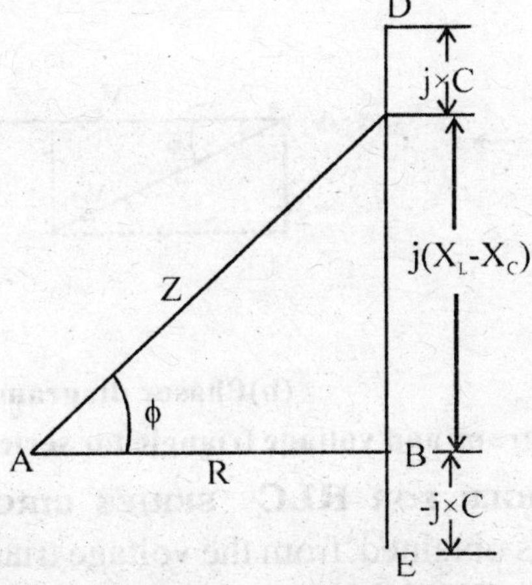

Fig 3.28 (d)Impedance triangle for $X_L > X_C$

Power factor of the circuit is given by $\cos \Phi = \dfrac{R}{Z} = \dfrac{R}{\sqrt{R^2 + (X_L - X_C)^2}}$

Power consumed in the circuit, $P = I^2 R$ or $VI \cos \Phi$

Case 2 $X_L < X_C$

In this case, Voltage drop across L i.e. V_L is less than Voltage drop across C , i.e. V_C.

Or $V_C > V_L$

The circuit is capacitive if **$X_L < X_C$ or $X_C > X_L$**

The applied voltage V is equal to the phasor sum of voltages V_R, V_L and V_C and therefore its magnitude can be written as :

$$V = I R + j I X_L - j I X_C$$
$$= I(R - j(X_C - X_L)) = IZ \angle \Phi$$

where $Z = R - j(X_C - X_L)$ is known as **impedance** of the circuit

$$|Z| = \sqrt{R^2 + (X_C - X_L)^2}$$

In complex form, it can be written as:

$$Z = R - j X_{eq}$$

Where $X_{eq} = (X_C - X_L)$

As this circuit is **Capacitive**, current will lead the vo... .e by an angle Φ which is given by :

$$\Phi = \tan^{-1}\frac{V_C - V_L}{V_R} = \tan^{-1}\frac{I X_C - I X_L}{I R} = \tan^{-1}\frac{X_C - X_L}{R} = \tan^{-1}\frac{X_{eq}}{R}$$

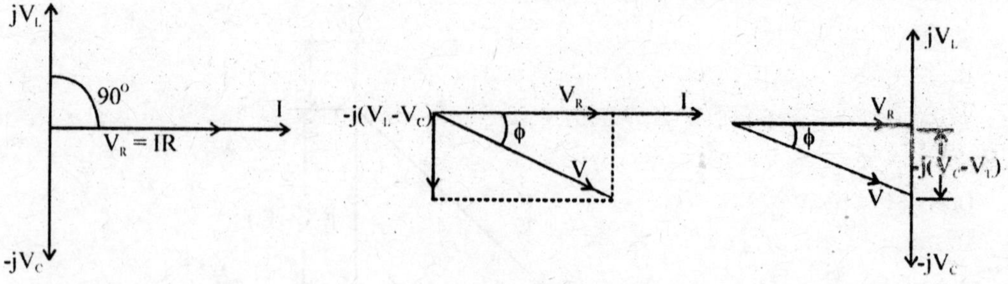

(a)$V_C > V_L$ **(b)Phasor diagram** **(c)Voltage triangle**

Fig 3.29 Phasor diagram and voltage triangle for series RLC circuit ($X_L < X_C$)

3.9.3.3 Impedance triangle for RLC series circuit ($X_L < X_C$)

The Impedance triangle is obtained from the voltage triangle by dividing its each side by current,I. and is as shown in Fig. 3.29 (d) .

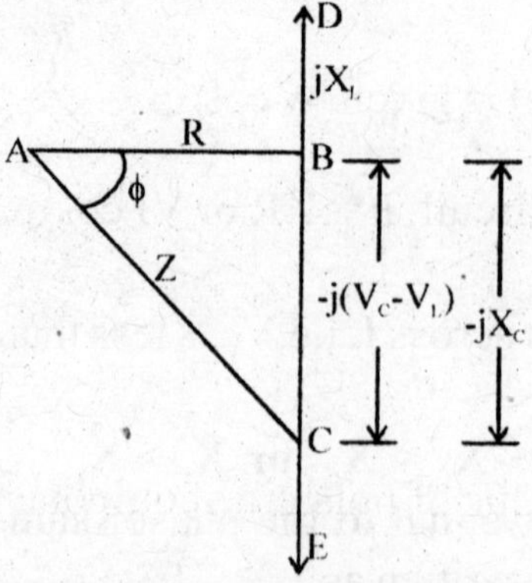

Fig 3.29 (d)Impedance triangle for $X_L < X_C$

Case 3. $X_L = X_C$

In this case, phasor V_L is equal to phasor V_C and these are 180° out of phase with each other. Hence their phasor sum is zero, Fig 3.30(a).

V_L 90° $V_R = IR$ I

V_C

$V_R = IR$ I

(a) $V_L = V_C$ (b) V and I are in phase

Fig 3.30 Phasor diagram for series RLC circuit ($X_L = X_C$)

Hence the resultant voltage, $V = V_R$ and is in phase with current 'I' i.e. angle ϕ is zero, as shown in the phasor diagram, Fig.3.30 (b).

i.e. $V = V_R = IR$

In this case, net impedance $Z = R$

Thus the net impedance is purely resistive. It is a **resistive** circuit

3.9.3.4 Power in RLC Series Circuit

(a) Instantaneous Power

Instantaneous power, $p = vi$

$$= V_{max} \sin \omega t \times I_{max} \sin(\omega t \pm \Phi)$$

$$= \frac{1}{2} V_{max} I_{max} [2 \sin \omega t . \sin(\omega t \pm \Phi)]$$

$\pm$Sign depends upon the condition whether the circuit is inductive or capacitive.

Using the Trigonometric Identity, $2 \sin A \sin B = \cos(A - B) - \cos(A + B)$, in the above equation, we get:

$$p = \frac{1}{2} V_{max} I_{max} [\cos\Phi - \cos(2\omega t \pm \Phi)]$$

i.e. Instantaneous power, $p = \frac{1}{2} V_{max} I_{max} \cos\Phi - \frac{1}{2} V_{max} I_{max} \cos(2\omega t \pm \Phi)$

(b) Average power

We know that average value of pulsating component, $\frac{1}{2} V_{max} I_{max} \cos(2\omega t \pm \Phi)$ over a complete cycle is zero, therefore, average power of the circuit :

$$P_{av} = \frac{1}{2} V_{max} I_{max} \cos \Phi = \frac{V_{max}}{\sqrt{2}} \frac{I_{max}}{\sqrt{2}} \cos \Phi$$

or $$P_{av} = V_{rms} I_{rms} \cos \Phi$$

Where V_{rms} and I_{rms} are the RMS values of voltage and current and Φ is the phase angle between applied voltage V and current I.

3.9.3.5 Power Triangle for RLC Series Circuit

The power triangle is obtained from the voltage triangle by multiplying each side by current ,I. and is as shown in Fig.3.32(a) and (b):

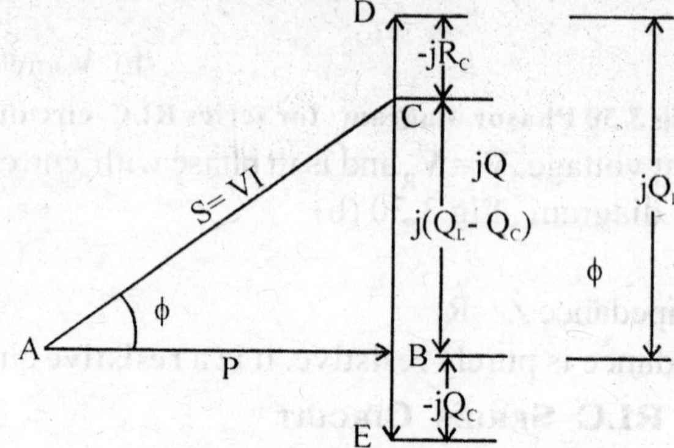

Fig 3.32 (a)Power triangle for $X_L > X_C$

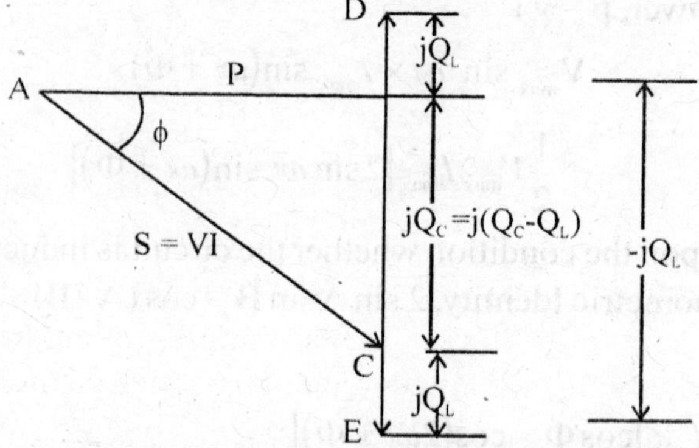

Fig 3.32 (b)Power triangle for $X_C > X_L$

3.9.3.6 Summary of Calculation Procedure RLC Series Circuit

1. Calculate the inductive reactance $X_L = \omega L = 2\pi fL$
2. Calculate the capacitive reactance $X_C = \frac{1}{\omega C} = \frac{1}{2\pi fC}$
3. Determine the equivalent reactance $X_{eq} = (X_L - X_C)$ or $(X_C - X_L)$
4. If there is more than one resistance in the circuit,

calculate the total resistance $R = R_1 + R_2 + ...$

5. Calculate the impedance $Z = \sqrt{R^2 + X_{eq}^2}$

6. Calculate the phase angle $\Phi = \tan^{-1} \frac{X_{eq}}{R}$

7. Calculate the current, $I = \frac{V}{Z}$

8. Determine the resistive voltage, $V_R = I R = V \cos\phi$

9. Determine the reactive voltage, $V_X = I X_{eq} = V \sin\phi$

10. Determine the voltage across inductance, $V_L = I X_L$

11. Determine the voltage across capacitance, $V_C = I X_C$

Example 3.26

An AC circuit consists of a resistance of 2 ohms, an inductance of 0.5 H, and a capacitance of 25μF, all in series.Determine:(i) total reactance , (ii)impedance. The supply frequency is 50 Hz.

Solution :

$R = 2\Omega$; $L = 0.5$ H; $f = 50$ Hz; $C = 25\mu F = 25 \times 10^{-6}$ F;

$$X_L = 2\pi \times 50 \times 0.5 = 157\,\Omega \text{ ; and } X_C = \frac{1}{2\pi \times 50 \times 25 \times 10^{-6}} = 127.38\,\Omega$$

(i) Total reactance, $X_{eq} = X_L - X_C = 157 - 127.38 = 29.62\Omega$ Ans

(ii) Impedance, $Z = \sqrt{R^2 + X_{eq}^2} = \sqrt{2^2 + 29.62^2} = \sqrt{881.34} = 29.687$ Ans

Example 3.27

A coil of resistance 10 ohms, and inductance 200 mH is connected in series with a capacitance of 100 μF across 200 V, 50 Hz AC supply. Calculate: (i) the magnitude of current: (ii) p.f. (iii) phase angle, and (iv) voltage across each element.

Solution :

$$X_L = 2\pi f L = 2\pi \times 50 \quad 200 \times 10^{-3} = 62.8\Omega$$

and $$X_C = \frac{1}{2\pi \times 50 \times 100 \times 10^{-6}} = 31.845\,\Omega$$

$$X_{eq} = X_L - X_C = 62.8 - 31.845 = 30.955\textbf{(Inductive)}$$

Impedance, $$Z = \sqrt{10^2 + 30.955^2} = \sqrt{100 + 958.21} = \sqrt{1058.21} = 32.53\,\Omega$$

(i) Current, $$I = V/Z = \frac{200}{32.53} = 6.148 \text{ amps}$$ **Ans**

(ii) p.f., cos = R/Z = $\frac{10}{32.53}$ = 0.307 Ans

(iii) Phase angle, = $\cos^{-1}$ 0.307 = 72.09 ^{0} (lag). Ans

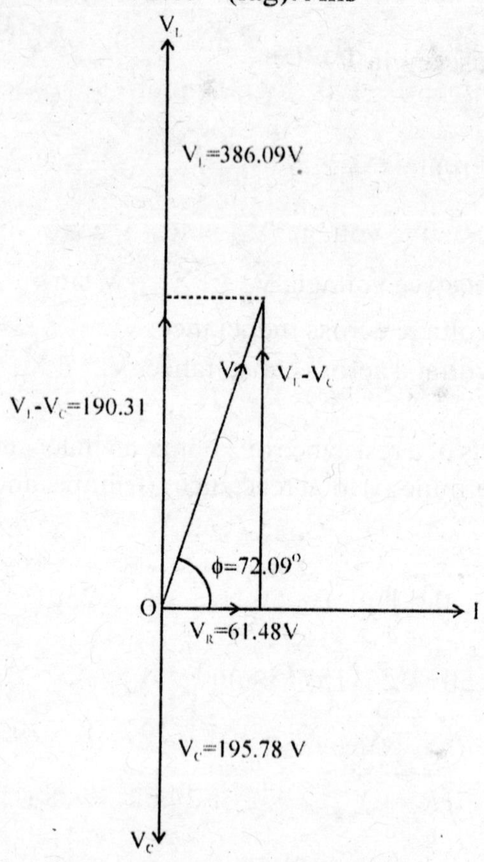

Phasor diagram for Example 3.27

(i) Voltage across resistor, V_R = I R = 6.148×10 = 61.48 volts Ans

Voltage across inductor, V_L = 1X_L = 6.148 × 62.8 = 386.09 volts Ans

Voltage across capacitor, V_C = I X_C = 6.148 × 31.845 = 195.78 volts Ans

Example 3.28

A resistance R, an inductance L = 0.05 H and a capacitance C are connected in series. A voltage v = 282 cos (1000 t – 10°) Volts is applied to the series combination, the current flowing is 5 $\sqrt{2}$ cos (1000 t – 60°) amp . Find R and C.

Solution:

$$Z = \frac{V_{max}}{I_{max}} = \frac{282}{5\sqrt{2}} = 40\Omega \ , 2\pi \ f = 1000 \ ,$$

$\therefore \quad X_L = 1000 \times 0.05 = 50\ \Omega$

$v = 282 \cos(1000\,t - 10^\circ)$

Converting cosine function in to sine function, we get:

$$v = 282 \sin(1000t - 10^\circ + 90^\circ) = 282 \sin(1000\,t + 80^\circ)$$

$$i = 5\sqrt{2} \cos(1000\,t - 60^\circ) = 5\sqrt{2} \sin(1000t - 60^0 + 90^\circ) = 5\sqrt{2} \sin(1000\,t + 30^\circ)$$

Phase angle, $= 80^\circ - 30^0 = 50^\circ$ lagging (current lags behind voltage)

$\cos\phi = \cos 50 = 0.642$

$\sin\phi = \sin 50 = 0.766$

$R = Z \cos\phi = 40 \times 0.642 = 25.68\Omega$ **Ans**

$X_{eq} = Z \sin\phi = 40 \times 0.766 = 30.64\ \Omega$

We know that, $X_{eq} = X_L - X_C$

Or $X_C = X_L - X_{eq}$

Substituting the values we get:

$X_C = 50. - 30.64 = 19.36\ \Omega$

$$X_C = \frac{1}{1000\ C} = \frac{1}{2\pi f C} = 19.36$$

$$\text{Or } C = \frac{1}{1000 \times 19.36} = 51.65\ \mu f$$ **Ans**

Example 3.29

A series RLC circuit is shown in Fig.3.33(a). Find applied voltage, frequency, power loss in the choke coil.

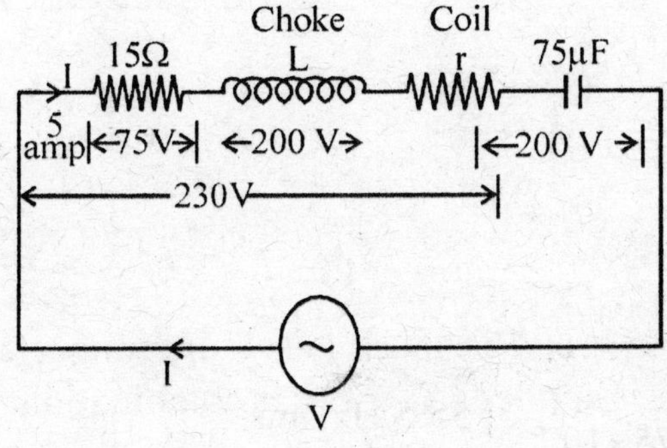

Fig. 3.33 : (a). A Series RLC circuit

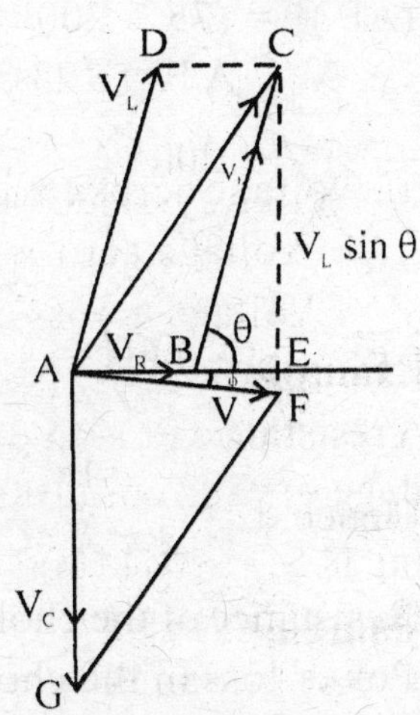

Fig. 3.33 : (b) Phasor diagram

Solution

Current, $I = \frac{V_R}{R} = \frac{75}{15} = 5$ amp

As explained in Example 3.21,

$$(V_R + V_L \cos \theta)^2 + (V_L \sin \theta)^2 = V^2 = 230^2$$

i.e. $$V_R^2 + V_L^2 + 2 V_R V_L \cos = V^2$$

or $$\cos \theta = \frac{V^2 - V_R^2 - V_L^2}{2 V_R V_L} = \frac{230^2 - 75^2 - 200^2}{2(75)(200)} = 0.2425$$

i.e. $$\theta = \cos^{-1}(0.2425) = 75.96^0$$

or $$\sin \theta = 0.97$$

Impedance of the choke coil, $Z_L = \frac{200}{5} = 40$ ohms

Inductive reactance of choke coil, $X_L = Z_L \sin\theta = 40 \times 0.97 = 38.8$ ohms

Resistance of choke coil, $r = Z_L \cos \theta = 40 \times 0.2425 = 9.7$ ohms

Voltage drop in $X_L = 5 \times 38.8 = 194.0$ volts

Voltage drop in resistance, $r = 5 \times 9.7 = 48.5$ volts

In right angled triangle, A E F

$$V_1^2 = (AF)^2 = (AE)^2 + (EF)^2 \quad \text{where, } EF = CF\text{-}CE$$
$$= (V_R + BC \cos\theta)^2 + (V_C - 194)^2$$

Substituting the values, we get :

$(AF)^2 = (75 + 200 \times 0.24)^2 + (200 - 194)^2 = (123.5)^2 + 36 = 15258.5$

i.e. $V_1 = AF = 123.5$ volts ans

$$X_C = \frac{200}{5} = 40 \text{ ohms}$$

i.e. $$\frac{1}{2\pi fC} = 40$$

Hence $$f = \frac{1}{2\pi X_C C} = \frac{1}{2\pi \times 40 \times 75 \times 10^{-6}} = 55.5 \text{ Hz ans}$$

Resistance of the choke coil = 9.7 ohms

Power loss in the choke = $5^2 \times 9.7 = 242.5$ watts **Ans**

3.10 IMPEDANCES IN SERIES

The series circuits discussed so far consisted of simple combinations of basic network elements. However several impedances may be connected in series in AC circuits. The computations in such cases become simple, if the impedances connected in series are replaced by a single equivalent impedance.

Figure 3.34 shows three complex impedances Z_1, Z_2 and Z_3 connected in series across an AC voltage source 'V', and a current 'I' passes through the impedances. Assume that V_1, V_2 and V_3 are the voltage drops across impedances Z_1, Z_2 and Z_3, respectively.

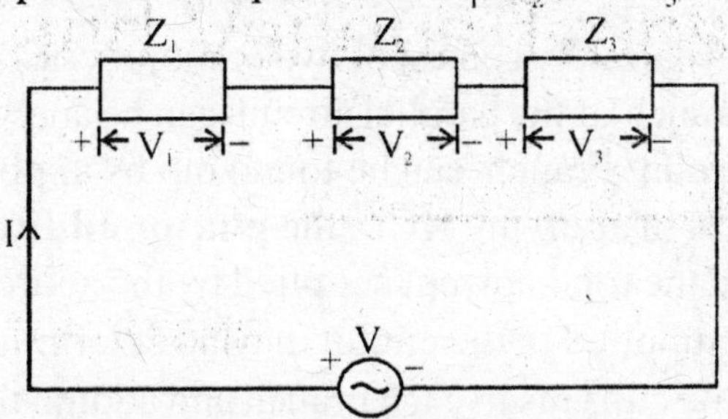

Fig.3.34 Impedances connected in Series

Applying KVL to the circuit shown in Figure3.34, the following expression results :

$$V = V_1 + V_2 + V_3 \qquad \text{(i)}$$

However,

$$V_1 = Z_1 \times I, \quad V_2 = Z_2 \times I, \quad V_3 = Z_3 \times I$$

Substituting the values of V_1, V_2 and V_3 in equation (i) given above, we get:

$$V = (Z_1 + Z_2 + Z_3) \times I = Z_{eq} \times I \qquad \text{(ii)}$$

Where, $Z_{eq} = Z_1 + Z_2 + Z_3$

Thus the equivalent complex impedance, Z_{eq} for a series circuit may be obtained by adding all the complex impedances connected in series.

Example 3.30

Three impedances are connected in series across a 200 V, 50Hz supply. Their impedances are: $Z_1 = 12 + j4$, $Z_2 = 10 + j2$ and $Z_3 = 3 - j6$

Calculate the current through the circuit and voltage drop in each impedance

Solution

Converting the three impedances in to polar form, we get:

$Z_1 = 12 + j4 = 12.64 \angle 18.41^0$, $Z_2 = 10 + j2 = 10.198 \angle 9.46^0$, $Z_3 = 3 - j6 = 6.7 \angle -63.43^0$

Total impedance $= Z_1 + Z_2 + Z_3 = 25 + j0 = 25 \angle 0$ ohms

Current through the circuit $= I = \dfrac{V}{Z} = \dfrac{200\angle 0}{25\angle 0} = 8 \angle 0$ amp. **Ans**

Voltage drops in different impedances are:

$V_1 = Z_1 \times I = (12+j4)8 = 96+j32 = 101.12$ volts Ans

$V_2 = Z_2 \times I = (10+j2)\,8 = 80+j6 = 81.58$ volts Ans

$V_3 = Z_3 \times I = (3 - j6)\,8 = 24-j48 = 53.6$ volts Ans

3.11 Single Phase Parallel Circuits

In practice, we come across the parallel AC circuits quite often, because all the electrical appliances or devices of different ratings are operated at the same supply voltage and are connected in parallel.

All parallel circuits normally consist of two or more series circuits connected to a common voltage source. Each branch of the parallel circuit can be analysed separately. The combined effect of each series circuit/ branch can be found out by applying KCL to the common node of the parallel branches. In applying KCL, the **phasor addition** of all the branch currents is found out to determine the total current supplied by the source.

The total current supplied to the circuit may be determined by the following methods.

(i) Determining the currents in each branch and adding them vectorially.

or

(ii) By determining an equivalent impedance of the whole circuit by series/parallel combinations and dividing the applied voltage by this equivalent impedance.

As all the individual branches are connected to the same voltage source, **the supply voltage phasor is taken as the reference phasor.**

Behaviour of AC circuits containing R & L,R & C and R,L,C elements connected in parallel is described below

3.11.1 RL in Parallel

Fig 3.35(a) shows a resistance R ohm and an inductance L connected in parallel to an AC voltage source of V volts and angular frequency rad/sec.

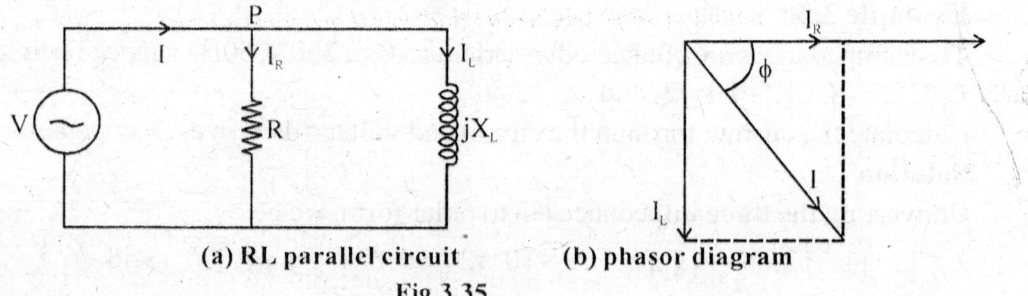

(a) RL parallel circuit (b) phasor diagram

Fig.3.35

The inductive reactance $X_L = \omega L = 2\pi f L$ ohm.

Let the voltage phasor **V** be the reference phasor. Hence, $\mathbf{V} = V\angle 0^\circ$.

Then ,

Current through the resistance, $I_R = \frac{V\angle 0^o}{R} = \frac{V}{R}$ ...(i)

Current through the inductance, $I_L = \frac{V\angle 0^o}{j\omega L} = -j\frac{V}{\omega L} = \frac{V}{\omega L}\angle -90^o$ ...(ii)

From Equations (i) and (ii), it is clear that the current I_R is in phase with the applied voltage V while the current I_L lags the supply voltage V by an angle 90°.

The phasor diagram for a parallel RL circuit is shown in Figure 3.35(b).

Applying KCL at node 'P' for the circuit, shown in Figure 3.35(a),

$$I = I_R + I_L \quad \text{...(iii)}$$

Substituting Equations (i) and (ii) into Equation (iii), we get:.

$$I = \frac{V}{R} - j\frac{V}{\omega L} = V\left(\frac{1}{R} + \frac{1}{j\omega L}\right) \quad \text{(iv)}$$

The angle ϕ between voltage phasor and current phasor is given by :

$$\tan\varphi = \frac{|I_L|}{|I_R|} = \frac{V/\omega L}{V/R} = \frac{R}{\omega L} = \frac{R}{X_L} \quad \text{(v)}$$

Example 3.31

A resistance of 10 Ω in parallel with a pure inductance of 100 mH is connected to a 200-V, 50-Hz supply. Calculate the total current drawn from the supply and the phase angle between the voltage and the total current.

Solution :

The parallel RL circuit is shown in Figure given below:

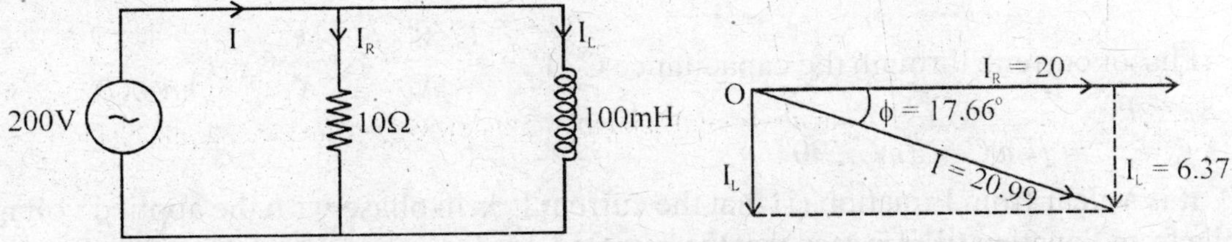

(a) RL parallel circuit **(b)Phasor diagram for Example 3.31**

Let the supply voltage be chosen as the reference phasor. Then $V = V\angle 0° = 200\angle 0°$.

$R = 10\ \Omega$

$jX_L = j\omega L = j314 \times 100 \times 10^{-3} = j31.4\ \Omega = 31.4\angle 90°$.

$$I_R = \frac{200\angle 0^o}{10} = 20.0\angle 0° = 20.0 + j0 \text{ amp.}$$

$$I_L = \frac{200\angle 0^o}{31.4\angle 90^o} = 6.369\angle -90^o = 0 - j6.369 \text{ amp.}$$

Then, total current drawn from the supply,

$I = I_R + I_L = 20.0 - j6.369$

$= 20.98 \angle -17.66°$ amp **Ans**

Phase angle between voltage and total current = 17.66° lagging Ans

The phasor diagram is drawn choosing proper scales for currents and is shown in Figure (b)given above.

3.11.2 R-C in Parallel

Figure 3.36(a) shows the R and C connected in parallel across an AC voltage source of V volts. The current flow in the network is also shown in the Fig

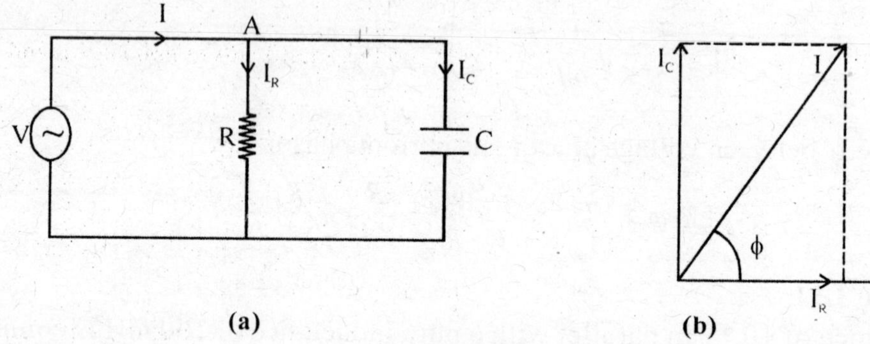

Fig.3.36

Let the voltage phasor V be the reference phasor. Then, $V = V\angle 0°$ and

Phasor current through the resistance R, $I_R = \frac{V\angle 0^o}{R} = \frac{V}{R}\angle 0$ ...(i)

Phasor current through the capacitance C, $I_C = \frac{V\angle 0^o}{-jX_C} = j\frac{V}{X_C} = j\frac{V}{1/(\omega C)}$

$= jV\omega C = V\omega C\angle 90^o$ ...(ii)

It is a clear from Equation (i) that the current I_R is in phase with the applied voltage V while from Equation (ii) it is seen that the current I_C leads supply voltage V by 90°. The phasor diagram for the parallel RC circuit is shown in Figure3.36(b).

Applying the KCL at node'A' for the circuit shown,

$I = I_R + I_C$ ...(iii)

Substituting Equations (i) and (ii) into Equation (iii).

$$I = \frac{V}{R} + jV\omega C = V\left(\frac{1}{R} + j\omega C\right)$$

From the phasor diagram , it can be seen that the angle ϕ between voltage phasor and current phasor is given by :

$$\tan\phi = \frac{|I_C|}{|I_R|} = \frac{V\omega C}{V/R} = R\omega C = \frac{R}{X_C}$$

Example 3.32

A circuit consisting of a resistance of 50 Ω in parallel with a pure capacitance of 100 μF is connected to a 200-V, 50-Hz supply. Calculate : (a) the branch currents (b) the supply current and the circuit phase angle.

Solution :

The parallel RC circuit is shown in Fig 3.37(a)

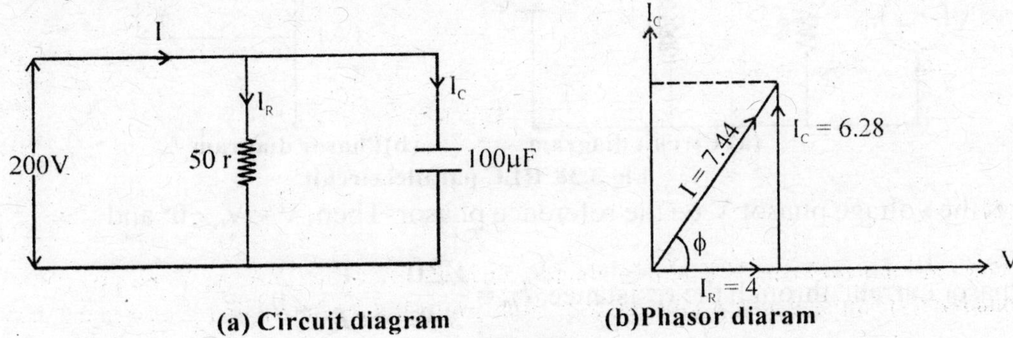

(a) Circuit diagram (b)Phasor diaram

Fig.3.37

Let the supply voltage be chosen as the reference phasor. Then V = 200 ∠ 0°.

R = 50 Ω

$$-jX_C = -j\frac{1}{\omega C} = \frac{-j}{314\times 100\times 10^{-6}} = -j31.85 = 31.85\angle -90^o$$

(a) Branch currents

$$I_R = \frac{200\angle 0^o}{50} = 4.0\angle 0^o = 4.0 + j0 \text{ amp}$$ **Ans**

$$I_C = \frac{200\angle 0^o}{31.85\angle -90^o} = 6.279\angle 90^o = j6.279 \text{ amp}$$ **Ans**

(b)Hence, Supply current, $I = I_R + I_C = 4.0 + j6.279 = \sqrt{4^2 + 6.279^2}\angle\tan^{-1}\frac{6.279}{4.0}$

= 7.445 ∠ 57.48° Ans

Phase angle = 57.48° leading. **Ans**

The phasor diagram is drawn choosing proper scales for currents and is shown in Fig 3.37(b).

3.11.3 RLC in Parallel

A circuit with resistance R, inductance L and capacitance C ,connected in parallel across an AC voltage source of V volt is shown in Figure3.38(a)

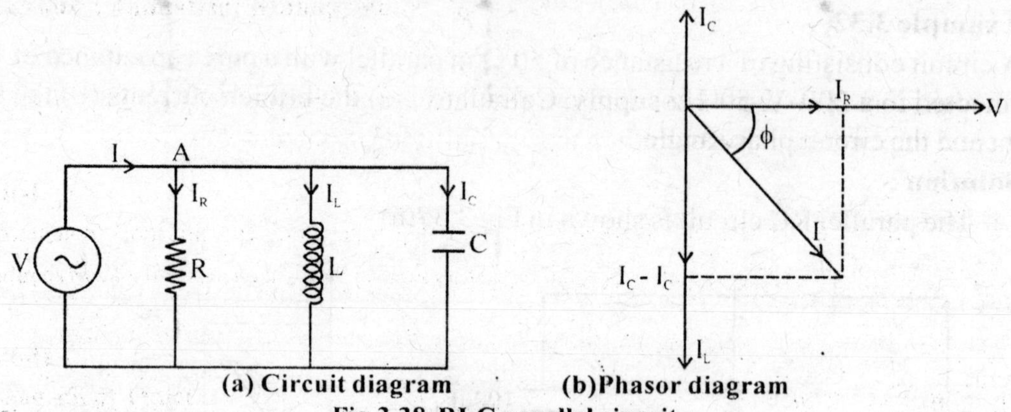

(a) Circuit diagram (b)Phasor diagram

Fig.3.38 RLC parallel circuit

Let the voltage phasor V be the reference phasor. Then, V = V $\angle$ 0° and

Phasor current through the resistance, $I_R = \dfrac{V\angle 0^o}{R} = \dfrac{V}{R}\angle 0$ (i)

Phasor current through the inductance, $I_L = \dfrac{V\angle 0^o}{j\omega L} = -j\dfrac{V}{\omega L} = \dfrac{V}{\omega L}\angle -90^o$ (ii)

Phasor current through the capacitance, $I_C = \dfrac{V\angle 0^o}{-jX_C} = j\dfrac{V}{X_C}$

$$= j\frac{V}{1/\omega C} = jV\omega C = V\omega C\angle 90^o \quad \text{(iii)}$$

It is evident that I_R is in phase with the voltage V, I_L lags the voltage V by 90°, while I_C leads the voltage V by 90°.

Applying the KCL at common node'A' , $I = I_R + I_L + I_C$ (iv)

Substituting Equations(i) , (ii) and (iii) into Equation (iv), we get:

$$I = \frac{V}{R} - j\frac{V}{\omega L} + jV\omega C$$

$$= V\left[\frac{1}{R} + j\left(\omega C - \frac{1}{\omega L}\right)\right]$$

The phasor diagram for the parallel RLC circuit, when $I_L > I_C$, is shown in Fig. 3.38(b)

It is observed that the current, I lags the applied voltage V by an angle ϕ. The angle ϕ is given by

$$\tan\phi = \frac{I_L - I_C}{I_R} = \frac{V[(1/\omega L) - \omega C]}{V/R} = R\left(\frac{1}{\omega L} - \omega C\right) \qquad = R\left(\frac{1}{X_L} - \frac{1}{X_C}\right)$$

It is evident from the phasor diagram that the current phasor I will lag or lead the voltage phasor V depending upon whether the magnitude of current phasor, $I_L > I_C$ or $I_C > I_L$, respectively.

Example 3.33

Three elements, a resistance of 50 Ω, an inductance of 0.2 H, and a capacitance of 300 μF, are connected in parallel to a 200-V, 50-Hz supply. Calculate (a) the current in each element, (b) the supply current, (c) the phase angle between the supply voltage and the supply current.

Solution :

The circuit diagram for the parallel branches is shown in Fig.3.39(a) Let the supply voltage be chosen as the reference phasor.

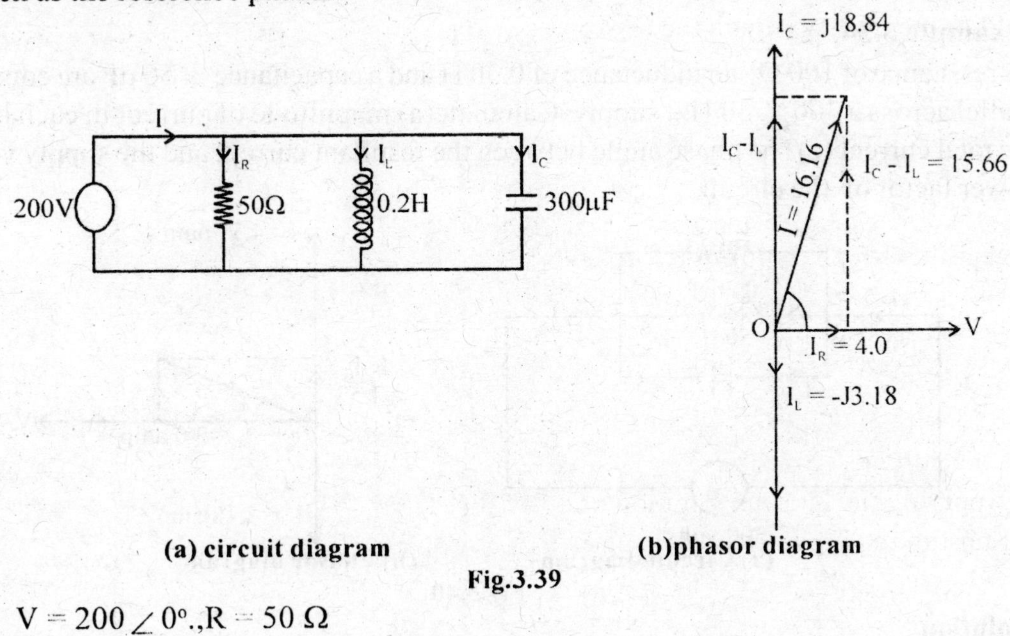

(a) circuit diagram (b)phasor diagram

Fig.3.39

V = 200 ∠ 0°., R = 50 Ω

$$jX_L = j\omega L = j314 \times 0.2 = j62.8\Omega = 62.8\angle 90^o$$

$$-jX_C = -j\frac{1}{\omega C} = \frac{-j}{314 \times 300 \times 10^{-6}} = -j10.615 = 10.615\angle -90^o$$

(a) Current in each element

$$I_R = \frac{200\angle 0^o}{50} = 4.0\angle 0^o = 4.0 + j0 \text{ amp Ans}$$

$$I_L = \frac{200\angle 0^o}{62.8\angle 90^o} = 3.1847\angle -90^o = -j\,3.1847 \text{ amp Ans}$$

$$I_C = \frac{200\angle 0^o}{10.615\angle -90^o} = 18.84\angle 90 = \text{j}18.84 \text{ amp} \quad \textbf{Ans}$$

(b) Supply current

$$I = I_R + I_L + I_C = 4.0 - j3.1847 + j18.84$$

$$= 4.0 + j15.66 = \sqrt{4^2 + 15.66^2}\angle \tan^{-1}\frac{15.66}{4} = 16.16\angle 75.67^o \text{ amp.} \quad \textbf{Ans}$$

(c) Phase angle between supply voltage and supply current

Phase angle = 75.67 leading. Ans

The phasor diagram is drawn choosing proper scales for currents and is shown in Fig.3.39(b)

Example 3.34.

A resistance of 100 Ω, an inductance of 0.30 H and a capacitance of 50 μF are connected in parallel across a 300 V, 50 Hz, supply. Calculate(a) magnitude of current in each branch (b)The total current (c)The phase angle between the resultant current and the supply voltage (d) Power factor of the circuit.

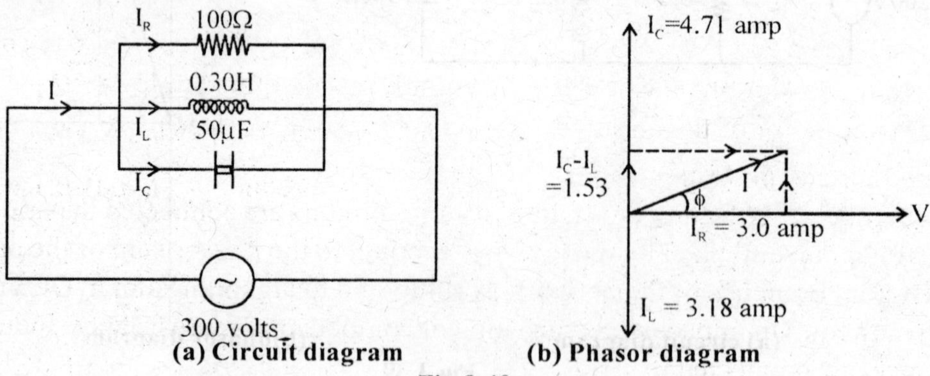

(a) Circuit diagram **(b) Phasor diagram**

Fig.3.40

Solution:

Resistance, R = 100 Ω

Inductive reactance, $X_L = 2\pi f L = 2\pi \times 50 \times 0.30 = 94.2\,\Omega$

Capacitive reactance, $X_C = \frac{1}{2\pi fC} = \frac{1}{2\pi \times 50 \times 50 \times 10^{-6}} = 63.6 \ \Omega$

(a) Current in each element

Current through resistance, $I_R = \frac{V}{R} = \frac{300}{100} = 3$ amp **Ans.**

Current through inductance, $I_L = \frac{V}{X_L} = \frac{300}{94.2} = 3.18$ amp

Current through capacitance, $I_C = \frac{V}{X_C} = \frac{300}{63.6} = 4.71$

Combining I_C and I_L which are in opposition and I_C I_L, we get $I_C - I_L = 4.71 - 3.18 = 1.53$ amp

(b) Supply current

Resultant current, which is phasor sum of I_R, I_C and I_L;

$I = \sqrt{(I_R)^2 + (I_C - I_L)^2} = \sqrt{(3)^2 + (1.53)^2} = 3.367$ amp. **Ans.**

(c) Phase angle

Phase angle $\Phi = \tan^{-1}\frac{I_C - I_L}{I_R} = \tan^{-1}\frac{4.71 - 3.18}{3} = \tan^{-1}\frac{1.53}{3} = \tan^{-1} 0.51 = 27^0$ **Ans.**

The phasor diagram is drawn choosing proper scales for currents and is shown in Fig.3.40(b)

(d) Power factor

It is seen that the current phasor leads the voltage phasor by an angle 27^0

Hence,Power factor of the circuit = cos = cos 27 ° = 0.89 (lead) **Ans.**

3.11.4 Impedances in Parallel

When a number of branches consisting of series circuits are connected in parallel in a network, the total current taken by the network is equal to the phasor sum of the currents drawn by different branches of the network as compared to algebraic sum in DC circuits. Each branch may contain impedances consisting of components like resistance, inductance and capacitance etc. connected in series.

The total current can be calculated by any of the following methods:

1. Phasor diagram Method/ Complex algebra Method using impedances.
2. Admittance Method .

The application of a particular method depends upon the problem under consideration

3.11.4.1 Phasor Diagram Method/Complex Algebra Method Using Impedances

Let us consider a parallel circuit consisting of two branches of impedances Z_1, Z_2 consisting of R,L,C elements connected in series respectively and connected in parallel across an alternating voltage of V volts, as shown in Fig3.41(a)

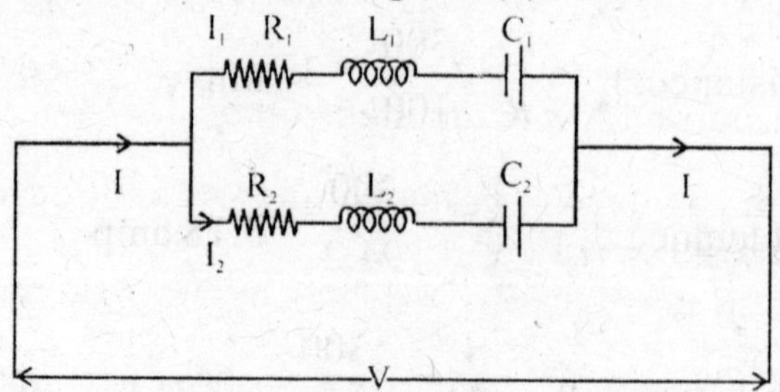

Fig 3.41(a)Parallel circuit

Since the branches are connected in parallel, therefore, the voltage across each branch is the same and is equal to supply voltage V, but currents through them will be different.

Let the current through branch '1' be I_1 and the current through branch '2' be I_2. The total current drawn from the supply is equal to the phasor sum of currents I_1 and I_2.

A **phasor diagram** can be drawn as shown in Fig.3.41(b), taking voltage 'V' as a reference phasorand choosing proper scales for the current phasors.

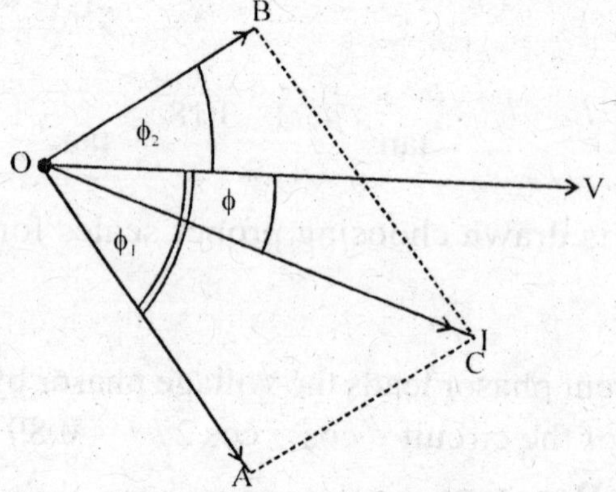

Fig 3.41(b)Phasor diagram

Branch currents I_1 and I_2 have been represented by the phasors OA at an angle, Φ_1 and OB at an angle, Φ_2 respectively with reference to voltage phasor 'V'

Phasor sum of OA and OB is given by phasor OC, which represents the total current ,'I' at an angle Φ w.r.t. voltage phasor'V'.

$\cos\Phi$ gives the value of p.f. of the circuit.

This method gives a clear insight in to the problem.

Branch currents I_1 and I_2 can be calculated as follws:

In Branch 1, (Assuming $X_{L1} > X_{C1}$)

Impedance of branch 1, $Z_1 = R_1 + j(X_{L1} - X_{C1}) = \sqrt{(R_1)^2 + (X_{L1} - X_{C1})^2}$

Branch Current , $I_1 = \dfrac{V}{Z_1}$

Phase angle of this current I_1 w.r.t. applied voltage is: $\Phi_1 = \tan^{-1}\dfrac{X_{L_1} - X_{C_1}}{R_1}$

In general , current I_1 will lead or lag behind the applied voltage V depending upon whether is +ve or – ve respectively.

In the phasor diagram shown in Fig.3.41(b), I_1 has been shown to be lagging the reference voltage V by an angle because X_{L1} has been assumed to be greater than X_{C1} and thus branch 1 is inductive.

Similarly in Branch 2 (Assuming $X_{C2} > X_{L2}$)

Impedance of branch 2, $Z_2 = \sqrt{(R_2)^2 + (X_{C_2} - X_{L_2})^2}$

Branch Current, $I_2 = \dfrac{V}{Z_2}$

Phase angle of this current I_2 w.r.t. applied voltage V is, $\Phi_2 = \tan^{-1}\dfrac{X_{C_2} - X_{L_2}}{R_2}$

In general , current I_2 will lead or lag behind the applied voltage V, depending upon whether is + ve or – ve respectively . I_2 has been shown to be leading the reference voltage V by an angle in the phasor diagram shown in Fig. 3.41(b),because $\mathbf{X_{C21}}$ has been assumed to be greater than $\mathbf{X_{L2}}$ and thus branch 2 is capacitive.

The phasor sum of currents can be obtained either graphically or by resolving branch currents I_1, I_2 in to horizontal and vertical components along X- axis and Y – axis and then determining the resultant of these components by combining them analytically as explained below.

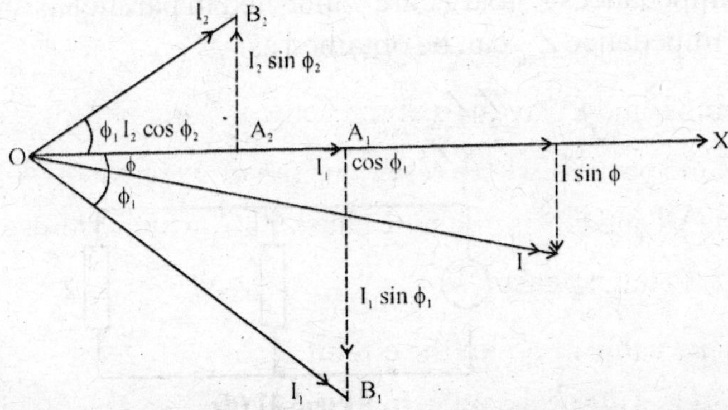

Fig 3.41(c)Resolution of branch currents along x-axis and y-axis

Component of resultant current I along X- axis, I cos Φ

= sum of components of branch currents I_1, I_2 along X – axis.

or $\quad I\cos\Phi = I_1 \cos\Phi_1 + I_2 \cos\Phi_2$ (3.30)

Similarly component of resultant current I along Y- axis, I sin

= sum of components of branch currents I_1, I_2 along Y – axis.

i.e. $\quad I\sin\Phi = -I_1 \sin\Phi_1 + I_2 \sin\Phi_2$ (3.31)

Resultant Current, $I = \sqrt{(I\cos\Phi)^2 + (I\sin\Phi)^2}$ (3.32)

Substituting equations (3.30) and (3.31)in equation (3.32),we get:

$$I = \sqrt{(I_1\cos\Phi_1 + I_2\cos\Phi_2)^2 + (-I_1\sin\Phi_1 + I_2\sin\Phi_2)^2}$$

and tan $\quad \Phi = \dfrac{I\sin\Phi}{I\cos\Phi}$

or (3.33)

If Φ is +ve , current I will lead the applied voltage V, if Φ is –ve , current I will lag behind the applied voltage V, Fig. 3.41(c)

Power factor of the overall circuit is:

$$\cos\Phi = \frac{\textit{Sum of active components of current}}{\textit{total resul} \tan t \textit{ current}}$$

or
$$\cos\Phi = \frac{I_1\cos\phi_1 + I_2\cos\phi_2}{I}$$

This method is fast and convenient but does not give clear insight in to the problem.

3.11.4.2 Simple Method When Only Two Impedances Z_1 and Z_2 are Connected in Parallel (Similar to DC Method for two Resistances Connected in Parallel)

If only two impedances Z_1 and Z_2 are connected in parallel as shown in Fig.3.41(d) then their equivalent impedance Z_{eq} can be obtained as:

$$Z_{eq} = \frac{Z_1 Z_2}{Z_1 \times Z_2}$$

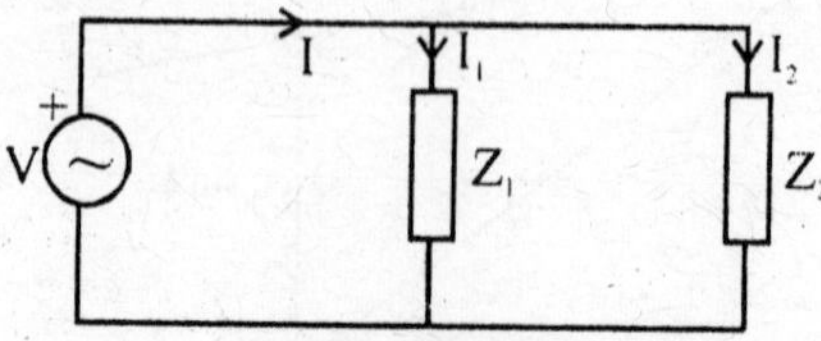

Fig.3.41 (d)

And the supply current ,I can be calculated as follows:

i.e. current, $I = \frac{V}{Z_{eq}}$

Example 3.35

Two circuits, the impedances of which are given by $Z_1 = (3 + j\,4)\,\Omega$ and $Z_2 = (6 - j\,8)\,\Omega$ are connected in parallel. If the potential difference across the impedances is $(230 + j\,0)$ V, Calculate (i) Total current and current supplied to each branch (ii) Total power consumed (iii) Overall p.f. and p.f. of each branch.

Solution.

$$Z_1 = (3 + j\,4) = 5, \text{ and } Z_2 = (6 - j\,8) = 10\angle -53.13^0$$

$$Z_1 + Z_2 = 9\text{-}j4 = 9.84\angle\text{-}23.96^0$$

(i)Branch currents and Total current

Current, $I_1 = \frac{V}{Z_1} = \frac{230\angle 0°}{5\angle 53.13} = 46\angle -53.13$ **Ans**

Resolving I_1 in to real and imaginary components, we get:

$$I_1 = 46(\cos 53.13\text{-}j\sin 53.13) = 46(0.59\text{-}j0.8) = 27.14\text{-}j36.8$$

Current, $I_2 = \frac{V}{Z_2} = \frac{230\angle 0°}{10\angle -53.13°} = 23\angle 53.13°$ **Ans**

Resolving I_2 in to real and imaginary components, we get:

$$I_2 = 23(\cos 53.13 + j\sin 53.13) = 46(0.59 + j0.8) = 13.57 + j18.4$$

Total current $I = I_1 + I_2 = 27.14\text{-}j36.8 + 13.57 + j18.4 = 40.71\text{-}j18.4$

$$I = \sqrt{1657.3 + 338.56} = \sqrt{1995.86} = 44.67 \text{ Amp}$$ **Ans**

Draw the phasor diagram as shown in Fig. 3.42, choosing proper scales and angles for the current phasors I_1 and I_2.

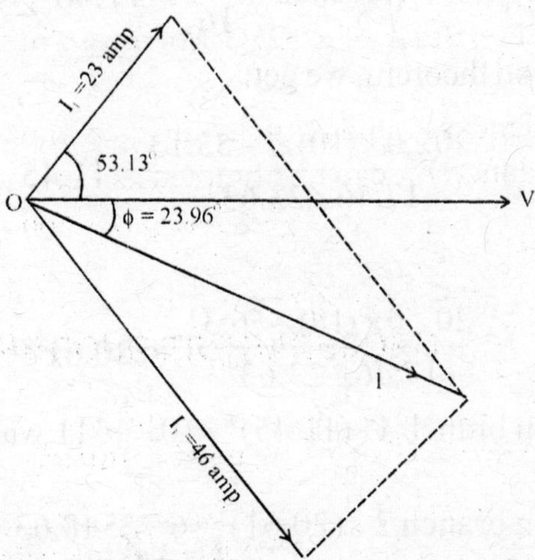

Fig.3.42 Phasor diagram for Example 3.20

Total current can be obtained either graphically as shown in the phasor diagram or analytically as shown below:

$$Z_{eq} = \frac{Z_1 Z_2}{(Z_1 + Z_2)} = \frac{(3+j4)(6-j8)}{(9-j4)} = \frac{(5\angle 53.13^0\ )(10\angle -53.13^0)}{9.84\angle -23.96} = 5.08\ \angle 23.96^0$$

$$\text{Total current, I} = \frac{V}{Z_{eq}} = \frac{230\angle 0}{5.08\angle 23.96°} = 45.27\angle -23.96° \text{ amp. Ans}$$

(ii) Total Power consumed

Total power, P = $I_1^2 R_1 + I_2^2 R_2 = (46)^2 \times 3 + (23)^2 \times (6) = 6348 + 3174 = 9522$ Watts Ans

(iii) Power factor

p.f. of branch 1 = cos (– 53.13°) = 0.60 (lag) Ans

p.f. of branch 2 = cos (53.13°) = 0.6 (lead) Ans

Overall p.f. = cos (–23.96°) = 0.91 (lag) Ans

Example 3.36

Two impedances $Z_1 = (10 + j\,15)\,\Omega$ and $Z_2 = (6\text{-}j8)\,\Omega$ are connected in parallel and supply current is 20 amp. What is the current and power dissipation in each branch.

Solution :

$$Z_1 = (10 + j\,15) = \sqrt{325}\ \tan^{-1}\frac{15}{10} = 18\ \angle 56.31^0$$

$$Z_2 = (6\text{-}j8) = \sqrt{100}\ \tan^{-1}\frac{-8}{6} = 10\angle \text{-}53.13^0$$

$$Z_1 + Z_2 = 16 + j\,7 = \sqrt{305 \tan^{-1}\frac{7}{16}} = 17.46\ \angle 23.63^0$$

Using current division theorem, we get:

$$I_1 = \frac{I Z_2}{Z_1 + Z_2} = \frac{20\angle 0 \times (10)\angle -53.13}{17.46\angle 23.63^0} = 11.45\ \angle 76.76^0 \quad \textbf{Ans}$$

$$I_2 = \frac{I Z_1}{Z_1 + Z_2} = \frac{20\angle 0 \times (18)\angle 56.31}{17.46\angle 23.63^0} = 20.61\ \angle\ 32.68^0 \quad \textbf{Ans}$$

Power dissipation in branch 1=$(11.45)^2 \times 10$= 1311 watts **Ans**

Power dissipation in branch 2 =$(20.61)^2 \times 6$= 2548.63 watts **Ans**

Alternative method

$$Z_{eq} = \frac{Z_1 Z_2}{Z_1 + Z_2} = \frac{(18\angle 56.31)(10\angle -53.13)}{17.46\angle 23.63} = 10.309\angle -20.45^0$$

$$V = I\,Z_{eq} = 20\angle 0 \times 10.309\angle -20.45^0 = 200.309\angle -20.45^0$$

$$I_1 = \frac{V}{Z_1} = \frac{200.309\angle -20.45}{18\angle 56.31} = 11.128\ \angle -76.76^0 \text{ Amp}$$ **Ans**

$$I_2 = \frac{V}{Z_2} = \frac{200.309\angle -20.45}{10\angle -53.13} = 20.0309\ \angle 32.68 \text{ Amp}$$ **Ans**

Example 3.37

Impedances $Z_1 = (60 - j80)\Omega$ and $Z_2 = (100 + j100)\Omega$ are connected in parallel across 220 V, 50 Hz supply. Find: (i) Branch currents, (ii)Total current, (iii) Total power.

Solution :

(i) Calculation of branch currents

$$Z_1 = 60 - j\,80 = 100\angle -\tan^{-1} 80/60 = 100\ \angle -\tan^{-1} 4/3 = 100\angle -53.13^0$$

$$Z_2 = 100 + j\,100 = \sqrt{100^2 + 100^2}\ \angle \tan^{-1} 100/100 = 141.42\angle 45^o$$

$$\text{Branch current } I_1 = \frac{V}{Z_1} = \frac{220\angle 0°}{100\angle -53.3°} = 2.2\angle 53.3^o$$ **Ans**

$$\text{and Branch current } I_2 = \frac{V}{Z_2} = \frac{220\angle 0°}{141.42\angle 45°} = 1.55\ -45° \text{ amp}$$ **Ans**

Draw the phasor diagram as shown in Fig. 3.43, choosing proper scales and angles for the current phasors I_1 and I_2.

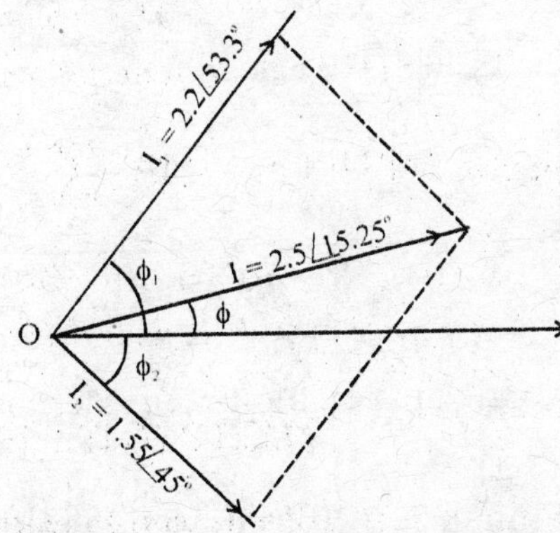

Fig.3.43 Phasor diagram for Example 3.21

(ii) Total current

Total current can be obtained either graphically as shown in the phasor diagram or analytically as shown below:

$$Z_1 + Z_2 = 160 + j\,20 = \sqrt{26000}\ \tan^{-1}\frac{20}{160} = 161.24\ \angle 7.12^0$$

$$\therefore \quad Z_{eq} = \frac{Z_1 Z_2}{Z_1 + Z_2} = \frac{(100\angle -53.13)(141.42)\angle 45^o}{161.24\angle 7.12^0}$$

$$\text{or} \quad Z_{eq} = \frac{(100)(141.42)\angle -8.13^o}{161.24\angle 7.12^o} = 87.707\ \angle -15.25^o$$

$$\therefore \quad \text{Total current, } I = \frac{V}{Z_{eq}} = \frac{220\angle 0^\circ}{87.707\angle -15.25^\circ} = 2.508\angle 15.25^\circ \text{ amp.}$$ **Ans**

Total current as seen from the phasor diagram = $2.508\angle 15.25^\circ$ **Ans**

(iii) Total power

p f = cos (15.25) = 0.964

$\therefore$ Total power, P = V I cos ϕ = 220 × 2.508 × cos 15.25 = 532.33 watts **Ans.**

3.11.4.3 Series Parallel Combination of Impedances

Several impedances may be connected in series - parallel combinations. In such cases, the complex impedances are expressed in both rectangular and polar forms. The rectangular form is used for addition and subtraction of impedances and the polar form is used for multiplication/division of impedances, as illustrated in the examples given below.

Example 3.39

For the following circuit, find: (i) the current delivered by the source, and (ii) the power factor, where, $Z_1 = 0 - j\,4\Omega$; $Z_2 = 0 - j\,4\Omega$; $Z_3 = 4 + j\,4$, V = 200 Volts

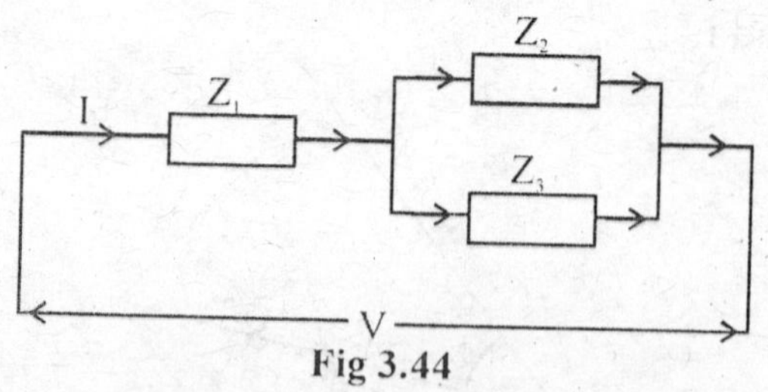

Fig 3.44

Solution

Impedances Z_2 and Z_3 are in parallel, their equivalent impedance Z_{eq} is:

$$\therefore \quad Z_{eq} = \frac{Z_2 Z_3}{Z_2 + Z_3} = \frac{(0 - j\,4)(4 + j\,4)}{0 - j4 + 4 + j4} = \frac{16 - j16}{4} = (4 - j4)$$

However, Impedances Z_1 and Z_{eq} are in series ,

$$Z_{total} = Z_1 + Z_{eq} = 0 - j\,4 + 4 - j\,4 = 4 - j\,8\ \sqrt{4^2 + 8^2}\ \tan^{-1} = (-8/4\,).$$

$$= 8.94\ \angle -63.43°$$

(i) Current, $I = V/\,Z_{total} = 200\ \angle 0°/8.94\ \angle -63.43 = 22.37\angle 63.43$ **Ans**

(ii) p.f., $\cos\phi = \cos 63.43 = 0.447$ (leading). **Ans**

Example 3.40

An inductive impedance BC is connected in series with a parallel combination of two branches across AB, consisting of a condenser, and non- inductive resistance. The circuit constants are so adjusted that the current in the two parallel branches of AB are equal, and that the voltage across AB is equal to, and in quadrature with the voltage across BC. When a voltage of 200 V is applied to AC, the total power absorbed is 1,200 W. Calculate the circuit constants R_1, X_C, R_2 and X_L

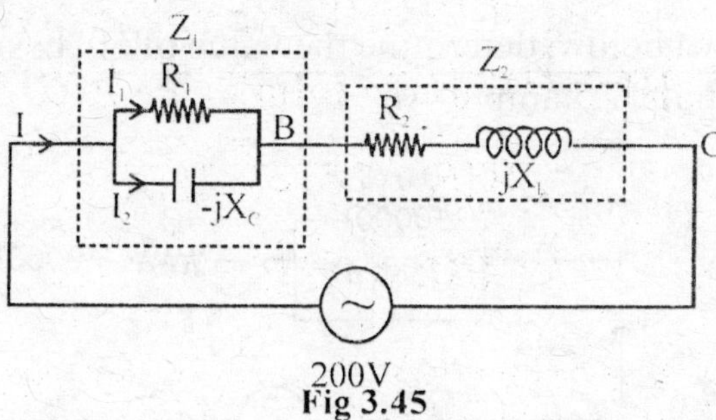

Fig 3.45

Solution:

As voltages across parallel elements are equal and it is given that currents in parallel branches are also equal,(i.e. $I_1 = I_2$)

$\therefore \quad R_1 = X_C.$

R_1 and $-jX_C$ are in parallel , hence their equivalent impedance Z_1 is :

$$\therefore \quad Z_1 = \frac{R_1 \times (-jX_C)}{R_1 - jX_C} = \frac{-jR_1^2}{R_1 - jR_1} = \frac{-jR_1^2(R_1 + jR_1)}{R_1^2 + R_1^2}$$

$$= \frac{R_1 - jR_1}{2} = \frac{R_1}{\sqrt{2}}\angle -45°\Omega$$

As voltage across BC is same and is in quadrature with the voltage across AB, therefore the two impedances are equal in magnitude but differ in phase by 90°

$$\therefore \quad Z_2 = \frac{R_1}{\sqrt{2}}\angle 45° = 0.5R_1 + j0.5R_1 = R_2 + jX_L$$

$R_2 = X_L = 0.5\ R_1$

$$Z = Z_1 + Z_2 = \frac{R_1 - jR_1}{2} + \frac{R_1 + jR_1}{2} = R_1$$

p.f., $\cos\phi = 1$, *and* $\phi = 0°$

$$\text{Current I} = \frac{P}{V\cos\phi} = \frac{1200}{200\times 1} = \mathbf{6\ amp.}$$

$I^2 R_1 = 1200$

$$R_1 = X_C = \frac{1200}{6\times 6} = 33.33\Omega$$ **Ans**

$X_L = 0.5\ R_1 = 16.66\ \Omega$ **Ans**

$R_2 = 16.66\Omega$ **Ans**

Example 3.41 .

In the circuit shown below, determine the value of 50Hz supply voltage that must be applied across AB so that I = 5 amp

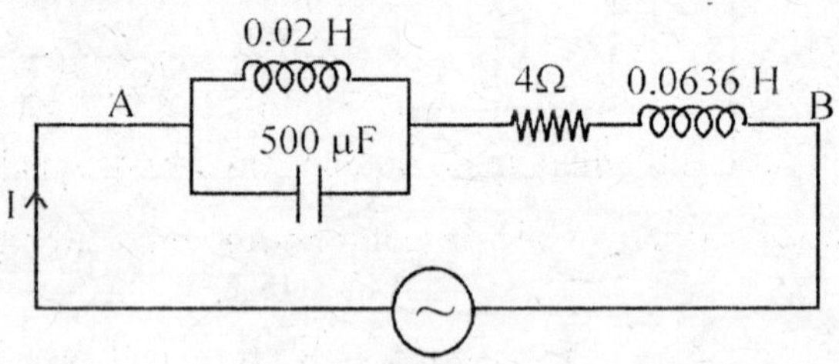

Fig 3.46

Solution:

Let $Z_1 = j\ X_L$ and $Z_2 = -j\ X_C$

Then , $Z_1 = j\,2\pi\times 50\times 0.02 = j6.28 = 6.28\angle 90°$

$Z_2 = -j/2\pi\times 50\times 500\times 10^{-6} = 6.366\angle -90°$

Z_1 and Z_2 are in parallel and their equivalent impedance Z_{eq} is:

$$\therefore Z_p = \frac{6.28\angle 90°\times 6.366\angle -90°}{j6.28 - j\,6.366} = \frac{39.97}{-0.086j} = j\,464.76 = 464.76\ \angle 90.$$

Let Z_3 be the series combination of resistance and inductance

Then $Z_3 = 4 + 2\pi\times 50\times 0.0636 = 4 + j\,20\ \Omega$

$\therefore Z_{eq} = j464.76 + 4 + j20 = 4 + j\,484.76 = 484.77\ \angle 89.5°\,\Omega$

Now current $I = 5\angle 0°$ amp

Hence, supply voltage,

$V = I Z_{total} = 5 \angle 0 \times 484.77 \angle 89.5 = 2423.85 \angle 89.5$ Volts **Ans**

Example 3.43

Determine the current in the circuit shown below, and also find the power consumed as well as power factor.

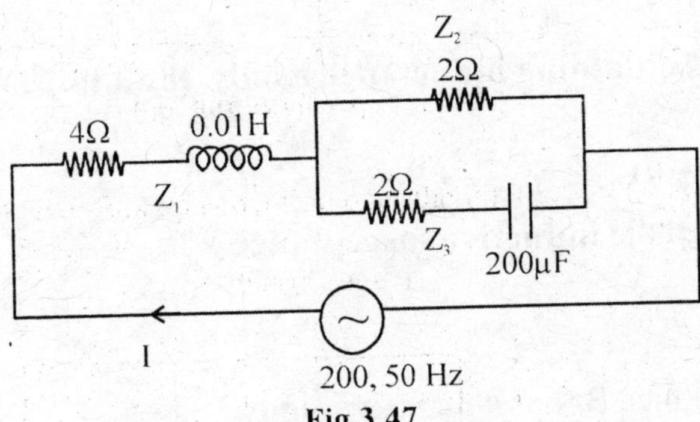

Fig 3.47

Solution.

$$Z_1 = 4 + j\,2\pi \times 50 \times 0.01 = (4 + j\,3.14) = 5.08 \angle 38.13^0 \Omega$$

$$Z_2 = 2\Omega$$

$$Z_3 = 2 - \left(j / 2\pi \times 50 \times 200 \times 10^{-6}\right) = 2 - j15.91\Omega = 16.04\angle 82.84°\Omega$$

$$\therefore\ Z_{eq} = \frac{Z_2 Z_3}{Z_2 + Z_3} = \frac{2 \times 16.04\angle -82.84°}{(2 + j0) + (2 - j15.91)} = \frac{32.08\angle -82.84}{4 - j15.91} = \frac{32.08\angle -82.84}{16.4\angle -75.88}$$

$$= 1.956\angle -6.96^0$$

$$\therefore\ Z_{total} = Z_1 + Z_{eq} = 4 + j\,3.14 + 1.936 - j0.234 = 5.936 + j2.906 = 6.60\angle 26.17°$$

(i) $I = \dfrac{V}{Z_{eq}} = 200\ \angle 0° / 6.6\angle 26.17° = 30.3\angle -26.17^0$ Ans.

(ii) p.f. = cos (-26.17°) = 0.897 (lagging) Ans.

(iii) Power= $V\,I\cos\phi = 200 \times 30.3 \times 0.897 = 5435.82$ watts Ans.

3.11.4.4 ADMITTANCE METHOD FOR IMPEDANCES CONNECTED IN PARALLEL

Most engineers think in terms of impedance. However, the use of admittance simplifies some problems. Hence, various forms of admittance and their interpretation is given below:

Concept of Admittance

The reciprocal of Impedance is known as Admittance and is represented /denoted by English capital letter **Y** and expressed in **Siemens (S).** It is a complex quantity

One siemen is defined as the admittance of a network having an impedance of one ohm.

An admittance has two components known as **Conductance, G and Susceptance, B** respectively. The unit of conductance and susceptance is **Siemens** (S)

Admittances of various circuit elements and their series combinations are as given below:

(i)For a circuit containing resistance R only, the admittance is :

$$Y_R = \frac{1}{Z_R} = \frac{1}{R} = G$$

(ii)For a circuit containing inductance L only, the admittance is :

$$Y_L = \frac{1}{Z_L} = \frac{1}{jX_L} = \frac{-j}{X_L} = -j\,B_L$$

Where B_L is called the **inductive susceptance.**

$$B_L = \frac{1}{X_L}$$

Inductive susceptance B_L has a **negative sign**

(iii)For a circuit containing capacitanceC, only, the admittance is :

$$Y_C = \frac{1}{Z_C} = \frac{1}{-jX_C} = \frac{j}{X_C} = j\,B_C$$

Where B_C is called the capacitive **susceptance.**

$$B_C = \frac{1}{X_C}$$

Capacitive susceptance B_C has a **positive sign**

(iv)Admittance of a Series RL circuit

$$Y = \frac{1}{Z} = \frac{1}{R + jX_L} = \frac{R - jX_L}{(R + jX_L)(R - jX_L)}$$

$$= \frac{R - jX_L}{(R^2 + X_L^2)}$$

$$= \frac{R}{(R^2 + X_L^2)} - \frac{jX_L}{(R^2 + X_L^2)}$$

$$= G - jB_L$$

Where conductance, $G = \frac{R}{(R^2 + X_L^2)} = \frac{R}{Z^2}$

Thus conductance is defined as the ratio of the resistance R and the square of the impedance Z^2 and is measured in Siemens.

and Susceptance $B_L = \frac{X_L}{(R^2 + X_L^2)} = \frac{X_L}{Z^2}$

Thus Susceptance is defined as the ratio of the reactance and the square of the impedance Z^2 and is measured in Siemens.

The susceptasnce can be inductive or capacitive depending on whether thre reactance is inductive or capacitive respectively.

(v)Admittance of a Series RC circuit

$$Y = \frac{1}{Z} = \frac{1}{R - jX_C} = \frac{R + jX_C}{(R - jX_C)(R + jX_C)}$$

$$= \frac{R + jX_C}{(R^2 + X_C^2)}$$

$$= \frac{R}{(R^2 + X_C^2)} + \frac{jX_C}{(R^2 + X_C^2)}$$

$$= G + jB_C$$

Where conductance, $G = \frac{R}{(R^2 + X_C^2)} = \frac{R}{(Z^2 + X_C^2)}$

and Susceptance $B_C = \frac{X_C}{(R^2 + X_C^2)} = \frac{X_C}{Z^2}$

Admittance triangle

The three sides of an admittance triangle shown in Fig. 3.48 indicate the conductance, suseptance and admittance. The conductance component is shown in the reference direction. The susceptance B can be either inductive, shown in the –ve direction or capacitive shown in the +ve direction.

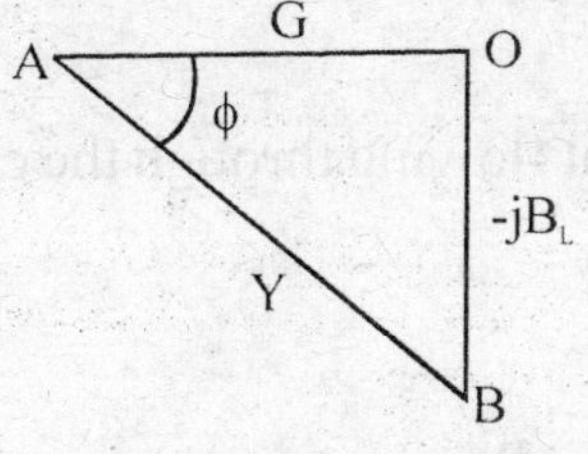

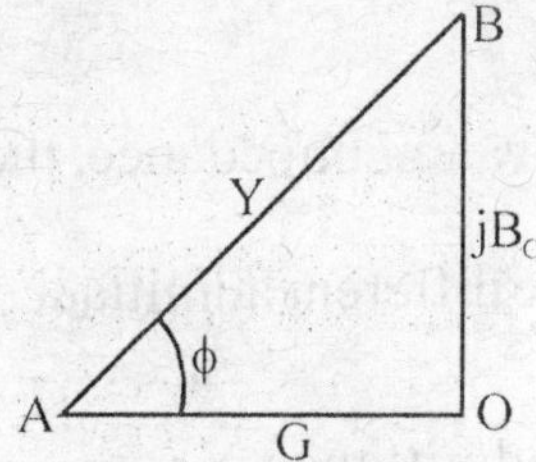

Admittance triangle **Admittance triangle**

(for series RL circuit) (for series RC circuit)

Fig 3.48

3.12 Step by Step Procedure to Solve Parallel Circuits Using Admittance Method

1. First determine conductance and susceptance of each branch from the relation

$$G_i = \frac{R}{Z^2} \text{ and } B_i = \frac{X}{Z^2},$$

Inductive susceptance, $B_L = \dfrac{X_L}{Z^2}$

Capacitive susceptance, $B_C = \dfrac{X_C}{Z^2}$

2. Find the algebraic sum of conductances and susceptances of individual branches respectively.

$$G = \Sigma G_i \qquad B = \Sigma B_i$$

3. Calculate the total admittance of the circuit, $Y = \sqrt{G^2 + B^2}$
4. Find the branch currents and total current in the circuit from the relation :

$I = V \times Y$ having phase angle $\phi = \tan^{-1}\dfrac{B}{G}$

Capacitive susceptance will be taken as +ve and inductive susceptance as –ve . If ϕ is + ve current will lead and if ϕ is –ve current will lag behind the applied voltage

Fig 3.48(b) shows two impedances connected in parallel.

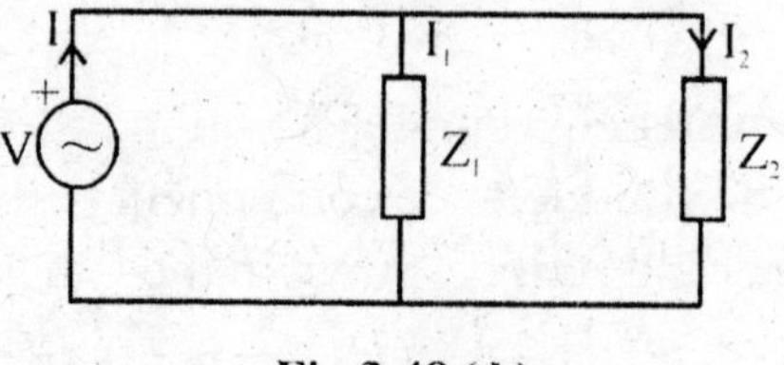

Fig.3.48 (b)

If Z_{eq} is equivalent impedance, then total current flowing through the circuit $I = \dfrac{V}{Z_{eq}}$

We can write different admittances as follows:

and Total admittance, $Y_{eq} = \dfrac{1}{Z_{eq}} = Y_1 + Y_2$ (Phasor sum)

Resultant current, $I = V \times \frac{1}{Z_{eq}} = V \times Y_{eq}$

Three impedances connected in parallel across an AC voltage source V are shown in Fig. 3.48(c)

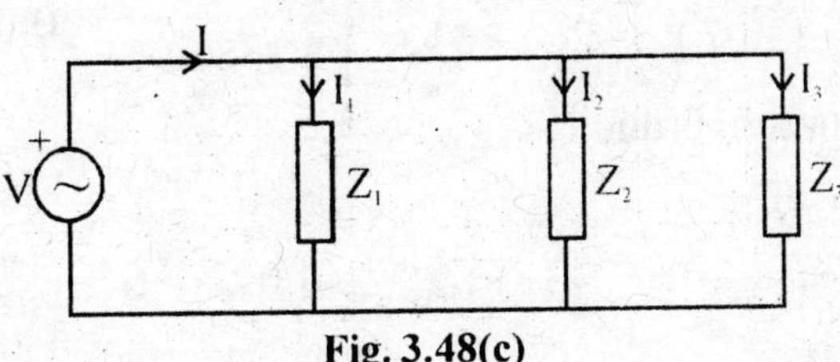

Fig. 3.48(c)

Applying KCL, we get

$I = I_1 + I_2 + I_3$ (i)

$I_1 = Y_1 \times V \quad I_2 = Y_2 \times V \quad I_3 = Y_3 V$ (ii)

where $Y_1 = 1/Z_1$, $Y_2 = 1/Z_2$

$Y_3 = 1/Z_3$ are the complex admittances

Substituting the values of I_1, I_2 and I_3 from equation (ii) in equation (i), we get:

$I = (Y_1 + Y_2 + Y_3)\, V$

i.e. $I = Y_{eq}\, V$

where $Y_{eq} = (Y_1 + Y_2 + Y_3)$

Examples given below illustrate the admittance method for the solution of parallel AC circuits

Example 3.38

For the parallel circuit shown in the Fig 3.49q given below, the parameter value are:

$X_C = 30$ (at 50 Hz) , $R_2 = 40$

Coil : $R_C = 20$; $L_C = 0.1$ H;

Determine (in magnitude and phase) the branch currents and the total current.

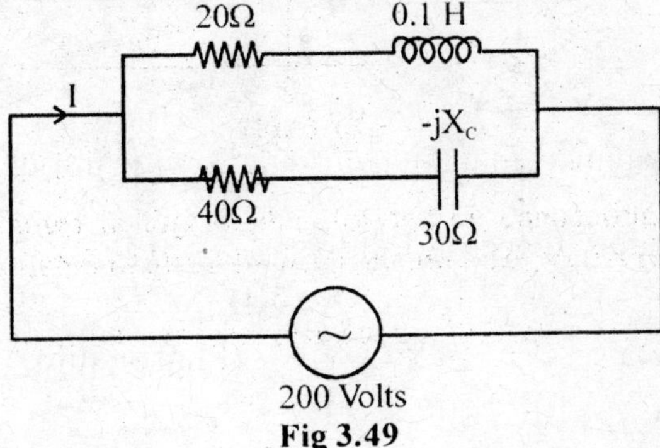

Fig 3.49

Solution:

Resistance of the inductive branch = 20 Ω

Reactance of inductive branch = 2 π f L = 2π × 50 × 0.1 = 31.4

Conductance of inductive branch,

$$G_1 = \frac{R_1}{(R_1)^2 + (2\pi fL)^2} = \frac{20}{20^2 + 31.4^2} = \frac{20}{1385.96} = 0.0144 \text{ Siemens}$$

Susceptance of inductive branch,

$$B_1 = \frac{2\pi fL}{(R_1)^2 + (2\pi fL)^2} = \frac{31.4}{1385.96} = -0.0226 \text{ S} \qquad \text{(-ve sign being inductive)}$$

Resistance of capacitive branch, $R_2 = 40$, Reactance of capacitive branch, $X_C = 30$

Conductance of capacitive branch, $G_2 = \dfrac{R_2}{R_2^2 + X_C^2} = \dfrac{40}{40^2 + 30^2} = 0.016$ S

Susceptance of capacitive branch, $B_2 = \dfrac{X_C}{R_2^2 + X_C^2} = \dfrac{30}{40^2 + 30^2} = 0.012$ S

(+ve sign, being capacitive)

Total conductance of circuit, $G = G_1 + G_2 = 0.0144 + 0.016 = 0.0304$ S

$B = B_1 + B_2 = -0.0226 + 0.0120 = -0.0106$ S

$$Y = \sqrt{G^2 + B^2} = \sqrt{(0.0304)^2 + (0.0106)^2} = \sqrt{0.00091 + 0.0001} = \sqrt{0.0010} = 0.031$$

I = VY = 200(0.031) = 6.32 amp. **Ans.**

Example 3.39 Find the current, and power supplied by the 200V source for the circuits shown below, using admittance method.

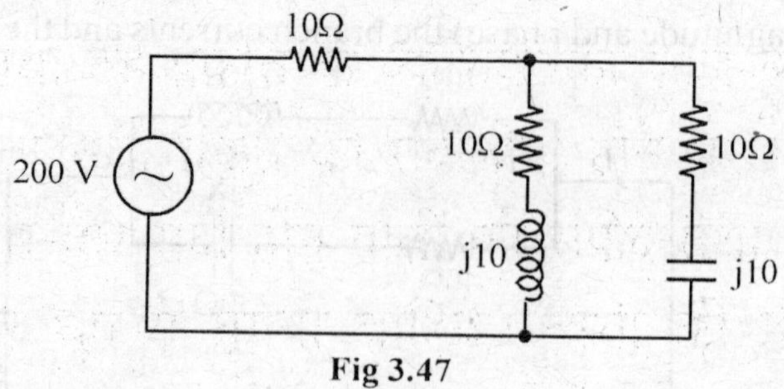

Fig 3.47

Solution:

$$Z_1^2 = R_1^2 + X_{L1}^2 = 10^2 + 10^2 = 200$$

$$G_1 = R_1/Z_1^2 = 10/200 = 0.05$$

$$B_1 = X_{L1}/Z_1^2 = 10/200 = 0.05$$

$$Z_2^2 = R_2^2 + X_{L2}^2 = 10^2 + 10^2 = 200$$

$$G_2 = R_2/Z_2^2 = 10/200 = 0.05;$$

$$B_2 = -X_{C2}/Z_2^2 = -10/200 = -0.05;$$

$$\therefore G_{12} = G_1 + G_2 = 0.05 + 0.05 = 0.10$$

$$B_{12} = B_1 + B_2 = 0.05 - 0.05 = 0$$

$$Y_{12} = \sqrt{G_{12}^2 + B_{12}^2} = \sqrt{0.10^2 + 0^2} = 0.10$$

or $$Z_{12} = 1/Y_{12} = 1/0.10 = 10\ \Omega$$

$$R_{12} = G_{12} \times Z_{12}^2 = 0.10 \times (10)^2 = 10\Omega$$

$$X_{12} = B_{12} \times Z_{12}^2 = 0 \times (10)^2 = 0$$

$$\therefore R_{eq} = 10 + 10 = 20\ \Omega$$

$$X_{eq} = 0 + 0 = 0\Omega$$

$$Z_{eq} = 20\ \Omega$$

(i) Current, $I = V/Z_{eq} = 200/20 = 10$ amp Ans

Power, $P = I^2 R = (10)^2\ 20 = 2000$ W. Ans

3.13 Resonance in AC Circuits

In circuits containing R,L,C elements , voltage and current outputs are quite different at different frequencies. These outputs may be greater over a small range of frequencies as compared to all other frequencies.

Resonance is a condition in a circuit, which is observed at a particular frequency , when the effects of inductive reactance and capacitive reactance cancel each other and the supply voltage and supply current are in time phase with each other in spite of the presence of energy storing elements, i.e. inductance and capacitance.

Resonance cannot take place when only one type of energy-storage element is present. There must exist at least two types of independent energy storing elements capable of interchanging energy between one another

3.13.1 Series Resonance/Resonance in Series R-L-C Circuits

A series R, L ,C circuit with a constant voltage and a variable frequency supply is shown in Fig 3.51

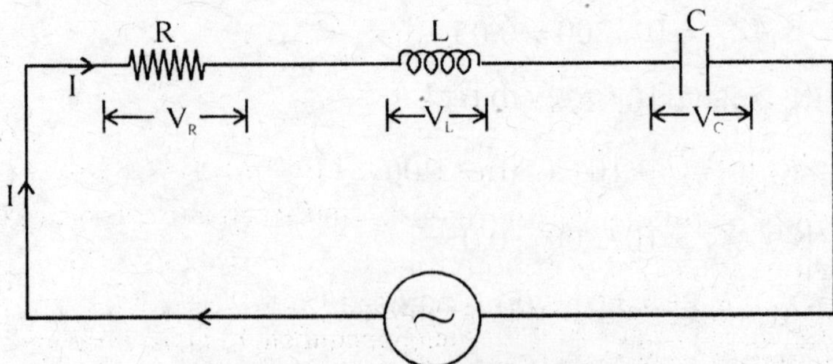

Fig.3.51: A Series RLC circuit

3.13.1.1 Expression for Resonant frequency

When the supply frequency is varied, the inductive reactance and capacitive reactance vary. At a certain frequency , the inductive reactance and capacitive reactance may become equal and thus cancel each other, making the circuit purely resistive.

As the impedance of the circuit changes with the change in frequency, the current in the circuit changes.

For an applied sinusoidal voltage **V**, having 'ω' as the angular frequency (in radians / second.), the phasor current 'I', is given by :

$$I = \frac{V}{Z} = \frac{V}{R + j\left(\omega L - \frac{1}{\omega C}\right)} \tag{3.34}$$

At resonance, the applied voltage and current must be in phase. This requires that impedance Z should be equal to resistance, R, i.e. imaginary part of the impedance should be zero.

i.e. $$\omega_r L - \frac{1}{\omega_r C} = 0$$

where , ω_r is the resonant frequency.

or $$\omega_r L = \frac{1}{\omega_r C}$$

or $$\omega_r^2 = \frac{1}{LC}$$

i.e., At resonance, $$\omega_r = \frac{1}{\sqrt{LC}} \tag{3.35}$$

But $\omega_r = 2\pi f_r$

Therefore, Resonant frequency in Hz, $f_r = \dfrac{1}{2\pi\sqrt{LC}}$ (3.36)

This equation shows that in a series R L C circuit, resonance can be produced :

1. By varying the frequency for fixed values of L and C.
2. By varying either L or C or both for a fixed frequency.

For fixed values of L and C and by varying the frequency of supply, the condition for resonance is given by :

$$f_r = \frac{1}{2\pi\sqrt{LC}}$$

Thus **inductive reactance** $\mathbf{X_L}$ is equal to the **capacitive reactance** $\mathbf{X_C}$ **at resonant frequency.**

3.13.1.2 Reactance Curves

A plot of **reactance** versus **frequency** is called a **reactance curve.** In a series R L C circuit, three curves can be drawn for the following cases as shown in Fig 3.52(a), (b)and (c)

(i) Inductive reactance versus frequency. Fig3.52(a)

(ii) Capacitive reactance versus frequency and Fig3.52(b)

(i) The total net reactance versus frequency Fig 3.52(c).

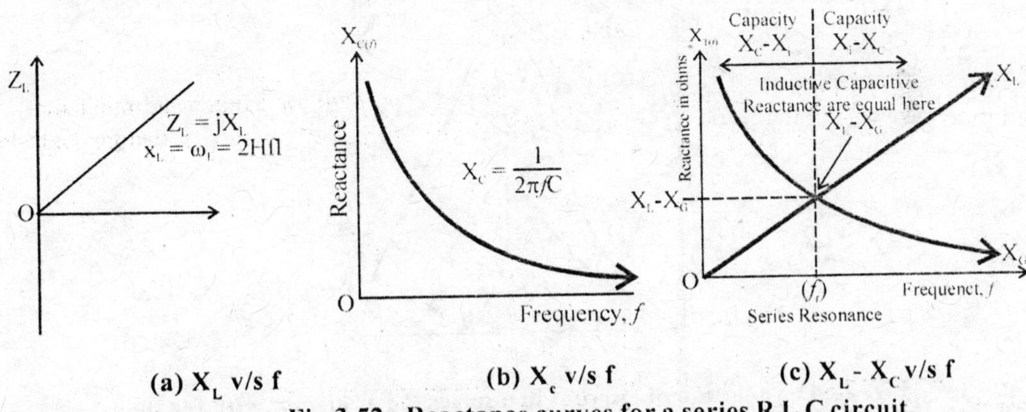

(a) X_L v/s f **(b) X_c v/s f** **(c) X_L - X_C v/s f**

Fig.3.52 : Reactance curves for a series R L C circuit.

It can be seen from Fig 3.52(c) that for any frequency f **lower** than f_r, inductive reactance $\mathbf{X_L}$ is less than capacitive reactance $\mathbf{X_C}$ i.e. capacitive reactance dominates over inductive reactance and hence the circuit behaves as a **capacitive circuit.**

For any frequency **higher** than resonant frequency f_r, inductive reactance is greater than capacitive reactance and the circuit behaves as an **inductive circuit.**

At the resonance frequency,inductive reactance and capacitive reactance cancel each other.

3.13.1.3 Phasor Diagram

Figure 3.53 shows phasor diagram for the series RLC circuit at $f = f_r$ with L and C kept constant.

Fig.3.53 Phasor diagram of RLC circuit at $f = f_r$

At resonance ($f = f_r$), the impedance of the circuit is equal to R. Consequently, the current is a maximum and is in time phase with voltage. The power factor is unity. The expression for the current phasor is :

$$I_r = \frac{V}{Z} = \frac{V}{R} \angle 0^o$$

Where I_r denotes the current at resonance.

Variation of current I with frequency for two different values of resistances R_1 and R_2(R_1 > R_2)is shown in Fig.3.54

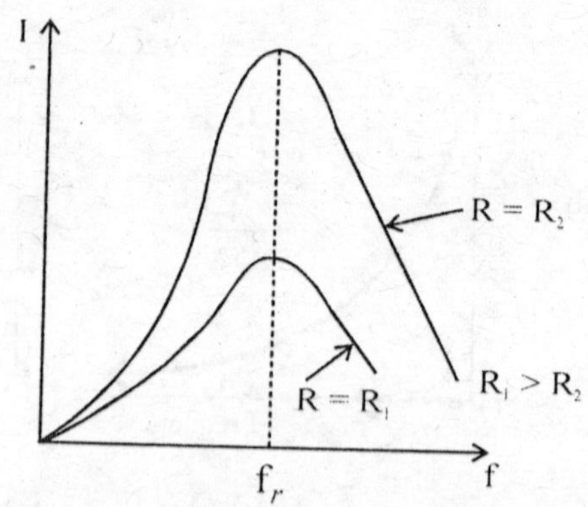

Fig. 3.54 : Variation of current in a series R L C circuit with frequency.

3.13.1.4 Quality factor of a Series Circuit, Q

It is a factor to determine the quality of a network and is defined as the ratio of reactive power to the resistive power.

$$\text{Q factor} = \frac{\text{reactive power}}{\text{resistive power}} = \frac{I^2 X_L}{I^2 R} = \frac{X_L}{R} \qquad (3.37)$$

It can also be defined as the ratio of inductive or capacitive reactance at resonance to the resistance of the circuit.

i.e. $Q = \frac{X_L}{R} = \frac{2\pi f_r L}{R}$ or $Q = \frac{X_C}{R} = \frac{1}{2\pi f_r C R}$ (3.38)

Q- factor remains constant over a wide range of frequencies. At high frequencies ,resistance changes due to skin effect and the ratio of X_L and R remains almost constant.

Q factor is also defined as **Voltage Amplification/ Magnification** at resonance

$$\textbf{Voltage Magnification} = \frac{V_C}{V} = \frac{V_L}{V} = \frac{\text{Voltage across L or C}}{\text{Supply voltage}}$$

$$= \frac{I_r X_L}{I_r R} = \frac{X_L}{R}$$

At the resonant frequency f_r, the voltage across an inductance is :

$V_L = I_r X_L$ (3.39)

However at resonance frequency, current in the circuit, $I_r = \frac{V}{R}$

Substituting this value of I_r in equation (3.39), we get:

Therefore, $V_L = \left(\frac{V}{R}\right) X_L = \frac{V}{R} \times 2\pi f_r L$

Substituting the value of f_r, we get:

$$V_L = \frac{V}{R}(2\pi L) . \frac{1}{2\pi\sqrt{LC}}$$

Or $V_L = \frac{V}{R}\sqrt{\frac{L}{C}}$ (3.40)

Similarly, the voltage across capacitance, C at resonance frequency f_r is :

$$V_C = I_r X_C = \frac{V}{R} . \frac{1}{2\pi f_r C}$$

Substituting the value of f_r, we get:

$V_C = \frac{V}{R} . \frac{1}{2\pi C} . \frac{2\pi\sqrt{LC}}{1} = \frac{V}{R}\sqrt{\frac{L}{C}}$ (3.41) which is equal to V_L as given by equation (3.40)

Voltage magnification

It can be seen from equations(3.40) and (3.41) ,that

$$\frac{V_C}{V} = \frac{V_L}{V} = \frac{1}{R}\sqrt{\frac{L}{C}} = Q \quad (3.42)$$

At resonant frequency $V_L = V_C = V \times$ Q- factor (3.43)

Hence voltage across L and C can be extremely large

Q factor is a measure of selectivity or sharpness of tuning of the series RLC circuit.

Since the impedance offered by a series resonant circuit is minimum , therefore it is called an **acceptor** circuit

3.13.1.5 Bandwidth of a Series Resonant circuit (B)

The concept of resonance is useful in discriminating a range of frequencies from the other frequencies. The circuit gives maximum response in a range around the resonant frequency and this range of frequencies is known as a **useful band** of frequencies.

An important parameter related with resonant circuits is **Bandwidth** and is denoted by **B**.

Band width of a series resonant circuit is a band of frequencies , which lies between two frequencies on either side of resonant frequency , where the current falls to or 0.707 of its maximum value at resonance and for which the power delivered to resistance R is greater than or equal to $P_r/2$ where P_r is the power delivered to R at resonance

As the current at resonance is maximum, the power dissipated in the resistance is maximum at resonance.

Consider the current-versus frequency curve shown in Fig. 3.55. It is clear from the curve that the current reaches a maximum level (I_r) at resonance and at frequencies close to resonance, the current level is only a little below its maximum value. Thus, the resonant circuit is said to select a band of frequencies rather than just one frequency. The lowest and highest frequencies of the band are identified as f_1 and f_2, respectively, and the bandwidth is B = . (Fig. 3.55).

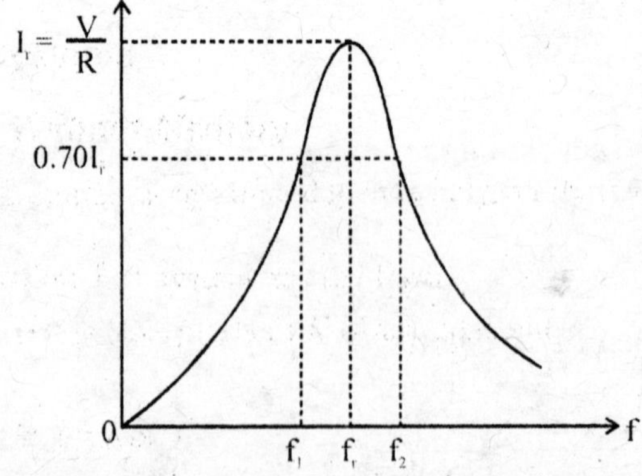

Fig.3.55 : Band width of a series resonant circuit

The frequencies f_1 and f_2 correspond to those frequencies at which the power delivered to the circuit is half the value of power which is delivered at resonance.

Hence, f_1 and f_2 are termed as **half- power points**. The power P_r delivered to the circuit at resonance is

$$P_r = I_r^2 R$$

Thus, value of half- power at resonance is :

$$\frac{1}{2}P_r = \frac{I_r^2 R}{2} \tag{3.44}$$

It can be rewritten as :

$$\frac{1}{2}P_r = \left(\frac{I_r}{\sqrt{2}}\right)^2 R \tag{3.45}$$

Equation (3.45) shows that at the half- power points, the current, I in the series RLC circuit is :

$$I = \frac{I_r}{\sqrt{2}} = 0.707 I_r \tag{3.46}$$

From equation (3.34), magnitude of the current in this circuit is:

$$|I| = \frac{V}{\sqrt{R^2 + (X_L - X_C)^2}}$$

Taking out 'R' from the denominator ,this equation can be rewritten as:

$$|I| = \frac{V}{R\sqrt{1 + \left(\frac{X_L - X_C}{R}\right)^2}}$$

We know that, $\frac{V}{R} = I_r$,

Substituting this in the above equation, we get:

$$|I| = \frac{I_r}{\sqrt{1 + \left(\frac{X_L - X_C}{R}\right)^2}} \tag{3.47}$$

From Equation (3.46), current I, at the half-power points is:

$$I = \frac{I_r}{\sqrt{2}}$$

Therefore, **at the half power frequencies (f_1 and f_2)**, equating the quantities under the square root in equations (3.46) and (3.47), we get: .

$$1 + \left(\frac{X_L - X_C}{R}\right)^2 = 2$$

or
$$\frac{X_L - X_C}{R} = \sqrt{2-1} = 1$$

i.e. $X_L - X_C = R$ at half power frequencies (3.48)

At f_r, $X_L - X_C = 0$. Therefore, when the frequency increases from f_r to f_2, X_L must increase by R/2 and X_C must decrease by R/2, to satisfy equation (3.48).

$$\text{i.e. } 2\pi f_2 L - 2\pi f_r L = \frac{1}{2} R$$

$$\text{or } 2\pi L(f_2 - f_r) = \frac{1}{2} R$$

$$\text{i.e. } f_2 - f_r = \frac{R}{4\pi L} \tag{3.49}$$

Similarly, when the frequency decreases from f_r to f_1, X_C must increase by R/2 and X_L must decrease by R/2. Consequently,

$$\text{i.e. } 2\pi f_r L - 2\pi f_1 L = \frac{1}{2} R$$

or
$$2\pi L(f_r - f_1) = \frac{1}{2} R$$

$$\text{i.e. } f_r - f_1 = \frac{R}{4\pi L} \tag{3.50}$$

Adding Equation (3.49) to Equation (3.50), we get:

$$f_2 - f_1 = \frac{R}{4\pi L} + \frac{R}{4\pi L} = \frac{R}{2\pi L}$$

Hence, the circuit bandwidth is :

$$B = f_2 - f_1 = \Delta f = \frac{R}{2\pi L} \tag{3.51}$$

or $\frac{1}{2}\Delta f = \frac{R}{4\pi L}$

From equations (3.49),(3.50) and (3.51),we can also write:

$$f_1 = f_r \times \frac{1}{2}\Delta f$$

and $f_2 = f_r \frac{1}{2} + f$

Multiplying the right- hand side of Equation (3.51) by f_r/f_r gives

$$B = \Delta f = \frac{R\, f_r}{2\pi\, f_r\, L}$$

Substituting the value of Q = in the equation given above, we get:

i.e. $$B = \Delta f = \frac{f_r}{Q} = \frac{resonant\ frequency}{quality\ factor} \qquad (3.52)$$

Q is defined as the ratio of the resonant frequency to the band width.i.e. $Q = \frac{f_r}{\Delta f}$

If Q is high, band width will be small for a given resonant frequency. i.e. if we want to select a particular frequency only and reject all other frequencies from a signal, then Q should have a high value.

3.13.1.6 Selectivity

The ability of a resonant circuit to select one particular frequency and to discriminate against other frequencies is termed the **selectivity** of the circuit.

Selectivity is defined as the ratio of band width to resonant frequency and is given by:

$$\text{Selectivity} = \frac{Band\ width}{Resonance\ frequency} = \frac{1}{Q}$$

Example 3.44

Calculate the half – power frequencies and the bandwidth for the series RLC circuit shown in Figure 3.56. Given, R = 10 ohm, L = 50×10^{-6} H and C = 1000×10^{-12} Farad

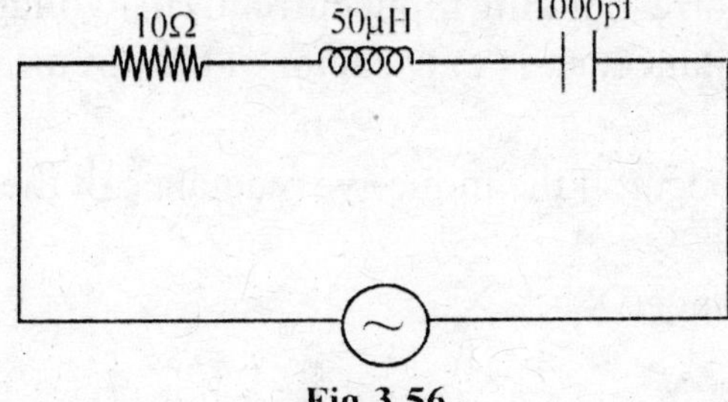

Fig.3.56

Solution :

$\therefore$ Resonant frequency, $f_r = \dfrac{1}{2\pi\sqrt{LC}} = \dfrac{1}{2\pi\sqrt{50\times10^{-6}\times1000\times10^{-12}}}$

i.e. $\mathbf{f_r = 711.7\ kHz}$ **Ans**

Q-factor $= \dfrac{\omega_r L}{R} = \dfrac{2\pi\times711.7\times10^3\times50\times10^{-6}}{10}$ (from equation 3.38)

i.e. $Q = 22.35$

Upper half power frequency, $f_2 = \dfrac{R}{4\pi L} + f_r$ (refer equation 3.49)

$= \dfrac{10}{4\pi\times50\times10^{-6}} + 711.7\ KHz = 15.91 + 711.7 = 727.61\ \text{kHz}$ **Ans**

Lower half power frequency $f_1 = f_r - \dfrac{R}{4\pi L}$ (refer equation 3.50)

or $f_1 = 711.7 - \dfrac{10}{4\pi\times50\times10^{-6}} = 711.7 - 15.91 = 695.79\ \text{kHz}$ **Ans**

Band width, $\Delta f = f_2 - f_1 = 727.61 - 695.79 = 31.82\ \text{kHz}$ **Ans**

Alternatively

Band width, $\Delta f = \dfrac{f_r}{Q}$ (refer equation 3.52)

Substituting the values of f_r and Q, we get:

Band width, $\Delta f = \dfrac{711.7}{22.35} = 31.82\ \text{kHz}$ **Ans**

Example 3.45

A circuit having a resistance of 4 Ω, an inductance of 0.2 H and a variable capacitance in series, is connected across a 100 V, 50 Hz supply. Calculate (i) The value of capacitance to produce resonance, (ii)Corresponding input current, (iii) Voltage across the inductance, (iv) Voltage across the capacitance, and (v) *Q*-factor of the circuit.

Solution :

(i) Resonance will occur, if the inductive reactance of the circuit equals the capacitive reactance,

i.e. At resonance, $X_L = X_C$

$$\text{or } 2\pi \times 50 \times 0.2 = \frac{1}{2\pi \times 50 \times C}$$

Capacitance required, $C = \dfrac{1}{(2\pi \times 50)^2 \times 0.2} = 50.6 \times 10^{-6}\,F = 50.6\ \mu F$ **Ans.**

(ii) At resonance, $X_L = X_C$, and the net impedance of the circuit, Z is:

$Z = R = 4\ \Omega$

Hence, Current drawn by the circuit at resonance, $I = \dfrac{100}{4} = 25$ Amp **Ans.**

(iii) Inductive reactance of the circuit, $X_L = 2\pi \times 50 \times 0.2 = 62.83\ \Omega$

Voltage across inductance, $V_L = I X_L = 25 \times 62.83 = 1570.75$ Volts

(iv) Capacitive reactance, $X_C = X_L = 62.83\ \Omega$

Thus, Voltage across the capacitor $= I X_C = 25 \times 62.83 = 1570.75$ Volts

(v) Q-factor of the circuit $= \dfrac{\textit{Voltage}\text{ across L or C}}{\textit{Supply}\text{ voltage}} = \dfrac{1570.75}{100} = 15.70$ **Ans**

Example 3.46

A coil of resistance 4.0 ohm and inductance 0.04 H is connected in series with a condenser across 220 V mains. What capacitance should the condenser have so that maximum current flows at (a) 20 Hz (b)50 Hz. ? Find also the current and the voltage across the condenser in each case.

Solution :

Part(a) $f_r = 20$ Hz

$\omega_r = 2\pi \times f_r = 2\pi \times 20 = 40\pi$ rad/sec

At resonance, $\omega_r = \dfrac{1}{\sqrt{LC}}$

or $C = \dfrac{1}{\omega_r^2\, L}$

Substituting the values of known parameters, we get:

Or Capacitance, $C = \dfrac{1}{(40\pi)^2 \times 0.04} = 1583.1\ \mu F$ **Ans**

Current at resonance, $I = \dfrac{220}{4} = 55$ amp **Ans.**

Voltage across the capacitor, $V_C = I X_C = \frac{I}{C\omega} = \frac{55}{1583.1 \times 10^{-6} \times 40\pi} = 276.4$ Volts**Ans**

Part (b) $f_r = 50$ Hz

$w_r = 2\pi \times 50 = 100\pi$ rad/sec

$$C = \frac{1}{\omega_r^2 L} = \frac{1}{(100\pi)^2 \times 0.04} = 253.5\ \mu F$$

I = 55 amp.

Voltage across the capacitor, $V_C = \frac{I}{C\omega} = \frac{55}{253.5 \times 10^{-6} \times 100\pi} = 690$ volts **Ans**

3.13.2 Resonance in Parallel Circuits/Parallel Resinance

Consider a parallel RLC circuit shown in Fig 3.57, supplied with a constant current, I.

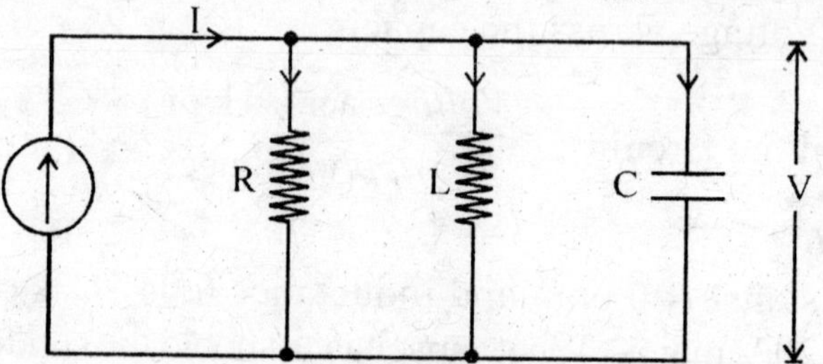

Fig.3.57 : Circuit configuration

3.13.2.1 Expression for Resonant Frequency

A parallel circuit can be analyzed easily in terms of admittances as compared to impedances.

The total admittance, Y for this circuit is:

$$Y = Y_R + Y_L + Y_C$$

$$= \frac{1}{R} + \frac{1}{j\omega L} + \frac{1}{-jX_C}$$

$$= \frac{1}{R} + \frac{1}{j\omega L} + j\omega C = \frac{1}{R} + j\left(\omega C - \frac{1}{\omega L}\right) \tag{3.53}$$

The total input current, I is :

$$I = VY$$

Hence, $$I = V\left[\frac{1}{R} + j\left(\omega C - \frac{1}{\omega L}\right)\right] \tag{3.54}$$

or

$$V = \frac{I}{Y} = \frac{I}{G + j(\omega C - \frac{1}{\omega L})}, \text{ Where } G = \frac{1}{R}$$

If current, I and conductance, G are assumed to be constant, hence the voltage will be maximum when the susceptance part or imaginary part of the denominator i.e. Y, is zero.

Under this condition, the output voltage will be in time phase with the current, which is the basic condition of resonance. Hence, the radian or angular frequency, ω_r at which resonance occurs can be found from:

$$\omega_r C - \frac{1}{\omega_r L} = 0$$

i.e. $$\omega_r C = \frac{1}{\omega_r L}$$

or $$\omega_r = \frac{1}{\sqrt{LC}}$$

or The resonant frequency, $$f_r = \frac{1}{2\pi\sqrt{LC}} \quad (3.55)$$

This is identical to the expression for resonance applicable to the series RLC circuit.

3.13.2.2 Practical Parallel Resonant Circuit

A practical parallel circuit frequently encountered in practice, consisting of resistance, inductance and capacitance is shown in Fig 3.58

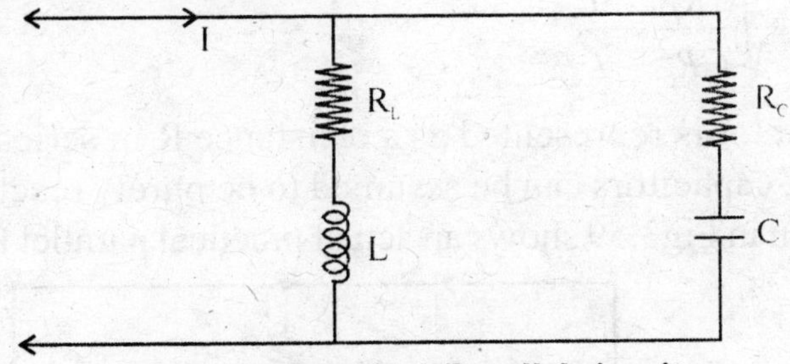

Fig.3.58 : Practical Parallel circuit

Admittance of the circuit shown in Fig 3.58 is given by:

$$Y = Y_L + Y_C = \frac{1}{R_L + j\omega L} + \frac{1}{R_C - \frac{j}{\omega C}} \quad (3.56)$$

Rationalizing equation (3.56) and collecting the real and imaginary terms,

$$Y = \frac{R_L}{R_L^2 + \omega^2 L^2} + \frac{R_C}{R_C^2 + \frac{1}{\omega^2 C^2}} - j\left(\frac{\omega L}{R_L^2 + \omega^2 L^2} - \frac{\frac{1}{\omega C}}{R_C^2 + \frac{1}{\omega^2 C^2}}\right)$$

At resonance, the imaginary part of admittance, Y must be zero.

i.e. $$\frac{\omega_r L}{R_L^2 + \omega_r^2 L^2} = \frac{\omega_r C}{R_C^2 \omega_r^2 C^2 + 1}$$

or $$\frac{L}{R_L^2 + \omega_r^2 L^2} = \frac{C}{R_C^2 \omega_r^2 C^2 + 1}$$

Cross multiplying the two sides, we get:

$$R_L^2 C + \omega_r^2 L^2 C = R_C^2 L C^2 + L$$

Rearranging the terms and taking out, we get:

$$(L^2 C - R_C^2 L C^2) = L - R_L^2 C$$

i.e. $$\omega_r^2 = \frac{R_L^2 C - L}{R_C^2 LC^2 - L^2 C}$$

$$\omega_r^2 = \frac{CR_L^2 - L}{LC(CR_C^2 - L)}$$

Solving for $_r$, we get:

Resonant cyclic frequency of the given circuit, can be written as :

$$f_r = \frac{1}{2\pi\sqrt{LC}}\sqrt{\frac{CR_L^2 - L}{CR_C^2 - L}} \tag{3.57}$$

A practical inductor is represented as a resistance R in series with an inductance L. However, in general, capacitors can be assumed to be purely reactive i.e. resistance R_C is zero. The circuit given in Fig3.59 shows an actual practical parallel LC circuit.

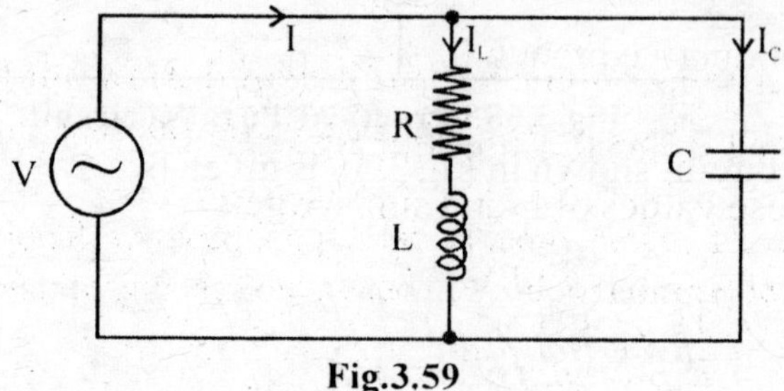

Fig.3.59

Substituting $R_C = 0$ in equation (3.57), we get the equation for resonant frequency for the circuit shown in Fig.3.59 as:

$$f_r = \frac{1}{2\pi\sqrt{LC}}\sqrt{1 - \frac{CR_L^2}{L}} = \frac{1}{2\pi}\sqrt{\frac{1}{LC} - \frac{R_L^2}{L^2}} \quad (3.58)$$

The phasor diagram in Fig3.60 represents the circuit conditions at resonance. It is seen that the presence of the resistive component in the inductive branch makes the I_L phase angle ϕ slightly less than 90°. If the reactive components of I_L and I_{-C} are equal, the resultant current , I will be in phase with the voltage. This is the condition of resonance in a parallel circuit. Hence, resonance is achieved when the reactive component of I_L is equal to the capacitor current I_{-C}.

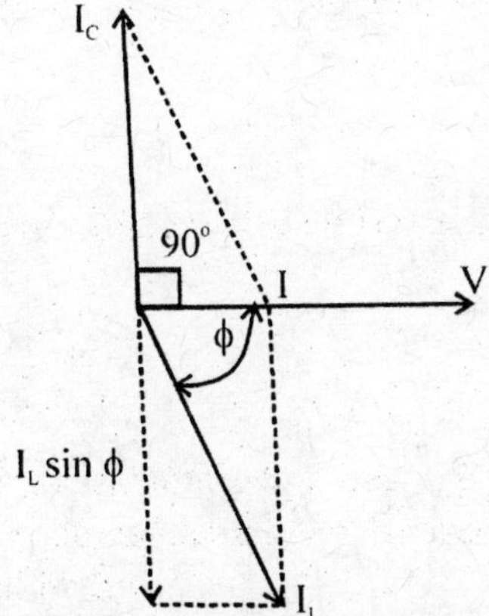

Fig.3.60 : Current phasor diagram for the circuit of Fig.3.59

3.13.2.3 Alternative Method-1 for Derivation of Equation for Resonant Frequency for the Circuit Shown in Fig. 3.59

For resonance in a parallel circuit,

$$I_C = I_L \sin\phi \quad \text{(i)}$$

Current in the capacitive branch, $I_C = \dfrac{V}{X_C}$ (ii)

Current in the inductive branch $I_L = \dfrac{V}{Z_L}$

We know that in a series R-L circuit, sin =

Substituting these values of I_L and sin , we get:

$$I_L \sin = \left(\frac{V}{Z_L}\right)\left(\frac{X_L}{Z_L}\right) = \left(\frac{VX_L}{Z_L^2}\right) \quad \text{(iii)}$$

Substituting the values of I_C and I_L sin from equations (ii) and(iii) in equation(i), we get:

$$\frac{V}{X_C} = \frac{VX_L}{Z_L^2}$$

On simplification, we get:

$$Z_L^2 = X_L X_C$$

or $R^2 + (X_L)^2 = (\omega L)(\frac{1}{\omega C})$

At resonance, this equation can be rewritten as:

$$R^2 + (\omega_r L)^2 = \frac{2\pi f_r L}{2\pi f_r C} = \frac{L}{C}$$

i.e. $$(\omega_r L)^2 = \frac{L}{C} - R^2$$

Keeping on the LHS, we get:

or $$\omega_r^2 = \frac{1}{LC} - \frac{R^2}{L^2}$$

or Resonant frequency, $$\omega_r = \sqrt{\frac{1}{LC} - \frac{R^2}{L^2}} \tag{3.59}$$

Resonant cyclic frequency $$f_r = \frac{1}{2\pi}\sqrt{\frac{1}{LC} - \frac{R^2}{L^2}} \tag{3.60}$$

3.13.2.4 Dynamic impedance of a practical parallel circuit

The admittance of the parallel circuit shown in Fig. 3.59 is:

$$Y = \frac{1}{R + jX_L} + j\frac{1}{X_C} \tag{3.61}$$

rationalising the first term on RHS of equation (3.61) gives :

$$Y = \frac{R}{R^2 + X_L^2} - j\frac{X_L}{R^2 + X_L^2} + j\frac{1}{X_C} \tag{3.62}$$

At resonance, the imaginary components of Y should be zero.

$$\frac{1}{X_C} = \frac{X_L}{R^2 + X_L^2}$$

Or $$X_C = \frac{R^2 + X_L^2}{X_L} \tag{3.63}$$

Which, when substituted into Equation (3.61), gives the admittance at resonance, Y_r as:

$$Y_r = \frac{1}{R + jX_L} + \frac{jX_L}{R^2 + X_L^2}$$

$$= \frac{R - jX_L + jX_L}{R^2 + X_L^2}$$

Or $$Y_r = \frac{R}{R^2 + X_L^2} \quad (3.64)$$

Thus at resonance, mpedance, $Z_r = \frac{1}{Y_r} = \frac{R^2 + X_L^2}{R}$ (3.65)

However from Equation (3.63), $R^2 + X_L^2 = X_L X_C$

Hence, $Z_r = \frac{X_L X_C}{R} = \frac{2\pi f L}{2\pi f C R}$

Or At resonance, impedance offered by a practical resonant parallel circuit,

$$Z_r = \frac{L}{CR} \quad (3.66)$$

This impedance is called **dynamic impedance** of the circuit.

3.13.2.5 Q Factor for Parallel LC Circuits

The total current drawn from the supply at resonance is I, which is in phase with supply voltage V and is much smaller than I_C and I_L. Thus, in the parallel resonant LC circuit, there is a current magnification, which is analogous to the voltage magnification that occurs in a series resonant circuit.

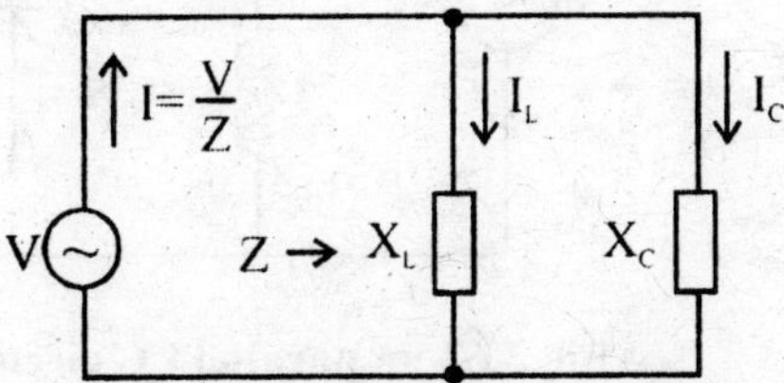

Fig.3.61 : Parallel LC circuit

In a parallel resonance circuit, I_L and I_C are much larger than the supply current (I). The circuit Q factor is the ratio I_L/I or I_C/I.

Since the input current at resonance is minimum, a parallel resonant circuit is called a **rejector circuit**. In other words , the circuit almost rejects the current at resonance.

As illustrated in Fig.3.61

$$I = \frac{V}{Z}$$

From Equation (3.66), $Z_r = \frac{L}{CR}$

so, $$I = \frac{V}{Z_r} = \frac{V}{L/(CR)} \quad (3.67)$$

Also,

$$I_L = I_C = \frac{V}{X_C} = V\omega C$$

Hence Q factor , $$Q = \frac{I_C}{I} = V\omega C \times \frac{L/(CR)}{V} = \frac{\omega L}{R} \quad (3.68)$$

This equation is exactly the same as the Q factor equation for a series resonant circuit

Example 3.47

A parallel LC circuit, Fig 3.62 has an inductance L = 200 μH with a coil resistance of 10 Ω. The capacitor C is adjustable over the range 100 pF to 200 pF. Determine the maximum and minimum resonance frequencies for the circuit. Also, calculate the Q factor of the circuit at the two resonance frequency extremes .

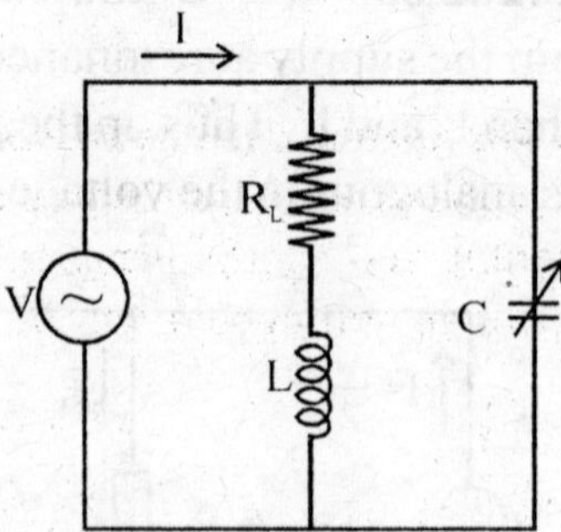

Fig.3.62 : A parallel LC circuit

Solution :

In a parallel resonance circuit with Q > 10, the resonance frequency and bandwidth are calculated as in a series resonance circuit.

(i) Calculation of maximum and minimum resonant frequencies

If $\quad Q > 10, \quad f_r = \frac{1}{2\pi\sqrt{LC}}$

When C = 100 pF, $f_r = \dfrac{1}{2\pi\sqrt{200\times10^{-6}\times100\times10^{-12}}} = \dfrac{10^7}{2\pi\sqrt{2}}$

Maximum resonant frequency , $f_{r(max)} = \dfrac{10^7}{8.8857} = 1.125$ MHz **Ans**

When C = 200 pF,

Minimum resonant frequency, $f_r = \dfrac{1}{2\pi\sqrt{200\times10^{-6}\times200\times10^{-12}}}$

$= \dfrac{10^7}{4\pi} = 0.795$ MHz **Ans**

(ii) Calculation of Q- factor

$$Q = \frac{\omega L}{R}$$

At $f = 1.125$ MHz,

$Q = \dfrac{2\pi\times1.125\times10^6\times200\times10^{-6}}{10} = 40\,\pi\times 1.125 = 125.664\times1.125 = 141.372$ **Ans**

At $f = .795$ MHz,

$Q = \dfrac{2\pi\times0.795\times10^6\times200\times10^{-6}}{10} = 99.902$ **Ans**

Example 3.48

The Circuit shown in Fig.3.63 has L=200 μ H and is to be tuned to resonate at frequencies ranging from 250 KHz to 500 KHz. Calculate the range of adjustment of the tuning capacitor.

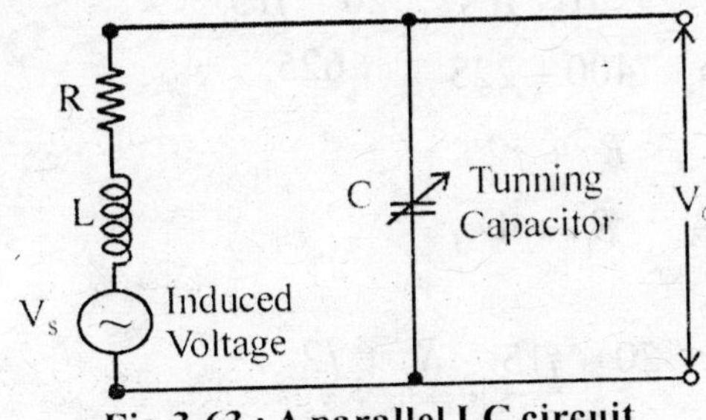

Fig.3.63 : A parallel LC circuit

Solution :

We know that:

Resonant frequency, $f_r = \dfrac{1}{2\pi\sqrt{LC}}$

$$\therefore \quad f_r^{\,2} = \frac{1}{4\pi^2 LC}$$

and $$C = \frac{1}{4\pi^2 f_r^{\,2} L}$$

At $\mathbf{f_r = 500\ kHz}$

$$C = \frac{1}{4\pi^2 (500\times10^3)^2 \times 200\times10^{-6}} = \frac{1000\times10^{-12}}{2} = 500\ \text{pf}$$ **Ans**

At $\mathbf{f_r = 250\ kHz}$

$$C = \frac{1}{4\pi^2 (250\times10^3)^2 \times 200\times10^{-6}} = \frac{1000\times10^{-12}}{0.5} = 2000\ \text{pf}$$ **Ans**

Hence the range of adjustment of capacitor is from 500 pf to 2000 pf **Ans**

Example 3.49

Determine the value of R in the network shown in Fig 3.64 to yield resonance.

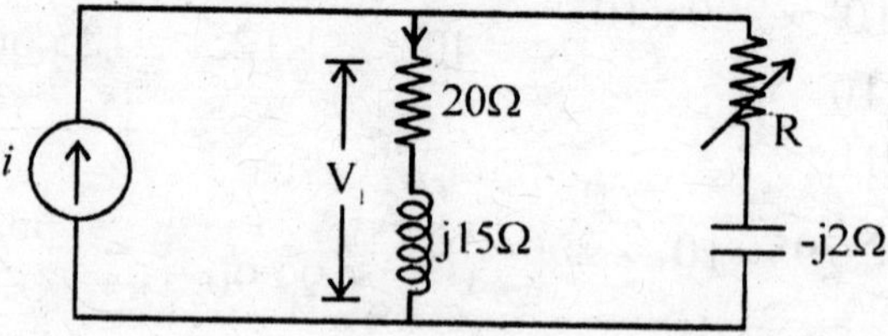

Fig.3.64 : A parallel LC circuit

Solution :

Here $$Y_1 = \frac{1}{20 + j15} = \frac{20 - j15}{400 + 225} = \frac{20 - j15}{625}$$

and $$Y_2 = \frac{1}{R - j2} = \frac{R + j2}{R^2 + 4}$$

Then $$Y_T = Y_1 + Y_2 = \frac{20 - j15}{625} + \frac{R + j2}{R^2 + 4}$$

Separating the real and imaginary components , we get :

$$Y_T = \left(\frac{20}{625} + \frac{R}{R^2 + 4}\right) + j\left(\frac{2}{R^2 + 4} - \frac{15}{625}\right)$$

For resonance, the imaginary part of Y_T should be zero.

i.e. $$\frac{2}{R^2+4} = \frac{15}{625} = \frac{3}{125}$$

Or $3(R^2+4) = 250$ or $3R^2 + 12 = 250$

or Value of resistance to yield resonance is: $R = \sqrt{\frac{238}{3}} = \sqrt{79.3} = 8.88$ ohms Ans

Example 3.50

Find the resonance frequency for the circuit shown in Figure.3.65.L=10mH , R_L=5 Ω ,C=40 μ F ,R_C =2 Ω

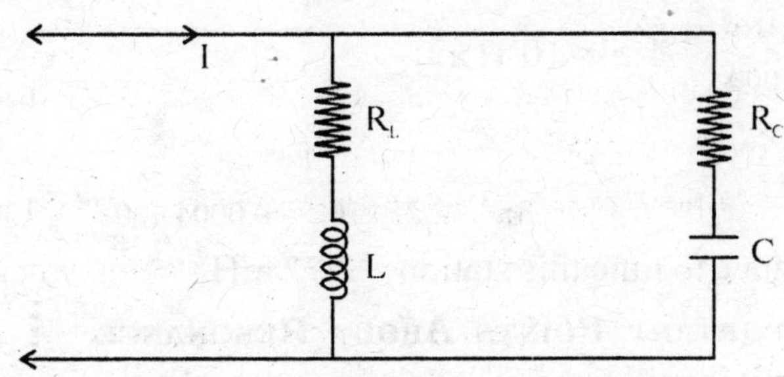

Fig.3.65 : A parallel LC circuit

Solution :

$$Y = Y_L + Y_C = \frac{1}{R_L + j\omega L} + \frac{1}{R_C - \frac{j}{\omega C}}$$

Rationalizing and collecting the real and imaginary terms,

$$Y = \frac{R_L}{R_L^2 + \omega^2 L^2} + \frac{R_C}{R_C^2 + \frac{1}{\omega^2 C^2}} - j\left(\frac{\omega L}{R_L^2 + \omega^2 L^2} - \frac{\frac{1}{\omega C}}{R_C^2 + \frac{1}{\omega^2 C^2}}\right)$$

For resonance, the imaginary part of Y must be zero.

Solving for ω, resonant frequency of the given circuit is:

$$f_r = \frac{1}{2\pi\sqrt{LC}}\sqrt{\frac{CR_L^2 - L}{CR_C^2 - L}}$$ (refer equation 3.63)

Substituting the values of parameters, we get:

$$f_r = \frac{1}{2\pi\sqrt{10\times10^{-3}\times40\times10^{-6}}} \times \sqrt{\frac{40\times10^{-6}\times5^2 - 10\times10^{-3}}{40\times10^{-6}\times2^2 - 10\times10^{-3}}} = 263$$

Resonant frequency f_r = 263 Hz. Ans

Example 3.51

A broadcasting station transmits signals on a wave length of 2000meters at a speed of 3 10^8 m/sec. A capacitor of 0.0004 F is connected in parallel with a variable inductor to form the tuning circuit. Find the value of inductance to tune this station.

Solution

Velocity of propagation= wave length × frequency

Where velocity is in m/sec , wave length is in meters, freq. is in Hz.

$$\text{Frequency} = \frac{3\times10^8}{2000} = 1.5\times10^5 \text{Hz}$$

$$\text{At resonance, } L = \frac{1}{4\pi^2 f_r^2 C} = \frac{1}{4\pi^2\times2.25\times10^{10}\times0.0004\times10^{-6}} = \frac{1}{36\pi^2} = 2.77 \text{ mH}$$

Value of inductance to tune this station = **2.77** mH **Ans**

3.13.3 Some Important Points About Resonance

1. An AC circuit is said to be in resonance when the applied voltage and the resulting current are in phase.
2. At resonance, the equivalent complex impedance of the circuit consists of only resistance and the power factor of the circuit is unity.
3. Resonance in series circuits is referred to as **series resonance** or simply **resonance**, while resonance in parallel circuits is referred to as **parallel resonance** or simply **anti-resonance**.

Tuning of radio or television involves changing the resonant frequency of one or more resonant circuits by varying the capacitance.

3.13.4 Comparison of Series Resonance and Parallel Resonance

Comparison of Series Resonance and Parallel Resonance is given in the Table 3.1 given below.

Table 3.1 Comparison of Series Resonance and Parallel Resonance

S.No.	Parameter	Series Resonance	Parallel Resonance
1	Current magnitude	Maximum $I_r = \frac{V}{R}$	Minimum $I_r = \frac{V}{(L/CR)}$
2	p.f.	Unity	Unity
3	Power	Low	High
4	Impedance	Minimum, $Z_r = R$	Maximum $Z_r = \frac{L}{CR}$
5	Q Factor	Depends on voltage amplification $Q = \frac{1}{R}\sqrt{\frac{L}{C}}$	Depends on current amplification $Q = \frac{1}{R}\sqrt{\frac{L}{C}}$
6	Type of circuit	Acceptor	Rejector
7	Nature of circuit at resonance	Resistive	Resistive
8	Resonant Frequency	$f_r = \frac{1}{2\pi\sqrt{LC}}$	$f_r = \frac{1}{2\pi\sqrt{LC}}$
9	Signal Magnification	Voltage signal Magnification	Current signal Magnification

3.14 SUMMARY

In this chapter, the terminology related to AC quantities and phasor representation of alternating quantities has been thoroughly explained.

1. **Basic concepts, theory and techniques for the analysis of single phase AC series and parallel circuits consisting of resistance, inductance and capacitance elements have been given.**
2. **Resonance in series and parallel AC circuits has been thoroughly discussed.**

3.15 SHORT QUESTION AND ANSWERS

1. What is an alternating quantity?

Ans. A quantity whose magnitude and direction changes periodically at regular intervals is called alternating quantity.

2. What is a cycle?

Ans. One complete set of positive and negative values of an alternating quantity constitutes one cycle.

3. What is time period?

Ans. It is the time taken by an alternating quantity to complete one cycle. It is denoted by 'T'. Its unit is second.

4. What is frequency?

Ans. The number of cycles completed by an alternating quantity per second is known as frequency. It is denoted by 'f' and its unit is hertz.

5. What is amplitude?

Ans. The maximum value of an alternating quantity, either positive or negative is called amplitude.

6. What is R.M.S. value?

Ans. R.M.S. value of an alternating current is given by that equivalent value of direct current which when flowing through a given circuit for a given time, produces the same heating effect as that produced by an alternating current when flowing through the same circuit for the same time.

7. Define Form factor and Peak factor.

Ans. 1. $\text{Form Factor} = \dfrac{R.M.S.\ \text{Value}}{Average\ \text{Value}}$

2. $\text{Peak Factor} = \dfrac{Maximum\ \text{Value}}{R.M.S.\ \text{Value}}$

10. What is the significance of peak factor?

Ans. Knowledge of peak factor is essential to determine the dielectric strength of an insulating material since the dielectric stress developed is proportional to the peak value of the voltage applied to it.

11. What is the difference between a vector and a phasor?

Ans. Vector is a quantity that has magnitude as well as direction.

In addition to above, phasor rotates at a constant angular speed.

12. What is the significance of the phasor representation of an alternating quantity?

Ans. It enables us to understand its magnitude and position with respect to reference line.

13. What do you mean by phase of an alternating quantity?

Ans. The phase of an alternating quantity (voltage or current) at any instant is defined as the fractional part of a cycle through which the quantity has advanced.

14. What do you mean by phase difference?

Ans. The phase difference may be defined as the angular displacement between the maximum positive values of the two phasors representing the two quantities having the same frequency

15. What do you understand by power factor?

Ans. Power factor may be defined as the cosine of the phase angle between voltage and current. It may also be defined as the ratio of resistance to impedance or the ratio of true power to apparent power.

16. What do you understand by active and reactive components of current ?

Ans. The current component which is in phase with circuit voltage (i.e. I cos ϕ) and contributes to active or true power of the circuit is called the active (wattfull or in-phase) component of current.

The current component which is in quadrature (or 90° out of phase) to the circuit voltage (i.e. I sin ϕ) and contributes to reactive power of the circuit is called the reactive (or wattles) component of current.

17. What are active and reactive powers?

Ans. The power which is actually consumed in an ac circuit is called the **true or active power**. Power is consumed only in resistance. It is expressed in watts.

A pure inductor and a pure capacitor do not consume any power, as in a half cycle whatsoever power is drawn from the supply source, the same is returned to the supply source in the other half cycle. This power which reacts itself is called the **reactive power**. This is also known as wattless power.

18. In an R–L–C circuit on what factors does it depend whether the phase angle between voltage V and current I is lagging or leading?

Ans. Phase angle in an R–L–C circuit is given by:

$$\phi = \tan^{-1}\frac{X_L - X_C}{R}$$

i.e. the phase angle between voltage V and current I will depend upon whether the inductive reactance is greater or smaller than capacitive reactance. When X_L exceeds X_C, applied voltage V will lead current I and when X_C exceeds X_L applied voltage V will lag behind the current I.

19. What is meant by resonant frequency?

Ans. The frequency of supply at which inductive reactance and capacitive reactance are equal is called resonant frequency.

20. What is series resonance?

Ans. When in an R – L – C series circuit the inductive reactance equals capacitive reactance, the circuit is said to be in **series resonance**.

21. Why is the series resonance called the voltage resonance?

Ans. Since in series resonance the voltage across inductance and voltage across capacitance is maximum, it is called the **voltage resonance**.

22. Why is the series R – L – C circuit called the acceptor circuit?

Ans. Series R – L – C circuit accepts currents at one particular frequency but rejects currents of other frequencies and therefore, series R – L – C circuit is called the **acceptor circuit.**

23. What do you understand by the term bandwidth of a resonant circuit?

Ans. The frequency band over which the magnitude of current is equal to $\frac{1}{\sqrt{2}}$ times the peak current is known as bandwidth of the circuit.

24. Why is parallel R – L – C circuit also called the rejector circuit?

Ans. Parallel R – L – C circuit is called the **rejector** circuit because at resonant frequency the line current is minimum or it almost rejects it.

25. What is meant by Q-factor of series resonant circuit?

Ans. The Q-factor of a series resonant circuit is given as the voltage magnification that the circuit produces at resonance.

$$\text{Q-factor at resonance, } Q_r = \frac{\omega_r L}{R} = \frac{2\pi f_r L}{R} = \left(\frac{2\pi L}{R}\right)\frac{1}{2\pi\sqrt{LC}} = \frac{1}{R}\sqrt{\frac{L}{C}}$$

26. What is meant by Q-factor of a parallel resonant circuit?

Ans. Q-factor of a parallel resonant circuit is defined as the ratio of the circulating current to the line current at resonance or as the **current magnification**.

3.16 Problems

1. For AC single phase 230 V, 50 Hz, assuming sinusoidal waveform find (i) maximum value (ii) rms value (iii) frequency (iv) instantaneous value voltage equations (v) voltage 0.003 second after starting from zero and rising in +ve direction.

Ans. (i) 325.3 V, (ii) 230 V, (iii) 50 Hz, (iv) 325.3 sin 314 t, (v) 263.17 V

2. Calculate the maximum value, and the root mean square value of the following quantities :

(i) 40 sin ωt (ii) (A + B) sin $\left(\omega t - \frac{\pi}{2}\right)$; (iii) 10 sin ωt – 17.3 cos ωt

Ans. (i) $40, 20\sqrt{2}$ **, (ii)** $A+B, \frac{A+B}{\sqrt{2}}$ **, (iii)** $20, 10\sqrt{2}$

3. The currents flowing in the three branches of an ac parallel circuit are :

$i_1 = 20 \sin \omega t$; $i_2 = 25 \sin\left(\omega t - \frac{\pi}{6}\right)$; $i_3 = 15 \sin\left(\omega t + \frac{\pi}{6}\right)$. Determine the total current drawn from the supply.

Ans. 54.87 sin (ωt - .029 π)

4. Three sinusoidal voltages acting in series are given by :

$v_1 = 10 \sin 440\,t$

$v_2 = 10\sqrt{2}\ \sin (440\,t - 45^o)$

$v_3 = 20 \cos 440\,t$

Determine :

(i) an expression for the resultant voltage.

(ii) The frequency and rms value of the resultant voltage.

Ans.22.36sin(440*t*+26.56^0),70.6Hz,15.81 volts

5. A 50-Hz voltage of 115 V (rms) is impressed on a 50 Ω resistance. Write the time equations for the voltage and resultant current. Let the zero point of the voltage wave be at $t = 0$.

Ans. 162.6 sin 314 *t* and 3.25 sin 314*t*

6. Find the current through a coil whose inductance is 60 mH and resistance is negligible, if it is connected to a 230volts , 50c/s supply.

Ans. 12.2 amp.

7. Find the current through a capacitotr whose capacitance is 100 μ F and resistance is negligible , if it is connected to a 200volts , 50c/s supply

Ans 6.29 amp.

8. A resistance of 15 Ω, choke coil of 0.05 H and the capacitor of 300 μF are connected in series. If the circuit is connected to 250 V, 50 Hz, power supply, find (i) circuit impedance (ii) power factor (iii) absorbed power (iv) resonant frequency and (v) Q-factor of the coil.

Ans.(i) 15.84 Ω; (ii) 0.947 lag; (iii) 3,735 W; (iv) 41 Hz; (v) 1.056

9. A coil having a resistance of 6 Ω and an inductance of 0.03 H is connected across a 50 V, 60 Hz supply. Calculate (i) the current (ii) the phase angle between the current and the applied voltage, (iii) the power factor, (iv) the volt-amperes and (v) the power.

Ans.(i) 3.905 A; (ii) 62.05°; (iii) 0.4686 lag; (iv) 195.25 VA; (v) 91.5 W

10. An inductive coil takes a current of 10 amp lagging 30° from 200 V, 50 Hz supply.Calculate the resistance, reactance and inductance of the coil.

11. $v = 100 \sin 100\,\pi t$ and $i = 8 \sin (100\,\pi t + 60^o)$.From these expressions , find the impedance and the power factor of the circuit.

Ans. 12.5 Ω; 0.5 lead

12. A voltage of 125 volts is applied across a resistance connected in series with a capacitor. The circuit current is 2.2 amp. Power lost in the resistance is 96.8 watts. Calculate the resistance and capacitance . Assume supply frequency as 50c/s.

Ans.R=20 ohm,C=59.85μF

13. A resistance of 50 ohms , an inductance of 0.15 H and a capacitance of 100 μF are connected in parallel across a 100 volt , 50Hz supply. Calculate the current in each branch.

Ans. 2amp, 2.12 amp , 3.14 amp

14. A circuit comprises an inductance of 24 μH associated with resistance of 0.1 Ω in parallel with a capacitance of 0.16 μF. Find the resonant frequency?

15. A coil of 20 ohm resistance has an inductance of 0.2 H and is connected in parallel with a condenser of 100 μF capacitance. Calculate the frequency of resonance.

Ans. 31.83 Hz

16. Two impedances $Z_1 = 10-j15$ ohms and $Z_2 = 4+j8$ are connected in parallel. The supply voltage is 100 volts, 25 Hz. Calculate(i) admittance, conductance, susceptance of the total circuit (ii) total current drawn.

Ans [0.097,0.081,0.054, 9.7, 0.83]

❑❑❑

Chapter 4

Three Phase A.C. Circuits

OBJECTIVE

- To learn the basic concepts of three phase voltages and currents, double subscript notation and phase sequence
- To understand the star and delta connections of three phases
- To understand the relationships between line and phase values of voltages and currents in three phase star and delta connections
- To learn the methods for the measurement of power in three phase circuits

4.1 INTRODUCTION

AC circuits considered so far were single phase circuits, which are widely used for the operation of domestic and commercial electrical devices, in electronics engineering and are most familiar to the consumers. However, single phase circuits have limitations for generation, transmission and distribution of electrical energy. Three phase systems are quite common for these applications and industrial use . For certain jobs, greater number of phases is also used. Mercury arc rectifiers are either six phase or twelve phase and most of the rotary converters are six phase. Some experimental six phase systems and integrated three phase and six phase systems are under study and research for proper applications.

All modern generators are three phase and for transmitting large amounts of power, three phase transmission lines are used. In this section , star and delta connected three phase systems are discussed.

An AC system having a set of two or more than two equal voltages/currents of same frequency arranged to have equal phase difference between adjacent voltages/currents is called a poly phase system.

4.2 Advantages of Poly Phase Systems

Poly phase systems have many advantages. Some of them are as follows :

1. Three phase machines produce about 45% higher output as compared with single phase machines.
2. Size of three phase machines is smaller than that of single phase machines for the same output. For the same frame size, machine rating/output of a three phase machine can be 1.5 times the rating of the single phase m/c.
3. Three phase induction motors are self starting.
4. More power can be transmitted over a three phase system using conductors of same size i.e. same cross sectional area as compared to single phase system.
5. Cross sectional area of conductors is small for a 3-phase system for the transfer of same amount of power at a given voltage, resulting in saving of copper .
6. The power delivered by a single phase system pulsates. In a poly phase system, power delivered is almost constant.
7. Power factor of three phase induction motors is better than single phase induction motors.
8. Direction of rotation of a three phase induction motor can be easily changed.

4.3 Generation of Three Phase Voltages and Currents

A coil rotating in a uniform magnetic field generates an alternating emf. Three phase voltages are generated by three phase generator.

A three phase generator consists of two main parts i.e. (i) stator (ii) rotor.

Stator is a slationary part rotor produces a constant rotating magnetic field.

In a three phase generator, three coils/ windingsa1 a_1, a_2, b_1 b_2, and c_1 c_2, displaced from each other by 120^0 are put in the stator as shown in Fig.4.1(a)

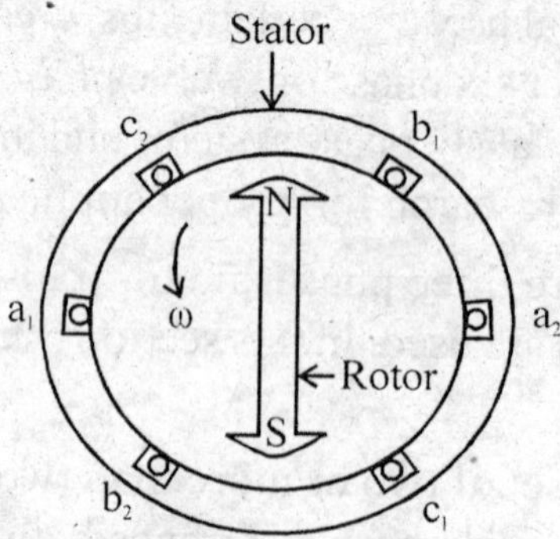

Fig.4.1 : (a)Generation of three phase voltages

For a generator operation to take place, the rotor is driven by a prime -mover at its rated speed N. DC current supplied to the rotor field system sets up alternate North and South pole magnetic fields. As the rotor is rotated, the flux linking stator coils changes and hence emf is induced in them according to Faraday's Law of electromagnetic induction. The emf induced in the coils is alternating emf and has a frequency f , given as

$$f = \frac{PN}{120}$$

Where P - Number of poles in the rotor

N - Speed of rotor in RPM

The magnitude of the voltage induced in each phase depends on the rotor flux , the number of conductors and the speed of the rotor. As three sets of coils are placed in the stator core, physically 120 electrical degrees apart from one another, three-phase alternating voltage , with a 120° phase angle displacement will be available on the terminals of AC generator.

The instantaneous values of these three sinusoidal voltages induced in the coils, assuming phase R as the reference phase, are expressed as:

$$V_a = V_{max} \sin \omega t$$
$$V_b = V_{max} \sin (\omega t - 120^\circ)$$
$$V_c = V_{max} \sin (\omega t - 240^\circ) \quad (4.1)$$

It can be seen that these voltages/emf's are displaced from each other by an angle of 120°. These three voltages are taken out from stator terminals.

The phasor diagram showing these three voltages is shown in Fig 4.1(b)

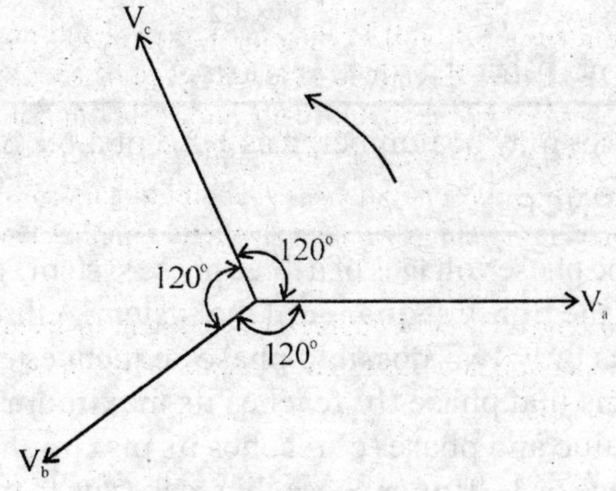

Fig4.1 : (b)Phasor diagram

A set of three sinusoidal voltages or currents which are equal in magnitude but have a phase difference of 120^0 , is known as a balanced three phase system.

The graphical reprcsentation of these three emf waveforms is shown in Fig 4.1(c

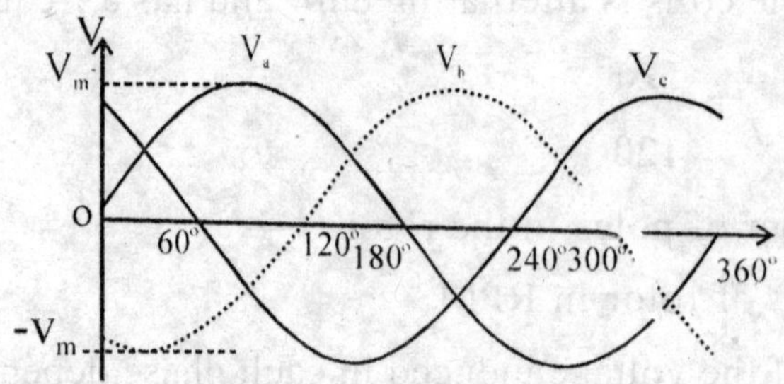

Fig.4.1 : (c) Waveforms of three phase voltages

4.4 Double-Subscript Notation

It is convenient to describe polyphase voltages and currents using **Double-Subscript Notation.** Subscript 'ab' used with voltage V i.e. V_{ab} indicates that point *'a' is positive* with respect to point *'b'*. Thus, the plus sign is located at *'a'*, as indicated in Figure 4.2

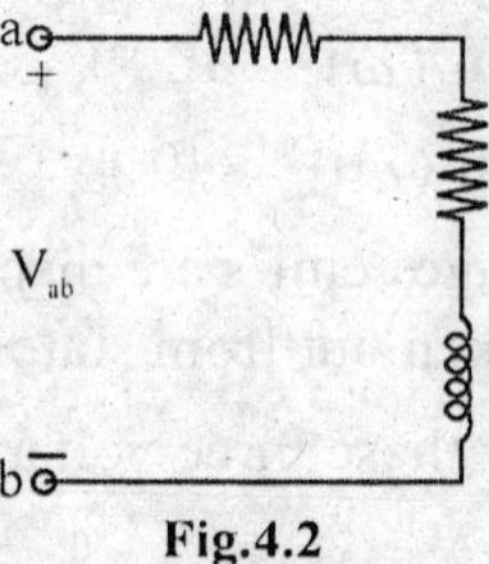

Fig.4.2

4.5 Numbering of Different Phases

Three different phases may be numbered as 1,2,3 or a,b,c or red ,yellow and blue etc.

4.6 Phase Sequence

The order in which the phase voltages of a three phase system attain their peak or maximum positive values is called the **phase sequence** of the system. A three phase system consisting of three phases a,b,c has only two possible phase sequences : **abc** and **cba.**If the phase sequence is abc , it means that phase 'b' reaches its maximum value 120^0 after phase 'a' reaches its maximum value and phase 'c' reaches its maximum value 240^0 after phase 'a reaches its maximum value or 120^0 after phase 'b' reaches its maximum value.

If one of the phase voltages ,say V_a is taken as reference voltage, then for the phase sequence 'abc', we can write three different phase voltages as follows:

$$V_a = V\angle 0^0$$
$$V_b = V\angle -120^0$$
$$V_c = V\angle -240^0 = V\angle 120^0$$

Phase sequence in any system can be changed by interchanging any pair of lines i.e. by interchanging the connections of any two phases.

By convention , sequence RYB is taken as positive phase sequence and RBY as negative phase sequence.

4.6.1 Effect of Change of Phase Sequence

In the case of three phase unbalanced loads, the effect of change of phase sequence is to cause a completely different set of values of the currents. Reversing the phase sequence of a generator which is to be paralleled with a similar generator can cause damage to the machines. When electric power is connected to a three phase induction motor, the direction of rotation of motor depends upon the phase sequence.

4.7 Interconnection of Three Single Phases

Three single phases may be interconnected to form a three phase system.

Three phase systems are of two types:

(i) Balanced three phase systems(ii) Unbalanced three phase systems

(i) Balanced three phase systems : A three phase supply system is assumed to be balanced if all the phase voltage magnitudes are equal and all are separated from each other by an angle of 120^0. In addition to having balanced supply voltages, the load impedances of all the three phases should be balanced. i.e. load impedances in all the three phases are equal in magnitude and have the same impedance angle.

A balanced three phase system can be analysed as a single phase circuit.

(ii) Unbalanced three phase systems : A three phase supply system is assumed to be unbalanced if either the phase voltage magnitude or the phase angles or both are different for different phases.

A three phase system is said to be unbalanced ,if:

(i) Loads are unbalanced, supply system is balanced.

(ii)Supply system is unbalanced , loads are balanced

(iii) If both supply system and loads are unbalanced.

Three phases are generally interconnected in such a way that it leads to a change in the output voltage and current levels and thus to a considerable saving of copper in the conductors. The general methods of interconnection of three phases are:

(a) Star or Wye connection (Y)

(b) Mesh or Delta connection(Δ)

4.7.1 Star Connected Supply Ststem (Y)

In this connection, similar ends i.e. starting ends or finishing ends of coils/windings are joined together at a point, usually called as **star point or neutral point,** 'N'

The three isolated phases are connected as shown in Fig.4.3(a) in which terminals R_2, Y_2 and B_2 are joined together to form a neutral point. Such interconnection of the three phases is called **star connection** and is designated by a symbol Y.

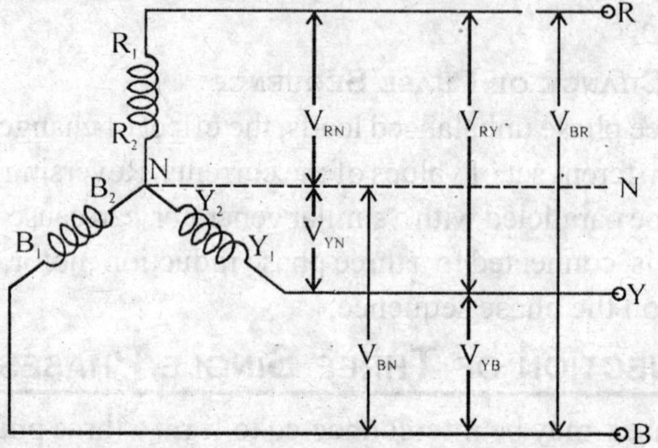

Fig. 4.3 : (a) Three phase star connection (three phase three wire)

This system is known as **Three phase Three wire** supply system

The system shown in Fig 4.3(b) is known as **Three phase Four wire** supply system. Here the fourth neutral wire is also available in addition to the three wires.

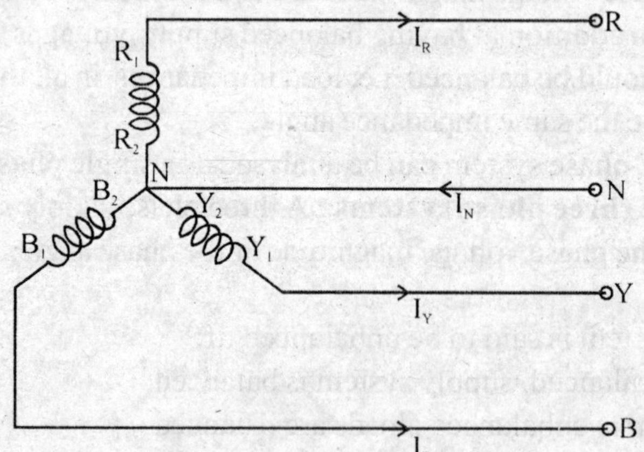

Fig. 4.3 : (b) Three phase star connection with neutral wire (three phase four wire)

All the three windings in a three phase alternator are identical and produce the same level of voltage, known as **phase voltage** V_{ph}. V_{R1R2}, V_{Y1Y2}, V_{B1B2} are the phase voltages as shown in Fig 4.3 (a) and(b). The **phase current** I_{ph} is the current supplied by each of the windings. The directions of these phase currents I_R, I_Y, I_B i.e out of terminals R,Y and B, are shown in Fig 4.3(b)

Let i_R, i_Y and i_B be the instantaneous values of the sinusoidal currents in the three phases.These currents may be represented by the following equations for a 3- phase symmetrical balanced system.

$$i_R = I_{max} \sin \omega t$$
$$i_Y = I_{max} \sin(\omega t - 120)$$
$$i_B = I_{max} \sin(\omega t - 240^\circ) \qquad (4.2)$$

It can be proved that their sum is equal to zero, as follows:

$$\text{i.e. } i_R + i_Y + i_B = I_{max} [\sin \omega t + \sin(\omega t - 120^\circ) + \sin(\omega t - 240^\circ)] = 0 \qquad (4.3)$$

Using the trigonometric identities, the above equation can be written as :

$$= I_{max} [\sin t + \sin t \cos 120 - \cos t \sin 120 + \sin t \cos 240 - \cos t \sin 240]$$

Substituting the values of different trigonometric functions, we get:

$$i_R + i_Y + i_B = I_{max}[\sin \omega t + \sin \omega t(-\frac{1}{2}) - \cos \omega t(\frac{\sqrt{3}}{2}) + \sin \omega t(-\frac{1}{2}) + \cos \omega t(\frac{\sqrt{3}}{2})]$$

$$= I_{max}[\sin \omega t + \sin \omega t(-\frac{1}{2}) + \sin \omega t(-\frac{1}{2})]$$

$$= I_{max}[\sin \omega t - \sin \omega t]$$

$$= 0$$

Thus, in a balanced 3- phase star connected circuit, the sum of the instantaneous currents in the three phases is always zero or the current in the neutral conductor is zero at every instant.

4.7.1.1 Line and Phase Quantities in Star-Connected Circuits

The conductors or wires that connect a three phase supply system to a load are referred to as lines and the current carried by these conductors is known as the line current I_L. It is seen from Fig.4.3(b) that line currents and phase currents are the same.

The neutral conductor is the return conductor for all the three windings . Therefore the neutral current I_N is the phasor sum of all the three line/phase currents. It is seen that when all the three windings have identical loads i.e. a balanced load, neutral current I_N is zero. However, when load is unbalanced , current flows in the neutral wire.

In a Y connected system, the **phase voltage** is the voltage between any one line and neutral.The **line voltage** is the potential difference between any two lines. Thus the line voltage is the phasor difference of two phase voltages.

Phasor diagram of a star connected system shown in Fig. 4.4 shows V_{RN}, V_{YN} and V_{BN} as the RMS values of the phase voltages.

The phase sequence of the system has been assumed as RYB and V_{RN} has been taken as the reference phasor.

The line voltages across the lines, V_{RY}, V_{YB} and V_{BR} are obtained from the phase voltages as explained below.

Line voltage across terminals R and Y,

$$V_{RY} = V_{RN} + V_{NY} = V_{RN} + (-V_{YN})$$

$$= V_{RN} - V_{YN} \text{ (i.e. Phasor difference of two phase voltage phasors)} \quad (4.4)$$

Similarly, line voltage across terminals Y and B, $V_{YB} = (Y_{YN} - V_{BN})$ and the line voltage across terminals B and R, $V_{BR} = (V_{BN} - V_{RN})$.

Thus the line voltages in a star – connected system are obtained by subtracting vectorially, the appropriate phase voltages.

Line voltages so obtained are equal and are displaced from each other by 120°. i.e. Line voltage V_{YB} is lagging the Line voltage V_{RY} by 120° and V_{BR} is lagging V_{YB} by 120°, thus maintaining the phase sequence of the system as RYB.

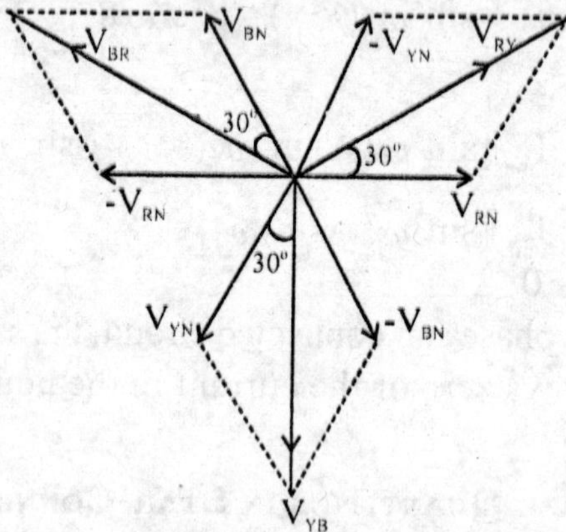

Fig 4.4 : Phasor Diagram of Astar Connected Circuit

Referring to Fig. 4.4, it is quite clear that the line voltages are 30° ahead of the phase voltages.

Line voltage magnitude can be calculated as follows:

Line voltage $V_{RY} = V_{RN} - V_{YN}$

It can be seen from the phasor diagram that by resolving the components along the diagonal, we get:

$$V_{RY} = |V_{RN}| \cos 30° + (|-V_{YN}| \cos 30°)$$

However, $|V_{RN}| = |-V_{YN}| = V_{ph}$

Hence, $V_{RY} = 2 V_{ph} \cos 30°$

$$= 2 V_{ph} \frac{\sqrt{3}}{2} = \sqrt{3}\, V_{ph} \quad (\because \cos 30° = \frac{\sqrt{3}}{2})$$

Similarly, line voltage, $V_{YB} = \sqrt{3}\ V_{YN} = \sqrt{3}\ V_{ph}$

and the line voltage, $V_{BR} = \sqrt{3}\,V_{BN} = \sqrt{3}\,V_{ph}$

Hence, in a star- connected 3- phase system,

Line voltage, $V_L = \sqrt{3} \times$ phase voltage, V_{ph} (4.5)

Line current, I_L = phase current, I_{ph} (4.6)

Example 4.1 : In a 3-phase star connected system, the three phase voltages are given as:

$V_{RN} = 10\angle 0^o$ V, $V_{YN} = 10\angle -120^o$ V and $V_{BN} = 10\angle -240^o$ V

Find the line voltages.

Solution :

The line voltage V_{RY} may be found as , $V_{RY} = V_{RN} - V_{YN}$

$$= 10\angle 0^o - 10\angle -120^o \text{ V} = 10-(-5.0 - j8.66) \text{ V} = 15.0 + j8.66$$

$$= \sqrt{(15)^2 + (8.66)^2}\ \tan^{-1}\frac{8.66}{15} = \sqrt{300}\ \tan^{-1}0.577 = 17.32\angle 30^o \text{ V}$$

The three given phase voltages and the construction of the line voltage phasor V_{RY} are shown in the phasor diagram of Figure 4.5

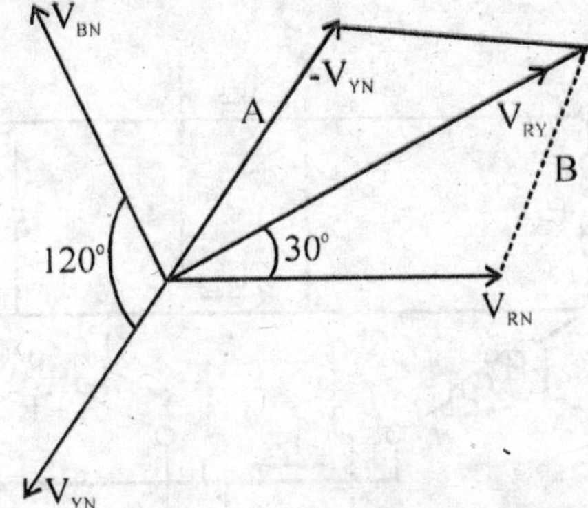

Fig.4.5 : Phasr diagram for Example 4.1

Alternative method of solution

It is seen that voltage $V_{NY} = -V_{YN} = 10\angle 60^o$

$$V_{RY} = V_{RN} + V_{NY} = 10\angle 0^o + 10\angle 60^o \text{ Volts} = 10 + 10\,(0.5 + j0.866) \text{ Volts}$$

$$= 10 + 5 + j8.66 = 15 + j8.66$$

$$= 17.32\angle 30^o \text{ volts Ans}$$

4.7.1.2 Three Phase Y-Connected Loads

It is possible to operate three phase loads from either the **phase voltages**(connecting them between any phase and neutral) or the **line voltages**(connecting them between any two lines). If the loads are to be connected to the phase voltages, fourth wire i.e. a **neutral wire** has to be provided.

If the three phase load is balanced i.e. three phase currents are equal , then the current in the neutral wire is zero.

Let the three phase currents be:

$$I_R = I\angle 0^{\circ}$$

$$I_Y = I\angle -120^{\circ}$$

and

$$I_B = I\angle -240^{\circ}$$

By applying KCL at the neutral point, the current in the neutral I_N is equal to the sum of three incoming phase currents.

$$I_N = I_R + I_Y + I_B = I\angle 0 + I\angle -120 + I\angle -240$$

$$= I(1+j0) + I(-0.5-j0.866) + I(-0.5+j0.866) = 0$$

Hence in a three phase star system having a balanced load, the neutral current is zero and the neutral wire is not usually needed.

A 3-phase Y- connected load connected with 3-phase Y- connected supply is shown in Fig 4.6

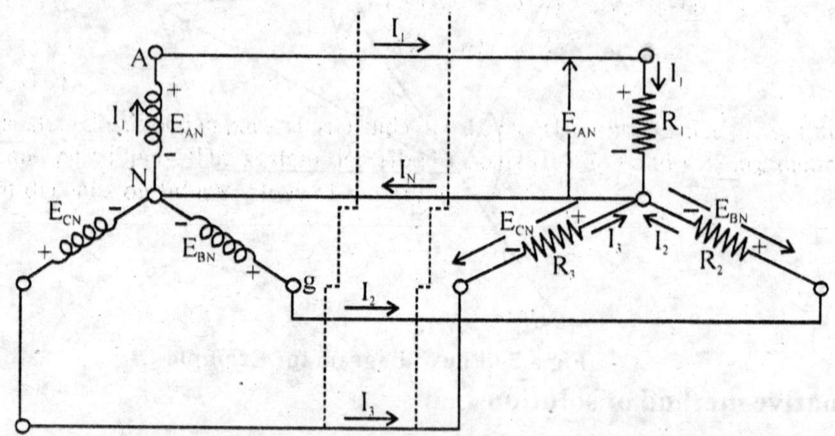

Fig.4.6 : Y–Connected Load With Y–Connected Supply

The voltages across the individual loads are equal to the supply phase voltages.

Line voltage = $\sqrt{3}$ load voltage

Individual load currents are the same which flow in the lines.

∴ **Line current = Individual load current**

4.7.2 Delta Connection

In this type of three phase connection, the three phases are connected such that, starting end of one phase is joined to the finishing end of another phase to form a closed delta or mesh as shown in Fig.4.7

This type of arrangement of three phases is called **Delta or Mesh connection** and is designated by the **symbol,** Δ.

There is no neutral conductor in such a connection.

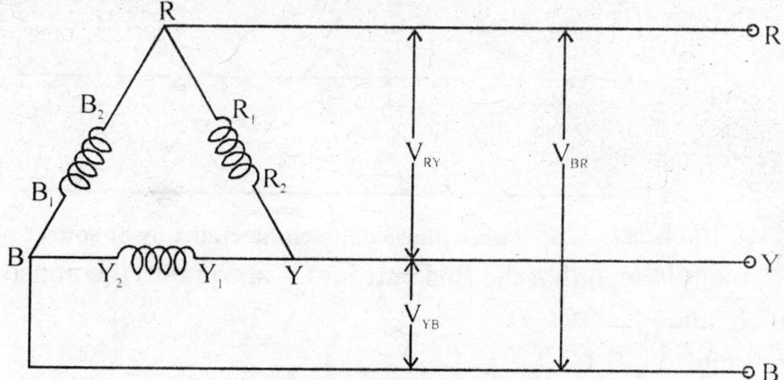

Fig.4.7 : Three phase Delta connection

Let the instantaneous emf's of the three phases be given by :

$$e_R = E_{max} \sin \omega t$$
$$e_Y = E_{max} \sin (\omega t - 120^0)$$
$$e_B = E_{max} \sin (\omega t - 240^o)$$

For a symmetrical, balanced circuit, the circulating current in the closed delta is always zero, i.e. the sum of the instantaneous values of emfs of the three phases is zero at every instant . It can be proved as follows:

i.e. $e_R + e_Y + e_B = E_{max} [\sin \omega t + \sin (\omega t - 120^o) + \sin (\omega t - 240^o)] = 0$

$$= E_{max} \angle 0^o + E_{max} \angle 120^o + E_{max} \angle 120^o$$
$$= E_{max} (1) + E_{max} (-0.5 - j0.866) + E_{max} (-0.5 + j0.866)$$
$$= 0$$

4.7.2.1 Line And Phase Quantities In Delta-Connected Circuit

Figure 4.8 shows a delta- connected system , in which the phase currents I_R, I_Y and I_B. are equal in magnitude and displaced from each other by 120°. Currents in the line conductors are now different than the phase currents.

Let the currents in the three line conductors be I_{L1}, I_{L2} and I_{L3}

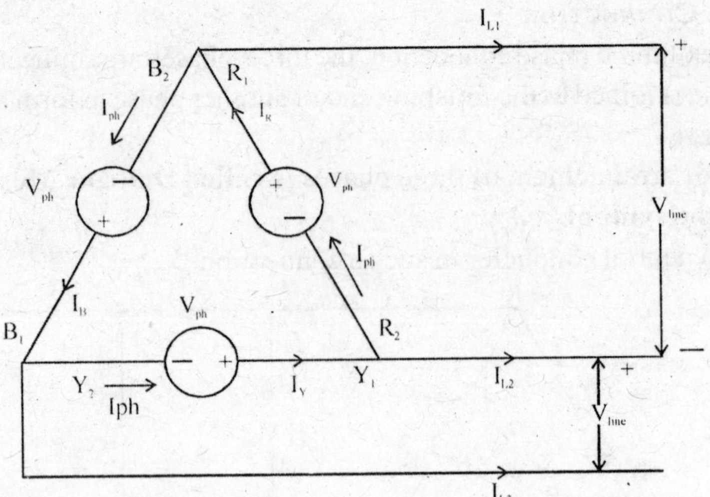

Fig. 4.8 : Three phase delta connected power source

From Fig.4.8, it is seen that the line current I_{L1} is equal to the phasor difference of the phase currents I_R and I_B.

i.e. Line current $I_{L1} = I_R - I_B$

$$= 2\, I_{ph} \cos 30^o = \sqrt{3}\, I_{ph}$$

Similarly, Line current

$$I_{L2} = I_Y - I_R = \sqrt{3}\, I_{ph}$$

Line current $\qquad I_{L3} = I_B - I_Y = I_{ph}$

Hence, in a balanced 3- phase delta- connected system.

Line Current $\qquad I_L = I_{ph}$ (4.7)

Phasor diagram of a three phase delta connected circuit showing phase currents and line currents is shown in Fig.4.9

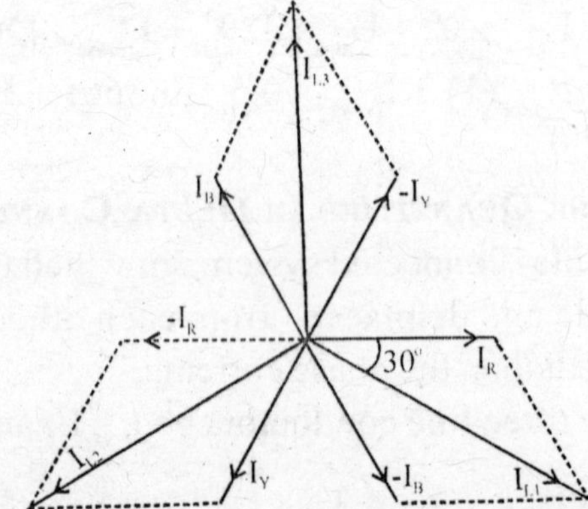

Fig 4.9 : Phasor Diagram of a Delta Connected Circuit

The three line currents I_{L1}, I_{L2} and I_{L3} are equal in magnitude and are displaced from each other by 120°.

Referring to Fig. 4.7 and 4.8, it is obvious that the line voltages are equal to the phase voltages in a delta connected system.

Line Voltage, V_L = Phase Voltage, V_{ph} (4.8)

4.7.2.2 Three Phase Delta Connected Loads

A three phase delta connected load connected to a three phase delta connected supply is shown in Fig. 4.10

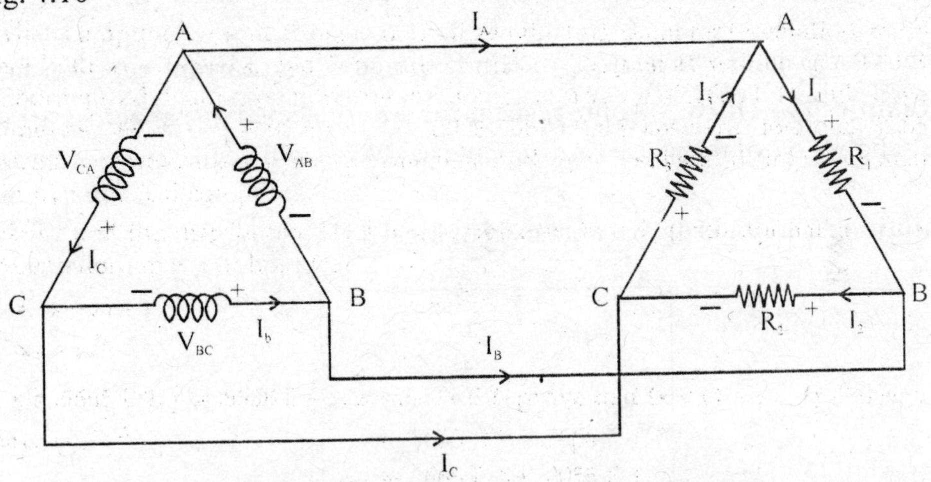

Fig.4.10 : Δ connected load with Δ connected supply

The delta connected load voltages are equal to the line voltages.

Load voltage = Line voltage

Individual load currents combine to produce line currents and,

Line current $I_L = \sqrt{3}$ load current

4.8 Power In Three-Phase System With Balance Load

Power input to a single-phase AC circuit is $P = V_{ph} I_{ph} \cos\phi$.

Thus, power per phase in a 3- phase system is $V_{ph} I_{ph} \cos\phi$,

where V_{ph} and I_{ph} are the rms values of phase voltage and phase current respectively and $\cos\phi$ is the power factor of the load, ϕ is the angle between the phase voltage and the corresponding phase current.

Hence, the total power fed to a 3- phase system with balanced load.

$$= 3 \times \text{power per phase} = 3\, V_{ph}\, I_{ph} \cos\phi \quad (4.9)$$

4.8.1 Power in Three Phase Star Connected Circuit

In star- connected circuit,

Line voltage, $V_L = \sqrt{3}\, V_{ph}$

Line current, $I_L = I_{ph}$

Substituting the phase values in Eq. (4.9) in terms of line values of voltage and current, we get an expression for power in terms of line voltage and line current.

$$\text{Total 3-phase power} = 3\frac{V_L}{\sqrt{3}} \times I_L \cos\varphi = \sqrt{3}\, V_L \times I_L \cos\phi$$

i.e. Total 3-phase power = $\sqrt{3}$ × line voltage × line current × power factor (4.10)

Example 4.2 : Three 100 ohm resistances are connected in star across a 440 V , 50 Hz, 3 phase supply. One of the phases is disconnected. What will be the line current and absorbed power?

Solution : Circuit is drawn as shown in Fig 4.11

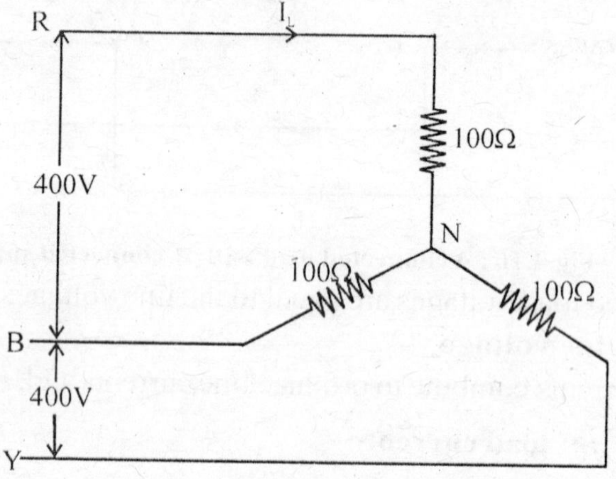

Fig.4.11

If one of the phases , say phase 'Y' is disconnected, then two resistances between terminals RN and terminals NB get connected in series across a supply of 440 volts.

$$\text{Hence line current}, I_L = \frac{440}{100+100} = \frac{440}{200} = 2.2 \text{ amp.}$$ **Ans**

$$\text{Power absorbed, P} = \frac{V^2}{R} = \frac{(440)^2}{200} = 968 \text{ Watts}$$ **Ans**

Example 4.3 : Three inductive coils, each having a resistance of 10 Ω and an inductance of 0.02 H are connected in star to a 3- phase, 410 V, 50 Hz supply. Calculate (i) Phase current and line current and (ii) Total power absorbed by the star connected load.

Solution : Three inductive coils connected in star are shown in Fig. 4.12

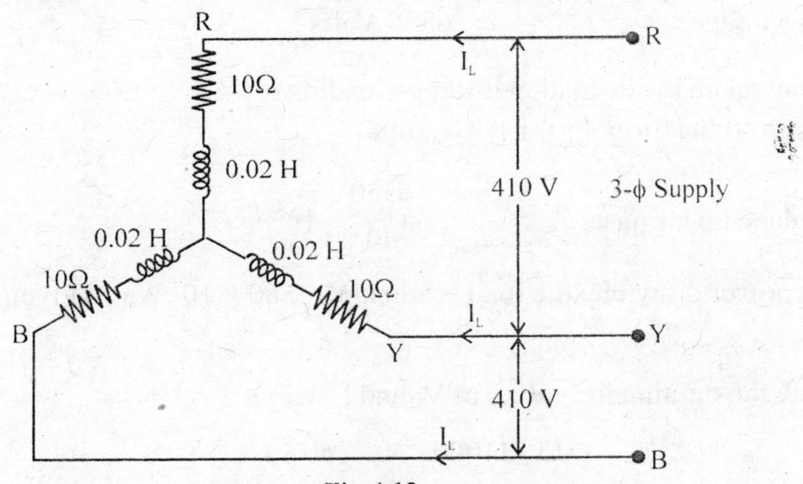

Fig.4.12

Line voltage applied to star connected circuit, V_L = 410 Volts.

$$\text{Phase voltage} = \frac{V_L}{\sqrt{3}} = \frac{410}{\sqrt{3}} = 236.72 \text{ Volts}$$

Inductive reactance of each coil, $X_L = 2\pi \times 50 \times 0.02 = 6.28\ \Omega$

Impedance per phase $= 10 + j\,6.28 = \sqrt{(10)^2 + (6.28)^2} = 11.8$ ohms

(i) Hence, Current per phase $= \dfrac{\textit{Phase voltage}}{\textit{Phase impedance}} = \dfrac{236.72}{11.8} = 20.06$ amp. **Ans**

In star- connected circuit, Line current = phase current
Thus, line current = 20.06 Amp **Ans**

$$\text{Power factor of the circuit, } \cos\phi = \frac{R_{ph}}{Z_{ph}} = \frac{10}{11.8} = 0.847 \text{ (lag)}$$

Total Power absorbed $= \sqrt{3}\, V_L \times I_L \cos\phi$

$= \sqrt{3} \times 410 \times 20.06 \times 0.847 = 12065.52$ watts

$= 12.065$ Kw **Ans**

Example 4.4 : A balanced 3- phase star connected load of 500 kW takes a leading current of 40 amps , when connected to a 3- phase, 11000 V, 50 Hz supply. Find the circuit parameters of the load i.e.resistance, R and capacitance, C per phase.

Solution : Line voltage across the load, V_L = 11000 Volts

Phase voltage, $V_{ph} = \dfrac{11000}{\sqrt{3}} = 6350$ Volts

Current taken by the load = 40 amps(leading)
In a star connection, $I_{ph} = I_L = 40$ amps

Impedance per phase, $Z_{ph} = \dfrac{V_{ph}}{I_{ph}} = \dfrac{6350}{40} = 158.8\Omega$

Total power drawn by the load = 500 kW = 500 × 10^3 Watts(given)
$\sqrt{3} = V_L I_L \cos\phi$
Hence, substituting the values of V_L and I_L we get:

$$500 \times 10^3 = \sqrt{3} \times 11000 \times 40 \times \cos\phi$$

Thus power factor of the load, $\cos\phi = \dfrac{500 \times 10^3}{\sqrt{3} \times 11000 \times 40} = 0.656$

But, power factor, $\cos\phi = \dfrac{R_{Ph}}{Z_{Ph}}$

i.e. $R_{ph} = Z_{ph} \times \cos\phi$
Thus resistance per phase, $R_{ph} = 158.8 \times 0.656 = 104.17\Omega$ **Ans**
As the power factor of the load is leading , hence the load is capacitive.

i.e. $Z_{ph} = \sqrt{R_{ph}^2 + X_C^2}$

$$X_C = \sqrt{Z_{ph}^2 - R_{ph}^2} = \sqrt{(158.8)^2 - (104.17)^2}$$

$$= \sqrt{25217.44 - 10851.389} = \sqrt{14366.05}$$

$$= 119.858\Omega$$

But $X_C = \dfrac{1}{\omega C} = \dfrac{1}{2\pi f C}$

Thus the capacitance/ phase, $C = \dfrac{1}{2\pi f X_C} = \dfrac{1}{2\pi \times 50 \times 119.858} = 26.55 \times 10^{-6}$ Farad

or Capacitance per phase = 26.55 μF **Ans**

4.8.2 Power in Three- Phase Delta Connected Circuits

In a delta- connected circuit,

$$V_L = V_{ph}$$

and $I_L = \sqrt{3}\, I_{ph}$

Substituting these relations in Equation (4.9), we get:

$$\text{Total 3-phase power} = 3 \times V_L \times \frac{I_L}{\sqrt{3}} \times \cos\phi = \sqrt{3}\, V_L \times I_L \cos\phi$$

$$= \sqrt{3} \times \text{line voltage} \times \text{line current} \times \text{p.f} \qquad (4.11)$$

Hence in general, it can be stated that the expression for power in a 3- phase system with balanced load remains the same irrespective of the connection of three phases i.e. star or delta and is given by :

Total 3-phase power = $\sqrt{3}\, V_L I_L \cos\phi$ (**in terms of line values)**

= $3V_{ph} I_{ph} \cos\phi$ (**in terms of phase values)**

Example 4.5 : Three inductive coils, each with a resistance of 10Ω and an inductance of 0.02 H are connected in delta to a 3- phase, 410 V, 50 Hz supply. Calculate (i) Phase current and line current and (ii) Total power absorbed by the delta connected load.

Solution : Figure 4.13 shows the three inductive coils connected in delta

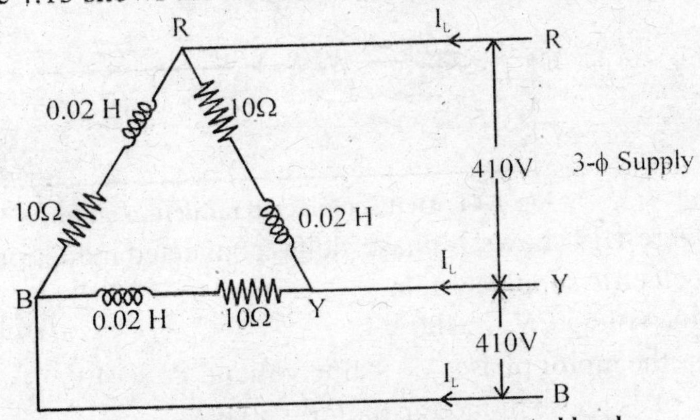

Fig.4.13 : Three Phase Delta connected load

Phase voltage = line voltage = 410 Volts

Inductive reactance of each coil,

$$X_L = 2\pi \times 50 \times 0.02 = 6.28\ \Omega$$

Impedance of each coil or impedance per phase

$$= 10 + j\,6.28 = \sqrt{(10)^2 + (6.28)^2} = 11.8 \text{ ohms}$$

(i) Hence, Phase current

$$= \frac{V_{ph}}{Z_{ph}} = \frac{410}{11.8} = 34.745 \text{ amps}$$ **Ans**

In a delta connection, Line current,

$$I_L = \sqrt{3} \times I_{ph} = \sqrt{3} \times 34.745 = 60.179 \text{ amps}$$ **Ans**

(i) Total three phase Power absorbed,

$$P = \sqrt{3} V_L I_L \cos\phi = \sqrt{3} \times 410 \times 60.179 \times 0.847$$
$$= 36195.96 \text{ W} = 36.196 \text{ kW}$$ **Ans**

Example 4.6 : A 3- phase 50 kW, 50 Hz, delta- connected induction motor is supplied from 3- phase, 400 V, 50 Hz supply system. The efficiency and the power factor of 3- phase induction motor are 90 percent and 0.85 respectively. Calculate (i) Current in each motor phase, (ii)The line current, (iii) Active and reactive components of phase current

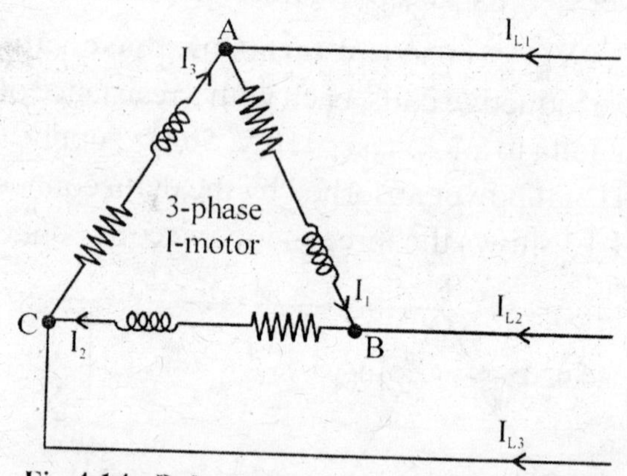

Fig.4.14 : Delta connected induction motor.

Solution : Figure 4.14 shows 3- phase, delta- connected induction motor connected to a 3- phase, 400 V, electric supply system.

(i) Supply Line voltage, $V_L = 400$ V

Voltage across the motor phase, V_{ph} = line voltage $V_L = 400$ V

Output power of 3- phase induction motor = 50 kW

Input power to the motor

$$= \frac{output}{efficiency} = \frac{50 \times 10^3}{0.90} = 55555.55 \text{ Watt}$$

$$= 3 \, V_{ph} \, I_{ph} \cos\phi$$

Substituting the values of V_{ph} and cosϕ, we get :

$$55555.55 = 3 \times 400 \times I_{ph} \times 0.85$$

Thus current drawn by each phase of motor,

$$I_{ph} = \frac{55555.55}{3 \times 400 \times 0.85} = 54.46 \text{ Amp.}$$ **Ans**

(ii) Line current, $I_L = \sqrt{3} \times I_{ph} = \sqrt{3} \times 54.46 = 94.33$ amp. **Ans**

(iii) Active component of phase current $= I_{ph} \cos\phi = 54.4 \times 60.85 = 46.291$ amp **Ans**

Reactive component of phase current $= I_{ph} \sin\phi = 54.4 \times 60.526 = 28.645$ amp **Ans**

Example 4.7 : A 3- phase delta connected AC motor, when connected to a 3- phase 420 V, 50 Hz system is developing 20 kW at an efficiency of 90 percent and a power factor of 0.85. Calculate (i) the line current and (ii) the phase current.

Solution : Output power of the 3- phase AC motor = 20kW = 20×10^3 Watt

Efficiency of the motor = 0.9

Thus, input power to the motor $= \dfrac{output}{\eta} = \dfrac{20\times10^3}{0.9} = 22222$ Watt

Line voltage, $V_L = 420$ V , Power factor, $\cos\phi = 0.85$

However ,Power input to 3-phase AC motor $= \sqrt{3} \times V_L \times I_L \times \cos\phi$

Substituting the values , we get:

$$22222 = \sqrt{3} \times 420 \times I_L \times 0.85$$

(i) Line current drawn by the motor, $I_L = \dfrac{22222}{\sqrt{3} \times 420 \times 0.85} = 35.94$ Amp **Ans**

(ii) As the motor is delta connected, phase current $= \dfrac{1}{\sqrt{3}} \times$ line current

Current flowing in each phase of the motor $= \dfrac{35.94}{\sqrt{3}} = 20.75$ amp **Ans**

4.9 Unbalabnced Load Circuits

In a balanced load, the impedances in all the phases are same. In unbalanced load circuits , the impedances of one or more phases are different from the other phases.

4.9.1 Unbalanced Star- Connected Load Circuits

When an unbalanced star- connected load is connected to a balanced 3- phase 4- wire supply system as shown in Fig 4.15, the line currents are unbalanced and due to this unbalance, the neutral wire carries some current. The voltage across each impedance is same irrespective of the fact that the load is balanced or unbalanced.

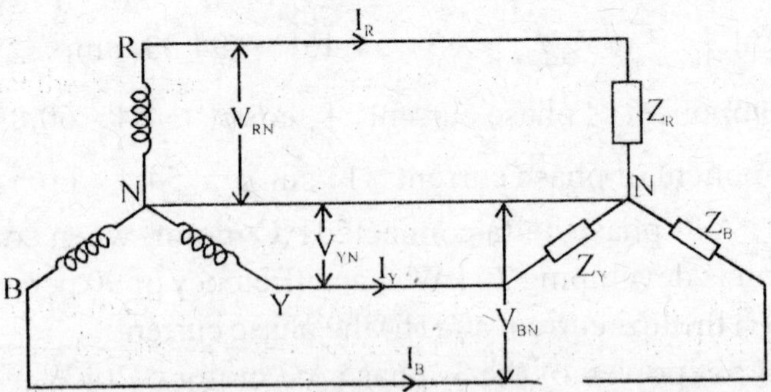

Fig.4.15 : Unbalance Star-Connectec Load

The three phase currents are determined by dividing the phase voltages by the impedance of the concerned phase, i.e.

$$I_R = \frac{V_{RN}}{Z_R},\ I_Y = \frac{V_{YN}}{Z_Y}$$

$$I_B = \frac{V_{BN}}{Z_B} \tag{4.12}$$

where V_{RN}, V_{YN} and V_{BN} are phase voltages and Z_R, Z_Y and Z_B are phase impedances of the load.

Current in the neutral wire, $I_N = I_R + I_Y + I_B$ (Phasor sum) (4.13)

Total three phase power $= V_{RN} I_R \cos\phi_R + V_{YN} I_Y \cos\phi_Y + V_{BN} I_B \cos\phi_B$ (4.14)

Where ϕ_R is the angle between V_{RN} and I_R, ϕ_Y is the angle between V_{YN} and I_Y and ϕ_B is the angle between V_{BN} and I_B

Example 4.8 : Three resistive loads of 2 kW, 4 kW and 6 kW are connected between the three line conductors and neutral of a 3 phase, four wire system with line voltage of 400 V as shown in Figure 4.16. Find (i) Current in each line and (ii) Current in neutral conductor.

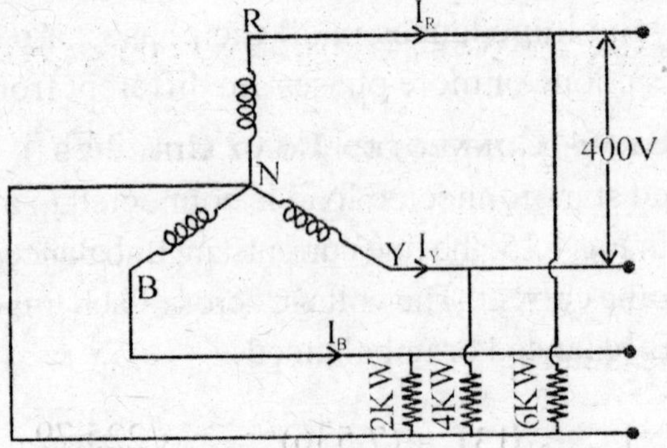

Fig.4.16 : Unbalanced Star- Connected Load

Solution:

(i) Line voltage,

$$V_L = 400 \text{ V}$$

$$\text{Phase Voltage} = \frac{1}{\sqrt{3}} \times V_L = \frac{400}{\sqrt{3}} = 231 \text{ volts}$$

Current in each line can be calculated as follows:

Magnitude of current in R-phase,

$$I_R = \frac{6 \times 10^3}{231} = 26.0 \text{ amps} \qquad \textbf{Ans}$$

Magnitude of Current in Y-phase,

$$I_Y = \frac{4 \times 10^3}{231} = 17.30 \text{ amps} \qquad \textbf{Ans}$$

Magnitude of Current in B-phase,

$$I_B = \frac{2 \times 10^3}{231} = 8.65 \text{ amps} \qquad \textbf{Ans}$$

(ii) Current in the neutral conductor is obtained by calculating the phasor sum of the three currents.

The loads on the three phases are resistive, therefore the current in a particular phase will be in phase with its phase voltage.

Let the currents in all the phases be represented with respect to the current in R-phase as a reference. Assume the phase sequence as RYB.

Phasor current, $I_R = 8.65 \angle 0° = 8.65 + j0$

Phasor current, $I_Y = 17.30 \angle -120°$

$= 17.30 [\cos(-120°) + j \sin(-120°)] = 17.30 (-0.5 - j\,0.866)$

$= -8.65 - j14.981$

Phasor current, $I_B = 26.0 \angle -240°$

$= 26.0 [\cos(-240°) + j \sin(-240°)] = 26(-0.5 + j0.866)$

$= -13 - j\,22.516$

Current in the neutral conductor, $I_N = I_R + I_Y + I_B$ (phasor sum)

Thus, $I_N = 8.65 + j0 + -8.65 - j14.981 + (-13 - j\,22.516) = -13 + j7.536$

or Current in the Neutral,

$$I_N = \sqrt{(13)^2 + (7.536)^2} = \sqrt{225.79} = 15.02 \text{ amp.} \qquad \textbf{Ans}$$

4.9.2 Unbalanced Delta Connected Load

In unbalanced delta- connected load, the voltage across, each impedance is same, independent of the load and is equal to the line voltage. Individual phase currents are calculated by dividing the phase or line voltages by the respective phase impedances. Considering the impedances connected in Fig.4.17 to be unbalanced, the various currents can be calculated as follows:

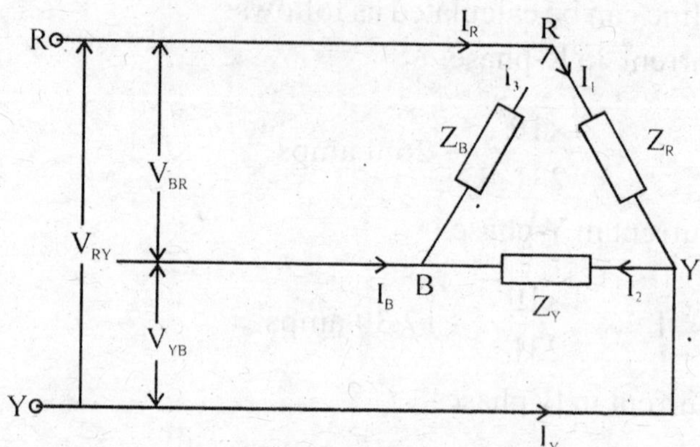

Fig.4.17 : Unbalanced Delta Connected Load

Phase current, $I_1 = \dfrac{V_{RY}}{Z_R}$, $I_2 = \dfrac{V_{YB}}{Z_Y}$

$$I_3 = \frac{V_{BR}}{Z_B}$$

Line currents are the phasor difference of two phase currents and are given by:

$$I_R = I_1 - I_3$$

$$I_Y = I_2 - I_1$$

$$I_B = I_3 - I_2$$

3-phase power is the sum of powers of all the three phases

Hence , 3-phase power which is the sum of powers of each phase is

$$= V_{RY} I_1 \cos\phi_1 + V_{YB} I_2 \cos\phi_2 + V_{BR} I_3 \cos\phi_3$$

Where ϕ_1 is the angle between V_{RY} and I_1 , ϕ_2 is the angle between V_{YB} and I_2 and ϕ_3 is the angle between V_{BR} and I_3

Example 4.9 : An unbalanced delta connected load is connected across a 400 V, 50 Hz, 3- phase supply system as shown in Fig.4.18 . Calculate the phase currents. I_1 ,I_2 and I_3

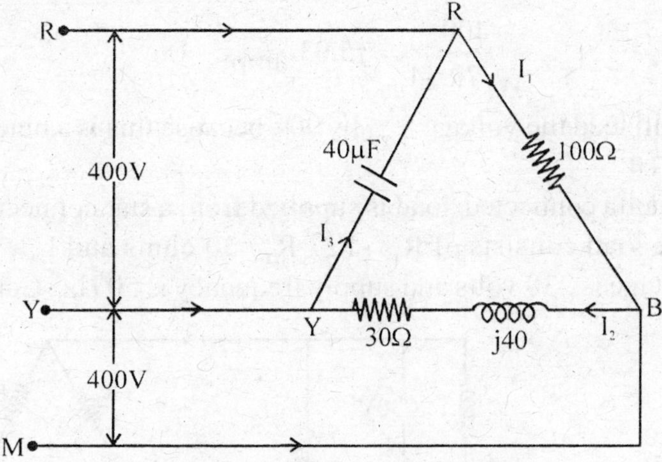

Fig.4.18 : A delta connected load

Solution :

Voltage across the lines,

$$V_L = 400 \text{ V}$$

Phase voltage,

$$V_{ph} = 400 \text{ V} = \text{Line voltage}$$

Impedance of the phase between terminals R and B = 100 Ω (resistive)

Thus phase current

$$I_1 = \frac{400}{100} = 4 \text{ amp}$$

As this load is resistive, the current I_1 will be in phase with the voltage V_{RB}. Impedance of the phase between terminals B and Y = 30 + j40 = 50 Ω

Thus phase current,

$$I_2 = \frac{400}{50} = 8 \text{ amp}$$

Current I_2 will lag the voltage V_{BY} by the impedance angle ϕ_2, which is given by,

$$\phi_2 = \tan^{-1}\frac{40}{30} = 53.12^0$$

Thus I_2 will lag the voltage V_{BY} by an angle 53.12^0

Impedance X_C of the phase between terminals Y and

$$R = \frac{1}{\omega C} = \frac{1}{2\pi fC} = \frac{1}{2 \times 3.143 \times 50 \times 40 \times 10^{-6}} = 79.54 \text{ ohms}$$

Current, $I_3 = \frac{400}{-j79.54} = j5.03$ amp.

Current I_3 will lead the voltage V_{YR} by 90°, because this is a purely capacitive branch.

Example 4.10

A balanced delta connected load is supplied from a star connected generator as shown in Fig. 4.19 .The load consists of $R_1 = R_2 = R_3 = 30$ ohms and $L_1 = L_2 = L_3 = 500$mH. The supply phase voltage is 230 volts and supply frequency is 50 Hz . Calculate the line current.

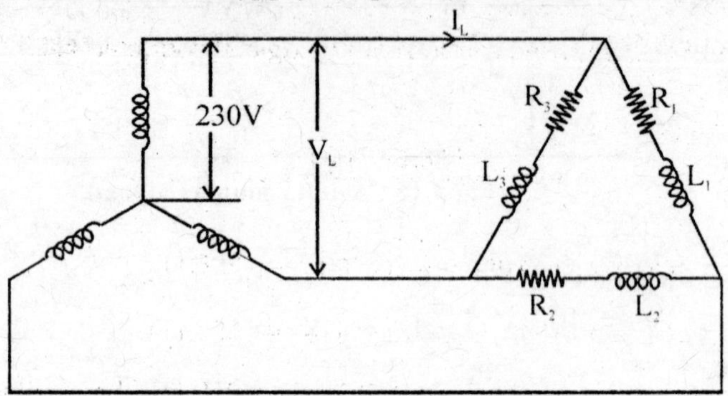

Fig.4.19 : Balanced Delta Connected Load

Solution:

$$V_L = \sqrt{3}\ V_{Ph} = \sqrt{3} \times 230 = 398.36 \text{ volts}$$

$$X_L = 2\pi\ fL = 2\ \pi\ 50\ 500\ 10^{-3} = 157.08 \text{ ohms}$$

$$Z_1 = Z_2 = Z_3 = R + j\,X_L = 30 + j157.08 \text{ ohms}$$

Magnitude of impedance of each phase, $|Z| = = 160.036$ ohms

$$I_{Ph} = I_1 = I_2 = I_3 = \frac{V_L}{Z} = \frac{398.36}{160.036} \times 2 = 2.489 \text{ Amp}$$

$$I_L = \sqrt{3}\, I_{Ph} = \sqrt{3} \times 2.489 = 4.31 \text{ amp}$$ **Ans**

Example 4.11 : A 3- phase, 420 V, star- connected alternator supplies a 3- phase 100 kW mesh- connected induction motor of efficiency 90% and power factor 0.85. Find the current (i) in each motor phase, (ii) in each generator phase.

Solution:

(i) Line voltage across the star connected alternator = 420 V

Line voltage across the motor, $V_L = 420$ V

As the motor is delta connected, voltage across each phase of the motor (Phase voltage),

$$V_{ph} = 420 \text{ V}$$

Output power of the motor = 100 kW = 100000 watt

Efficiency of the motor = 90%

Thus input power to the motor $= \dfrac{100000}{0.9} = 111111$ watt

Power factor of the motor = 0.85 (given)

We know that input power to the motor = $\sqrt{3}\, V_L I_L \cos\phi$

Substituting the values of V_L and cos ϕ, we get:

$$111111 = \sqrt{3} \times 420 \times I_L \times 0.85$$

Hence line current I_L drawn by the delta- connected motor,

$$I_L = \frac{111111}{\sqrt{3} \times 420 \times 0.85} = 179.697 \text{ amp}$$

As the motor is delta connected, phase current $= \dfrac{1}{\sqrt{3}} \times$ line current

(i) Thus current in each phase of motor,

$$I_{ph} = \frac{179.697}{\sqrt{3}} = 103.75 \text{ amp}$$ **Ans**

(ii) Alternator is star connected and supplies a 3- phase motor.

Therefore current in the alternator phase will be the same as the line current drawn by the motor.

Thus current in each phase of alternator = 179.697 amp. **Ans**

4.10 Power Measurement in Three Phase Circuits

The measuring instrument used for the measurement of power is a **wattmeter**. A wattmeter consists of two coils (i) current coil (ii) potential coil. The current coil is connected in series with the load and carries the load current. Potential coil is connected in parallel with the load, just like a voltmeter and it carries current proportional to the voltage across the load.

The power input to a three phase circuit can be measured by the following methods:

(i) One wattmeter method

(ii) Two wattmeter method

(iii) Three wattmeter method

4.10.1 One Wattmeter Method

This method can be used for measuring power of balanced three phase loads only. A wattmeter is connected in one of the lines as shown in Fig 4.20 .

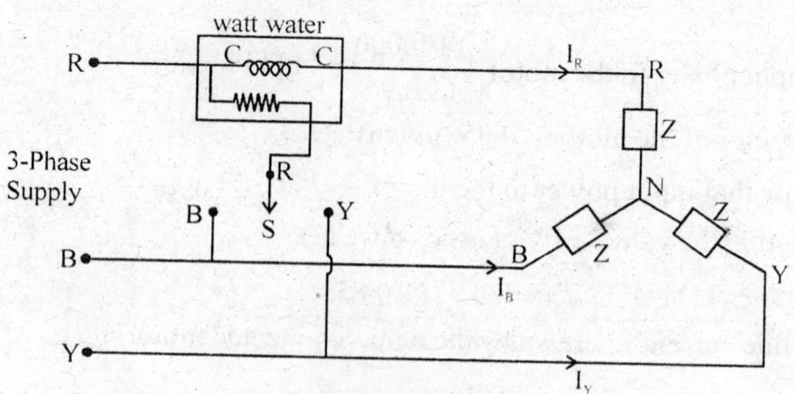

Fig.4.20 : One wattmeter method

The current coil of the wattmeter is connected in series with one of the lines .Using a two way switch 'S', at first potential coil is connected between lines R and Y and let the reading of the wattmeter be W_1 and later between R and B and let the reading of the wattmeter be W_2.

Sum of the two wattmeter readings i.e. $W_1 + W_2$ gives the total power input to the load.

One wattmeter method can be used for balanced three phase four wire systems i.e. where fourth wire or neutral point is available and the load is balanced as follows:

The current coil of the wattmeter is connected in series with one of the lines and potential coil is connected between this line and the neutral point.

The wattmeter reading gives the power input per phase.

Since the system is balanced, the total power input to the three phase system is given by:

Total power input = 3 × wattmeter reading

4.10.2 Two Wattmeter Method

Power in a three phase balanced or unbalanced system can be measured by using two wattmeters.

In this method, the current coils of the two wattmeters are connected in two of the three lines of the three phase circuits. The pressure coils are connected between the two lines in which the current coils are connected and the third line as shown in Fig. 4.21.

It can be proved that the sum of readings indicated by the two wattmeters gives the total power absorbed by the three phase load.

A wattmeter will give a negative reading if the phase angle between the voltage and current exceeds 90^0. In such cases , current coil connections of the wattmeter are reversed and the reading is treated to be negative.

(i) Star connected load

Circuit diagram for measurement of power in a three phase star connected circuit using two wattmeters is shown in Fig 4.21(a)

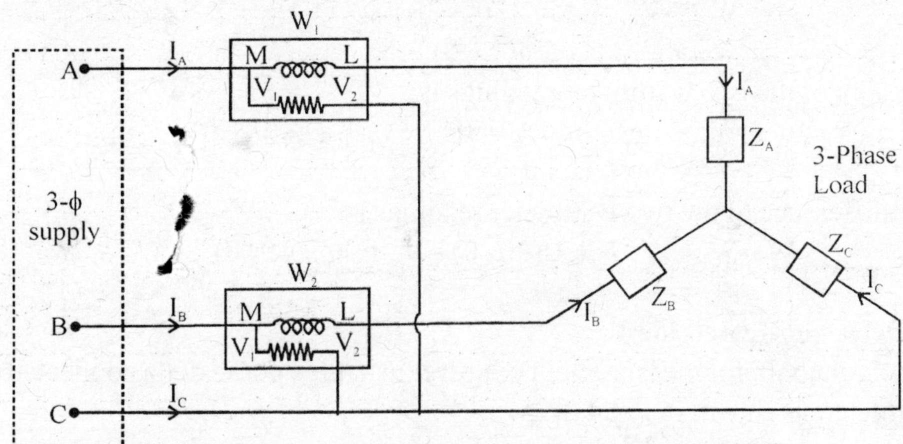

Fig.4.21 : (a) Two wattmeter method Star connected load

Fig. 4.21(b)shows the phasor diagram, where it has been assumed that the load pf is lagging.

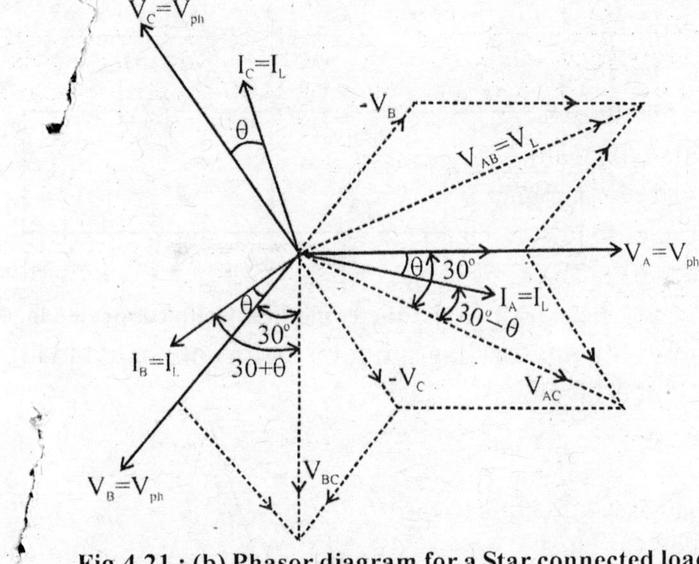

Fig.4.21 : (b) Phasor diagram for a Star connected load

From the phasor diagram , it can be seen that the current I_A leads the voltage V_{AC} by 30-θ and the current I_B leads the voltage V_{BC} by 30 + θ

Power measured by wattmeter W_1 is due to current I_A and voltage V_{AC} because the current through its current coil is I_A and the voltage across its pressure coil is V_{AC}

$$\therefore \quad W_1 = V_{AC} I_A \cos(30^0 - \theta)$$
$$= V_L I_L \cos(30^0 - \theta)$$

Similarly, Power measured by wattmeter W_2 is due to current I_B and voltage V_{BC} because the current through its current coil is I_B and the voltage across its pressure coil is V_{BC}

$$W_2 = V_{BC}\, I_B \cos(30^0 + \theta)$$

or $$W_2 = V_L\, I_L \cos(30^0 + \theta)$$

The sum of the two wattmeter readings is:

$$W_1 + W_2 = V_L\, I_L \cos(\theta - 30^0) + V_L\, I_L \cos(\theta + 30^0)$$
$$= V_L\, I_L \cos\theta$$

The difference of the two wattmeter readings is:

$$W_1 - W_2 = V_L\, I_L \cos(30 - \theta) - V_L\, I_L \cos(30 + \theta)$$
$$= V_L\, I_L \sin\theta$$

(ii)Delta connected load

Circuit diagram for measurement of power in a three phase delta connected circuit using two wattmeters is shown in Fig 4.22(a)

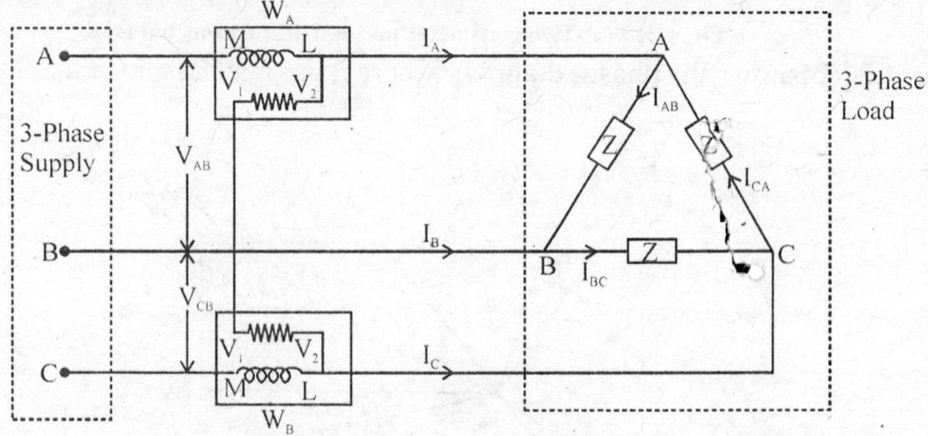

Fig. 4.22 : (a) Two wattmeter method-Delta connected load

Fig. 4.22(b) shows the phasor diagram for a Delta connected load, where it has been assumed that the load pf is lagging.

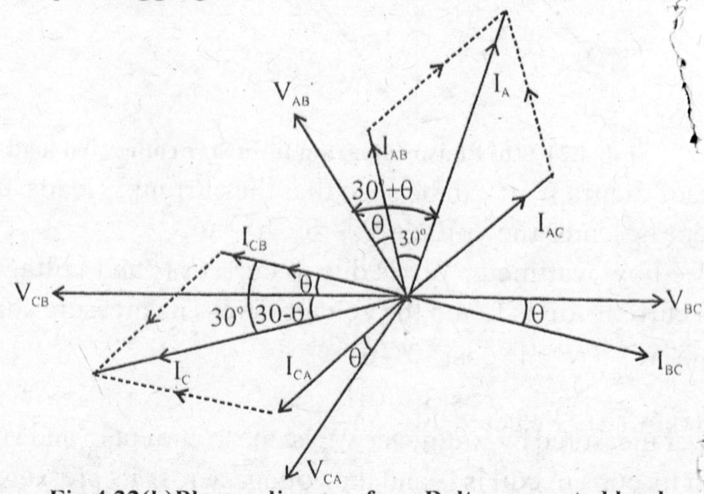

Fig.4.22(b)Phasor diagram for a Delta connected load

When three equal impedances are connected in delta, the resultant line currents make 30 degree angles with their phase currents

Phasor diagram of Fig 4.22(b) corresponds to circuit diagram given in Fig 4.22(a), assuming phase sequence as ABC

Applying KCL at nodes A and C, we get line currents in terms of phase currents:

$$I_A = I_{AB} + I_{AC}$$
$$I_C = I_{CA} + I_{CB}$$

Applying KVL,

$$V_{AB} - V_{CB} = V_{AC}$$

It can be seen from the phasor diagram that : I_A lags V_{AB} by $\theta + 30^0$

Power W_A in phase AB of the load is due to voltage V_{AB} and current I_A:

$$W_A = V_{AB} I_A \cos(\theta + 30^0)$$

I_C lags V_{CB} by $30^0 - \theta$

Power in phase BC of the load is:

$$W_C = V_{CB} I_C \cos(\theta - 30^0)$$

because the current through its current coil is I_C and the voltage across its pressure coil is V_{CB}

In general, the relative order in the voltage sequence of the two lines in which the wattmeters are connected is not known, hence we can write:

$$W_1 = W_A = V_L I_L \cos(\theta + 30^0)$$
$$W_2 = W_C = V_L I_L \cos(\theta - 30^0)$$

The sum of the two wattmeter readings is:

$$\begin{aligned} W_1 + W_2 &= V_L I_L \cos(\theta + 30^0) + V_L I_L \cos(\theta - 30^0) \\ &= V_L I_L [(\cos(\theta + 30^0) + \cos(\theta - 30^0)] \\ &= V_L I_L [\cos\theta \cos 30 - \sin\theta \sin 30 + \cos\theta \cos 30 + \sin\theta \sin 30] \\ &= 2 V_L I_L \cos\theta \cos 30 \\ &= 2 \frac{\sqrt{3}}{2} V_L I_L \cos\theta \\ &= \sqrt{3} V_L I_L \cos\theta \qquad \text{(i)} \end{aligned}$$

The difference of the two wattmeter readings is:

$$\begin{aligned} W_1 - W_2 &= V_L I_L \cos(30 - \theta) - V_L I_L \cos(30 + \theta) \\ &= V_L I_L [(\cos(30 - \theta) - \cos(30 + \theta)] \\ &= V_L I_L [\cos\theta \cos 30 + \sin\theta \sin 30 - \cos\theta \cos 30 + \sin\theta \sin 30] \\ &= V_L I_L \sin\theta \qquad \text{(ii)} \end{aligned}$$

Hence dividing(i) by (ii) , we get :

$$\frac{W_1 - W_2}{W_1 + W_2} = \frac{V_L I_L \sin\theta}{\sqrt{3} V_L I_L \cos\theta} = \frac{\tan\theta}{\sqrt{3}}$$

or $$\tan\theta = \sqrt{3}\frac{(W_1 - W_2)}{(W_1 + W_2)}$$

or $$\theta = \tan^{-1}\sqrt{3}\frac{(W_1 - W_2)}{(W_1 + W_2)}$$

Thus power factor $\cos\theta$ or impedance angle can be found from the two wattmeter readings.

4.10.3 Effect of Load on Wattmeter Readings

For a lagging pf load , the wattmeter readings are given by:

$$W_1 = V_L I_L \cos(30^0 + \theta)$$
$$W_2 = V_L I_L \cos(30^0 - \theta)$$

It is seen that wattmeter readings depend on the load power factor

(a) When $\theta = 0$, ie. pf is unity, then

$$W_1 = W_2 = V_L I_L \cos(30^0)$$

i.e. both wattmeters give equal readings.

(b) When pf is 0.5 (lagging), then

$$W_2 = V_L I_L \cos(30^0 + 60) = 0$$

Hence one wattmeter gives a zero reading and the other wattmeter gives the total power.

(c) When 90 › θ › 60 ⁰, ie. pf is 0.5 › pf ›0, then

W_1 is positive and the reading W_2 will be reversed because the phase angle between current and voltage becomes more than 90^0

Hence total power is obtained by subtracting the reading of W_2 from W_1

(d) When θ =90 ⁰, ie. pf is 0.0 (pure inductive or capacitive load), then

$$W_1 = V_L I_L \sin(30^0)$$
$$W_2 = -V_L I_L \sin(30^0)$$

i.e. two readings are equal but are of opposite sign.

$$W_1 + W_2 = 0$$

It means that the active power consumed by an inductive or a capacitive load is zero.

4.10.4 Three Wattmeter Method

This method is used only for three phase unbalanced loads.

In this method, three wattmeters are used such that one wattmeter is connected in each phase separately, as shown in Fig.4.23.

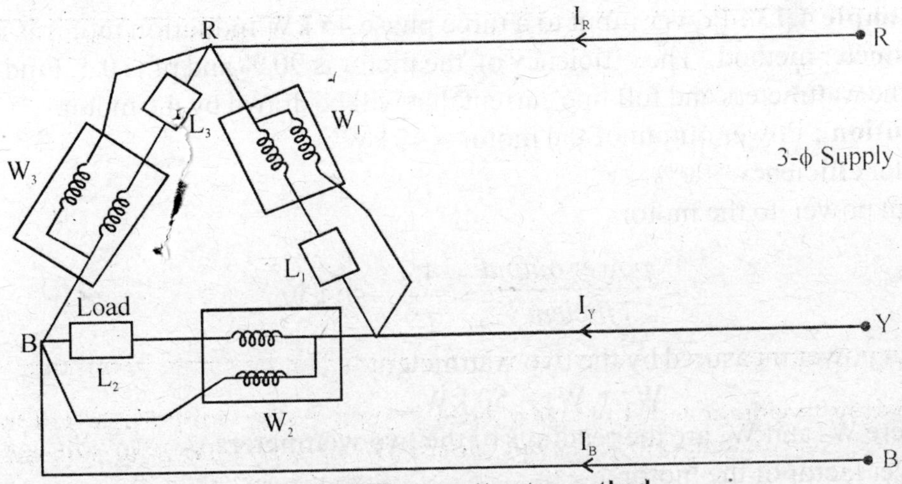

Fig. 4.23 Three wattmeter method

If W_1, W_2 and W_3 are the readings of three wattmeters, then the total power is equal to the algebraic sum of the three wattmeter readings i.e. $(W_1 + W_2 + W_3)$ watts.

The drawback of this method is that it may not be always possible to get the neutral point in the case of a star connected load and it may not be feasible to break in to the different phases of a delta connected load as shown in Fig.4.23.

Some examples are given below to illustrate these concepts.

Example 4.12 : Two wattmeters connected to measure three phase power of a load give the readings as $W_1 = 4kW$ and $W_2 = 2kW$. Reading of the second wattmeter is obtained after reversing the connection to the current coil of the wattmeter. Calculate the power and power factor of the load.

Solution :

$$W_1 = 4kW = 4000 \text{ watts}$$

and $$W_2 = -2kW = -2000 \text{ watts}$$

Power of the load = Power output = sum of the readings of two wattmeters

$$= W_1 + W_2 = 4000 - 2000 = 2000 \text{ watts}$$ **Ans**

Difference of the readings of two wattmeters,

$$W_1 - W_2 = 4000 - (-2000) = 6000 \text{ watts}$$

$$tan\theta = \sqrt{3}\frac{(W_1 - W_2)}{(W_1 + W_2)} = \sqrt{3}\frac{(6000)}{(2000)} = \sqrt{3}(3) = 5.196$$

$$\theta = \tan^{-1} 5.196 = 79^o$$

Hence power factor of the load, $\cos\theta = \cos 79^0 = 0.19$ **Ans**

Example 4.13 : Power input to a three phase 45 kW induction motor is measured by two wattmeter method. The efficiency of the motor is 90 % and pf is 0.5. Find the readings of both the wattmeters and full line current that will be drawn by the motor.

Solution : Power output of the motor = 45 kW

Motor efficiency =90%

Input power to the motor

$$= \frac{power\ output}{efficiency} = \frac{45}{0.9} = 50 \text{ kW}$$

Input power measured by the two wattmeters

$$= W_1 + W_2 = 50 \text{ kW} \qquad \text{(i)}$$

Where W_1 and W_2 are the readings of the two wattmeters.

Power factor of the motor

$$= \cos\theta = 0.5$$

∴ Power factor angle,

$$\theta = \cos^{-1} 0.5 = 60°$$

However, $$\tan\theta = \sqrt{3}\left(\frac{W_1 W_2}{W_1 + W_2}\right) = \sqrt{3}\left(\frac{W_1 - W_2}{50}\right)$$

Substituting the values of θ and($W_1 + W_2$), we get:

$$\tan 60 = \sqrt{3}\left(\frac{W_1 - W_2}{50}\right)$$

Hence, $$W_1 - W_2 = \frac{50 \times 1.732}{\sqrt{3}} = 50 \text{ kW} \qquad \text{(ii)}$$

Adding equations (i) and (ii),we get

$$2W_1 = 100 \text{ kW}$$

i.e. $W_1 = 50$ kW **Ans**

and W_2 = Zero **Ans**

i.e. one of the wattmeter gives a zero reading.

Example 4.14 : Two wattmeters are connected to measure power input to a three phase balanced load . If both the wattmeters give equal reading , calculate the load power factor.

Solution :

$$W_1 = W_2$$

$$\tan\theta = \sqrt{3}\frac{(W_1 - W_2)}{(W_1 + W_2)} = \sqrt{3}\left(\frac{0}{W_1 + W_2}\right) = 0$$

$$\theta = \tan^{-1} 0 = \text{zero}$$

Hence $\quad pf = \cos\theta = \cos(0) = 1$ **Ans**

i.e. load pf is unity.

It means that the load is resistive and the two wattmeters give equal readings for such resistive loads.

Example 4.15 : The total power input to a three phase circuit is measured to be 100 kW by using two wattmeter method. Assume the power factor is 0.8. Find the wattmeter readings.

Solution :

$$\cos\theta = 0.8$$

Hence $\quad \theta = \cos^{-1} 0.8 = 36^0$

$$\tan\theta = \tan 36^0 = \sqrt{3}\left(\frac{W_1 - W_2}{W_1 + W_2}\right)$$

$$\tan 36^0 = 0.7265$$

$$W_1 + W_2 = 100 \text{ KW} \qquad \text{(i)}$$

Hence, $\quad W_1 - W_2 = \dfrac{0.7265 \times 100}{\sqrt{3}} = \dfrac{72.65}{\sqrt{3}} = 41.94 \qquad \text{(ii)}$

Adding equations (i) and (ii), we get :

$$2W_1 = 141.94 \text{ kW}$$

or $\quad W_1 = 70.97 \text{ kW}$ **Ans**

and $\quad W_2 = 100 - 70.97 = 29.03 \text{ kW}$ **Ans**

4.11 SUMMARY

In this chapter concepts of three phase voltages and currents, phase sequence , star and delta connections have been discussed. Relationships between line and phase values of voltages and currents in three phase star and delta connections have been given. Methods for the measurement of power in three phase circuits have been explained.

4.12 SHORT QUESTIONS AND ANSWERS

1. What is meant by a balanced system?

   Ans. It is a system in which currents in the three phases are equal in magnitude and are displaced from each other by equal angles and also all the phase voltages are equal in magnitude and are displaced from each other by equal angles.

2. Give the relation between line and phase values of voltage and current in a star connection.

Ans. In a star connection :

Line voltage = $\sqrt{3}$ × Phase voltage

Line current = Phase current

3. What do you mean by three-phase balanced load?

Ans.Any three-phase load will be balanced when the loads (impedances) connected in three phases are same in magnitude as well as in phase.

4. What is three-phase unbalanced load?

Ans.Any three-phase load will be unbalanced if impedances in one or more phases differ from the impedance (s) of the remaining phase (s).

5. What is phase sequence in a 3-phase system? How is given phase sequence reversed?

Ans.Phase sequence is the order or sequence in which the currents or voltages in different phases attain their maximum values one after the other.

The given phase sequence can be reversed by interchanging any two terminals of the supply.

6. What are the advantages of a three phase system over a single phase system?

Ans. Advantages of three phase systems are :

1. Three phase machines produce about 45% higher output as compared with single phase machines.
2. Size of three phase machines is smaller than that of single phase machines for the same output. For the same frame size, machine rating/output of a three phase machine can be 1.5 times the rating of the single phase m/c.
3. Three phase induction motors are self starting.
4. More power can be transmitted over a three phase system using conductors of same size i.e. same cross sectional area as compared to single phase system.
5. Cross sectional area of conductors is small for a 3-phase system for the transfer of same amount of power at a given voltage, resulting in saving of copper .
6. The power delivered by a single phase system pulsates. In a poly phase system, power delivered is almost constant.
7. p.f. of three phase induction motors is better than single phase induction motors.
8. Direction of rotation of a three phase induction motor can be easily changed.

Review Questions

1. Three impedances each having a resistance of 30 ohms and an inductive reactance of 40 ohms are connected in star across a 400 V , 3 phase supply. Calculate (i) line current(ii) pf (iii) total power in kw.

Ans

2. A 400 V , three phase voltage is applied to a balanced delta connected load having a phase impedance of (3+j4) ohms. Find
(i) Line current in each line
(ii) Power consumed per phase
(i)Phasor sum of three line currents

Ans

3. Draw waveforms showing the instantaneous values of voltages of a three phase balanced system. Assume 'R' as the reference phase having a voltage $V_m = 50$ volts.
4. Three identical impedances are connected in star to a 400 volts three phase supply. the line current is 35 amperes and the total three phase power consumed is 15 kw. Calculate the per phase resistance and reactance

Ans R_{ph} = 4.08 ohm X_{ph} = 5.18 ohm

5. The total power input to a three phase circuit is measured to be 100 KW by using two wattmeter method. Assume the power factor is 0.66 lagging. Find the wattmeter readings

Ans 82.85 kw , 17.15 kw

6. Two wattmeters connected to measure three phase power of a load give the readings as
W_1 = 5 KW and W_2 = 3 KW . Reading of the second wattmeter is obtained after reversing the connection to the current coil of the wattmeter. Calculate the power and power factor of the load.
Ans 8 kw , 0.91

❑❑❑

UNIT-3

Chapter 5

Electromagnetic Induction and Transformations

Chapter 6

Electrical DC Machines

Chapter 5

Electromagnetic Induction and Transformers

OBJECTIVE

- To learn the basic concepts of electro magnetic induction , statically and dynamically induced emf ,self induced emf , mutually induced emf , self and mutual inductance.
- To learn the construction and principle of operation of single phase transformers

5.1 INTRODUCTION

The link between electricity and magnetism was discovered by Oersted in 1824. He discovered that a magnetic field exists around a current carrying conductor i.e. magnetism can be created by means of an electric current. A few years later, in 1831, Faraday, the famous English scientist discovered that a magnetic field can create an electric current in a conductor. When the magnetic flux linking a conductor changes, an emf is induced in the conductor. This phenomenon is known as electromagnetic induction. This important effect has brought a revolution in the engineering world. Most of the electrical devices are based on this principle. This chapter deals with the various aspects of electromagnetic induction.

5.2 ELECTROMAGNETIC INDUCTION

The phenomenon of induction or generation of voltage in a conductor placed in a magnetic field is called **electromagnetic induction**. The fact that came as a surprise to the scientists was that a magnetic field can produce an electric field that can drive a current . This link between a magnetic field and the electric field it produces or induces, is known as Faraday's law of induction. Electric guitars which revolutionized the early rock music, are based on the principle of electromagnetic induction.

5.2.1 FARADAY'S LAWS OF ELECTROMAGNETIC INDUCTION

(a) Faraday's First Law : Whenever the magnetic lines of force (flux) linking with a coil or circuit change, an emf always gets induced in the coil. This emf exists till the change of flux linking the coil takes place.

Or

Whenever a conductor cuts across the magnetic field , an emf is induced in the conductor.

If a conductor is moved up or downward i.e. across the magnetic field , emf is induced in the conductor. However if the conductor is moved along the magnetic field i.e. parallel to the magnetic field, emf is not induced in the conductor.

(b) Faraday's Second Law : The magnitude of such an emf induced in the coil is proportional to the rate at which the conductor cuts or is cut by the magnetic field i.e. it is proportional to the rate of change of flux linkages. Flux linkages mean the product of the number of turns of the coil and the flux linking it.

5.2.2 Mathematical Expression for Magnitude of Induced EMF

Let N = number of turns in the coil

ϕ_1 = Initial flux linking the coil in Webers

ϕ_2 = Final flux linking the coil in Webers

t = time taken in seconds for changing the flux from ϕ_1 to ϕ_2

Then, initial flux linkages of the coil = Nϕ_1

and Final flux linkages of the coil = Nϕ_2

According to Faraday's laws of electromagnetic induction, magnitude of the average induced emf in the coil,

$$E = \text{Rate of change of flux linkages}$$

$$= \frac{\text{Change of flux linkages}}{\text{time}} = \frac{N\phi_2 - N\phi_1}{t} \text{ volts}$$

i.e.

$$E = \frac{N(\phi_2 - \phi_1)}{t} \text{ volts}$$

Putting the above expression in differential form, we get the emf induced at any instant as

$$E = \frac{d}{dt}(N\phi) = N\frac{d\phi}{dt} \text{ volts} \quad ...(5.1)$$

To indicate that emf is induced in such a direction, such that it opposes the cause(i.e. change in flux) that produces it(Lenz's law), a minus sign is used for the induced emf as shown below:

$$E = -N\frac{d\phi}{dt} \text{ volts}$$

5.2.3 Direction of Induced EMF

There are two rules by which direction of induced emf can be established :

1. Fleming's Right Hand Rule
2. Lenz's law

5.2.3.1 Fleming's Right Hand Rule

The direction of the induced emf can be easily found out by applying Fleming's Right Hand Rule (Fig. 5.1), which is stated as follows:

Hold the thumb , the fore finger/first finger and the middle finger/second finger of the right hand mutually perpendicular to each other. If the thumb points to the direction of motion of conductor and the fore finger or first finger to the direction of magnetic field , then the middle or second finger will point to the direction of the induced emf in the conductor.

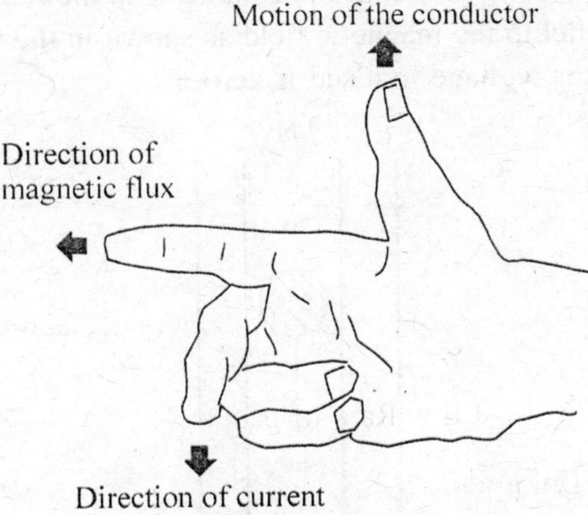

Fig 5.1 Fleming's Right Hand Rule

5.2.3.2 Lenz's Law

Faraday's law gives the magnitude of an induced voltage. The polarity of the induced voltage is determined by Lenz's law.

The law states that the polarity of the electro magnetically induced voltage is such as to oppose the change which produces it. The direction of the induced emf is always such that it tends to cause a current flow in a coil or circuit in such a direction that the magnetic field set up by it opposes the cause which produces it.

5.3 Nature of Induced EMF

An emf is induced in a coil or conductor whenever there is a change in flux linkages. The change in flux linkages of the coil/conductor can be brought about in the following two ways:

(i) If out of the conductor and the magnetic field, one of them is moved physically and the other is kept stationary ,in such a way that conductors cut across the magnetic field, then the induced emf, is known as **Dynamically Induced emf.**

(ii) If emf is induced due to change in flux which is due to the increase or decrease of current in the conductor or coil without moving either the conductor or magnetic field, then the induced emf , is known as **Statically induced emf.**

5.3.1 Dynamically Induced EMF

Consider a stationary magnetic field of flux density B wb/m^2. Let the direction of the magnetic field be as shown in the Fig 5.2. Let a conductor with circular cross section be placed in this magnetic field .

(a) Let 'l' be the effective length of the conductor in the field. in 'm'. Let the conductor be moved parallel to the magnetic field as shown in the Fig 5.2(a). As no flux is cut by the conductor ,voltage induced is zero.

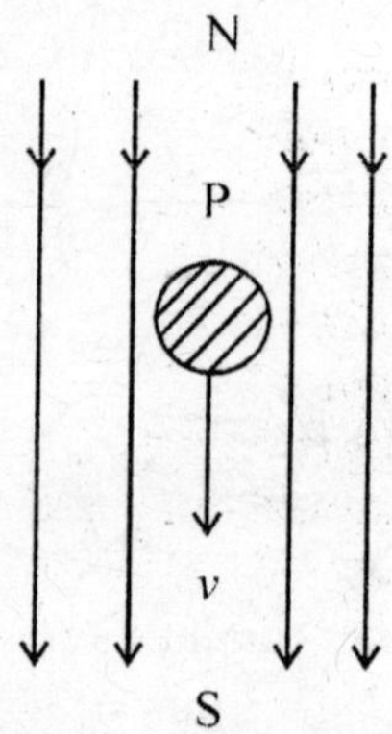

Fig 5.2(a) Conductor moving parallel to the field

(b) Let the conductor be moved at right angles to the magnetic field as shown in the Fig 5.2(b). In a time dt seconds, the distance moved is 'dx' meters.

$\therefore$ Area swept by the moving conductor, A in time 'dt' = l dx m^2

Magnetic flux linked/ cut by the conductor , $d\phi$ = flux density $\times$ area = B A = B l dx wb

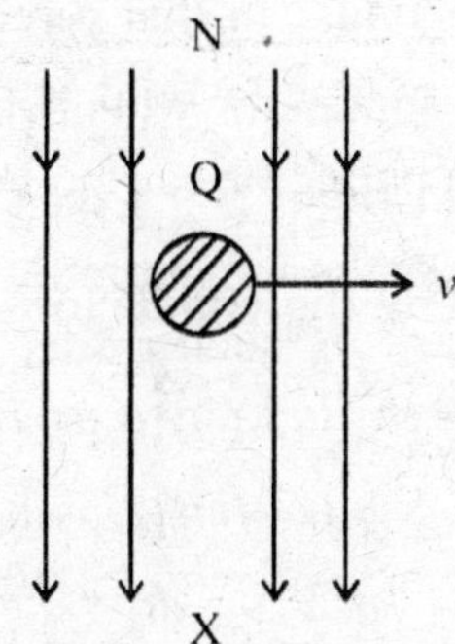

Fig. 5.2(b) Conductor moving perpendicular to the field

If the conductor has one turn, the flux linkage = B l dx

Rate of change of flux linkages

$$= \frac{Bldx}{dt}$$

According to Faraday's law of electro magnetic induction , the emf induced in the conductor is:

$$E = \text{Rate of change of flux linkages}$$

$$= \frac{Bldx}{dt} \text{ volts}$$

i.e.
$$E = B\,l\,v \text{ volts} \qquad ...(5.2)$$

where,
$$v = \text{linear velocity} = \frac{dx}{dt}$$

(c) Let the conductor be moved with the same velocity 'v' m/sec in an inclined direction making an angle 'θ' to the direction of the magnetic field as shown in Fig. 5.2 (c).

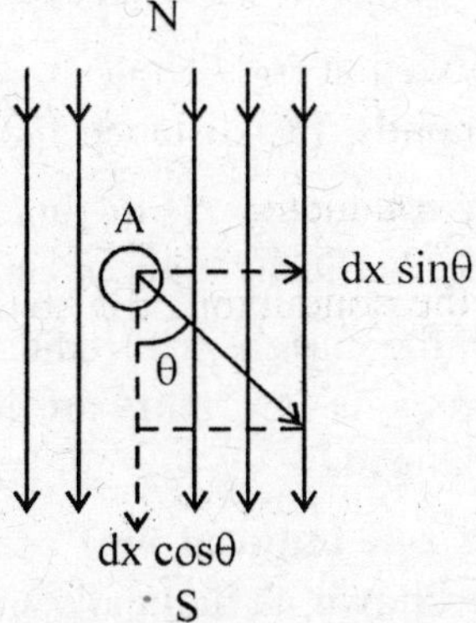

Fig 5.2(c) Conductor moving at an angle θ to the field

Let the distance covered in dt seconds in that direction be dx. Then the component of distance which is **perpendicular** to the magnetic field is dx sin θ.

$\therefore$ Area swept by the conductor in time 't' = l dx sin θ.

Hence, Magnetic flux linked/ cut by the conductor , $d\phi$ = B A =B l dx sinθ wb

The emf induced in the conductor is :

$$E = \frac{d\phi}{dt} = \frac{Bldx}{dt} \sin\theta \text{ volts}$$

i.e. The induced emf in the conductor is :

$$E = B\, l\, v \sin\theta \text{ volts} \qquad ...(5.3)$$

It can be seen that the magnitude of induced emf has been reduced by a factor equal to sinθ.

5.3.2 Statically Induced EMF

In this case, both the conductor and the magnetic field are held stationary and the magnetic field is varied by changing the magnitude of the current that produces the changing magnetic field and the emf may be **Self induced emf or Mutually induced emf.**

Consider two coils A and B wound over a magnetic core as shown in Fig 5.3. Coil A is energized by a voltage of V volts. If switch 'S' is closed, then a steady current of 'I_1' amp. will flow through the coil A. It will produce a flux of ϕ wb. Let us assume that this entire flux links coils A and B.

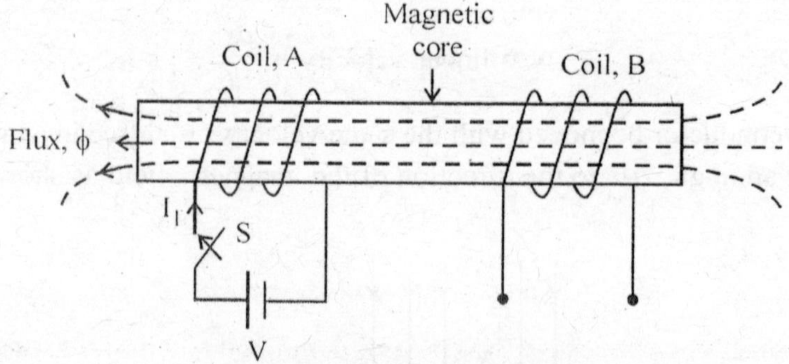

Fig. 5.3 Statically induced emf

If the switch is suddenly opened, the current will reduce to zero and the flux linking both the coils will become zero. As per Faraday's law, emfs are induced in both the coils A and B, because flux linking these coils has changed.

EMF induced in coil A is called **self induced emf** and emf induced in coil B is called **mutually induced emf**. Such emfs are known as Statically induced emfs. These are also called **"transformer emfs."**

Statically induced emf can be classified into two categories, namely, self-induced emf and mutually induced emf.

5.3.2.1 Self Induced EMF

If a single coil carries a current, a flux will be set up in it. If the current in the coil changes, the flux will change. This change in flux will induce an emf in the coil. This kind of emf is known as "**self induced emf.**"

In other words, self-induced emf is the emf induced in a coil or circuit when the magnetic flux linking it changes , the flux being produced by current in the same coil or circuit.

The magnitude of this self induced emf is $E = N\dfrac{d\phi}{dt}$, where 'N' is the number of turns in the coil.

The direction of this induced emf would be such as to oppose any change of flux which is the very cause of production of emf..

Hence, it is also known as the **opposing** or **counter emf** of self-induction.

Mathematically it can be expressed as:

$$E \propto \frac{di}{dt}$$

or,

$$E = L\frac{di}{dt} \qquad \text{...(5.4)}$$

where L is coefficient of self-induction and is known as **Inductance.**

5.3.2.2 Mutually Induced EMF

It is the emf induced in a coil/circuit , due to change of flux linking it , flux produced by current flowing in another coil or circuit.

Referring to Fig 5.3 , if current through coil A is changed, the flux-linking coil B will change. Hence, an emf will be induced in coil B, which is called the mutually induced emf. The same phenomenon can be observed in coil A when current in coil B is changed.

Example 5.1 : A magnetic flux of 0.9 m Wb passing through a coil of 500 turns is reversed in 0.2 sec. Calculate the average value of emf induced in the coil.

Solution : The magnetic flux is first decreased from its value of 0.9 mWb to zero and then increased in the reverse direction to - 0.9 mWb.

Hence the change of flux is

$$= \phi_2 - \phi_1 = -0.9 - 0.9 = -1.800 \text{ mWb.}$$

The emf induced in the coil,

$$E = \frac{d\phi}{dt} N$$

Number of turns in the coil, $N = 500$,

Change in flux, $d\phi = -1.800$ mWb

Time in which change of flux takes place,

$$dt = 0.2 \text{ sec.}$$

Thus average emf induced in the coil

$$= \frac{500 \times (-1.8 \times 10^{-3})}{0.2} = -4.5\text{V}$$ **Ans.**

Example 5.2 : A wire whose length is 100 cm moves at right angles to its length at a velocity of 50m/s in a uniform magnetic field of flux density 1.0 Wb/m^2. Calculate the electromotive force induced in the conductor when the direction of motion is (a) perpendicular to the field, (b) inclined at 30° to the direction of field.

Solution : Length of the conductor,

$$l = 100\text{cm} = 1.0 \text{ m}$$

Flux density of uniform magnetic field,

$$B = 1.0 \text{ Wb/m}^2 \text{ (given)}$$

Velocity of motion of conductor,

$$v = 50\text{m/s (given)}$$

(a) EMF induced in the conductor, when the direction of motion of the conductor is perpendicular to the uniform field :

$$E = Blv = 1.0 \times 1.0 \times 50 = 50.0 \text{ V}$$ **Ans.**

(b) EMF induced when the direction of motion of the conductor is inclined at 45° to the direction of magnetic field:

$$E = Blv \sin\theta = 50.0 \sin 30^\circ = 25.0 \text{ V}$$ **Ans.**

Example 5.3 : A coil having 100 turns is placed in a magnetic field of 1.0 m Wb. Calculate the average emf induced in the coil when it is moved in 0.05 sec. from the given field to a field of 0.5 mWb.

Solution : Number of turns in the coil,

$$N = 100 \text{ (given)}$$

Change in flux due to motion of conductor,

$$d\phi = \phi_2 - \phi_1$$

$$= 0.5 - 1.0 = -0.5 \text{ mWb} = -0.5 \times 10^{-3} \text{ Wb}$$

Time, $dt = 0.05$ secs

Induced emf, $e = N \dfrac{d\phi}{dt}$

$$= 100 \times \frac{(-0.5)\times 10^{-3}}{0.05} = -1.0 \text{ V}$$ **Ans.**

5.4 Self Inductance (L)

A magnetic circuit is the heart of every energy conversion device. The magnetic circuit consists of a coil or coils wound around a highly magnetic material. Since the coil is made of a conducting material such as copper , it has a finite resistance. In the design of an electrical machine , an attempt is always made to keep the resistance of the coil as small as possible in order to minimize the power loss in the coil. The resistance of the coil always forms part of the equivalent circuit.

The other part of the equivalent circuit is due to the flux in the magnetic core of the device and its interaction with the coils wound around it.

In this section mathematical relationships are developed to represent a magnetic circuit by its equivalent self and mutual inductances.

5.4.1 Definition

Self Inductance is the property of a circuit element due to which a voltage is induced in the element by changing the current in it. When the voltage is induced in the element by the current changing in its own circuit , the property is called the self inductance or simply inductance.

The property of the coil to oppose any change in current flowing through it , is known as the **self inductance** or **inductance** of the coil.

Self inductance of a circuit is the flux linkages per unit current in it.

Its unit is Henry.

i.e. $$L = \frac{N\phi}{I} \text{ Henry} \qquad ...(5.5)$$

5.5 Relationship Between Self Induced Emf and Self Inductance

We know that induced emf is given by

$$E = N \frac{d\phi}{dt}$$

and, Inductance $L = \dfrac{N\phi}{I}$

or $$L\,I = N\phi$$

For a small increment of current 'd i' let the incremental flux produced be equal to 'dϕ'.

i.e. $$L\,di = N d\phi$$

$$\therefore \quad L\frac{di}{dt} = N\frac{d\phi}{dt} = E$$

Hence we can write :

$$E = L\,\frac{di}{dt}$$

L is the constant of the circuit and is called the **Self Inductance (or) Coefficient of Self-induction**.

This is the form of Faraday's law of electromagnetic induction which is commonly used in the analysis of electric circuits.

The concept of inductance is very useful because it provides a simple expression to express the induced emf in the coil directly in terms of the current producing it. Without this concept , first flux has to be determined and then faraday's law is to be applied to obtain the induced emf.

5.6 Mutual Inductance (M)

When two coils or circuits are placed near each other and a change in current in one coil produces a change in magnetic flux which links both the coils , these coils are said to be mutually coupled.

When the flux from a coil cuts another coil , an emf is induced in the second coil. This emf induced sets up a flux which opposes the original flux from the first coil. The induced emf effect in such cases is referred to as due to mutual inductance.

Mutual inductance is the property of two circuits , when a voltage is induced in one circuit due to a change of current in the other circuit. In other words, Mutual inductance may be defined as the ability of one coil to produce an emf in a nearby coil by induction, when the current in the first coil changes. This ability is reflected in terms of Coefficient of Mutual Induction, M.

5.6.1 Definition

Mutual inductance between two circuits/ coils is defined as the flux linkages of one circuit/ coil per unit current in the other circuit/coil. Its unit is Henry.

If a current I_1 flows through coil1 having N_1 turns , produces a flux ϕ_1 wb and if this flux ϕ_1 links with coil2 having N_2 turns, then the flux linkage will be $N_2\,\phi_1$ and hence mutual inductance of coil 2 due to current in coil 1 is:

Mutual inductance between coil 1 and coil 2,

$$M_{12} = \frac{N_2\phi_{12}}{I_1} \qquad ...(5.7)$$

where is the flux linking with coil 2 due to Current I_1 in coil 1, and

Mutual inductance between coil 2 and coil 1,

$$M_{21} = \frac{N_1\phi_{21}}{I_2} \qquad ...(5.8)$$

where is the flux linking with coil 1 due to current I_2 in coil 2

5.6.2 Relationship Between 'M' and Induced EMF

Referring to Fig. 5.5, mutually induced emf in coil 2 due to change in current of coil 1 is,

$$E_{m2} = M\frac{di_1}{dt} \qquad ...(5.9)$$

Similarly, mutually induced emf in coil 1 due to change in current of coil 2 is,

$$E_{m1} = \frac{di_2}{dt}M \qquad ...(5.10)$$

Thus, the mutually induced emf in a circuit is directly proportional to the rate of change of current in another circuit. The constant of proportionality (M) is called the **Mutual Inductance or the Coefficient of Mutual Induction.**

5.7 Coupling Coefficient Between Two Magnetically Coupled Circuits (K)

The two coils are said to be magnetically coupled if the flux produced by one coil links with the other fully or partially.

Normally only a part of flux produced by one coil links with the other coil because some flux leaks and does not link with the other coil.

Let I_1 = Current in coil 1, ϕ_1 = Flux caused by current I_1

Let I_2 = Current in coil 2, ϕ_2 = Flux caused by current I_2

There will be mutual flux linkages when both the coils are energized simultaneously.

Let a fraction of flux ϕ_1, say $K_1\phi_1$, link coil 2

The, Flux linkages of coil 2 is,

$$\psi_{21} = N_2(K_1\phi_1)$$

Mutual Inductance, $$M = \frac{\psi_{21}}{I_1} = \frac{N_2 K_1 \phi_1}{I_1} \quad \text{...(i)}$$

We know that, $$\phi_1 = \frac{N_1 I_1}{\mathfrak{R}} \quad \text{...(ii)}$$

Substituting the value of in equation (i), we get:

$$M = \frac{N_2 K_1 N_1 I_1}{I_1 \mathfrak{R}} = \frac{N_1 N_2 K_1}{\mathfrak{R}} \quad \text{...(iii)}$$

Where $\mathfrak{R}$ = Reluctance of the magnetic circuit = $\frac{l}{\mu_0 \mu_r a}$

Let a fraction of flux ϕ_2, say $k_2 \phi_2$, link coil 1.

Flux linkages of coil 1 is

$$\psi_{12} = N_1 (K_2 \phi_2)$$

Mutual inductance,

$$M = \frac{\psi_{12}}{I_2} = \frac{N_1 K_2 \phi_2}{I_2} \quad \text{...(iv)}$$

$$\phi_2 = \frac{N_2 I_2}{\mathfrak{R}}$$

Substituting the value of ϕ_2, we get:

$$M = \frac{N_1 K_2}{I_2} \left(\frac{N_2 I_2}{\mathfrak{R}} \right)$$

i.e. $$M = \frac{N_1 N_2 K_2}{\mathfrak{R}} \quad \text{...(v)}$$

Multiplying equations (iii) and (v), we get:

$$M^2 = K_1 K_2 \frac{N_1^2 N_2^2}{\mathfrak{R}^2}$$

Substituting $K_1 K_2 = K^2$, we get:

$$= K^2 \left(\frac{N_1^2}{\mathfrak{R}} \right) \left(\frac{N_2^2}{\mathfrak{R}} \right)$$

$$M^2 = K^2 \; L_1 L_2 \qquad ...(5.11)$$

Where L_1 and L_2 are the self-inductances of the coils 1 and 2 respectively

$$\therefore \qquad K^2 = \frac{M^2}{L_1 L_2};$$

or

$$K = \frac{M}{\sqrt{L_1 L_2}} \qquad ...(5.12)$$

Or

$$M = K \sqrt{L_1 L_2} \qquad ...(5.13)$$

Where 'K' is coupling coefficient and is given by:

Coupling coefficient, $$K = \frac{M}{\sqrt{L_1 L_2}}.$$

K is always ≤ 1

The coupling between the two coils is said to be a **tight** coupling , if K=1 and it is called **loose** coupling if K is less than one.

Example 5.4 : Two identical coils A and B consisting of 1000 turns each lie in parallel planes such that 60 percent the flux produced by the current in coil A links the coil B. Current of 2 amp flowing in the coil A produces in it a flux of 0.05 mWb. Calculate (i) Self – inductance of each coil and (ii) Mutual inductance between them.

Solution : (i)Self-inductance of a coil 'A' is given by,

$$L_A = \frac{N_A \phi_A}{I_A}$$

Number of turns in coil A,

$$N_A = 1000 \text{ (given)}$$

Flux linking coil A,

$$\phi_A = 0.05 \text{ mWb} = 0.05 \times 10^{-3} \text{ Wb}$$

Current in coil A,

$$I_A = 2 \text{ Amps}$$

Thus, self-inductance of coil A,

$$L_A = \frac{1000 \times 0.05 \times 10^{-3}}{2}$$

$$= 25 \times 10^{-3}\, Henry$$

$$= \mathbf{25\ mH}$$ **Ans.**

As the coils are identical, coil B will also have the same self- inductance.

∴ Self-inductance of coil B = **25 mH** **Ans.**

(ii) Mutual inductance between the coils A and B is given by :

$$M = \frac{N_B K \varphi_A}{I_A}$$

Number of turns in coil B,

$$N_B = 1000 \text{ (given)}, K = 0.6$$

Flux linking coil B = $K\phi_A$

$$= 0.6 \times 0.05 = 0.030 \text{ mWb} = 0.030 \times 10^{-3} \text{ Wb}$$

Mutual Inductance,

$$M = \frac{1000 \times 0.030 \times 10^{-3}}{2}$$

$$= 15 \times 10^{-3} \text{ H} = \mathbf{15\ mH}$$ **Ans.**

Example 5.5 : A solenoid consists of 800 turns of wire wound on a length of 100 cm. A search coil of 200 turns enclosing a mean area of 20cm^2 is placed centrally in the solenoid and mutual inductance of the arrangement is 0.402 mH, find emf induced in the search coil when the current in the solenoid is changing uniformly at the rate of 100 A/s.

Solution : EMF induced in the search coil due to a change in current in the solenoid

$$= M \frac{di}{dt}$$

Current in the solenoid is changing uniformly at a rate of 100 A/s,

i.e. $$\frac{di}{dt} = 100$$

Thus, emf induced in search coil

$$= 0.402 \times 10^{-3} \times 100 = 0.04 \text{ V}$$ **Ans.**

5.8 Introduction of Transformers

A Transformer transforms electrical energy from one level of voltage and current to another level of voltage and current without any change in frequency. The energy transfor-mation takes place via magnetic field.

As energy handled on both input and output sides is electrical energy, transformer has no moving parts and hence mechanical losses are absent. Therefore, its full-load efficiency is higher than any other rotating energy converting machine, and is around 98%.

Transformer is one of the most useful electrical devices and has made possible the widespread use of alternating current for generation, transmission, distribution and utilization of electrical energy. In this chapter, fundamental concepts of transformers will be illustrated.

5.9 Classification of Transformers

Transformers may be broadly classified into three groups according to the area of application

(a) Power transformers

- For transmission and distribution of electrical power.

(b) Communication transformers

- For impedance matching of loads and sources to achieve maximum transfer of power to the loads.
- For isolation of different circuits/systems.

(c) Instrument transformers

- Used for measuring high voltages and high currents by means of normally available small range measuring instruments
- Used with relays for protective equipment

Transformers can also be classified according to the following different criteria :

(i) Duty they Perform /service conditions:

1. Power transformer–for transmission. and distribution of power.
2. Current Transformers
3. Potential Transformers
4. Welding Transformers
5. Furnace Transformers
6. Radio Transformers

(ii) Construction /Relative position of winding and core

1. Core Type
2. Shell type
3. Berry type

(iii) Methods/Type of Cooling

1. Dry Type
   - (i) AN Air natural
   - (ii) AB Air blast
2. Oil Immersed
   - (i) ON Oil natural
   - (ii) OB Oil blast
   - (iii) OF Forced oil cooling
   - (iv) OW Oil and water cooled
   - (v) OFW Forced oil and water cooled

(iv) Number of Phases / windings

1. Single Phase
2. Three Phase – star-star, delta-delta, star-delta , delta-star, open-delta, scott connection etc.
3. Auto transformer

5.10 CONSTRUCTIONAL FEATURES OF A TRANSFORMER

A single-phase transformer consists of a laminated steel magnetic core and two wind-ings. These windings are called Low-Voltage (LV) winding and High-Voltage (HV) winding.

The transformer core is made up of silicon steel laminations, which have high permeability and offer a low reluctance path for the magnetic flux .Hence the magnetic field produced in the core is very strong. As the magnetic flux is alternating in nature, the flux reversals cause the magnetic losses in the core i.e., Hysteresis and Eddy current losses. To reduce Hysteresis loss, silicon is added to steel to manufacture core stampings. Stacking the core by thin insulated stampings , reduces Eddy current loss.

The laminations are generally of 0.35 to 0.5 mm thickness. The core is assembled such that it provides a continuous path for the magnetic flux. The cross section of the limb of the core is rectangular for small transformers. However, the cross section is either square or stepped for large size transformers.

The LV and HV windings are wound with copper wires.

There are two basic types of transformer construction:

(i) Core type Transformer

(ii) Shell type Transformer

5.10.1 Core Type Transformer

In this type, core is in the form of a rectangular frame and windings surround a considerable part of core .It has only one path for magnetic flux as shown in Fig 5.4 (a) and(b).the laminations are cut in the form of L shaped strips.

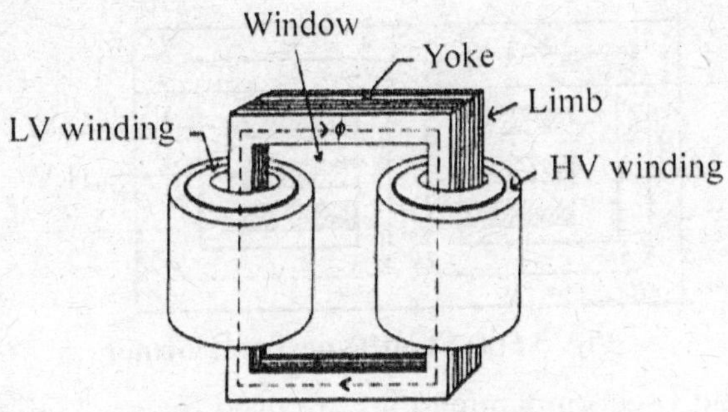

Fig. 5.4(a) Core type transformer

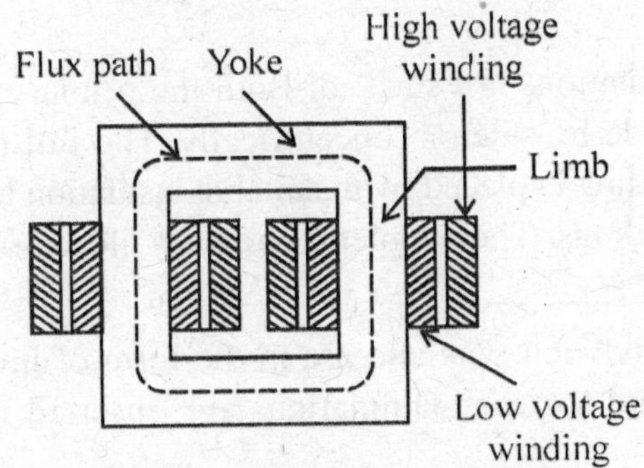

Fig. 5.4(b) Core type transformer

Single-phase core type transformer has two limbs, around which windings are uniformly distributed . Both the windings are of cylindrical shape.

Initially laminations are stacked. After placing sufficient insulation around the limbs, pre wound LV windings are inserted around the limb near the core. LV winding on each limb has half the required number of turns. These two half LV windings on each limb are connected in series and the resultant two terminals are taken out for making external connections. After providing suf-ficient insulation around the LV winding, half HV coils are placed on each limb over the LV windings as shown in Fig. 5.4(b). The two half HV coils are connected in series and resultant two terminals are taken out for making ex-ternal connection. The windings are form wound in cylindrical shape and the core laminations are inserted later on.

5.10.2 Shell Type Transformer

The shell type Transformer has three limbs- two outer limbs, one middle limb, two yokes and two windows. There are two flux paths. Full flux, ϕ passes through the middle limb and in the remaining parts of the core, the flux is less than $\phi/2$. Hence, the width of middle or central limb is twice as that of outer limbs. This type of core provides a better mechanical support and protection for the windings.

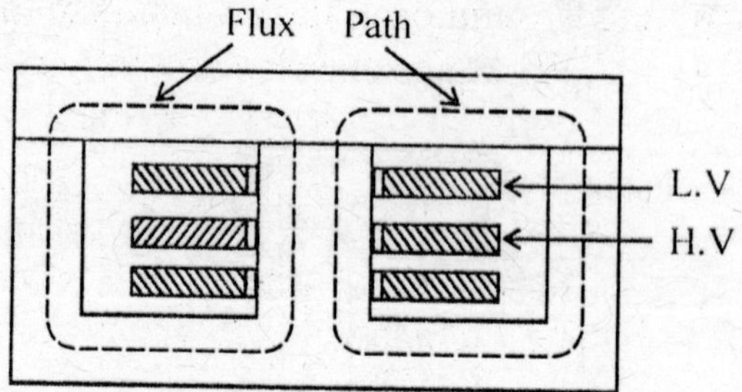

Fig. 5.5(a) Shell Type Transformer

Long strips of E and I section laminations are used for stacking the shell type transformer core. In order to avoid high reluctance at the joints, the alternate layers are stacked differently to eliminate continuous joints.

Initially E section laminations are stacked. Both the primary and secondary windings are placed on the central limb side by side or concentrically. The full LV winding is placed nearer the core and over this HV winding is placed, with suitable insulation between limb and LV winding and between LV and HV windings. The magnetic circuit is closed with I section lamination, Fig 5.5 (a) and (b)

The windings are of sandwich type and are in the form of interleaved pancakes, are form wound in cylindrical shape and the core laminations are inserted later on..

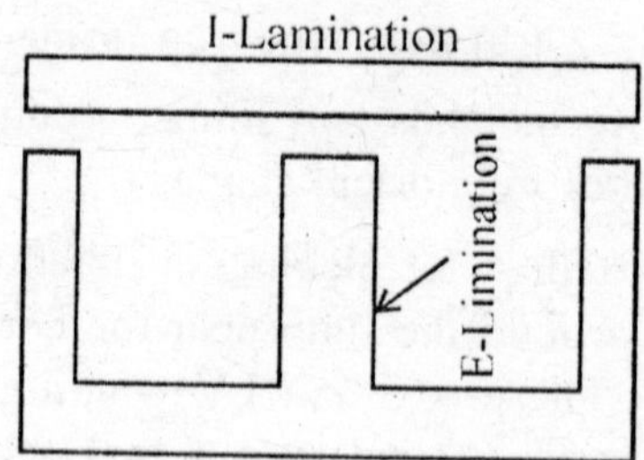

Fig. 5.5(b) E and I section laminations

Transformers of smaller capacity are provided with a simple enclosure with a provision for natural air circulation. Large size transformers are put inside a leak proof tank, filled with insulation oil.

5.10.3 Comparison Between Core Type and Shell Type Transformers

Comparison between core type and shell type transformers is given below in Table 5.1

Table 5.1 : Comparison between core type and shell type transformers

S.No.	Point of Comparison	Core type	Shell type
1	Location of coils	½ LV and ½ HV coils around each limb , Windings encircle the core	Both windings in center limb of Core
2	Length of mean turn of coil	Shorter	Longer
3	Insulating the coils	Easy	More difficult
4	Number .of windows	One window	Two windows
5	Shape of winding	Cylindrical windings	Sandwich type windings
6	Cooling surface	Better cooling since more surface is exposed	Cooling not very effective
7	Length of magnetic path	Shorter – has1 path only	Longer – has 2 paths
8	Mechanical support	Good	Stronger
9	Repairing at site	Easy access	More difficult
10	Cost	Cheap	Costlier

5.10.4 Primary and Secondary Windings

From the operating point of view, the two windings are called primary and secondary.

The winding connected to the supply, which is the prime source of energy, is called the **primary winding**. The second winding which gives out electrical energy to the load is called the **secondary winding**.

The terms primary and secondary in a transformer are really not rigid ; any winding may be called as primary, the other being secondary. Thus a transformer having windings labeled as 100V/400V with 100V source connected to winding labeled 100V acts as a step up transformer; whereas with 400V source connected to winding labeled 400V acts as a step down transformer.

5.11 EMF Equation of a Transformer

EMF equation of a transformer can be derived as follows :

5.11.1 Derivation of EMF Equation of a Transformer (Method 1)

The primary winding is connected across an AC supply. Let this supply voltage be v_1, which is sinusoidal and is given by:

$$v_1 = V_m \cos \omega t \qquad ...(5.14)$$

The sinusoidal magnetizing current i_m and the flux ϕ set up by it lag the applied voltage by 90^0

$\therefore$

$$i_m = I_m \sin \omega t \qquad ...(5.15)$$

The sinusoidal magnetising current passing through the primary winding causes an alternating flux ϕ to be set up in the core and is in phase with i_m

i.e.

$$\phi = \phi_m \sin \omega t \qquad ...(5.16)$$

Where ϕ_m is the maximum value of the instantaneous flux.

The alternating flux ϕ linking the primary winding having N_1 turns induces an emf 'e_1' in it.

i.e.

$$e_1 = N_1 \frac{d\phi}{dt}$$

$$= N_1 \frac{d}{dt}\phi_m \sin \omega t$$

$$= N_1 \phi_m \omega \cos$$

or

$$e_1 = E_{1m} \cos \omega t \qquad ...(5.17)$$

where,

$$E_{1m} = N_1 \phi_m \omega$$

The induced emf in the primary winding e_1 is in the opposite direction to the supply voltage V_1. The RMS value of induced emf in the primary winding, E_1 is given as:

$$E_1 = \frac{E_{1m}}{\sqrt{2}}$$

Substituting the value of E_{1m}, we get:

$$E_1 = \frac{N_1\phi_m\omega}{\sqrt{2}}$$

Substituting the value of ω, we get:

$$E_1 = \frac{N_1\phi_m(2\pi f)}{\sqrt{2}} = \sqrt{2}\,\pi\,\phi_m f N_1$$

or

$$\mathbf{E_1 = 4.44\ \phi_m\ f\ N_1\ Volts} \qquad ...(5.18)$$

Further, the same alternating flux φ links the secondary winding also. Repeating the above steps for secondary winding, the RMS value of induced emf in the secondary winding, E_2 is given as:

i.e.

$$\mathbf{E_2 = 4.44\phi_m\ f\ N_2\ Volts} \qquad ...(5.19)$$

5.11.2 Derivation of Emf Equation (Method 2)

Applied voltage to the primary winding of a transformer is sinusoidal. Hence flux set up in the core of the transformer is also sinusoidal as shown in Fig.5.6

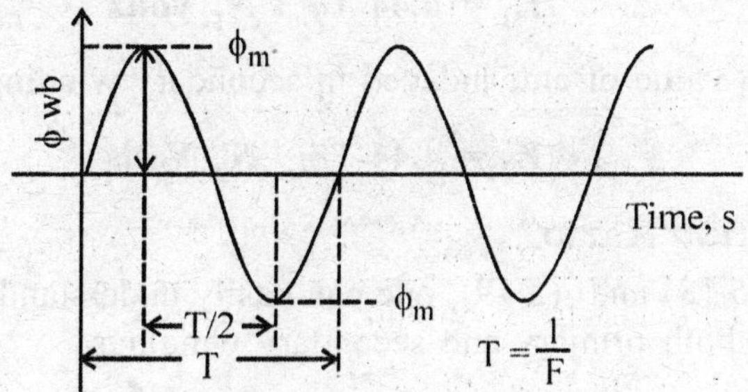

Fig. 5.6 Flux wave in a transformer

If 'T' is the time period i.e. time taken to complete one cycle.

and $$T = \frac{1}{f}$$

It can be seen from Fig. 5.6, Flux changes from 0 to ϕ_m as time changes from 0 to $\frac{T}{4}$ seconds.

Average rate of change of flux,

$$= \frac{\phi_m - 0}{\frac{T}{4} - 0} = \frac{4\phi_m}{T} \text{ wb/sec}$$

Substituting the value of 'T', in terms of 'f' ,we get:
Average rate of change of flux

$$= \frac{\phi_m}{1/4f} = 4f\phi_m \text{ wb/sec}$$

If it is assumed that the coil has a single turn, then according to Faraday's law of electromagnetic induction, Avg. value of emf induced/turn = 4 fϕ_m volts.

Form factor(for sinusoidal waveform)

$$= \frac{RMS Value}{Avg Value} = 1.11 \qquad ...(5.20)$$

Using this factor, we get:

RMS value of emf induced/turn = (1.11) (4 fϕ_m) = 4.44 fϕ_m volts

RMS value of emf induced in primary winding having N_1 turns

$$\mathbf{E_1 = 4.44\ f\phi_m\ N_1\ volts}$$

and similarly,RMS value of emf induced in secondary winding. having N_2 turns ,

$$\mathbf{E_2 = 4.44\ f\phi_m\ N_2\ Volts}$$

5.11.3 Transformation Ratio

From equations (5.18) and (5.19), one can easily understand that the emf induced per turn E_t, is the same for both primary and secondary windings.

i.e.
$$E_t = \frac{E_1}{N_1} = \frac{E_2}{N_2} = 4.44\ \phi_m f \qquad ...(5.21)$$

Equation (8.8) can be rearranged as:

$$\frac{E_2}{E_1} = \frac{N_2}{N_1} = K \qquad ...(5.22)$$

Where '*K*' is called the **transformation ratio** i.e. ratio of number of turns of secondary winding to number of turns primary winding .

For an ideal Transformer, voltage drops in the two windings are neglected

i.e. $E_1 = V_1$ and $E_2 = V_2$

Hence in an **ideal Transformer**

$$\frac{E_2}{E_1} = \frac{N_2}{N_1} = K = \frac{V_2}{V_1} \qquad ...(5.23)$$

If **K > 1**, Transformer is called a **Step-up Transformer.**

If **K < 1**, Transformer is called a **Step-down Transformer.**

5.11.4 Current Ratios

For an ideal transformer, i.e. neglecting voltage drops, and losses,

Primary VA = Secondary VA

i.e.
$$V_1 I_1 = V_2 I_2$$

$$\therefore \qquad \frac{V_2}{V_1} = \frac{I_1}{I_2} = \frac{E_2}{E_1} = K = \frac{N_2}{N_1}$$

Hence, we can deduce the following current ratio.

i.e. $$\frac{I_1}{I_2} = \frac{N_2}{N_1} = K \qquad ...(5.24)$$

or $$I_1 N_1 = I_2 N_2$$

Or Amp turns of primary winding = Amp turns of secondary winding

Equation (5.24) gives us the windings currents ratio.

5.11.5 Turns Ratio of a Transformer

It is defined as ratio of number of primary turns to the number of secondary turns.

i.e. $$\text{Turns ratio} = \frac{N_1}{N_2} \qquad ...(5.25)$$

Example 5.6 : A Transformer with 60 turns on the HV windings is to be used to step down the voltage from 200V to 100V. Find the number of turns in the LV winding.

Solution : We know that:

$$\frac{V_L}{V_H} = \frac{N_L}{N_H}$$

∴ Number of turns in the LV winding,

$$N_L = N_H \times \frac{V_L}{V_H} = 60 \times \frac{100}{200} = 30 \text{ turns} \qquad \textbf{Ans.}$$

Example 5.7 : A transformer supplies a load of 30 amp at 400 V. If the primary voltage is 2000 V, find (i) the secondary volt-amperes (ii) the primary current and (iii) primary volt-amperes. Neglect losses and magnetizing current.

Solution : Let S_1 and S_2 be the primary and secondary volt amperes.

(i) Secondary volt amperes

$$S_2 = V_2 \ I_2$$

$$= 400 \times 30 = 12000 \text{ VA} = 12 \text{ KVA} \qquad \textbf{Ans.}$$

(ii) Primary current

We know that:

$$\frac{I_1}{I_2} = \frac{V_2}{V_1}$$

$$I_1 = \frac{V_2}{V_1} \times I_2$$

$$= \frac{400}{2000} \times 30 = 6 \text{ amp.}$$ **Ans.**

(ii) Primary volt amperes

$$S_1 = V_1 I_1$$

$$= 2000 \times 6 = 12000 = 12 \text{ KVA}$$ **Ans.**

Example 5.8 : A 44 KVA, 4440V / 444 V, 50 Hz single-phase transformer has an effective cross-sectional area of core 0.01 m^2 and has 200 turns in its LV winding. Determine (i) The maximum value of flux density, (ii) The number of turns on the HV winding and (iii) Full load current in each winding.

Solution : Let S be the KVA rating of the transformer.

(i) Maximum value of flux density

From emf equation of a transformer, we know that:

$$\phi_m = \frac{E}{4.44 \times f \times N}$$

$$= \frac{444}{4.44 \times 50 \times 200}$$

$$\phi_m = 0.01 \text{ wb}$$

Max. flux density,

$$B_m = \frac{\phi_m}{A_i} = \frac{0.01}{0.01}$$

or

$$B_m = 1 \text{ wb/m}^2$$ **Ans.**

(ii) Number of turns on HV winding

$$\frac{V_H}{V_L} = \frac{N_H}{N_L}$$

$$N_H = N_L \times \frac{V_H}{V_L} = 200 \times \frac{4440}{444}$$

Number of turns, $N_H = 2000$ turns.

(iii) Full load currents in each winding

Current in the HV winding,

$$I_H = \frac{S}{V_H} = \frac{44\times10^3}{11\times10^3} = 4.0 \text{ amp} \qquad \textbf{Ans.}$$

Current in the LV winding,

$$I_L = \frac{S}{V_L} = \frac{44\times10^3}{444} = 99.09 \text{ amp} \qquad \textbf{Ans.}$$

5.12 Ideal Transformer

A transformer is called an ideal transformer if it satisfies the following conditions:

(1) Iron and copper losses are zero i.e. efficiency is 100%

(2) Primary and secondary winding resistances are zero.

(3) The leakage flux is zero, hence the flux linking both the windings is the same

(4) Voltage applied to the primary winding V_1 is equal to primary induced emf E_1

(5) Voltage induced in the secondary winding E_2 is equal to the voltage supplied to the load, V_2.

(6) Voltage regulation is 0%.

5.12.1 Principle of Operation of an Ideal Transformer on No Load

Consider that the secondary winding of an ideal transformer is open i.e. load is not connected to it, as shown in Fig 5.7

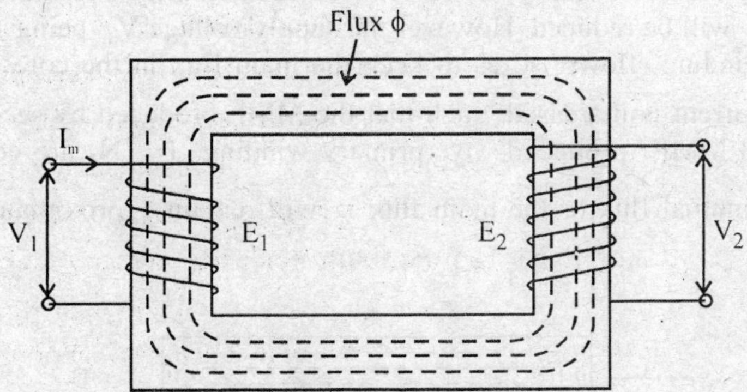

Fig. 5.7 Ideal transformer on no load

When a supply voltage V_1 having frequency 'f' is applied to the primary winding, it draws a magnetizing current I_m, lagging the supply voltage V_1 by 90º, as shown in the phasor diagram , Fig. 5.8, because the winding is purely inductive. This current produces an alternating flux ϕ, which is proportional to and is in phase with it.

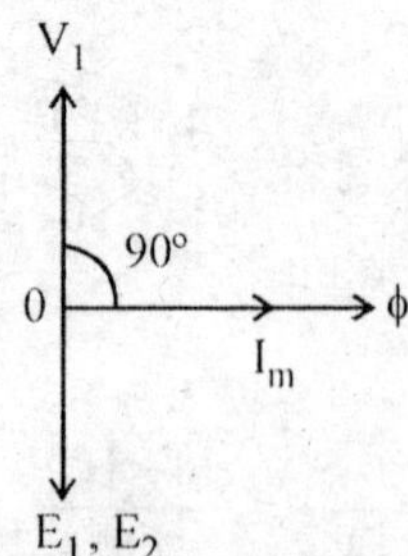

Fig. 5.8 Phasor diagram of an ideal transformer on No load

This alternating flux ϕ links with both the primary and secondary windings and induces emf's E_1 and E_2 in them respectively, both of which lag the flux ϕ by 90^0 according to Faraday's law of electromagnetic induction. The magnitude of induced emf's depends on the number of turns in the windings.

It may be seen in the phasor diagram that E_1 is equal to V_1 in magnitude but it is 180° out of phase with V_1

In the phasor diagram, flux phasor φ has been taken as the reference phasor , because it is common to both the windings.

5.12.2 Ideal Transformer on Load

Let us now consider the case when the ideal transformer secondary winding is connected to a load . A current I_2 will flow in the closed circuit consisting of secondary winding and load. Due to this current , an MMF equal to I_2N_2 is set up by the secondary winding. This MMF will produce a flux which tends to oppose the main flux produced by primary winding, according to Lenz's law. Hence the main flux will be reduced momentarily and the induced emf in the primary winding E_1 will be reduced. However the supply voltage V_1 being constant , a greater current in primary winding flows so as to keep the main flux in the core constant.

The primary current is increased such that the MMF produced by secondary winding $I_2 N_2$ and the increased MMF produced by primary winding $I_2' N_1$ are equal and opposite.

However, the mutual flux or the main flux ϕ will remain approximately the same.

Thus, $$I_2' N_1 = I_2 N_2 \quad ...(5.26)$$

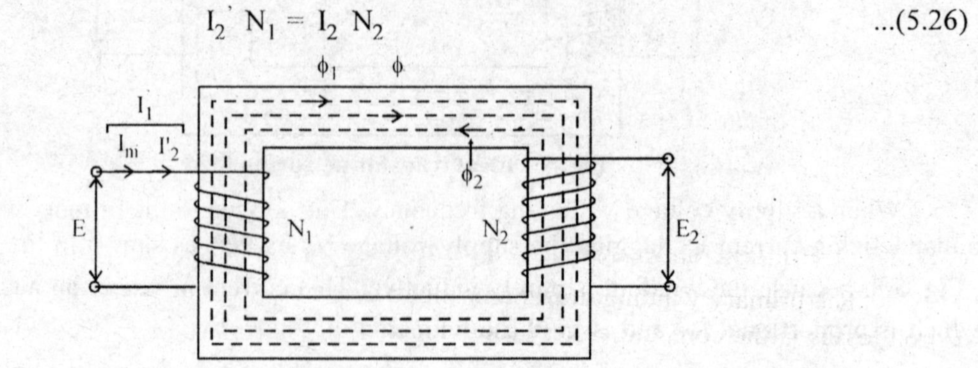

Fig. 5.9 Ideal Transformer on load

Hence, as and when the secondary winding delivers current I_2 to the load, the primary immediately draws additional current I_2' from the supply. Total primary current I_1, then will be the phasor sum of magnetizing current, $I\mu$ and I_2'. However, in an ideal transformer, magnetizing current is negligible and may be assumed as zero.

Combining equations (5.23) and equation (5.26), we get:

$$\frac{E_2}{E_1} = \frac{N_2}{N_1} = K = \frac{I'_2}{I_2} \qquad ...(5.27)$$

Fig.5.10 shows the complete phasor diagram of a transformer under different types of loads.

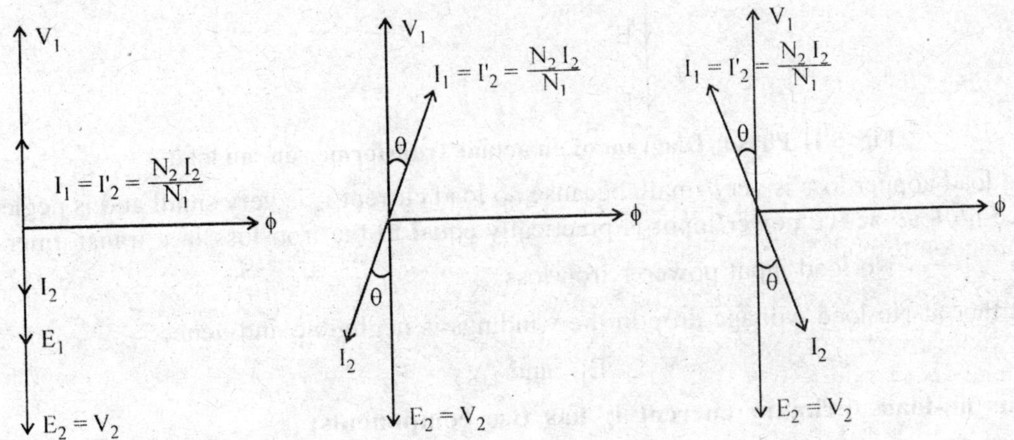

Fig 5.10 Phasor diagrams of ideal transformer under load conditions

5.13 Actual/Practical Transformer on No Load

To study an actual transformer, following points are to be considered :

(i) Resistances of primary and secondary windings are considered because windings consist of copper conductors, which have some resistance.

(ii) Magnetic leakage is not neglected and therefore leakage reactances have to be taken into account.

(iii) Although Hysteresis and Eddy current losses are small, these have to be considered because iron core is subjected to alternating flux.

5.13.1 Phasor Diagram of a Transformer on No Load

The primary winding draws a no- load current I_0 to magnetise the core and to supply the iron losses in the core and a very small amount of copper losses in the primary winding. Hence

I_o is not exactly 90° behind the supply voltage but lags the supply voltage by an angle ϕ_0, which is slightly less than 90° as shown in phasor diagram in Fig. 5.11. In the phasor diagram , Flux ϕ is taken as the reference phasor as it is common to both the windings. Induced emf's E_1 and E_2 lag the flux by 90°.

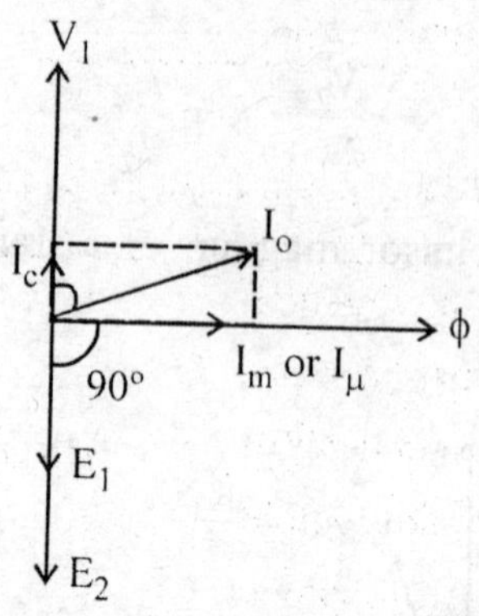

Fig. 5.11 Phasor Diagram of an actual Transformer on no load

No load copper loss is very small, because no load current I_o is very small and is neglected and hence no load active power input is practically equal to the iron loss in a transformer.

i.e. No load input power = iron loss

Further at No load, voltage drop in the windings is negligible and hence,

$$V_1 = E_1 \text{ and } V_2 = E_2$$

This no-load primary current I_o has two components:

(i) Active or power component or working component or iron loss component , I_W. It is in phase with supply voltage V_1 and supplies the iron loss in the core and a small amount of copper loss in the primary winding.

i.e.
$$I_W = I_0 \cos \varphi_o \qquad ...(5.28)$$

(ii) Magnetizing component or reactive component. $I\mu$. It lags the phasor V_1 by 90° and is in phase with the flux φ. It produces the mutual or useful flux in the core.

i.e.
$$I\mu = I_o \sin \varphi_o \qquad ...(5.29)$$

Thus, no load current I_o is a phasor sum of the currents I_W and I.

i.e.
$$I_o = \sqrt{I_W^{\,2} + I_\mu^{\,2}} \qquad ...(5.30)$$

and no load pf,
$$\cos \phi_0 = \frac{I_W}{I_0} \qquad ...(5.31)$$

The no load current component I_W is small as compared to component $I\mu$.

5.13.2 Equivalent Circuit of a Transformer on No Load

Under no load conditions, equivalent circuit of a transformer can be drawn as shown in Fig 5.12

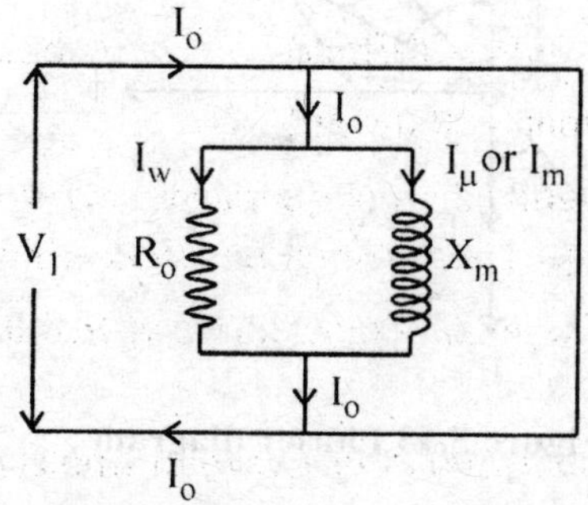

Fig 5.12 Equivalent circuit of a transformer on No Load

It can be seen from the equivalent circuit that :

$$R_0 = \frac{V}{I_W}$$

$$X_m = \frac{V}{I_\mu}$$

Where, $I_W = I_0 \cos\phi_0$

and $I = I_o \sin\phi_0$

Example 5.9 : A 300/200V transformer takes a current of 0.50 amps on open circuit at a pf of 0.2 . Find the magnetizing and iron loss components of current.

Solution : $E_1 = 300V,\ E_2 = 200V$

$$I_0 = 0.50 \text{ amps}$$

$$\cos\phi_0 = 0.2$$

$$\therefore \quad \sin\phi_0 = \sqrt{1-\cos^2\phi_0} = \sqrt{1-0.04} = \sqrt{0.96} = 0.9797$$

$$\therefore \quad I_w = I_0 \cos\phi_0 = 0.50\ (0.2) = 0.1 \text{ amps}$$

$$I_\mu = I_0 \sin\phi_0 = 0.50\ (.9797) \text{ amps}$$

$$= 0.4898 \text{ amp.}$$

Ans.

Primary Current in an Actual Transformer when Secondary is Loaded

When secondary of an actual transformer is loaded, the phasor diagram can be drawn as shown in Fig 5.13

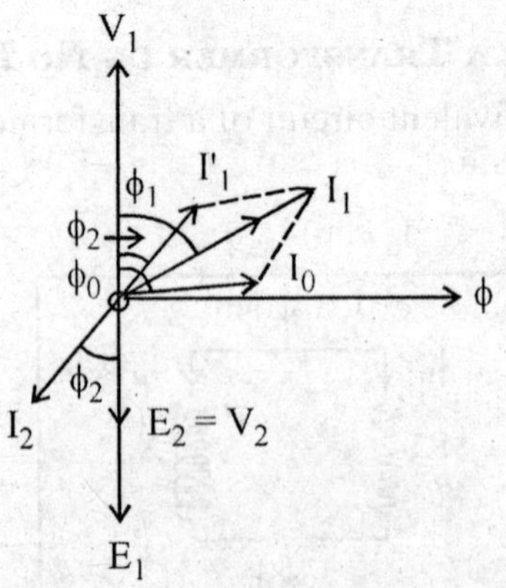

Figure 5.13 Phasor diagram

Consider that the secondary winding draws a load current I_2 at a lagging pf angle φ_2, i.e. current I_2 lags the secondary voltage V_2 by an angle φ_2. This current when referred to the primary side is:

$$I_2' = I_2 \left(\frac{N_1}{N_2}\right)$$

In addition to I_2', no load current I_0 also flows in the primary winding. Hence the total primary current I_1 is the phasor sum of I_0 and I_2' as shown in the phasor diagram, Figure 5.13

Example 5.10 : A single-phase transformer has 1000 turns on the primary and 200 turns on the secondary winding. The no-load current is 2.0 amp and the no-load power factor is 0.2 lagging. Calculate the current and power factor of the primary winding when the secondary draws a current of 200 amp at a power factor of 0.8 lagging. The voltage drop in the windings may be neglected.

Solution : We know that:

$$I_2' N_1 = I_2 N_2$$

i.e.

$$I_2' = \frac{200 \times 200}{1000} = 40 \text{ amp.}$$

$$I_o = 2.0 \text{ Amp}$$

$$\cos \phi_0 = 0.2 \qquad \sin \phi_0 = 0.98$$

$$\cos \phi_2 = 0.8 \qquad \sin \phi_2 = 0.6$$

From Figure (5.7), resolving the currents in to vertical and horizontal components, we get:

$$I_1 \cos \phi_1 = I_2' \cos \phi_2 + I_o \cos \phi_o$$

$$= 40 \times 0.8 + 2.0 \times 0.2 = 32.4$$

and $$I_1 \sin\phi_1 = I_2' \sin\phi_2 + I_o \sin\phi_o$$

$$= 40 \times 0.6 + 2.0 \times 0.98 = 25.96 \text{ amps}$$

Combining the horizontal and vertical components, we can get the primary current I_1 as:

i.e. $$\text{current } I_1 = \sqrt{(I_1\cos\phi_1)^2 + (I_1\sin\phi_1)^2}$$

$$= \sqrt{(32.4)^2 + (25.96)^2} = 41.51 \text{ amp.}$$

Primary Current = 41.51 amp. **Ans.**

$$\tan\phi_1 = \frac{I_1\sin\phi_1}{I_1\cos\phi_1} = \frac{25.96}{32.4} = 0.80$$

$$\phi_1 = \tan^{-1} 0.80 = 38.7^0$$

Primary power factor,

$$\cos\phi_1 = \cos(38.7) = 0.78$$ **Ans.**

5.14 Leakage Flux and Leakage Reactance

The flux ϕ which links both the windings is called the **useful flux or mutual flux**. However, whole of the flux which is produced by the primary current does not link the secondary winding. A small portion of it, say Φl_1 links only the primary winding. This flux Φl_1 which links primary winding only is called the **leakage flux**. Fig.5.14(a)

As the path of leakage flux is through the air, it has high reluctance. Φl_1, the leakage flux produced by the primary current I_1, is in phase with I_1 and induces a voltage Φl_1 in the primary winding.

Φl_2, the leakage flux produced by the secondary current I_2, is in phase with I_2 and induces a voltage el_2 in the secondary winding. The primary supply voltage V_1 is opposed by el_1 to produce E_1 and E_2 has to overcome el_2 to produce V_2

The induced voltages el_1 and el_2 due to leakage fluxes are considered as the voltage drops across the assumed fictitious reactances X_1 and X_2 placed in series with the primary and secondary windings. These reactances , X_1 and $X_2$are called leakage reactances and are shown in Fig 5.14(b) given below.

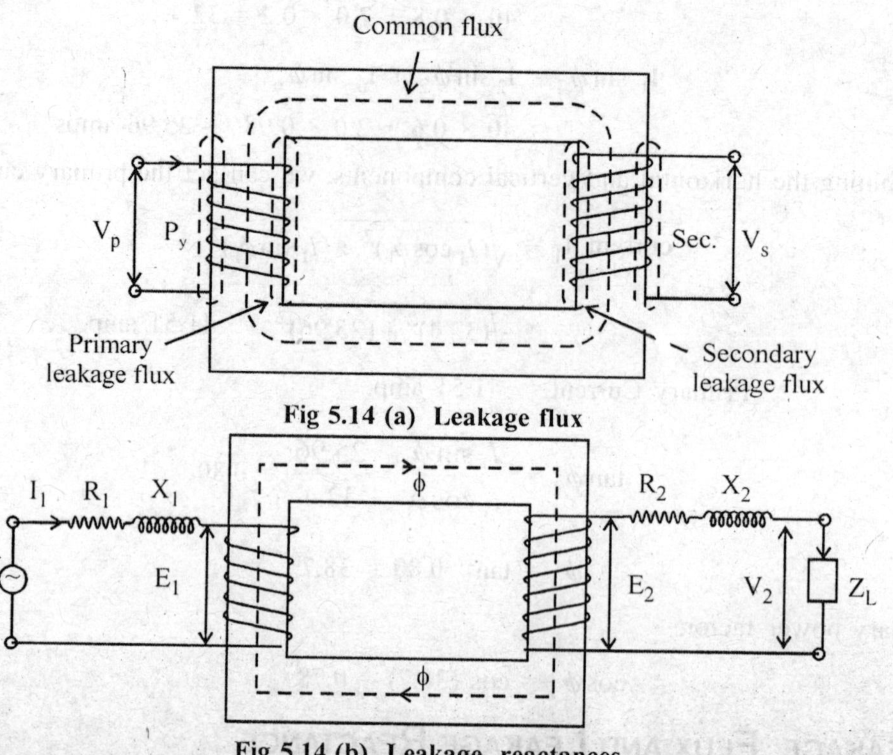

Fig 5.14 (a) Leakage flux

Fig 5.14 (b) Leakage reactances

5.15 EQUIVALENT RESISTANCE AND REACTANCE

In order to simplify the calculations, resistance and reactance on one side of a transformer can be transferred to the other side i.e. these parameters of the primary side can be transferred to the secondary side and vice versa.

Let R_1 and R_2 be the resistances of the primary and secondary windings, and X_1 and X_2 be the leakage reactances of the primary and secondary windings.

and $$K = \frac{N_2}{N_1}, \text{ be the transformation ratio.}$$

If the resistance of secondary winding R_2 is referred to primary side, then the equivalent total resistance R_{1eq} on the primary side will be :

$$R_{1eq} = R_1 + R_2'$$

where $$R_2' = R_2 \left(\frac{N_1}{N_2}\right)^2 \quad \text{...(5.32)}$$

If the primary resistance is referred to secondary side, then the equivalent total resistance R_{2eq} on the secondary side will be:

$$R_{2eq} = R_2 + R_1'$$

where, $$R_1' = R_1 \left(\frac{N_2}{N_1}\right)^2 \qquad ...(5.33)$$

Similarly, $$X_{1eq} = X_1 + X_2 \qquad ...(5.34)$$

and $$X_{2eq} = X_2 + X_1 \left(\frac{N_2}{N_1}\right)^2 \qquad ...(5.35)$$

Example 5.11 : A 100-kVA transformer has 400 turns on the primary winding and 100 turns on the secondary winding. The primary and secondary resistances are 0.4 and 0.02 Ω respectively, and the corresponding leakage reactances are 1.0 and 0.04 Ω respectively. Calculate the equivalent impedance referred to the primary circuit.

Solution :

Equivalent impedance Z_e of the transformer referred to the primary side is given by:

$$Z_e = \sqrt{(R_1 + R_2')^2 + (X_1 + X_2')^2}$$

$$R_2' = R_2 \left(\frac{N_1}{N_2}\right)^2 = \left(\frac{400}{100}\right)^2 \times 0.02 = 0.32 \ \Omega$$

$$X_2' = X_2 \left(\frac{N_1}{N_2}\right)^2 = \left(\frac{400}{100}\right)^2 \times 0.04 = 0.64 \ \Omega$$

$$Z_e = \sqrt{(0.4 + 0.32)^2 + (1.0 + 0.64)^2}$$

$$= \sqrt{(0.72)^2 + (1.64)^2} = \sqrt{0.5184 + 2.6896}$$

Equivalent impedance Z_e of the transformer referred to the primary side

$$= \sqrt{3.2080} = 1.791 \text{ ohms}$$ **Ans.**

5.16 Transformer Losses

A transformer is a static device . It does not have any rotating or mechanical losses. It has only electrical losses i.e. iron losses in the magnetic core and copper losses in the primary and secondary windings.

5.16.1 Iron Losses

A magnetic core when subjected to an alternating magnetization , has two types of losses i.e. Hysteresis loss and Eddy Current loss. As flux in the core is constant, these losses are constant. Sum of these two losses is called **core loss or iron loss.**

Iron losses remain constant for all loads put on the transformer.

5.16.2 Copper Losses

These losses are due to the resistance of copper in the primary and secondary windings. Total approx copper loss = $I_1^2 R_1 + I_2^2 R_2$, where I_1 and I_2 are the currents in the primary and secondary windings respectively.

It is clear that copper loss is a variable loss and varies with the square of the primary and secondary currents. Thus copper loss at half load i.e. half secondary current, will be one fourth of the copper loss that will occur at full load.

If this loss is very high, it can damage the insulation of the windings.

5.16.3 Mechanical Losses

Transformer has no rotational or mechanical losses like friction and windage losses , because it does not have any rotating part and is a static device.

5.17 Efficiency

5.17.1 Definition

It is defined as the ratio of output power to input power.

Input power = output power + iron loss + copper loss

$$\therefore\ \eta \text{ at F.L.} = \frac{\text{Output power at full load}}{\text{Output powerat full load} + \text{iron loss} + \text{Copper Loss at F.L}} \qquad ...(5.36)$$

If 'x' is the fraction of the full load, at which transformer is operating, then efficiency at the fractional load, x,

$$= \frac{\text{output power at fractional load}, x}{\text{output power at fractional load}, x + \text{iron loss} + x^2\left(\text{Copper loss at F.L}\right)} \qquad ...(5.37)$$

e.g. x = 0.5 means, 0.5 times the full load i.e. half of full load and so on.

5.17.2 Condition For Maximum Efficiency of a Transformer

Let us consider the secondary side of the transformer to obtain the condition for maximum efficiency.

Let R_{2eq} be the total equivalent resistance on the secondary side.

∴ Copper loss on secondary side

$$= I_2^2 R_{2eq}$$

Substituting this into Equation 5.36,we get,

$$\eta = \frac{V_2 I_2 \cos\phi_2}{V_2 I_2 \cos\phi_2 + I_2^2 R_{2eq} + P_i} \quad ...(5.38)$$

Where P_i = iron losses , I_2 is the full load current in secondary winding.

Dividing the numerator and denominator of equation 5.38, by I_2 we get,

$$\eta = \frac{V_2 \cos\phi_2}{V_2 \cos\phi_2 + \frac{P_i}{I_2} + I_2 R_{2eq}} \quad ...(5.39)$$

If the secondary terminal voltage V_2 is assumed to be constant, then for a constant power factor cos ϕ_2, we will get maximum efficiency when the denominator of Equation 5.39, is at its minimum value.

Denominator is minimum when its derivative with respect to I_2 is zero .

i.e. when,

$$\frac{d}{dI_2}\left[V_2 \cos\phi_2 + \frac{P_i}{I_2} + I_2 R_{2eq}\right] = 0$$

i.e

$$-\frac{P_i}{I_2^2} + R_{2eq} = 0$$

$$-P_i + I_2^2 R_{2eq} = 0$$

or

$$P_i = I_2^2 R_{2eq} \quad ...(5.40)$$

Thus the transformer efficiency is maximum at that value of current I_2 when the iron loss P_i becomes equal to the total copper loss $I_2^2 R_{2eq}$.

Thus the condition for maximum efficiency is: iron loss = copper loss

Example 5.12 : Determine the efficiency of a single phase, 50 KVA, 2500/250V , transformer at full load ,0.80 pf., F.L. copper loss = 800 watts and iron loss = 400 watts

Solution :

Output at full load = 50 × 1000 × 0.80 = 40000 watts

Total loss at full load = iron loss + copper loss

= 400 + 800 = 1200 watts

$$\therefore \quad \eta = \frac{40000}{40000+1200} = \frac{40000}{41200} = 97.08\ \% \quad \textbf{Ans.}$$

Example 5.13 : Calculate the efficiency of a 75 kVA transformer for (i) unity pf full load (ii) 0.8 pf lagging load at full load . It has iron loss = 1000 watts, copper loss at full load = 1.2 kW.

Solution :

(i) **Unity pf full load**

Output = 75 × 1 = 75 kW

Iron loss = 1000 Watts

Copper loss = (1200) (1) = 1200 Watt

$$\therefore \quad \eta = \frac{75000}{75000+1000+1200} = \frac{75000}{77200} = 97.15\% \quad \textbf{Ans.}$$

(ii) **0.8 pf lagging full load**

Output = 75 × 0.8 = 60.0 kW

Iron loss = 1000 Watts

Copper loss = 1200 Watts

$$\%\eta = \frac{60000}{60000+1000+1200} = \frac{60000}{62200} = 96.46\% \quad \textbf{Ans.}$$

Example 5.14 : The efficiency of a 1000 KVA, 110/220 V , 50 Hz, single phase transformer is 98.5% at half load at 0.8 pf leading and 98.8% at full load unity pf. Find

(i) Iron loss

(ii) FL copper loss

Solution :

$$\eta_{FL} = \frac{V_2 I_2 \cos\phi_2}{V_2 I_2 \cos\phi_2 + P_{cu\,FL} + P_i}$$

$$0.988 = \frac{1000\times10^3\times1}{(1000\times10^3\times1) + P_{cu\,FL} + P_i} \quad ...(i)$$

Simplifying equation (i), we get:

$$P_i + P_{cu} = 12145.75 \quad \text{...(ii)}$$

Copper losses at half full load $= \frac{1}{4} P_{cu}$

$$0.985 = \frac{500 \times 10^3 \times 0.8}{(500 \times 10^3 \times 0.8) + \frac{1}{4} P_{cu} + P_i} \quad \text{...(iii)}$$

Simplifying equations (iii), we get:

$$P_i + \frac{1}{4} P_{cu} = 6091.37 \quad \text{...(iv)}$$

Solving equations (ii) and (iv), we get:

$$\frac{3}{4} P_{cu} = 6054.38$$

$$P_{cu} = 8072.5 \text{ watts}$$ **Ans.**

$$P_i = 12145.75 - 8072.5 = 4073.24 \text{ watts}$$ **Ans.**

5.18 Voltage Regulation

The voltage available at a load i.e. at the secondary of a transformer is expected to be constant except with certain prescribed permissible variations. Too much voltage drop may not allow certain electrical devices to operate properly or operate at all and an abnormally high voltage can damage the equipment.

If the secondary voltage V_2 is to be kept constant at its rated value, the primary supply voltage under loading condition will be different from the rated nominal voltage. Hence the primary or the secondary voltage needs to be properly regulated. Regulation of secondary or primary voltage is done by either increasing or decreasing the effective number of turns.

Mathematically, Percentage voltage regulation is defined as :

$$\% \text{ Voltage Regulation} = \frac{|V_1| - |V_2|}{|V_2|} \times 100$$

5.19 Rating of Transformers

Generally the rating of a machine indicates the power supplied by it. However as the electrical load and its power factor on a transformer keeps on changing, its power output also changes. Hence the rating of a transformer is given in terms of the product of voltage and current, because the design of the magnetic circuit i.e. core and the electrical winding is based on these two parameters.

5.20 Applications of Transformers

Transformers are used for the following applications:

(i) Stepping up and stepping down of voltage

(ii) Electrical isolation, providing electrical isolation in medical equipments

(iii) Impedance matching

(iv) Link between AC and DC system

(v) Providing power supplies in TVs, Radios

(vi) Extending AC voltmeter, ammeter and wattmeter ranges

(vii) Welding purposes

5.21 Summary

Concepts of electromagnetic induction ,Faradays laws along with the concepts of magnitude and direction of induced emf have been discussed. Expressions for the dynamically induced emf, self and mutual induced emf have been derived .The concepts of self, mutual inductance and coupling coefficient have been explained.

Construction and principle of operation of a single phase ideal transformer have been discussed.

5.22 Important Formulae

1. Faraday's law of electromagnetic induction gives the induced emf

$$E = -N \frac{d\phi}{dt} \text{ volts}$$

2. The direction of the induced emf and hence current may be found either by Fleming's Right Hand Rule or Lenz's law.

3. The induced emf may be subdivided as follows:

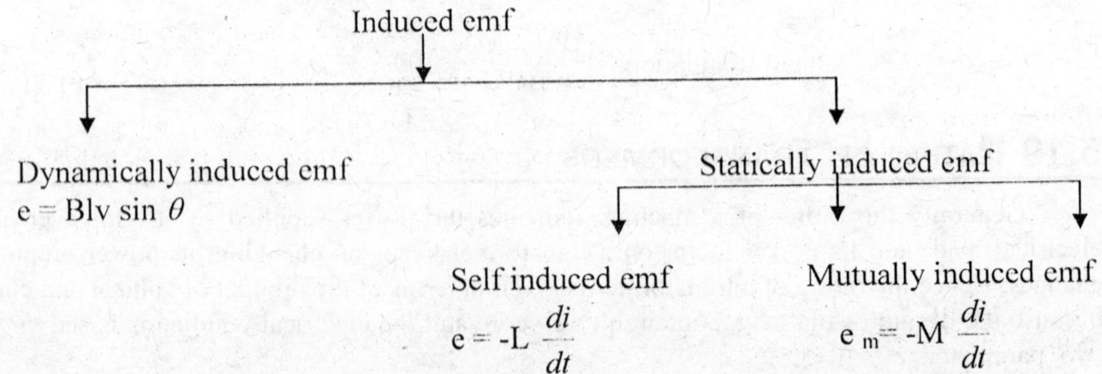

4. Flux linkages = $N\phi$

5. Value of the coefficient of self inductance of a solenoid is given by

$$L = \frac{N\phi}{I} \text{ henry}$$

Also, $$L = \frac{\mu_0 \mu_r a N^2}{l} \text{ henry}$$

6. Value of coefficient of mutual inductance between two coils is given by

$$M = \frac{N_2 \phi_2}{I_1} \text{ henry}$$

$$= \frac{\mu_0 \mu_r a N_1 N_2}{l} \text{ henry}$$

7. Coefficient of coupling between two magnetically coupled coils is given by

$$K = \frac{M}{\sqrt{L_1 L_2}}$$

Let E_1, E_2 are induced emfs in volts in the primary and secondary windings N_1, N_2 are primary and secondary number of turns Φ_m is maximum value of flux in webers B_m is maximum value of flux in density in weber /meter^2 I_1 and I_2 full load primary and secondary currents in amperes

(i) $E_1 = 4.44\ f\Phi_m N_1$ volts

(ii) $E_1 = 4.44\ f\ B_m\ A\ N_1$ volts

(iii) $E_2 = 4.44\ f\Phi_m N_2$ volts

(iv) $E_2 = 4.44\ f\ B_m\ A\ N_2$ volts

(v) $\frac{N_2}{N_1} = \frac{E_2}{E_1} = \frac{I_1}{I_2} = K$ = transformation ratio.

(vi) $I_o = \sqrt{I_W^2 + I_\mu^2}$

(vii) $I_w = I_o \cos\phi_0$

(viii) $I_\mu = I_o \sin\phi_o$

Where I_o = No load primary current in amperes, I_w = Active component of current in amperes, I_μ = Reactive component of current in amperes. Let R_1, R_2 are the resistance of primary and secondary windings in ohms.

(ix) $R_{01} = R_1 + \frac{R_2}{K^2}; R_{02} = R_2 + R_1K^2$ ohms

Where R_{o1} is equivalent resistance referred to primary in ohms ,R_{o2} is equivalent resistance referred to secondary in ohms.

(x) Efficiency = $\frac{Output}{Input}$

(xi) Output = Input – Total losses

(xii) Input = Output – Total losses

(xiii) Output at a fractional load, x = Rated VA × pf × x

(xiv) Copper loss at fractional load x = (x^2) (copper loss at F.L.)

where 'x' lies between 0 and 1.

5.23 PROBLEMS

1. A coil of 1250 turns is linking a flux of 5 mWb. The flux is reversed in an interval of 0.125 sec. Calculate the average value of the induced emf in the coil.

Ans. [100 V]

2. The field coils of a 4 pole dc generator ,each having 600 turns are connected in series,. With the field winding excited , they produce a flux of 0.05 Wb per pole. If the field is opened in 0.025 sec., calculate the average induced emf across the field terminals. The residual flux per pole is 0.001 Wb.

Ans. [4704 V]

3. Calculate the self inductance of an air cored toroidal coil with 50 cm mean dia and 6.25 cm^2 circular cross section wound uniformly with 1000 turns of wire. Also determine the induced emf in the coil when a current increasing at the rate of 200 A/s flows in the coil.

Ans [0.5mH, 0.1 V]

4. Two identical coils A and B having 800 turns lie nearby such that 80% of the flux produced by one coil links with the other. A current of 5 A in coil a produces in it a flux of 150×10^{-6}. The current in coil A changes from +20 amp to -20 amp in 20 ms. Calculate

(i) The self inductance of each coil

(ii) The mutual inductance and

(iii) The voltage induced in coil B.

Ans [0.04 H ,0.032 H , 64 V]

5. A 11000/400 V distribution transformer takes a no load primary current of 1 amp at a p.f. of 0.24 lagging. Find

(a) core loss current,

(b) magnetizing current

[Ans. (a) 0.24 A (b) 0.971 A]

6. A 6600/66 V single, phase, 50 cycle/sec core type transformer has an effective core section of 0.05 m^2 and maximum flux density is 1.2 Wb/m^2. Calculate the number of turns on the H.T. and L.T. sides.

Ans. [500, 5]

7. A 270/27 V, 5.4 KVA transformers has 50 turns in the secondary. Assuming the losses to be negligible, Calculate : Primary turns Primary and Secondary full load currents.

Ans. [500, 20, 200 Amp.]

8. A single phase transformer with a ratio of 230 / 115 V draws a no load current of 2 A at 0.28 p.f. lagging. The secondary winding supplies a current of 7 A at a p.f. of 0.8 lagging. Determine the current taken by the primary.

[Ans. 11.7 A]

9. A 1100/110 V, 22 KVA single phase transformer has primary resistance and reactance of 2 Ω and 5 Ω respectively. The secondary resistance and reactance are 0.02 Ω and 0.045 Ω respectively. Calculate :

(a) Equivalent resistance and reactance of secondary referred to primary.

(b) Total resistance and reactance referred to primary.

(c) Equivalent resistance and reactance of primary referred to secondary.

(d) Total resistance and reactance referred to secondary.

[Ans. (a) 4.5 Ω, (b) 9.5 Ω, (c) 0.05 Ω, (d) 0.095 Ω]

10. A single phase transformer has 350 turns on the primary winding and 1050 turns on the secondary winding. The primary winding is connected to a 400 volt, 50c/s supply. If the net cross sectional area of the core is 50 cm^2 , calculate (i) the maxm value of flux density in the core(ii)voltage induced in the secondary winding.

[Ans 1.0296 tesla, 1200 volts]

11. A 200 kVA , 3300/240 volt , 50 c/s , single phase transformer has 80 turns on the secondary winding . Assuming an ideal transformer , calculate

(i) primary and secondary current on full load

(ii) the maximum value of flux(iii) number of primary turns.

[Ans 60.6 amp ,833.3 amp,13.5 mwb , 1100 turns]

5.24 Short Questions and Answers

Q.1 State Faraday's Laws of electromagnetic induction.

Ans. First law: Whenever magnetic flux linking with a coil changes , an emf is induced in the coil.

Second law: The magnitude of the induced emf is equal to the product of number of turns of the coil and the rate of change of flux.

Q.2 What is Fleming's Left Hand Rule?

Ans. This is used to find the direction of the force acting on a current carrying conductor placed in the magnetic field. According to this rule, if we keep our thumb, first finger or forefinger and middle finger of the left hand at right angles to each other and if the first finger represents the direction of magnetic field, the middle finger indicates the direction of current then the thumb will indicate the direction of force acting on the conductor.

Q.3 What is Fleming's Right Hand Rule?

Ans. This is used to find the direction of induced emf. According to this rule, if we keep our thumb, first finger (or) forefinger and middle finger of the right hand at right angles to each other and if the first finger represents the direction of magnetic lines of flux, the thumb indicates the direction of motion of the conductor, then the middle finger will indicate the direction of induced emf.

Q.4 What is Self-Inductance?

Ans. The property of a coil due to which it opposes the change in the value of current or flux through it due to the production of self-induced emf in it , is called self-inductance.

Q.5 What is Mutual Inductance?

Ans. The ability of one coil to produce an emf in the nearby coil due to change of current or flux in the first coil is called Mutual Inductance.

Q.6 What is self induced emf?

Ans. The emf induced in a coil due to change in the value of its own flux linking with it, is called self-induced emf.

Q.7 What is mutual induced emf?

Ans. The emf induced in one coil due to change in value of flux linking in nearby coil is called mutual induced emf.

Q.8 How dynamically induced emf is produced?

Ans. If out of the conductor and the magnetic field , one of them is moved physically and the other is kept stationary ,in such a way that conductors cut across the magnetic field, then the induced emf , is known as Dynamically Induced emf.

Q.9 The inductance of a coil carrying a current of 1 amp. is 0.50 henrys. Find the energy stored in it.

Ans. Given Inductance,L = 0.50 henry , Current , I = 1 amperes

Energy stored = (1/2) L I^2

= (1/2) × 0.50 × 1 × 1 = 0.25 joules

Q.10 A coil of inductance 0.09 henry carrying a current of 10 amperes is wound with 1000 turns. Find the flux established $\phi = \frac{LI}{N} = \frac{0.05 \times 10}{1000}$ in the coil.

Ans. Given inductance, L = 0.05 henry , Current I = 10 amperes

No. of turns, N = 1000 turns

Flux established $\phi = \frac{LI}{N} = \frac{0.05 \times 10}{1000} = 0.5\,\text{m Wb}$

Q.11 What is the name of winding to which the supply is connected?

Ans. Primary winding.

Q.12 What is the name of the winding to which the load is connected?

Ans. Secondary winding

Q.13 Why the core is laminated?

Ans. To reduce eddy current loss in solid core which produces heat.

Q.14 How many magnetic paths are there in a core type transformer?

Ans. One only

Q.15 How many magnetic paths are there in a shell type transformer?

Ans. Two only.

Q.16 What do you understand by transformation ratio?

Ans. $\frac{E_2}{E_1} = \frac{N_2}{N_1}$ where E_2 and E_1 are secondary and primary voltages and N_2, N_1 are secondary and primary turns.

Q.17 What is formula for e.m.f. induced in a transformer?

Ans. E_1 = 4.44 ϕmax. × f × N_1 volts

Q.18 What will be the effect on the current of primary side, if the load on the secondary side is increased in a transformer?

Ans. Current will increase on the primary side.

Q.19 What are the different losses in a transformer.

Ans. Different losses in a transformer are:

1. Core losses or iron losses, which consist of Hysteresis and Eddy current losses.
2. Copper or winding losses in the primary winding and secondary winding.

Q.20 Why the efficiency of a transformer is high?

Ans. Because there is no moving part.

Q.21 Why do we use silicon steel for transformer core?

Ans. To reduce iron losses.

Q.22 What will happen if the primary of a transformer is connected to DC supply?

Ans The primary will produce constant flux. Consequently, no e.m.f. will be induced. The primary winding will draw excessive current due to low resistance of the primary and will overheat and get burn out.

Q.23 List four applications of transformers.

Ans. It can raise or lower the voltage or current in an a.c. circuit. It can increase or decrease the value of a capacitor, an inductor or resistance in an a.c. circuit. It can thus act as an impedance transferring device. It can isolate two circuits electrically. It can be used to prevent d.c. from passing from one circuit to the other.

Q.24 Where is a core- type construction suitable for a transformer?

Ans. The core type transformer affords better cooling surface for the winding than the shell-type core and is, therefore, more suitable for transformers which remain fully loaded.

Q.25 Why are iron losses constant at all loads in a transformer?

Ans. The flux in the core on load is the same as the flux on no load. Hence, the iron losses are constant and are independent of load.

Review Questions

1. What are the factors on which the self inductance of a coil depend?
2. Explain the concept of mutual inductance.
3. Define coefficient of coupling.
4. State the right hand screw rule.
5. State the right hand grip rule.

6. State and explain Faraday's laws of electromagnetic induction.
7. State and explain Lenz's law.
8. State Fleming's left hand rule.
9. State Fleming's right hand rule.
10. Explain the concept of induced emf.
11. Define mutually induced emf.
12. Define the terms tight coupling and loose coupling.
13. Explain the principle of operation of an ideal transformer under no load conditions
14. Explain the principle of operation of an ideal transformer under load conditions
15. Derive the emf equation of a transformer.
16. Draw the phasor diagrams of an ideal transformer under no load and load conditions.
17. Explain the constructional details of a transformer.

❑❑❑

Chapter 6

Electrical DC Machines

OBJECTIVES

- To learn the constructional details of DC machines
- To understand the principle of operation of DC generators and DC motors
- To develop EMF equation of DC machines
- To learn about various types of DC machines
- To understand the concept of Back emf in DC motors
- To develop Speed equation and Torque equation
- To learn the applications of DC generators and DC motors

6.1 INTRODUCTION

Generation of electrical energy, in general, is done by rotating electrical machines at the expense of mechanical energy. Further a large fraction of electrical energy is utilized by rotating electrical machines in which mechanical energy is developed at the expense of electrical energy, the machines being called electrical motors. In most of the cases electromechanical energy conversion process is reciprocal or reversible i.e. same machine can be used as a motor as well as generator. Hence due to its great importance, the study of rotating electrical machines is important.

Depending on the mode of operation, a DC machine is either called a DC generator or a DC motor. When the machine converts mechanical energy into electrical energy , which is DC in nature, it is called a DC generator. DC motor converts electrical energy which is DC in nature to mechanical energy.

6.2 Construction of a DC Machine

For emf to be induced in a DC machine, the following components are required:

(i) Magnetic Field System, which is capable of producing a suitable magnetic flux

(ii) Set of conductors/coils and

(iii) Relative movement between conductors and magnetic field system

Condition (iii) can be satisfied in two different ways : The magnetic field system can be placed in the stationary part of the machine called **stator** and the system of conductors in the rotating part of the machine, called **rotor.** Alternatively, the system of conductors can be placed in the stator and field system in the rotor.

In all electrical machines, the emf induced in the coil is always alternating in nature. This necessitates the system of conductors , called an **armature** to be placed in the rotor, along with the commutator assembly which converts AC induced emf into DC emf and supplies it to the external circuit. Fig. 6.1 shows the main parts of a DC machine.

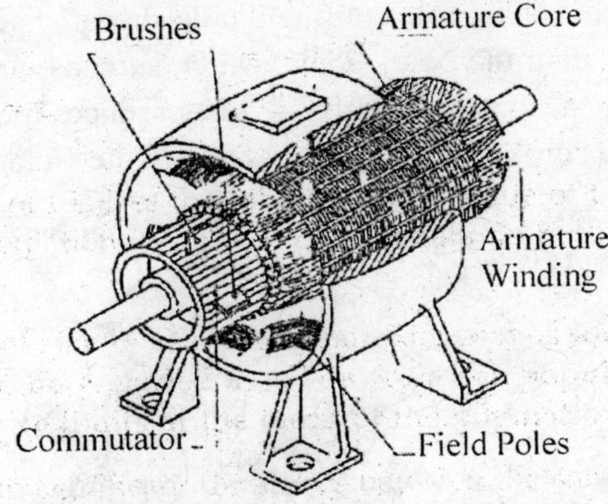

Fig 6.1 Constructional Details of a DC machine

The various parts of DC machines can be grouped into two categories :

Stationary Parts :

1. Frame or Yoke
2. Main poles
3. Main field winding
4. Interpoles/Commutating poles
5. Interpole windings

Rotating Parts

1. Armature core
2. Armature winding

3. Commutator
4. Brushes
5. Shaft

(i) **Frame/Yoke :** This part of the machine has two names as it serves two different functions. It gives mechanical support to the field poles and acts as a protecting cover for the whole machine and is therefore called as a **frame**. Secondly , it serves as a path for the lines of magnetic field and the magnetic circuit is completed through it . It is therefore also called as a **Yoke**.

In small machines, the frame/yoke is often made of cast iron in a single piece. In large sized machines, it is always made of cast steel or rolled steel either in one or more pieces. The feet and the terminal box are welded to the frame.

(ii) **Main poles :** These are salient poles and are bolted to the inside of the frame. Function of a field system is to produce a uniform magnetic field

The field magnets consist of pole cores and pole shoes. The portion of the poles facing the air-gap is wider than the pole body which spreads out the magnetic flux coming out of the pole in the air gap uniformly . It also, reduces the reluctance of the magnetic path, being of larger cross section. They support the field coils.

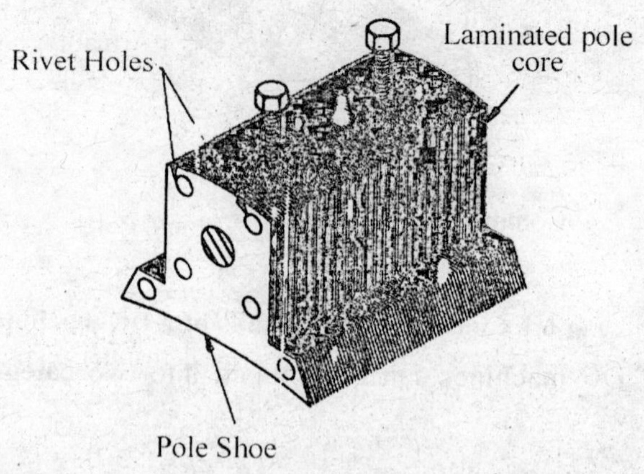

Fig 6.2 Laminated Pole Core

The pole core itself may be a solid piece made out of either cast iron or cast steel but the pole shoe is laminated and is fastened to the pole face by means of countersunk screws. In modern designs , the complete pole cores and pole shoes are built of thin laminations of annealed steel which are individually cut to the correct shape and are then stacked together as per required length and are riveted together under hydraulic pressure.

DC machines can be constructed with any even number of main poles i.e. 2 , 4, 6, 8 etc.

(iii) Main field windings : Field windings carry the dc exciting current. The main poles and their associated field windings are formed similar to an electromagnet. Former wound field coils of correct dimension are inserted around the main poles i.e. over the core,. Each pole may be provided with either shunt field coil or series field coil or both. Shunt field coils are wound with thin enamelled copper wire, which are ca-pable of carrying less current and have large number of turns. Series field coils are wound with thick enamelled copper wire capable of carrying large current and have small number of turns. Shunt field coils around all the poles are connected together in series and two field winding terminals are brought out on the terminal box.

Similarly series field coils form series field winding. If a DC current is passed through the field winding , a magnetic field having North and South pole polarity in alternate poles will be produced.

(iv) Interpoles : These poles are comparatively smaller in size than the main poles. They are fitted in between adjacent main poles and hence they are called **interpoles**. The flux provided by these poles gives spark less com-mutation and therefore they are also called as **commutating poles.**

Inter pole winding : All the exciting coils around interpoles are connected in series. This winding is normally connected in series with the armature circuit and carries the entire armature current I_a.

Air-gap : Radial spacing between main pole faces and surface of the ar-mature is called **air-gap**. The radial length of air-gap may be in the range of 0.5 mm and 1.5 mm. Number of turns of the field coils can be reduced by reducing the length of air gap.

(v) Armature core : The armature core houses the armature conductors and rotates in the magnetic field such that the conductors cut the magnetic flux. In addition to this, it provides a path of very low reluctance to the field flux and thus allows the magnetic circuit to complete its path through the yoke and the poles.

When the armature is rotated or rotating under the alternate North and South pole magnetic field, the flux reversal leads to iron losses in it. To reduce Hysteresis loss, the armature core is made up of low hysteresis steel containing few per cent silicon. To reduce eddy current loss, the core is assembled by stacking sheet steel laminations of the required length. The laminations are also known as stampings, as they are stamped out from steel plates. The stampings are slotted for insertion of arma-ture coils and to give smaller air gap. The assembled core is keyed on to the shaft.

(vi) Armature winding : It is the winding in which emf is induced. Properly connected coils of armature winding are inserted in the slots of the armature core. These are first wound in the form of flat rectangular coils and are then pulled into proper shape with a coil puller.

After insulating the armature slots, former wound, diamond shaped, enamelled copper coils are placed in them in two layers. If one side of a coil is placed in a slot lying under the North pole, then the other side of the same coil is placed in a slot lying under the next South pole and said to be 180 electrical degrees apart. A number of similar coils placed in the slots are then interconnected either in a **lap connected** manner or **wave connected** manner. Further, interconnecting ends of the consecutive coils are connected to the commutator segments. The total interconnected armature coils are called **Armature winding**.

(vii) Commutator : It is a mechanical rectifier. The functions of a commutator are (i) to facilitate collection of current from armature conductors and (ii) to rectify i.e. to convert AC emf/current induced in the armature coils to DC emf/current.

This is cylindrical in shape, fitted with the shaft on one side of the armature, namely the front side. The entire assembly is made of wedge shaped segments 0.5 to 1 mm thick of hard-drawn or drop-forged copper. The number of segments is equal to the number of coils in the armature winding. The segments are insulated from one another by a thin layer of mica sheet. The segments are held together by clamping flanges.

Each commutator segment is connected to the armature conductor by means of a copper lug or riser. The leads from the coils are soldered to commutator risers. To prevent them from flying out under the action of centrifugal forces, the segments have V grooves which are insulated by conical micanite rings.

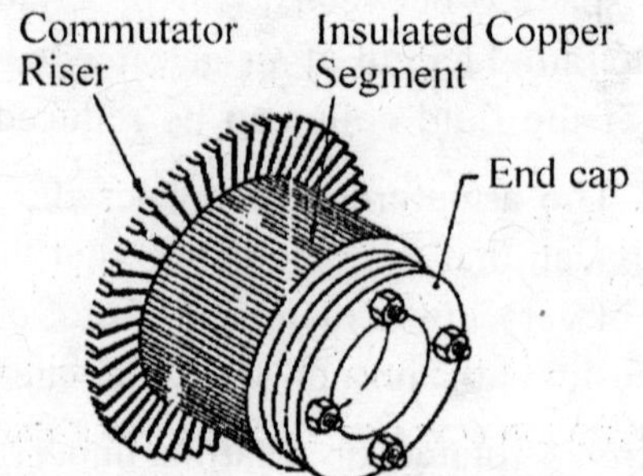

Fig 6.3 Complete Assembly of Commutator

(viii) Brushes : Function of the brushes is to collect current from the commutator and supply it to the external load circuit. These are usually made of carbon and are in the shape of a rectangular block. They may also be made of either copper gauze , copper - carbon compounds or graphite. Brushes are housed in brush holders, which have a box holder for the brush, a spring to maintain the brush tension and a hole to fix the holder to the rocker arm. The brushes can slide in the box open at both the ends. Brushes remain in contact with commutator segments by a spring , whose tension can be adjusted by changing the position of the tension lever in the notches. A flexible copper pigtail mounted at the top of the brush sends the current from the brushes to the holder.

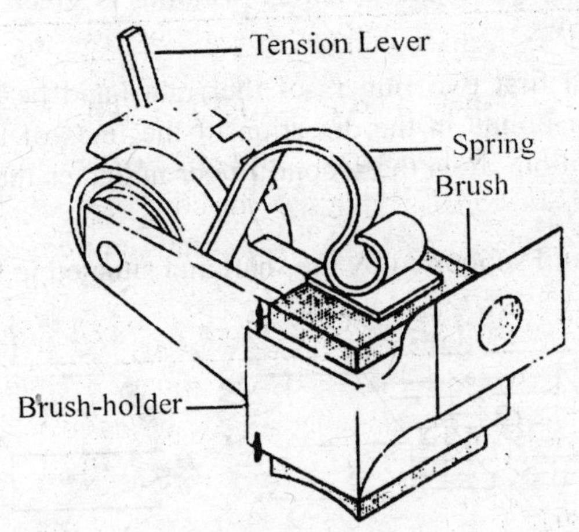

Fig 6.4 Brush Holder

Brush holders are fastened to a brush spindle which is attached to the front side end cover. The number of brush holders is equal to the number of field poles. Alternate $\frac{P}{2}$ positive brushes are connected together to form the armature terminal A_1 and the remaining negative brushes are connected together to form the other armature terminal A_2. The allowable current density varies from 5 amp. per sq. cm. in the case of carbon to 23 amps per sq. cm in the case of copper.

Brush rocker : The brush rocker may either be supported by a bearing cover in a small machine or by brackets attached to the yoke. The brush position to the neutral axis can be set by changing the position of the brush rocker.

(ix) **Bearings :** Because of their reliability, ball bearings are frequently used .For heavy duties roller bearings are preferred. These bearings are generally filled with hard oil for quiefer operation and reduced wear.

(x) **End plates :** The bearings are housed in the end plates and they are fixed to the yoke. They ensure the frictionless rotation of armature and to position the armature in the air gap.

(xi) **Shaft :** The steel shaft gives mechanical support to the armature and commutator. The rotor assembly is supported and allowed to rotate freely by two bearings fitted between the two end-covers and the shaft.

6.3 Principle of Operation of DC Generators

The DC generator operates on the principle of dynamically induced emf i.e. if the flux linking a coil changes, then an emf is induced in it. When an armature winding is rotated by the prime mover, the flux linking with it changes and an emf is induced in the armature winding. The direction of induced voltage in the armature winding is given by **Fleming's right hand rule,** which is stated as follows:

Let the thumb and first two fingers of the right hand be arranged at right angles to each other. If the first finger points in the direction of the lines of magnetic field, the thumb in the direction of the coil motion , then the second finger indicates the direction of the induced emf or current.

Consider a single coil supported by the shaft and situated in between a pair of poles as shown in Fig.6.5(a)

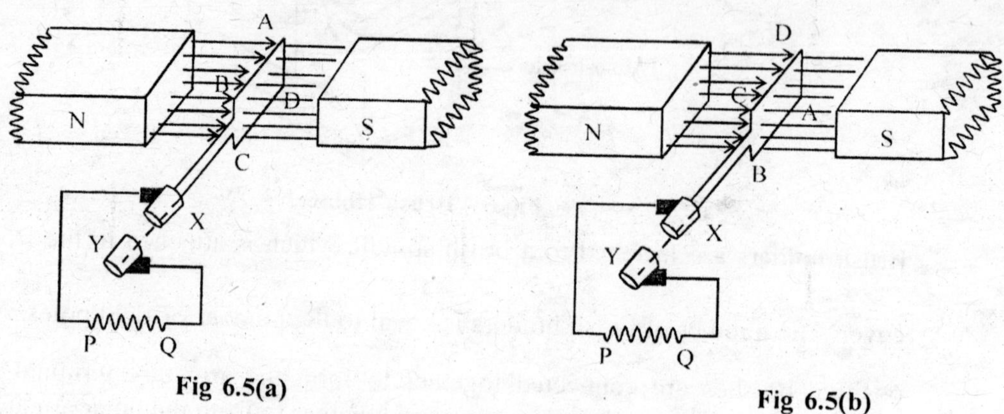

Fig 6.5(a) **Fig 6.5(b)**

Fig. 6.5: Principle of operation of DC Generators

Assume that the coil is rotated by a prime-mover, say, in the anticlockwise direction. When the coil side 'AB' moves under the N-pole and coil side 'CD' under the S-pole, the emf available to the external circuit across the brushes XY is say in +ve direction. After half rotation, coil side 'AB' moves under the S-pole and side 'CD' under the N-pole, as shown in fig. 6.5(b).

Fig. 6.5(b). Hence, emf collected from the coil, through the two slip rings and available across the brushes now is in -ve direction. Hence the use of slip rings makes an alternating emf available to the external circuit which is induced in the coil as shown in Fig. 6.5(c)

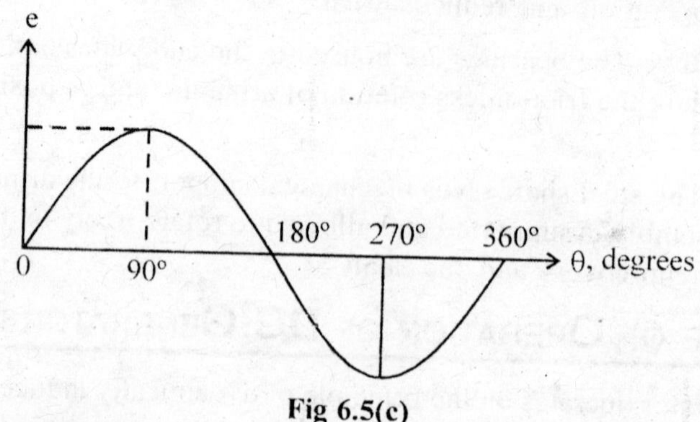

Fig 6.5(c)

To obtain the unidirectional current or direct current, the slip rings are replaced by split rings, as shown in Fig 6.6(a) and(b), which are splitted in to two equal segments a and b, separated from each other by insulation. This split ring is similar to the commutator assembly available in DC machines.

Coil side' AB ' is connected to one half of the split ring and side 'CD' to the other half. The coil side AB is always connected to the segment 'a' and CD to 'b'.

Brushes R and S are placed diametrically opposite to each other, instead of placing them side by side.

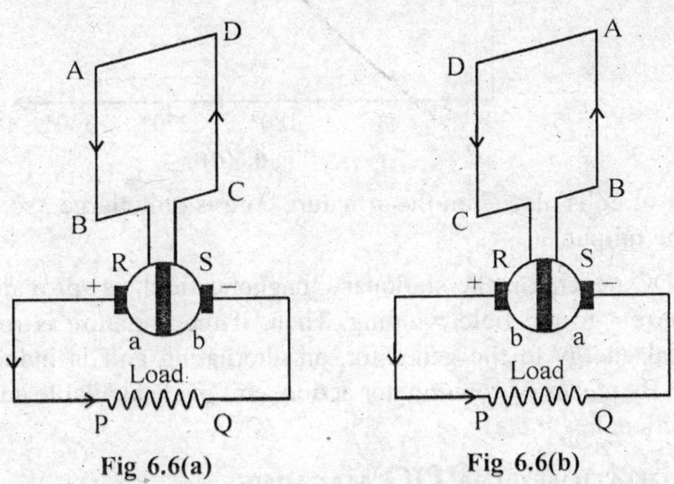

Fig 6.6(a) Fig 6.6(b)

It is seen that in the first half revolution, current flows along the path ABa RPQS bCDA, as shown in Fig 6.6(a) .In the next half revolution ,the direction of the induced current in the coil is reversed and the position of the segments 'a' and 'b' also gets reversed. Current flows in the path DCbRPQSaBAD, Fig 6.6(b). Hence brushes R and S contact positive polarity segment 'b' and negative polarity segment 'a' respectively i.e. direction of current in the external circuit PQ remains the same. The change over of segments takes place when the flux linkages or induced emf is minimum. Hence, emf available to the external circuit across brushes RS is always unidirectional as shown in Fig. 6.6(c).

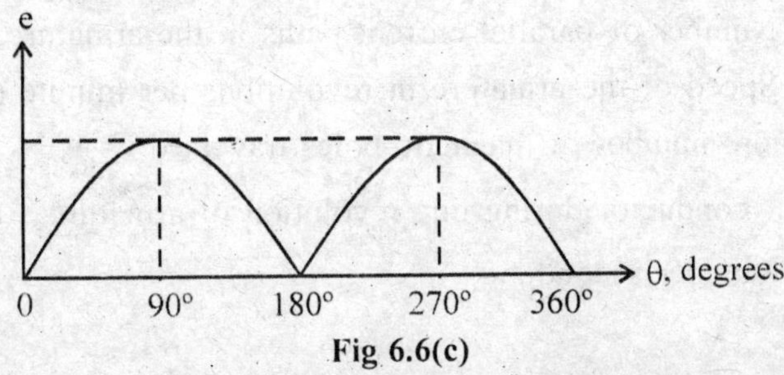

Fig 6.6(c)

A pulsating direct current given by a single coil is not suitable for actual applications. Hence, a large number of coils and commutator segments with the coils evenly distributed around the surface of the armature are used to make the brush voltage practically constant as shown in Fig.6.6(d).

The voltage generated by a single turn coil is small. Therefore, coils consist of several turns connected in series , so that generated emf is increased in proportion to the number of turns in the coil.

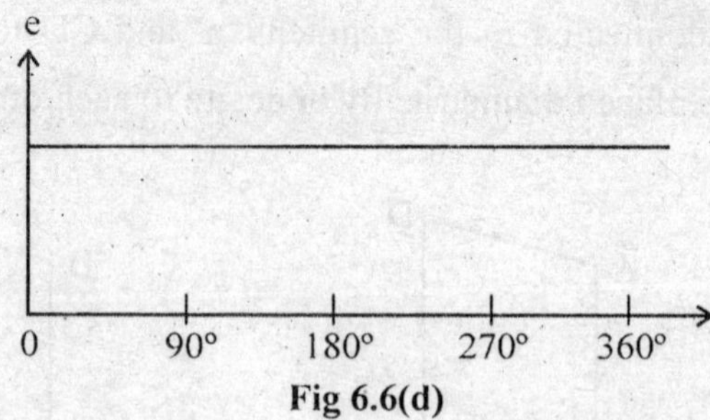

Fig 6.6(d)

The number of coils placed on the armature core is quite large , so that a pure direct current is obtained as the output.

Thus, in a DC generator, the stationary magnetic field is provided by the main poles by passing a DC current to the field winding. Then, if the armature is rotated by a prime-mover, giving mechanical energy to the generator, an alternating emf is induced in the armature coil (Faraday's Law), By means of commutator action, emf made available across arma-ture terminals becomes unidirectional.

6.4 EMF Equation of a DC machine

Amount of emf induced , E in the armature of a DC machine can be obtained using the Faraday's Second Law of Electromagnetic Induction.

Let

P Total number of main field poles

ϕ Useful flux per pole in Wb

Z Total number of conductors in the armature.

A Number of parallel current paths in the armature.

N Speed of the armature in revolutions per minute (RPM)

In one revolution, number of magnetic poles traversed = P

$\therefore$ Flux cut per conductor during one revolution of armature $= \phi$ P Webers

No. of revolutions per second

$$= \frac{N}{60}$$

Flux cut per conductor per second

$$= \phi P \frac{N}{60}$$

No of conductors in one parallel path of armature

$$= \frac{Z}{A}$$

∴ Flux cut by all the conductors in one parallel path per second

$$= \phi P \frac{N}{60} \frac{Z}{A}$$

∴ Total emf induced in all the conductors in one parallel path of the armature

$$\boxed{E = \frac{\phi ZNP}{60A} \text{ Volts}} \qquad ...(6.1)$$

For **Wave Winding** , Number of parallel paths in the armature , A = 2

Hence, $$E = \frac{\phi Z N P}{120} \text{ Volts} \qquad ...(6.2)$$

For **lap winding** , Number of parallel paths in the armature , A = No. of poles, P

Hence, $$E = \frac{\phi ZN}{60} \text{ volts} \qquad ...(6.3)$$

As the voltage across parallel paths is the same, hence this is the total emf induced by all the conductors in the armature

Example 6.1 : A4 pole, lap connected DC machine armature has 60 slots with 6 conductors per slot. The useful flux per pole is 50 mWb. Find the EMF available on the terminals when the armature is rotated at 1500 rpm

Solution :

Lap connection i.e. No. of parallel paths , A = P = 4;

Total conductors , Z = 6 × 60 = 360 conductors

We know that,

$$E = \frac{\phi ZNP}{60A} \text{ Volts}$$

EMF available at the terminals,

$$E = \frac{(50\times10^{-3})(360)(1500)(4)}{(60)(4)} = 450.0 \text{ Volts}$$ **Ans.**

Example 6.2 : A 4-pole, **lap wound** armature has 84 slots with 4 conductors per slot. Its useful flux per pole is 20 m Wb. At what speed should the armature be rotated in order to obtain an emf of 200 Volts?

Solution :

Total conductors, Z = 4 × 84 = 336, Speed, $N = \dfrac{60EA}{\phi ZP}$

Substituting the values, we get:

Speed at which the armature should be rotated,

$$N = \frac{60 \times 200}{20 \times 10^{-3} \times 336} \times \frac{4}{4} = 1785 \text{ rpm.}$$ **Ans.**

6.5 Armature Reaction

The effect of magnetic field set up by the armature current on the distribution of field flux under the main poles is called **'armature reaction'**

When DC machines are not supplying any current, flux in the magnetic circuit is set up only by the poles i.e.field winding. However, when the generator supplies current to a load, the armature current also sets up an MMF which distorts and weakens the main flux. The **armature reaction** poses problems for the proper operation of D.C. generators.

Since the magnetic circuit operates in its saturated region, there is a net demagnetization, which causes the pole flux to reduce.

The Armature reaction is present in DC motors also.

6.6 Commutation

The reversal of current in a coil as the coil passes the brush axis is called **commutation.** When commutation takes place, the coil undergoing commutation is short circuited by the brush.

The brief period during which the coil remains short circuited is known as **Commutation Period, T_c.**

(I) Ideal commutation : If the current reversal is completed by the end of commutation period, it is called **ideal commutation.** If the current reversal is not completed by that time, then sparking occurs between the brush and the commutator which results in damage to both.

(II) Methods of improving commutation : Improving commutation means making it as sparkless as possible. There are two methods of improving commutation:

(i) Resistance commutation

(ii) EMF commutation

(i) Resistance commutation : If contact resistance between the brush and the commutator is made large, then current would divide in the inverse ratio of contact resistances. This is achieved by using carbon brushes instead of copper brushes, which have high contact resistance. The other advantages of carbon brushes are:

(a) The carbon brushes lubricate and polish the commutator.

(b) If sparking occurs, damage to the commutation is less as compared to copper brushes.

(ii) EMF Commutation : In this method, the reactance voltage is neutralized by producing a reversing voltage in the coil undergoing commutation. The reversing voltage can be produced in the following two ways:

(a) By brush shifting

(b) By using Interpoles or Compoles.

6.7 Classification/Types of DC Machines

A DC machine consists of two windings i.e. a field winding which produces the magnetic flux and an armature winding which carries the main current.

The magnetic flux is provided by coils wound around the pole pieces. If the machine consists of any even number of poles , the number of turns on each pole piece is the same and the coils are connected in a manner so as to produce poles of alternate polarity around the periphery. The combination of all these coils which are generally connected in series is called **field winding.**

The simplest method of passing current through the field winding i.e. to excite it , is to connect it across a separate DC source , such as a battery. This type of generator is known as **Separately Excited DC Generator.**

For generating machines , there is another method of exciting the field windings , where no separate DC source is required . The field windings are connected in such a way with the armature winding , so that field exciting current is supplied by the emf induced in the armature itself. This class of generators is called **Self Excited DC Generators.**

Depending on the type of connection of the field winding with the armature , generators are classified as **Shunt** , where field winding and armature are connected in parallel, as **Series** , where field winding and armature are connected in series and **Compound** , where two field windings are used , one is connected in parallel to and the other is connected in series with the armature.

DC motors can also be similarly classified as per the connection of the field winding with the armature winding.

Classification of DC Machines is given below as shown in Fig 6.7

- DC Machine Motor or Generator
  - Separately Excited
  - Self Excited
    - Shunt
    - Series
    - Compound
      - Cumulative
      - Differential
        - Long shunt
        - Short shunt

Fig. 6.7 Classification of DC Machines

6.8 REPRESENTATION OF A DC MACHINE

The field winding and the armature windings of a DC machine are represented symbolically as shown in Fig 6.8 given below

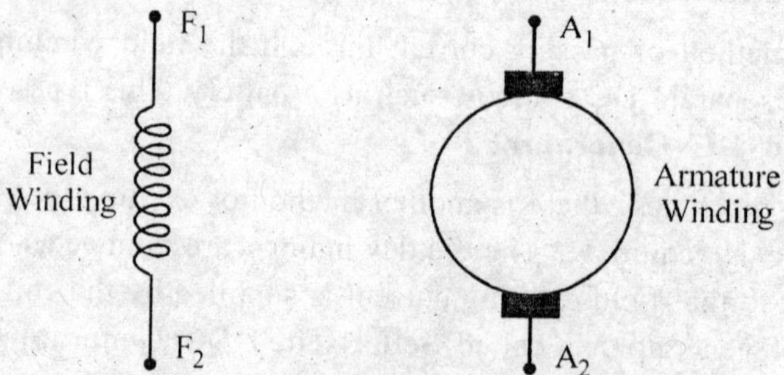

Fig. 6.8 Representation of a DC machine

Field windings are shown like coils and its terminals are marked as F_1 and F_2. Armature winding is shown like a drum with two brushes resting on them. Armature winding is connected to the external circuit through the brushes. Terminals of armature windings are marked as A_1 and A_2 , Fig. 6.8

Different types of DC machines will be discussed using these representations/ symbols.

6. 9 Terminologies Used for the Study of all types of DC Machines

Following terminologies will be used for the study of all types of DC machines.

I_a = Current in the armature winding in amperes.

I_f = Current in the shunt field winding in amperes

I_{se} = Current in the series field winding in amperes

I_L = Current taken by the load in amperes

R_a = Armature resistance in ohms

R_f = shunt field winding resistance in ohms

V = Terminal voltage in volts

E = Induced emf. in volts

BCD = Brush contact drop in volts

E_g = Generated emf. in volts

E_b = Back emf. in volts

6.10 Separately Excited DC Generator

In this type of generator, the exciting current to the field winding is supplied from a separate DC source. Fig. 6.9 shows the connec-tion diagram and equivalent circuit of a Separately Excited DC Generator. There is no electrical connection between the armature and field windings.

The prime mover rotates the armature winding mechanically. The magnetic flux produced by the field winding is cut by the rotating armature winding and hence emf is induced in the armature. The mechanical energy supplied by the prime mover is converted in to electrical energy.

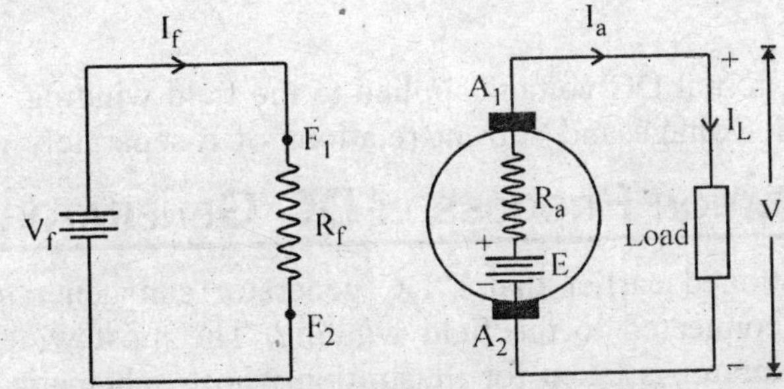

Fig 6.9 Equivalent circuit of Separately Excited DC Generator

6.10.1 Equivalent Circuit of Separately Excited DC Generator

Field winding is represented by its equivalent resistance R_f and armature winding is represented by an induced emf E in series with the resistance of armature winding R_a.

Field current I_f i.e. current in the field winding is constant and is given by:

$$I_f = \frac{V_f}{R_f}$$

where V_f is the external DC voltage applied to the field winding. If V_f is constant, I_f will be constant.

6.10.2 EMF, Voltage, Current Relations of Separately Excited DC Generator

For a DC generator, the terminal voltage is calculated by subtracting voltage drops from the induced emf.

When the load current is I_L, we can easily see that armature current, I_a is equal to I_L because the load and armature are in series.

i.e. $I_a = I_L$ ...(6.4)

The induced emf at no load is E, where

$$E = \frac{\phi ZNP}{120} \text{ Volts.}$$

It decreases to Voltage V at load , since the pole flux is reduced due to armature reaction. The voltage drops in the series circuit are the armature winding resistance drop and a brush contact drop at the moving contact between brush and commutator. This contact drop is either taken separately as 1 to 1.5 volts per brush, or alternatively , this drop is considered as equivalent to an increase in effective armature resistance.

Thus the terminal voltage,

$$V = E - I_a R_a \quad \text{...(6.5)}$$

Where R_a is armature circuit resistance (including resistance due to brush contact drop)

Field current,

$$I_f = \frac{V_f}{R_f} \quad \text{...(6.6)}$$

where V_f is the external DC voltage applied to the field winding. Equations(6.4), (6.5) and (6.6), describe the emf, voltage and current relations of a separately excited DC generator.

6.11 Self Excitation Process in DC Generators

It has been mentioned earlier that a DC generator can generate a voltage without any external source being connected to the field winding. The most widely used self excited DC generator i.e. shunt generator is taken for illustration whose schematic diagram is given below:

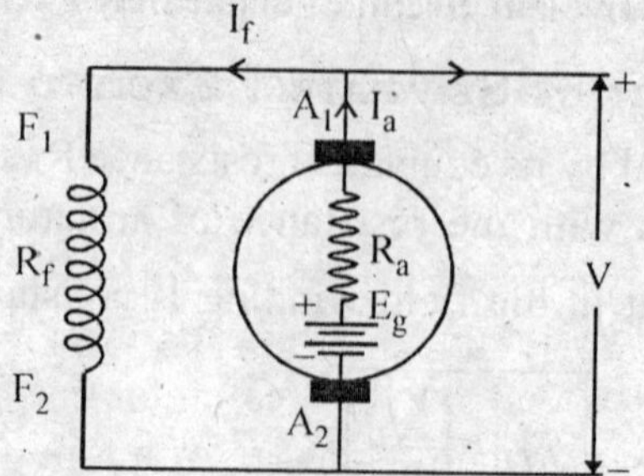

Fig.6.10 Schematic diagram of Self excited DC generator

For the sake of simplicity, let us consider that there is no load on the armature. The machine is driven at a constant speed, say at its rated speed.

A magnetic material has some residual magnetism in it even after the magnetizing force is reduced to zero. Hence,the poles of the shunt generator have some residual flux , even if the field current is zero , provided that the poles had been magnetized earlier. Due to this residual flux, (a small quantity), a small emf is generated across armature terminals A_1 A_2 according to emf equation (6.1). This emf finds a closed path A_1 F_1 F_2 A_2 and a current flows in it. If this current produces a flux in the magnetic circuit in the same direction as the residual flux, the net flux increases and the emf E_g across armature terminals increases, which further increases field current. This cycle continues and the emf across armature terminals A_1 A_2 goes on increasing.

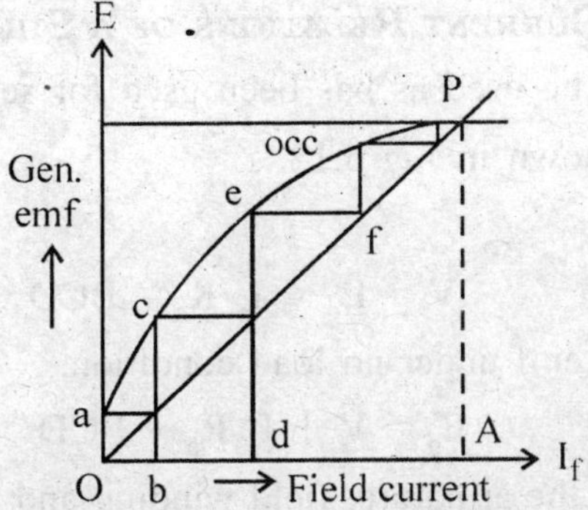

Fig. 6.11 Self Excitation in Generators

The generated emf v/s. field current, 'I_f' is shown in Fig. 6.11. It is seen from Fig 6.11 that when I_f is zero, the generated emf is 'oa', due to residual magnetism and it causes field current 'ob' to flow. This field current aids the residual flux and hence produces a large emf 'bc' . This increased emf causes an even larger field current 'od', which creates more flux for a larger induced emf.'de', therefore I_f further increases ,flux increases and as a consequence emf E increases further . This process of voltage build up continues up to point 'P'. This is the point of stable operation; the field current and emf being OA and AP. Based on the above logic, the essential conditions for self excitation are :

1. There must be a residual flux.
2. The connection between field and armature should be such that any increase in flux produced by increase in field current acts in the same direction as the residual flux.
3. Field Resistance line must intersect the E vs I_f curve (known as open circuit characteristic or O.C.C.).

6.12 Shunt Generator

In a Shunt Generator ,the field winding is connected in parallel with the armature and the current to the field winding is provided by the gen-erator itself .Hence, it comes under the category of self-excited genera-tor. Schematic diagram of a Shunt Generator is shown in Figure 6.12.

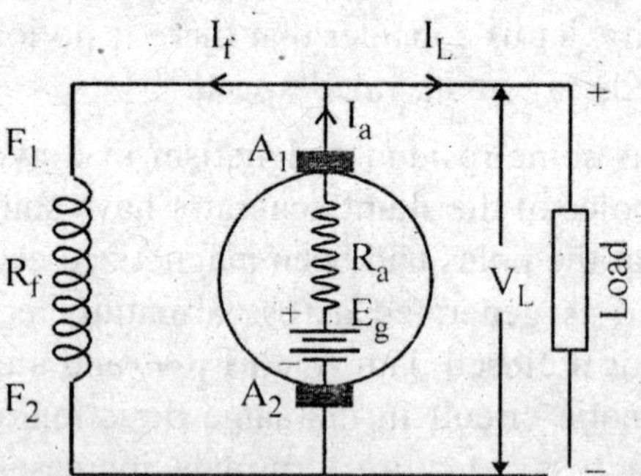

Fig. 6.12 Schematic diagram of Shunt Generator

6.12.1 EMF, Voltage, Current Relations of a Shunt Generator

Same terminologies will be used as has been used for separately excited generators.

For shunt generator, shown in Fig 6.12.

The terminal voltage

$$V = E_g - I_a R_a - BCD \quad ...(6.7)$$

Where E_g is the induced emf under no load condition.

Or

$$E_g = V + I_a R_a + BCD$$

However, by considering the armature, field winding and load current, we find that:

$$I_a = I_L + I_f \quad ...(6.8)$$

Field current,

$$I_f = \frac{V_L}{R_f} \quad ...(6.9)$$

Load current,

$$I_L = \frac{P_L}{V} \quad ...(6.10)$$

where P_L is the power consumed by the load. Power developed by the armature

$$= E_g \times I_a \quad ...(6.11)$$

Equations(6.6) to (6.11) describe the emf , voltage and current relations of a DC shunt Generator.

Example 6.3 : A D.C. shunt generator supplies a load of 4 kW at 200V. Calculate the induced emf, if the armature resistance is $0.5\,\Omega$ and the shunt field resistance is $100\,\Omega$.

Solution :

Load current,

$$I_L = \frac{4\times 10^3}{200} = 20 \text{ amperes}$$

$$I_f = \frac{V}{R_f} = \frac{200}{100} = 2.0 \text{ amperes}$$

$$\therefore \quad I_a = I_L + I_f = 20 + 2.0 = 22.0 \text{ amperes}$$

$\therefore$ Induced emf, $\quad E_g = V + I_a R_a = 200 + (22.0)\, 0.5 = 211$ Volts. **Ans.**

Example 6.4 : A 5 kW, 200V D.C. shunt generator has armature and field resistance of 0.10 and 50 respectively. Determine the total armature power developed

Solution:

$$I_f = \frac{V}{R_f} = \frac{200}{50} = 4 \text{ amperes}$$

$$I_L = \frac{P_L}{V} = \frac{5\times10^3}{200} = 25 \text{ amperes}$$

$$I_a = I_L + I_f = 25 + 4 = 29 \text{ amperes}$$

$$E_g = V + I_a R_a = 200 + (29 \times 0.1) = 202.9 \text{ Volts}$$

$\therefore$ Armature power developed,

$$E_g \times I_a = 202.9 \times 29 = 5884.1 \text{ Watts}$$ **Ans**

6.13 Series Generator

The series generator is a self excited generator with armature, field winding and load, all connected in series . Its schematic diagram is shown in Figure 6.13

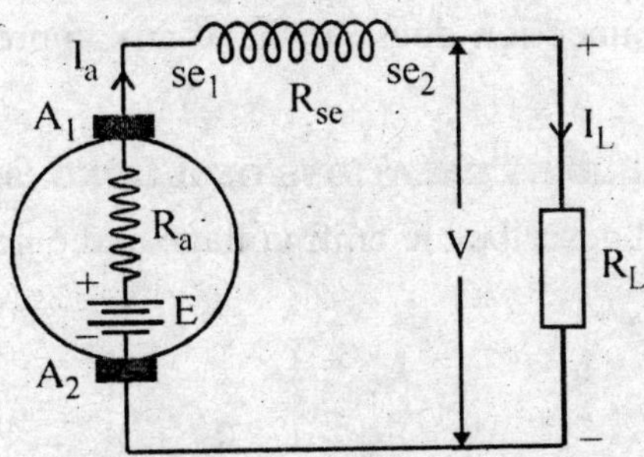

Fig. 6.13 Series Generator

6.13.1 EMF, Voltage, Current Relations of Series Generator

Following equations show the relations of EMF, voltage and current for series generator.

$$I_a = I_{se} = I_L \qquad \text{...(6.12)}$$

$$V = E - I_a (R_a + R_{se}) = I_L R_L \qquad \text{...(6.13)}$$

Where E is the emf generated when the generator is delivering load current I_L.

6.14 Compound Generator

Compound generators are of two types depending upon the connections of series and shunt field windings. These are:

1. Long Shunt Compound Generator
2. Short Shunt Compound Generator

The net flux set up in compound generators is combin. ·n of flux set up by shunt field winding on no load and the flux set up by series field winding d. ring load condition.

If the generator is to be connected for **Long Shunt**, the shunt field circuit is to be connected across the load terminals. Fig 6.14 shows the connection diagram of a **Lon, Shunt compound generator**

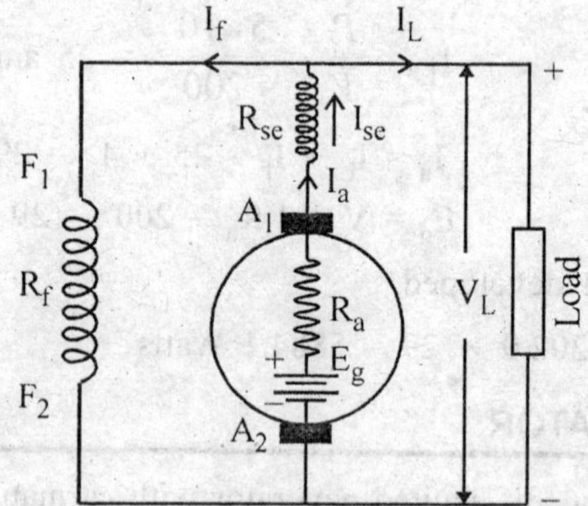

Fig 6. 14 Long Shunt Compound Generator

Short and Long Shunt connection does not have any appreciable change in characteristics of compound generators.

6.14.1 EMF, voltage ,current relations of a Long Shunt Compound Generator

Equations (6.14) to (6.15) describe the emf, voltage and current relations of a DC long shunt compound Generator.

$$I_L + I_f = I_{se} = I_a \quad \text{...(6.14)}$$

$$V = E_g - I_a (R_a + R_{se}) \quad \text{...(6.15)}$$

$$E_g = V + I_a R_a + I_{se} R_{se} + BCD$$

Fig. 6.15 shows the connection diagram of a **Short Shunt compound generator**, which is pref-erable to long shunt connection. **In this case , shunt field winding is connected across the armature only.**

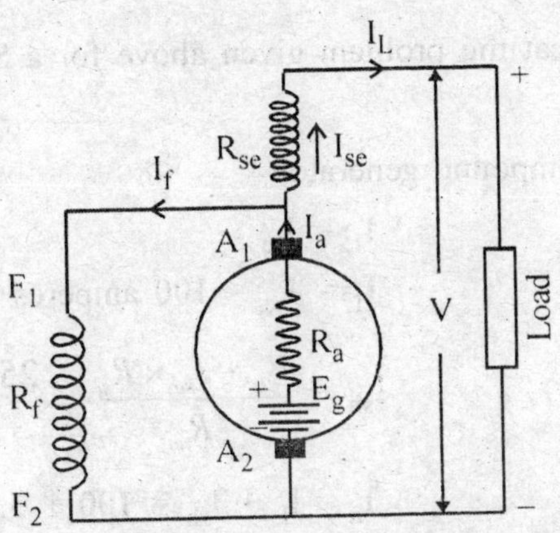

Fig 6.15 Short Shunt Compound Generator

6.14.2 EMF, Voltage, Current Relations of a Short Shunt Compound Generator

Equations (6.16) to (6.18) describe the emf , voltage and current relations of a DC shunt Generator.

$$I_L = I_{se} \quad ...(6.16)$$

$$I_a = I_{se} + I_f \quad ...(6.17)$$

$$V = E_g - I_a R_a - I_{se} R_{se} \quad ...(6.18)$$

$$E_g = V + I_a R_a + I_{se} R_{se} + BCD$$

Example 6.5 : A D.C. long shunt compound generator delivers a load current of 100 amperes at 250 volts. The resistance of the armature, series field and shunt field are 0.2 Ω, 0.01 Ω and 50 respectively. Calculate the emf induced in the armature. Allow a brush contact drop of 2V/Brush.

Solution :

B.C.D. = 2 Volts / brush = 2 × 2 = 4 volts.

$$I_{sh} = \frac{V_L}{R_{sh}} = \frac{250}{50} = 5 \text{ amperes}$$

$$I_a = I_L + I_{sh} = 100 + 5 = 105 \text{ amperes}$$

$$I_a = I_{se} = 105 \text{ amperes}$$

$$E_g = V + I_a R_a + I_{se} R_{se} + BCD$$

$$= 250 + (105 \times 0.2) + (105 \times 0.01) + 4$$

$$= 250 + 21.0 + 1.05 + 4$$

$$E_g = 276.05 \text{ Volts}$$

Ans.

Example 6.6 : Repeat the problem given above for a Short Shunt Compound Generator.

Solution :

For a short shunt compound generator

$$I_L = I_{se}$$

$$I_L = I_{se} = 100 \text{ amperes}$$

$$I_{sh} = \frac{V_L + I_{se} \times R_{se}}{R_{sh}} = \frac{250 + (100 \times 0.01)}{50} = 5.02 \text{ amperes}$$

$$I_a = I_L + I_{sh} = 100 + 5.02 = 105.02 \text{ amperes}$$

$$\therefore \quad E_g = V + I_a R_a + I_{se} R_{se} + BCD$$

$$= 250 + (105.02 \times 0.2) + (100 \times 0.01) + 4$$

$$= 250 + 21.004 + 1 + 4$$

$$E_g = 276.004 \text{ Volts.}$$

Ans.

6.15 Efficiency of a DC Generator

Efficiency is defined as the ratio of output power to the input power of a DC generator.

$$\eta_{gen} = \frac{output\ power}{input\ power} = \frac{P_o}{P_o + losses} \quad ...(6.19)$$

Variable copper losses = copper loss in armature winding + copper loss in field winding

$$= I_a^2 R_a + I_f^2 R_f$$

Constant iron loss = W_i

Output power = $V I_L$

$$\therefore \quad \eta_{gen} = \frac{VI_L}{VI_L + I_a^2 R_a + I_f^2 R_f + W_i} \quad ...(6.20)$$

Example 6.7 : A 200 V shunt generator has a full load current of 100 amp. Its armature resistance is 0.1 ohm, field resistance 100 ohm and constant losses and stray losses are 2000 watts. Calculate its efficiency at full load.

Solution :

$$I_f = \frac{V}{R_f} = \frac{200}{100} = 2 \text{ amperes}$$

Armature current at full load $= 100 + 2 = 102$ amp.

Armature Copper loss at full load $= (102)^2 \times 0.1 = 1040.4$ Watts

Total loss at full load $= 2000 + 1040.4 = 3040.4$ Watts

Output power at full load $= V\, I_L = 200 \times 100 = 20$ kW

Efficiency at full load $= \dfrac{2000}{23040.4}$

$= 86.8\%$ **Ans.**

6.16 Applications of DC Generators

(i) Separately Excited DC Generators :

Separately excited generators are preferred where the characteristics of DC shunt generator is not up to the expected level. Their applications have limitations because they need a separate source of excitation. Applications include refining of materials or electroplating

(ii) Shunt Generators :

(a) Battery charging in automobiles.

(b) As a pilot exciter for synchronous generator to excite the field magnets.

(c) Lighting and power supply purpose.

(d) Electroplating and electrolysis purposes.

(iii) Series Generators :

(a) As Boosters on DC feeders (railway services, compensating voltage drops in lines).

(b) Constant current source- welding-in- drooping characteristics

(c) Series arc lighting, series incandescent lighting

(d) Regenerative braking of D.C. locomotives

(iv) Compound Generators :

To supply DC power to railway circuits, elevator motors, welding.

6.17 DC Motors

DC motors are powered by DC sources. As nearly all electrical energy is distributed as AC, these motors have limited applications. However a rectifier or some converter is required where ever it is desirable to use DC motors only with the available AC supply, although it makes the system expensive.

In automotive applications , such as starting, wipers , power windows etc. DC motors are frequently used because DC is available readily from the battery.

An important advantage of dc motors has been that their speed and direction can be controlled more easily than ac motors. However with the advent of electronic systems that can vary the supply frequency, this advantage is disappearing.

6.17.1 Construction of DC Motors

A DC motor converts electrical energy in to mechanical energy i.e. input to a DC motor is electric power and its output is mechanical power.

The construction of DC motor is same as that of a DC generator. It also contains two windings i.e. field winding and an armature winding. The field winding or the field poles are stationary and the armature winding or the armature rotates. The armature is keyed to a shaft which couples the machine to the mechanical load. The shaft and tha armature are supported by bearings so that they can rotate freely.

In most of the DC motors , the direction of the current in the armature conductors on the rotor is reversed periodically during rotation. This is done with the help of a mechanical switch consisting of brushes mounted on the stator and a commutator mounted on the shaft. The brushes are in sliding contact with the commutator. As the commutator rotates , movement of the brushes from one segment to another changes the direction of current in the armature conductors.

6.17.2 Principle of Operation of DC Motors

The operating principle of a dc motor is that whenever a current carrying conductor is placed in a magnetic field, it experiences a force exerted on it.

The electrical energy, as the input to the motor, is supplied by a DC source voltage, V with a current I_f .

The current I_f flowing through the field winding causes a stationary magnetic field to be setup by the main poles i.e. alternate North and South poles. The current I_a entering into the armature winding via brushes and commutator gets distributed in such a way that all the conductors under the N-pole carry current in one particular direction, say outward and the conductors under S-pole carry current in opposite direction, say inward direction, as shown in Fig. 6.16.

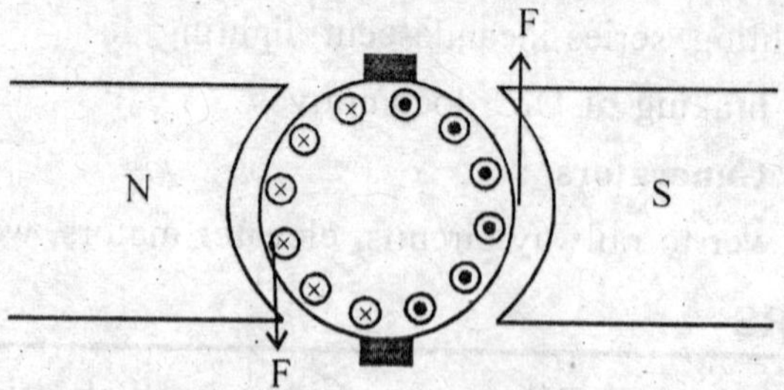

Fig. 6.16 Current Distribution

Each current carrying conductor of the armature winding within the magnetic field experiences a force given by the equation :

$$F = B\ I\ l \text{ Newtons}$$

The force developed by the conductor , acting on the shaft ,causes a torque T_d to develop. The direction of force and torque, is given by **Fleming's Left Hand Rule.**

The rule states that if the first three fingers of the left hand are held mutually at right angles to each other and if the first finger indicates the direction of the field , the centre finger indicates the direction of current, then the thumb indicates the direction of force exerted on the conductor.

If the torque developed is more than the load torque on the shaft and frictional torque on bearings, the arma-ture will start rotating in the same direction as that of the force .

Direction of the force depends upon the direction of magnetic field i.e. field current as well as, the direction of the current through the conductor i.e. armature winding.

Thus, direction of force exerted on the armature reverses , if the direction of either the field winding current or armature winding current is reversed. However , the direction of force will remain same, if we reverse the directions of both the field winding and armature winding currents simultaneously.

Therefore, if the direction of rotation of the motor is to be changed, **either** the direction of the current through the field winding **or** the direc-tion of the current through the armature winding should be changed.

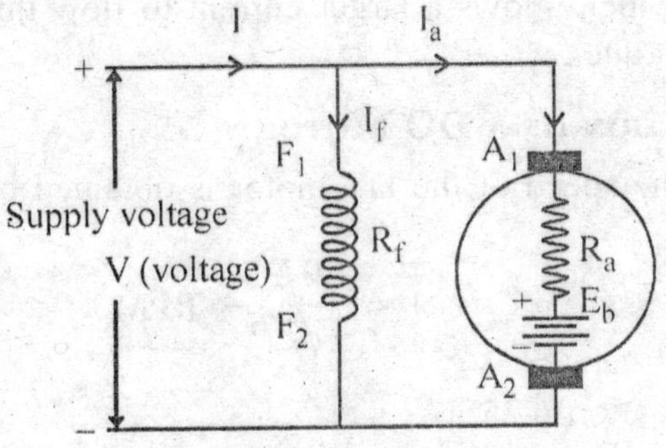

Fig. 6.17 D C Motor

6.17.3 Back EMF

As the armature starts rotating, the magnetic flux produced by the field winding and linking the armature conduc-tors changes. Hence, according to Faraday's First Law of electromagnetic induction, emf is induced in the armature con-ductors This emf opposes the supply voltage according to Lenz's law, hence it is called **back emf** and is denoted by E_b and is given by :

$$E_b = \frac{\phi ZN}{60} \cdot \frac{P}{A} \quad \text{...(6.21)}$$

This back emf opposes the applied voltage V. Back emf makes the motor self regulating motor i.e. motor draws as much armature current as is just sufficient to meet the load requirement

Fig. 6.17 shows the equivalent circuit of a DC motor. It can be seen that,

Armature current,

$$I_a = \frac{V - E_b}{R_a}$$

Or $$V = E_b + I_a R_a \qquad ...(6.22)$$

$$I = I_f + I_a \qquad ...(6.23)$$

Hence, the power equation is:

Power input to the armature

$$P = V I_a = E_b I_a + I_a^2 R_a \qquad ...(6.24)$$

Where,

$V I_a$ is electrical power input to the armature

$E_b I_a$ is mechanical power developed in the armature

$I_a^2 R_a$ is copper loss in the armature

If the load on the motor is increased , the motor will slow down i.e. motor speed decreases Hence back emf falls, which allows a larger current to flow through the armature, which leads to increase in driving torque .

6.17.4 Speed Equation of a DC Motor

The equation for the speed of the DC motor is obtained by rearrang-ing equation (6.6) as

$$N = \frac{60 E_b A}{\phi Z P} \text{ RPM} \qquad ...(6.25)$$

We can rewrite equation 6.25 as:

$$\boxed{N = K \frac{E_b}{\phi}} \qquad ...(6.26)$$

Where K is a constant $= \frac{60}{Z} \cdot \frac{A}{P}$, because A,Z,P are constants for a given machine.

The network equation for the motor armature is :

$$E_b = V - I_a R_a \qquad ...(6.27)$$

Hence, substituting the value of E_b in equation (6.22),we get:

$$\boxed{N = K \frac{(V - I_a R_a)}{\phi}} \qquad ...(6.28)$$

i.e. Speed of a motor is directly proportional to supply voltage and inversely proportional to the field flux or the field current.

6.17.5 Torque Equation of a DC Motor

The equation for the torque developed by a DC motor can be obtained by equating mechanical power developed by the armature with electrical power input to the armature.

Let P_m be the mechanical power developed by the armature and P_e is the electrical power input to the armature.

Then, $$P_m = P_e \quad \text{...(6.29)}$$

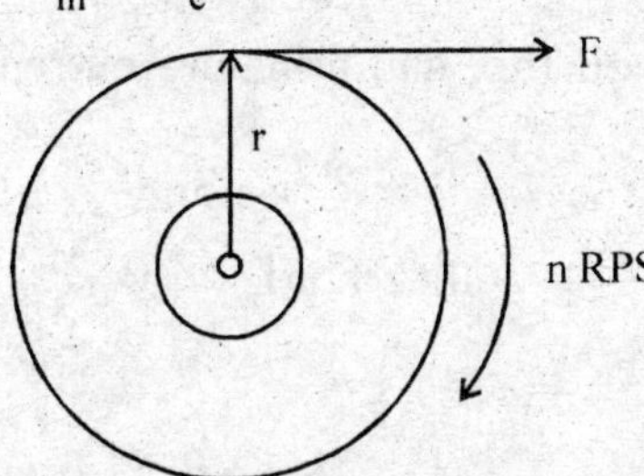

Fig 6.18 Torque Development

In Fig.6.18, a circumferential force of 'F' Newtons is exerted tangentially on a shaft of radius 'r' meters. The shaft is rotating at a speed of N revolutions per minute(RPM)

Distance covered in one revolution = circumference of the shaft = $2\pi r$

Work done by the force in one revolution = Force × distance

$$= F \times 2\pi\ r \text{ joules}$$

Work done/sec.

$$= \frac{F \times 2\pi \times r \times N}{60} \text{ joules/sec.}$$

Power developed = Work done/sec

i.e. Power $$P_m = F \times r \times \frac{2\pi N}{60} \text{ Watts}$$

Power developed $$= T \times \frac{2\pi N}{60}$$

Where Torque T, the mechanical torque developed = F × r

Substituting the value of P_m and P_e in equation(6.29), we get:

$$\frac{2\pi NT}{60} = E_b I_a \quad \text{...(6.30)}$$

Substituting the value of E_b , we get:

$$\frac{2\pi NT}{60} = \left(\frac{\phi ZN}{60} \times \frac{P}{A}\right).I_a \quad \text{...(6.31)}$$

Simplifying and rearranging equation (6.31) , we get:

$$T = \frac{1}{2\pi}\phi Z I_a . \frac{P}{A} \text{ Nm} \quad ...(6.32)$$

$$= 0.159\,\phi\, Z I_a \frac{P}{A}$$

Thus Torque, T depends on flux and armature current i.e.

$$T \propto \phi I_a$$

or $$T = K \phi I_a \quad ...(6.33)$$

Where

$$K = \frac{1}{2\pi} Z \frac{P}{A} \text{ is a constant.} \quad ...(6.34)$$

Thus, the motor develops mechanical energy with torque T and speed N.

Equation (6.33) is known as the **Torque Equation**.

Example 6.8 : A 4 pole, **wave wound** armature has 48 slots with 4 conductors per slot. Its useful flux per pole is 25m wb. Find the torque developed when the armature current is 50 amp.

Solution :

Wave wound i.e. $A = 2$;

Torque developed,

$$T_d = \frac{1}{2\pi}\phi Z I_a \frac{P}{A} \text{ Nw} - \text{m}$$

$$= \frac{1}{2\pi}(25\times10^{-3})\times48\times4\times50\times\frac{4}{2}$$

$$= \frac{240}{\pi} = 76.394 \textbf{ Nw – m}$$ **Ans.**

Example 6.9 : The armature of a 4-pole DC motor has a lap connected winding accommodated in 60 slots, each containing 10 conductors. If the useful flux per pole is 20 mwb ,calculate the torque developed when the armature current is 50 amps.

Solution :

Total conductors, $Z = 10 \times 60 = 600$

Torque, $$T_d = \frac{\phi Z I_a}{2\pi}\times\frac{P}{A}$$

$$= \frac{20 \times 10^{-3} \times 600 \times 50}{2\pi} \times \frac{4}{4} = 95.492 \text{ Nm.}$$ **Ans.**

6.17.6 Various Types of Torques in DC Motors

There are three torques in a DC motor:

(i) **Gross Torque, T_a :** The torque developed in the armature winding is called gross torque T_a. The power associated with this torque is gross mechanical output power. P_m.

(ii) **Frictional Torque, T_f :** The torque required to overcome the friction and windage losses is called **Frictional Torque T_f**

(iii) **Shaft Torque or Load torque, T_{sh} or T_L :** The torque available at the shaft to do useful work is called Shaft Torque or Load torque T_{sh} or T_L.

6.18 Types of DC Motors

DC motors may be classified depending upon the connections of field winding and armature as in the case of DC generators.

There are three types of DC motors :

(i) DC Shunt motor

(ii) DC Series motor

(iii) DC Compound motor

6.18.1 Shunt Motor

In DC shunt motors, the field winding is connected across the armature i.e. in parallel with the armature as shown in Fig. 6.19.

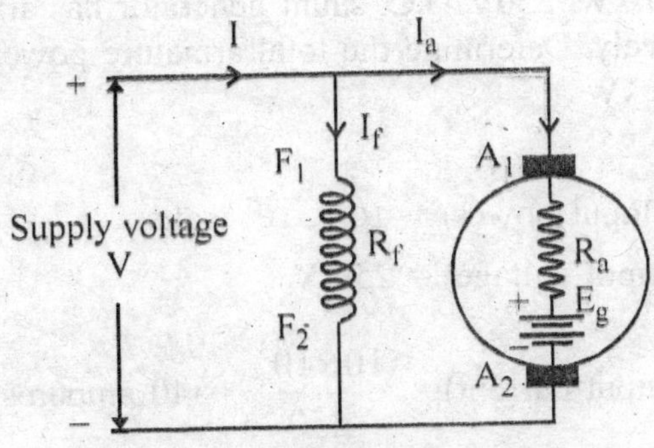

Fig. 6.19 Shunt Motor

6.18.1.1 EMF, Voltage, Current Relations in a DC Shunt Motor

For a motor, the load is replaced by a DC source and the voltage-current relations may be obtained by inspection. Further, the applied voltage in case of motor is the sum of back emf and all series voltage drops.

Field winding current,

$$I_f = \frac{V}{R_f}$$

Motor input current, $I = I_a + I_f$ ...(6.35)

Supply voltage, $V = E_b + I_a R_a$ ...(6.36)

Back emf, $E_b = V - I_a R_a$

Armature power developed

$$= E_b\, I_a$$

Equations (6.35) to (6.36) describe the emf , voltage and current relations of a DC shunt motor. Under normal operating conditions, the field current and hence the flux ϕ, remains constant.

Example 6.10 : A 250 V D.C. shunt motor takes 25 amperes at full load. The resistance of the armature and shunt field windings are 0.2 Ω and 100 Ω respectively. Determine the back emf on full load.

Solution :

$$I_f = \frac{250}{100} = 2.5 \text{ ampere}$$

$$I_a = I_L - I_{sh} = 25\text{-}2.5 = 22.5 \text{ amperes}$$

$$\therefore \quad E_b = V - I_a R_a$$

$$= 250 - 22.5 \times 0.2 = 245.5 \text{ volts.}$$ **Ans.**

Example 6.11 : A 10 kW, 250V D.C. shunt generator has armature and field resistance of 0.10Ω and 100Ω respectively. Determine the total armature power developed when working as a **shunt motor** taking 10 kW

Solution :

$$P_L \text{ (Input power)} = 10 \times 10^3 \text{ watts}$$

$$V_L \text{ (Input voltage)} = 250 \text{ V}$$

$$I_L \text{ (Input current)} = \frac{10 \times 10^3}{250} = 40 \text{ amperes}$$

$$I_{sh} = \frac{V_L}{R_{sh}} = \frac{250}{100} = 2.5 \text{ amperes}$$

$$I_a = I_L - I_{sh} = 40 - 2.5 = 37.5 \text{ amperes}$$

$$E_b = V_L - I_a R_a = 250 - (37.5 \times 0.10) = 246.25 \text{ volts}$$

Armature power developed,

$= 246.25 \times 37.5 = 9243.75$ watts **Ans.**

Example 6.12 : A 4 pole wave wound 220 V shunt motor has 400 armature conductors. The resistance of the armature is 0.05 Ω and that of shunt field is 55Ω. Find the torque developed when it takes a current of 40 amperes. The flux/pole is 25 milliwebers.

Solution :

$$I_{sh} = \frac{V_L}{R_{sh}} = \frac{220}{55} = 4 \text{ amperes}$$

$$I_a = I_1 - I_{sh} = 40 - 4 = 36 \text{ amperes}$$

$$A = 2$$

$$T = 0.159 \times \Phi \times Z \times I_a \times P/A \quad \text{N–m}$$

$$= \frac{0.159 \times 25 \times 10^{-3} \times 400 \times 36 \times 4}{2}$$

$$= 114.48 \text{ N–m}$$

Ans.

Example 6.13 : A 220 V, D.C. Shunt motor takes 4 amps on no load and runs at 1000 RPM. Armature resistance = 0.1 ohms and field resistance = 100 ohms. Calculate speed, when the motor is loaded and takes a current of 40 amps. Assume that the armature reaction weakens the field by 2%

Solution:

No load :

$$I_f = \frac{220}{100} = 2.2 \text{ amps}$$

$$I_a = I_L - I_f = 4 - 2.2 = 1.8 \text{ amps}$$

$$E_{b1} = 220 - (1.8 \times 0.1) = 219.82 \text{ volts.}$$

Load condition :

$$I_{a2} = 40 - 2.2 = 37.8 \text{ amps}$$

$$E_{b2} = 220 - (37.8 \times 0.2) = 220 - 7.56 = 212.44 \text{ volts}$$

$$\phi_2 = 0.98\phi_1$$

$$N_2 = \frac{Eb_2}{Eb_1} \times \frac{\phi_1}{0.98\phi_1} \times 1000 = \frac{212.44}{219.82} \times \frac{\phi_1}{0.98\phi_1} \times 1000$$

$$= 986.15 \text{ RPM.}$$

Ans.

6.18.1.2 Characteristics of DC Motors

In order to use a motor suitably, it is necessary to know its performance characteristics. These characteristics are of two types:

1. Electromechanical, showing the relationship between an electrical quantity and a mechanical quantity
2. Mechanical – showing the relationship between two mechanical quantities. In a DC motor, three important quantities are:
   (i) Current input to the armature
   (ii) speed
   (iii) torque

Characteristics relating current and speed, current and torque are **Electromechanical. Characteristics.** Characteristics relating torque and speed are **Mechanical Characteristics.**

In order to derive these characteristics, various equations of dc motors are rewritten as follows:

The induced emf is, $$E_b = \frac{\varphi Z N}{60} \cdot \frac{P}{A} \text{ volts}$$

The torque is, $$T = \frac{1}{2\pi} \phi Z I_a \frac{P}{A} \text{ Nm}$$

For a given machine, number of poles, number of conductors and number of parallel paths are constant.

Substituting, $$\frac{ZP}{60A} = K,$$

we get, $$E_b = K\phi N \quad ...(6.37)$$

$$T = K\phi I_a \quad ...(6.38)$$

6.18.2 Series Motor

In a series motor, the field winding is connected in series with the armature circuit as shown in Fig. 6.20.

As the field current I_{se} is either equal to or proportional to the armature current I_a, the flux setup by the field current, ϕ becomes $\propto$ I_a, when the motor is unsaturated and ϕ is constant when the motor is fully saturated.

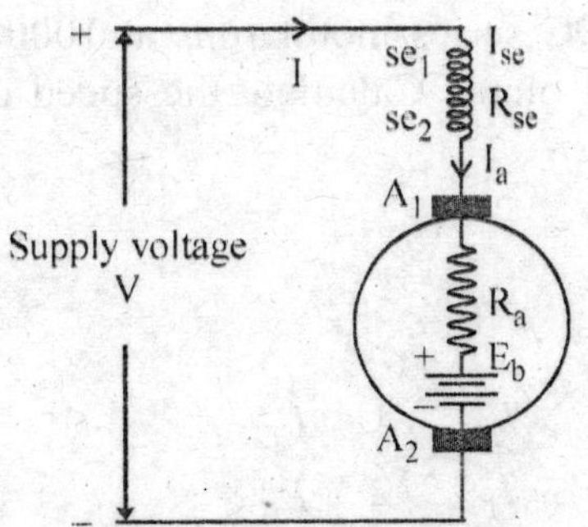

Fig 6.20 DC Series Motor

6.18.2.1 EMF, Voltage, Current Relations of DC Series Motor

For a motor, the load is replaced by DC supply source and the voltage - current relations may be obtained by inspection. The applied voltage in case of motor is the sum of back emf and all series voltage drops.

$$I = I_a = I_{se} \qquad ...(6.39)$$

$$V = E_b + I_a (R_a + R_{se}) \qquad ...(6.40)$$

Example 6.14 : A 250 D.C. **series motor** has an armature resistance of 0.10 Ω and series field resistance of 0.08 Ω. It is running at a speed of 1500 RPM, taking a current of 25 amperes. Determine its speed when it takes a current of 20 amperes.

Solution:

When the motor takes a current of 25 amperes

$$E_1 = V - I_a R_a - I_{se} R_{se}$$

$$I = I_a = I_{se} = 25 \text{ amp}$$

$$E_1 = 250 - (25 \times 0.1) - (25 \times .08) = 245.5 \text{ volts}$$

When the same motor takes a current of 20 amperes

$$I = I_{se} = I_a = 20 \text{ amperes} = I_2$$

$$\therefore \quad E_2 = 250 - (20 \times 0.1) - 20\,(.08) = 246.4 \text{ Volts}$$

We know that,

$$\frac{N_2}{N_1} = \frac{E_2}{E_1} \times \frac{I_1}{I_2}$$

$$N_2 = N_1 \times \frac{E_2}{E_1} \times \frac{I_1}{I_2}$$

$$= 1500 \times \frac{246.4}{245.5} \times \frac{25}{20} = 1882 \text{ r.p.m}$$

Ans.

Example 6.15 : A 220 V, DC series motor runs at 1000 RPM when taking 100 amps. Its armature circuit resistance is 0.2 ohms. Calculate the speed of the motor at half the torque.

Solution :

We know that, torque,

$$T \propto I_a^2$$

$$\frac{T_2}{T_1} = \frac{1}{2} = \frac{I_{a2}^2}{I_{a1}^2}$$

$$\therefore \quad I_{a2} = \frac{I_{a1}}{\sqrt{2}} = \frac{100}{\sqrt{2}} = 70.71 \text{ amp.}$$

$$E_{b1} = 220 - (100 \times 0.2) = 200 \text{ V}$$

$$E_{b2} = 220 - \left(\frac{100}{\sqrt{2}} \times 0.2\right) = 220 - (14.14) = 205.86$$

We know that,

$$\frac{N_2}{N_1} = \frac{E_{b2}}{E_{b1}} \times \frac{\phi_1}{\phi_2} = \frac{E_{b2}}{E_{b1}} \times \frac{I_{a1}}{I_{a2}}$$

$$\therefore \quad N_2 = \frac{205.86}{200} \times \frac{100}{70.71} \times 1000$$

$$= 1455 \text{ RPM}$$

Ans.

6.18.3 Compound Motor

6.18.3.1 EMF, Voltage, Current Relations of Compound Motor

Compound motor (Long shunt) : Fig. 6.21 shows the connection diagram of a **Long Shunt Compound Motor**.

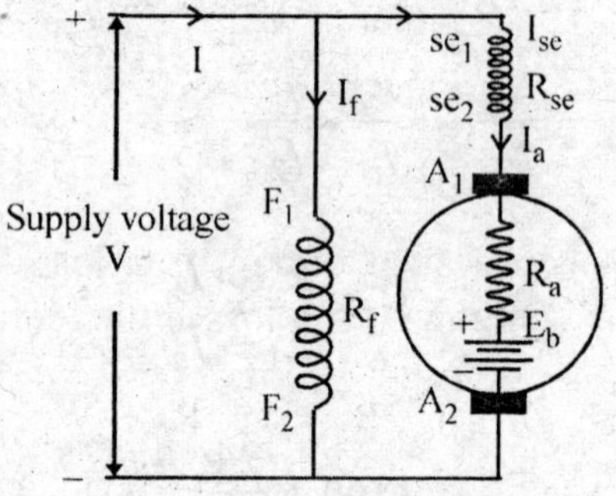

Fig 6.21 Long Shunt Compound Motor

Motor Input current, $I = I_f + I_{se}$ ...(6.41)

Series field winding current, $I_{se} = I_a$ ...(6.42)

Supply voltage, $V = E + I_a(R_a + R_{se})$ ...(6.43)

Compound motor (Short shunt) : Fig. 6.22 shows the connection diagram of a **Short Shunt Compound Motor.**

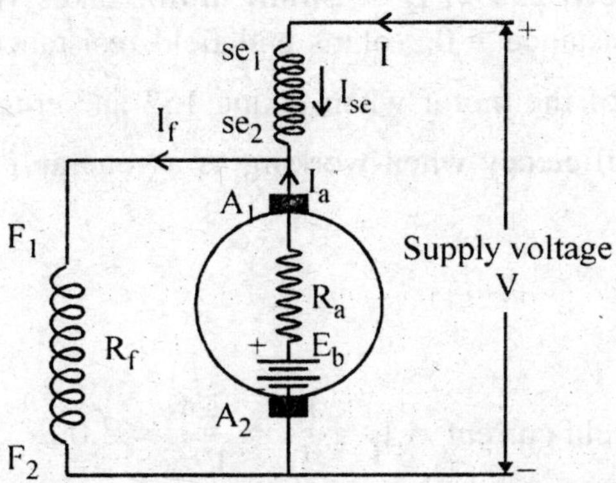

Fig 6.22 Short Shunt Compound Motor

$$I = I_{se} = I_a + I_f \quad \text{...(6.44)}$$

$$V = E + I_a R_a + I_{se} R_{se} \quad \text{...(6.45)}$$

On no load, as the armature current i.e. the series field current is small, most of the flux is set up by the shunt field winding only. Hence the compound motor starts on no-load as a shunt motor and a desired no-load speed N_0 can be set by suitable current I_f through the shunt field winding.

Classification of long and short shunt compound motor does not change the shape of the characteristic much.

6.19 Efficiency of a DC Motor

Efficiency is defined as the ratio of output power to the input power of a DC motor.

i.e. $$\eta_m = \frac{\text{output power}}{\text{input power}} = \frac{P_i - \text{losses}}{P_i}$$

where, Power input, $P_i = V\,I$

Total Losses = Variable copper losses + Constant iron loss = Copper loss in the armature winding + Copper loss in the field winding + Iron loss in the core.

$$= I_a^2\, R_a + I_f^2\, R_f + W_i$$

Hence, eff. of motor,

$$\eta_m = \frac{P_i - \left(I_a^2 R_a + I_f^2 R_f + W_i\right)}{P_i} \quad ...(6.46)$$

Example 6.16 : A 60 kW, 250 V, D.C. **Shunt motor** takes 16 amps on no load and runs at 1440 RPM. Armature resistance = 0.2 ohms and field resistance = 125 ohms

(i) Find the efficiency of the motor when taking 152 amperes.

(ii) Also estimate the efficiency when working as a generator and delivering 152 amperes at 250 volts.

Solution :

Efficiency as a motor

$$\text{Field current} = I_f = \frac{V}{R_f} = \frac{250}{125} = 2.0$$

Armature current,

$$I_a = I - I_f = 152 - 2 = 150 \text{ amperes}$$

Copper loss in armature, $I_a^2 R_a = (150)^2(0.2) = 4500$ watts

Copper loss in field, $I_f^2 R_f = (2)^2(125) = 500$ watts

Total losses, $= 4500 + 500 = 5000$ watts

Input power, $V I = 250(152) = 38000$ watts

$$\eta_m = \frac{P_i - \text{losses}}{P_i} = \frac{38000 - 5000}{38000} \times 100 = 86.84\%$$

Efficiency as a Generator

Output power, V $I_L = 250 \times 152 = 38000$ watts

$$I_a = I_2 + I_f = 152 + 2 = 154 \text{ amp.}$$

$$\text{Losses} = I_a R_a + I_f R_f$$

$$= (154)^2(0.2) + (2)^2(125) = 4743.2 + 500 = 5243.2 \text{ watts}$$

$$\eta_{gen} = \frac{P_o}{P_o + losses} = \frac{38000}{38000 + 5243.2} \times 100 = 87.86\%$$

6.20 Starters for DC Motors

If a dc motor connected to the dc supply directly, a heavy current will be drawn from the supply, because the back or counter emf is zero, when the speed is zero and the armature resistance is very small, generally less than one ohm. Hence an additional resistance is inserted in the armature circuit to control this starting current.. This additional resistance is cut off gradually as the speed increases and back emf builds up.

A starter for a dc motor consists of a starting resistance which is cut off in steps and also has some additional protective features, such as No Volt Release and Overload Protection. Three point and four point starters are used for starting dc shunt and compound motors.

6.21 Applications of DC Motors

DC motors offer a wide range of speed control and torque as well as excellent acceleration and deceleration.In some electronic control devices, a speed range of 40 : 1 is possible. Maximum torque in the case of DC machines is limited by commutation and not by heating as in other machines

Features like high flexibility and easier control of DC motors can not be matched by any other type of motor. Which type of motor suits a particular load depends on the characteristics of the load and characteristics of the motor. Application of different motors are as given below.

(i) **Shunt motors constant speed :** These are used essentially for constant speed applications i.e. where speed is required to be constant from no load to full load and requiring medium starting torque. Their starting torque is 30% to 40 % more than its rated torque. May be used for adjustable speed not greater than 2:1 range,such as: Wood working machines, battery operated fans, lathes, drills, boring mills, shapers; centrifugal pumps, blowers, conveyors, spinning and weaving machines.

(ii) **Shunt motors adjustable speed :** Same as above,for applications requiring adjustable speed control., either constant torque or constant output.

(iii) **Series motors :** These motors develop very high starting torque and has adjustable varying speed.These are variable speed motors and are used :

For drives requiring very high starting torque and where adjustable varying speed is acceptable and where the load is subjected to heavy fluctuations. Loads must be positively connected, not belted,.to prevent over speed. Lightest load should not be less than 15 to 20 % of full load torque.

Electric traction, elevators, conveyors, trains, fork lifting trucks, starter motor in automobiles, cranes, air compressors, vacuum cleaners, hair drier, sewing machines

(iv) **Compound motors :** Compound motor has considerably higher starting and maximum running torque as compared to DC shunt motor.

Ccompound motors are used for drives requiring high starting torque and fairly constant speed, pulsating loads,irregular loads or suddenly applied heavy loads, such as plunger pumps, shears, conveyors,crushers, hoists, Rolling mills, large Printing presses, heavy planers, reciprocating machines etc.

6.22 Revision Formulae

Let

- φ is flux per pole in webers
- N is speed of armature in R.P.M.
- P is number of poles
- Z is number of armature conductors
- A is number of parallel paths
- E_g is generated emf in volts
- I_a and R_a are armature current in amperes and armature resistance in ohms.
- I_{sh}, R_{sh} are shunt field current in amperes and shunt field resistance in ohms.
- I_L, R_L, P_L, V_L are load current in amperes, load resistance in ohms, load in watts and load voltage in volts.
- I_{se}, R_{se} are series field current in amperes and series field resistance in ohms.
- B.C.D is brush contact drop in volts.

For a shunt generator :

$$E_g = V_L + I_a R_a + \text{B.C.D. volts}$$

$$I_a = I_L + I_{sh} \text{ amperes}$$

$$I_{sh} = \frac{V_L}{R_{sh}} \text{ amperes}$$

$$I_L = \frac{P_L}{V_L} = \frac{V_L}{R_L} \text{ amperes}$$

For a series generator :

$$E_g = V_L + I_a R_a + I_{se} R_{se} + \text{B.C.D volts}$$

$$I_a = I_L = I_{se} = \frac{P_L}{V_L} \text{ ampere}$$

For a Long Shunt Compound Generator :

$$E_g = V_L + I_aR_a + I_{se}R_{se} + \text{B.C.D. volts}$$

$$I_L = \frac{P_L}{V_L} \; ; \; I_{sh} = \frac{V_L}{R_{sh}} \text{ amperes}$$

$$I_a = I_{se} = I_L + I_{sh}$$

For a Short Shunt Compound Generator :

$$E_g = V_L + I_aR_a + I_{se}R_{se} + \text{B.C.D volts}$$

$$I_{se} = I_L = \text{ampere}$$

$$I_a = I_L = I_{sh}$$

$$E_b = \text{volts}$$

Where E_b is back emf in volts.

DC MOTORS

For Shunt motor :

$$E_b = V_L - I_aR_a - \text{B.C.D volts}$$

$$I_L = \frac{P_L}{V_L} \text{ ampere}$$

$$I_{sh} = \frac{V_L}{R_{sh}} \text{ ampere}$$

$$I_a = I_L - I_{sh} \text{ ampere}$$

For Series motor :

$$E_b = V_L - I_aR_a - I_{se}R_{se} - \text{B.C.D volts}$$

$$I_L = I_{se} = I_a = \frac{P_L}{V_L} \text{ ampere}$$

For Long Shunt motor :

$$E_b = V_L - I_a\, R_a - I_{se}\, R_{se} - \text{B.C.D volts}$$

$$I_L = \frac{P_L}{V_L} \; ; \; I_{se} = I_a = I_L - I_{sh} \; ; \; I_{sh} = \frac{V_L}{R_{sh}} \text{ amperes}$$

For Short Shunt motor :

$$E_b = V_L - I_aR_a - I_{se}R_{se} - \text{B.C.D volts}$$

$$I_{se} = I_L \; \frac{P_L}{V_L} = \; ; \; I_{sh} = \frac{V_L - I_{se}R_{se}}{R_{sh}}$$

Where I_L is line current in amperes

P_L is the power input in watts

V_L is the voltage input in volts.

Armature torque, $$T_a = \frac{0.159ZP}{A}\phi I_a \text{ Nm}$$

6.23 Short Questions and Answers

Q.1 How can an e.m.f. be produced in a coil of wire? On What factors does its value and direction depend?

Ans. e.m.f. can be produced in a coil by rotating it in a magnetic field. Its value depends upon velocity and number of conductors. Its direction depends upon direction of rotation and direction of field.

Q.2 On what principle a D.C. generator works?

Ans. On the principle of Faraday's Law of Electro-magnetic induction.

Q.3 Is there any difference between D.C. motor and generator by construction?

Ans. No

Q.4 Why the armature is laminated?

Ans. To reduce eddy current losses.

Q.5 Can we use a wooden armature? If not, why?

Ans. No, because wood cannot withstand high speed rotation and the reluctance will be very high as it is a non-magnetic material.

Q.6 What are the types of D.C. generators with respect to their field excitation?

Ans. Two types ; separately excited and self-excited.

Q.7 What are the types of self excited generators?

Ans. Series, shunt and compound.

Q.8 What are the different types of compound generators?

Ans. Short shunt cumulative compound,

Short shunt differential compound,

Long shunt cumulative compound,

Long shunt differential compound.

Q.9 Where are the shunt generators used?

Ans. With regulators for ordinary lighting and power supply purposes. Mostly used for battery charging purpose.

Q.10 Where are the compound generators used?

Ans. Used for ordinary lighting and power supply systems.

Q.11 How a self-excited generator produces e.m.f. at starting?

Ans. Due to its residual magnetism.

Q.12 What is armature reaction?

Ans. It is the effect of magnetic field due to armature current on the distribution of main flux (poles).

Q.13 What are the effects of armature reaction?

Ans. (i) Main field is distorted.

(ii) Main flux weakens.

(iii) Less e.m.f. is induced.

Q.14 How to reduce the sparking on the commutator?

Ans. (i) By using high contact resistance brushes.

(ii) By using compensating winding.

(iii) By using Interpoles.

Q.15 On what principles does a D.C. motor work?

Ans. Whenever a current carrying conductor is placed in a magnetic field, conductor experiences a force and tends to move.

Q.16 What are the types of D.C. motor?

Ans. Series wound, shunt wound and compound wound.

Q.17 What are the further types of compound wound motor?

Ans. (i) Differential compound.

(ii) Cumulative compound.

Q.18 Why is the starter used for starting D.C. motors?

Ans. To give low supply voltage and to reduce heavy starting current.

Q.19 What are the parts of a D.C. machine?

Ans. Yoke, pole core, pole shoe, pole coil, armature, armature winding, commutator, brushes and bearings are the parts of a D.C. machine.

Q.20 What is the function of commutator?

Ans. The function of commutator is to convert the alternating emf induced in the armature into unidirectional current(DC) in the external circuit.

Q.21 What are the various losses in a D.C. machine?

Ans. The total losses in a D.C. machine are :

(i) Copper loss (includes armature copper loss, shunt field copper loss and series field copper loss).

(ii) Iron loss (includes hysteresis loss and eddy current loss).

(iii) Mechanical loss (includes friction loss and windage loss).

Q.22 What is back emf ?

Ans. When the motor rotates i.e., the conductors in the armature rotate, the magnetic flux is being cut by these conductors. Hence according to Faraday's Laws of electromagnetic induction, an emf is induced in the conductors acting in a direction opposite to the applied voltage and hence it is called back emf.

$$E_b = \Phi \frac{ZNP}{60A}$$

This back emf is always less than the applied voltage.

Q.23 Write the applications of a D.C. shunt motor.

Ans. Since D.C. shunt motor has approximately constant speed, it can be used for lathes, line - shaft drives, centrifugal pumps, reciprocating pumps, machine tools, blowers, fans, grinders, shapers, wood working machines, small printing presses etc.

Q.24 Write the applications of a D.C. series motor.

Ans. Since series motor has high starting torque, it can be used for electric traction, tram cars, railway cars, cranes, hoists, elevators, winches, conveyors etc.

6.24 PROBLEMS

1. A six pole DC generator armature has 00 conductors and is wave wound. If the flux per pole is 2×10^{-2} Wb and the speed is 500 rpm, calculate the emf generated.

   [Ans. 550 V]

2. An 8-pole, lap connected armature has 960 conductors, with flux per pole of 40 m Wb, running at a speed of 400 rpm. Calculate the emf generated on open circuit.

   [Ans. 256 V]

3. A 4 pole dynamo with wave wound armature has 5 slots each containing 20 conductors. The voltage induced is 357 V and the speed is 850 rpm. Calculate the flux per pole.

   [Ans. 1.2353 mWb.]

4. The armature of a 4 pole DC shunt motor has a lap connected winding, 60 slots and 20 conductors per slot. If the flux per pole is 23×10 wb, calculate the total torque developed, when the armature current is 50 amps.

   [Ans. 219.42 nm.]

5. A 4-pole, 500 V, D.C. shunt motor has 700 wave connected conductors on its armature. The F.L. armature current is 60 amps and the flux per pole is 30 mwb. Calculate the F.L. speed if the motor armature resistance is 0.2 ohms and the brush drop is 1V per brush.

[Ans. 694.3 RPM]

6. A 4-pole D.C. Shunt motor working on 250 V takes a current of 2 amps when running on 1000 RPM. What will be its back emf, speed, if the motor takes 51 amps at a certain load?Ra = 0.2 ohms, Rf = 250 ohms

[Ans. 240 V, 960.77 RPM]

7. A 30 Kw, 250 volts dc shunt generator has an induced voltage of 265 volts. If the field current when supplying rated load is 1.47 amp., find

(i) field resistance

(ii) armature resistanmce.

Neglect brush voltage drop.

[Ans. 170.07, 0.1235]

8. A short shunt compound generator supplies a current of 200 amp at 100 volts. The armature, series field and shunt field resistances are 0.04, 0.03, 60 ohm respectively. Calculate the generated emf.

[Ans 114 volts]

9. A 100 volt series motor takes 45 amp. When running at 750 rpm. Its armature fresistance is 0.22 ohm, series field resistance is 0.13 ohms. Find the back emf.

[Ans 84.25 volts]

□□□

UNIT-4

Chapter 7

Bipolar Junction Transistor

Chapter 8

Thyristors

Chapter 7

Bipolar Junction Transistor

7.1 INTRODUCTION

The branch of engineering which deals with flow of electrons through a vacuum or gas or semiconductor is known as electronics. Basically the word electronics derives its name from electron present in all electronic devices.

Electronics mainly touch with electronic devices and their utilization. An electronic device is that in which current conduction (i.e. flow of electron) take place through conductor (i.e. vacuum tubes, gas-filled tubes, semiconductor materials).

Electronics devices are being used in almost all the industries for quality control and automation. Electronics has gained much important due to its numerous applications in industries such as Amplification, control, conversion of light into electricity, conversion of electricity in to light generation, Rectification etc.

7.2 ENERGY BAND IN SOLIDS

The range of energies possessed by an electron in solids is known as energy band. This is shown in fig. 7.1

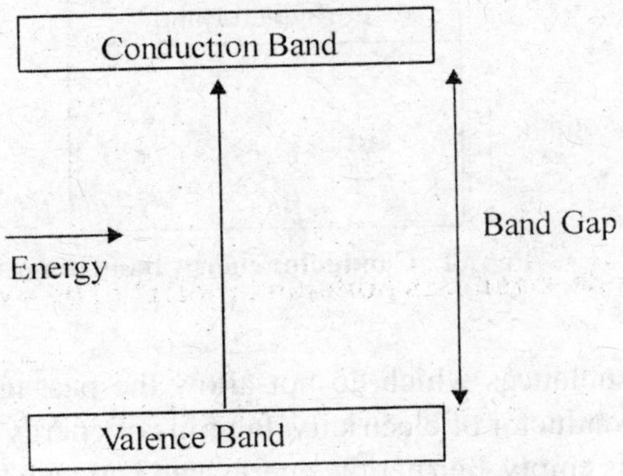

Fig.7.1 : Energy level diagrams

There is a number of energy bands in solids, the following are of particular importance.

Valance band : The range of energies (i.e. band) possessed by valance electrons is known as valance band. The electrons in the outermost orbit of an atom are known as valance electrons.

Conduction band : The range of energies (i.e. band) possessed by conduction band electrons is known as conduction band.Free electrons, which are responsible for the conduction of current in a conductor, are called conduction electrons.

Forbidden energy gap : The separation between conduction band and valance band on the energy level diagram is known as forbidden energy gap. No electron of a solid can stay in this gap.

7.3 Classification of Solids

This is well known that some solids are good conductors of electricity while others are insulators. There is also an intermediate class of semiconductos. According to behavoir of electricity, solids are classified into three categories.

(i) Conductor

(ii) Insulator

(iii) Semiconductor

7.3.1 Conductor

A solid which contains a partly filled band structure is called a conductor. In a conductor conduction band overlap with valence band, therefore, no forbidden gap is present. Electrons present in valence band can easily move into conduction band. Due to this, a slight potential difference across a conductor causes the free electrons to constitude electric current. Copper is an example of metal with 9×10^7 mho/m conductivity.

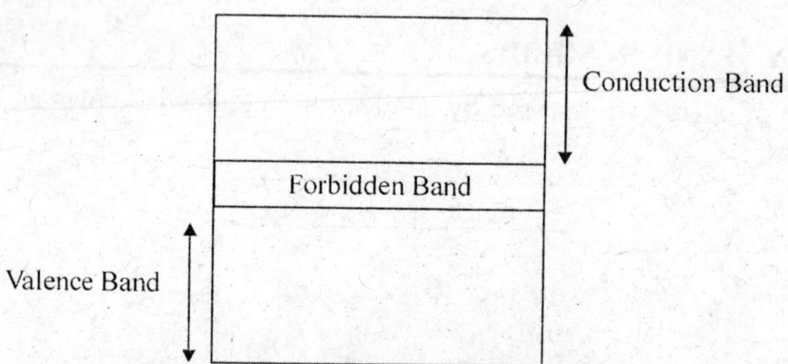

Fig.7.2 : Conductor energy band diagram

7.3.2 Insulator

Insulators are those subtances which do not allow the passage of electric current through them. Insulators are poor conductor of electricity. In terms of energy band, the valance bank is full which the cnduction band is empty. Further the energy gap between valance and conduction bands is very large. conduction band and valance band are separated by a forbidden gap of more than 6eV. This is shown in fig. 7.3

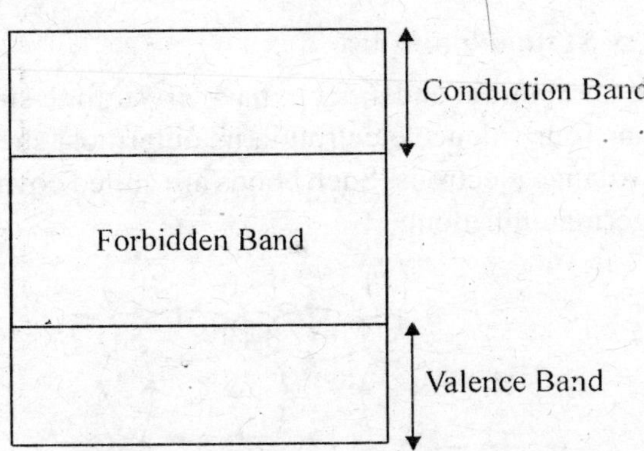

Fig.7.3 : Insulator energy band diagram

The energy which can be supplied to an electrion from an applied field is too small to carry the particle from the filled (valance band) into the vacant band (conduction band). Since the electron can not acquare sufficient applied energy, conduction is impossible. Diamond and glass are example of insulator.

7.3.3 SEMICONDUCTORS

A substance for which the electrical conductivity lies in between conductor and insulator is called a semiconductor. In this, the width of forbidden energy gap is relatively small. The forbidden energy gap semiconductor is nearly 1 eV.

Important practical semiconductors are germanium (ge) and silican (Si), which have 0.785eV and 1.21 eV forbidden energy gap respectively, at 0°k. Energy of this magnitude normally cannot be acquired from an applied field. Hence the valance band remains full, the conduction band empty, and these materials behave as insulators at low temperature. At room temperature, some valence band electrons acquire suffiicient energy to cross forbidden energy gap and move to the conduction band.

Once the valence band enectron jump into conduction band, it leaves vacancy in valence band. This vacancy is named as hole.

The semiconductor energy band diagram is shown in fig. 7.4.

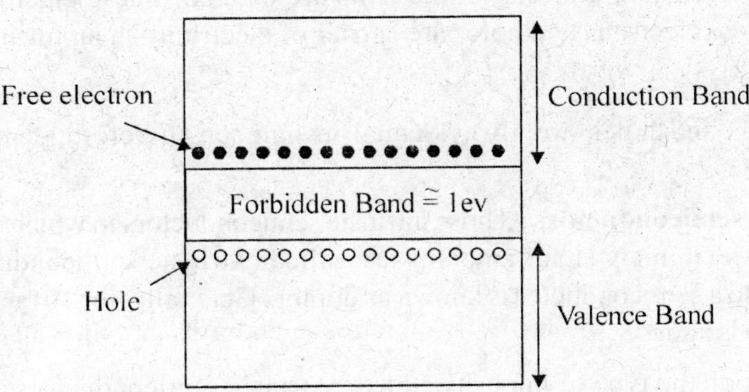

Fig. 7.4 : Semiconductor energy band diagram

7.3.3.1 Atomic Binding in Semiconductors

Semiconductors like germanium and silicon, have crystalline structure. Both these materials are tetravalent i.e. each has four valence electron in its outermost shell. In semiconductors, bonds are formed by sharing of valance electrons. Such bonds are called covalent bonds. Figure 7.5 shows the bonding diagram, of germanium atoms.

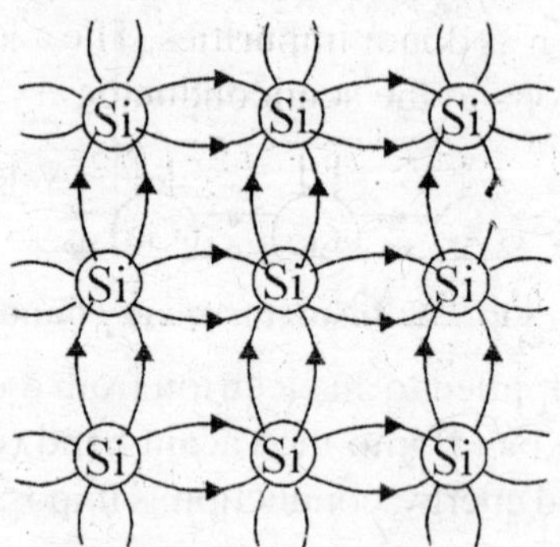

Fig.7.5 : Atomic binding in Si

Note :

1. A single germanium atom has 32 electrons out of which only four electron take part in electrical Properties of Ge, The remaining 28 electrons being tightly bound to the nucleus.
2. Forbidden energy gap for Si is 1.1eV & for Ge is 0.7 eV.

7.3.3.2 Types of Semiconductors

(i) Intrinsic semiconductors : An intrinsic semiconductor is one which is made of the semiconductor material in its extremely pure form. Examples of intinsic semiconductor are Pure Ge & Si.

These semiconductors are free from any bind of impurity. These substances are having lesser conductor than electrisic (impure) semiconductor. They have four electrons in their outmost shell. These four electrons forms covalent bond with neighboring atom. At high temperature some of covalent bond are broken and electron hole pairs are created. These electrons are named as free electrons. These free electrons and holes are carrier of electricity in an intrinsic semiconductor.

Note :

Free electrons and holes are always equal in number, therefore, semiconductor is always electrically neutral.

(ii) Extrinsic semiconductors : Those intrinsic semiconductors to which some suitable impurity has been added in extremely small amounts are called extrinsic semiconductors. The process of adding impurities to a semiconductor is known as doping. Generally, for 10^8 atoms of semiconductor, one impurity atom is added.

Depending upon the type of impurity added, extrinsic semiconductors are classified into two types.

(a) N –type semiconductor

(b) P-type semiconductor

(a) N-type semiconductor : This type of semiconductor is obtained when a pentavalent material like antimony (Sb) is added to pure Germanium (Ge) crystal. Typical examples of pentavalent impurities are antimony and arsenic (As) as shown in figure 7.6.Such impurities which produce n-type semiconductor are known as donor impurities. The addition of pentavalent impurities provides a large number of free electrons in the semiconductor crystal.

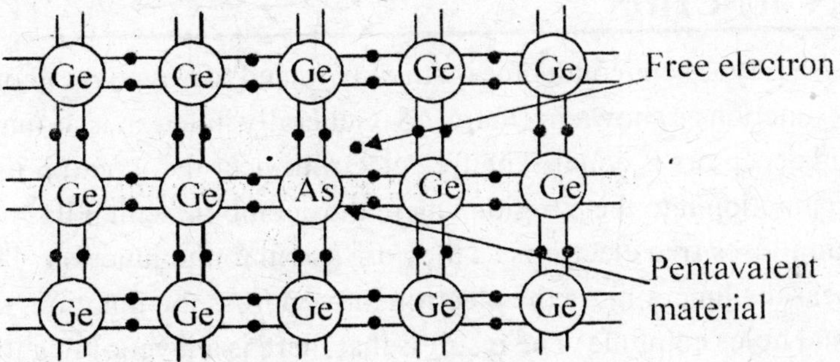

Fig.7.6 : N-type Semiconductor

(b) P-type semiconductor : When a small amount of trivalent impurity is added to a pure semiconductor, it is called p-type semiconductor. The addition of trivalent impurity provides a large number of holes in the semiconductor. Typical examples of trivalent impurities are gallium, indium and Boron as shown in figure 7.7. Such impurities which produce p-type semiconductor are known as acceptor impurities because the holes created can accept the electrons.

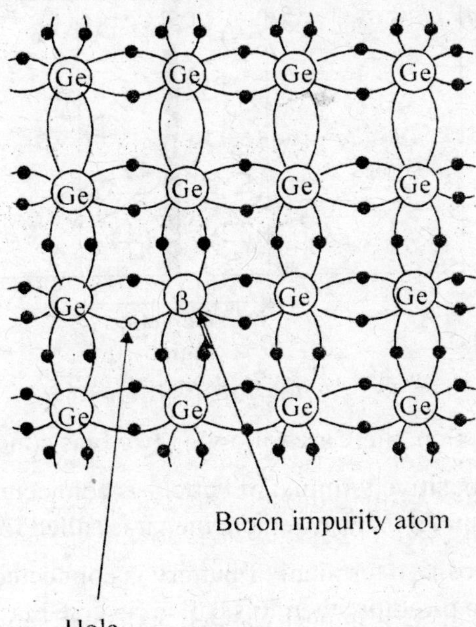

Fig.7.7 : P-type Semiconductor

We know that due to the effect of impurity, n-type material has a large number of free electrons whereas p-type material has large number of holes. Since the majority portion of current in n-type material is by the flow of free electrons. The free electron is known as majority carrier and the holes are the minority carriers.

Similarly in a p-type material, holes are the majority carriers and free electrons are the minority carriers.

7.4 P-N JUNCTION

When a p-type semiconductor is suitably joined to n-type semiconductor, the contact surface is called PN junction as shown in figure 7.8. Generally junction is assumed to lie where the density of donors and acceptors is equal. The free electrons near the junction in the n region begin to diffuse across the junction into the p region where they combine with holes near the junction. The result is that n region loses free electrons as they diffuse in to the junction. This creates a layer of positive charges near the junction. As the electron move across the junction, the p region loses holes as the electron and holes combine .The result is that there is a layer of negative charges near the junction. These two layers of positive and negative charges form the depletion region or layer.

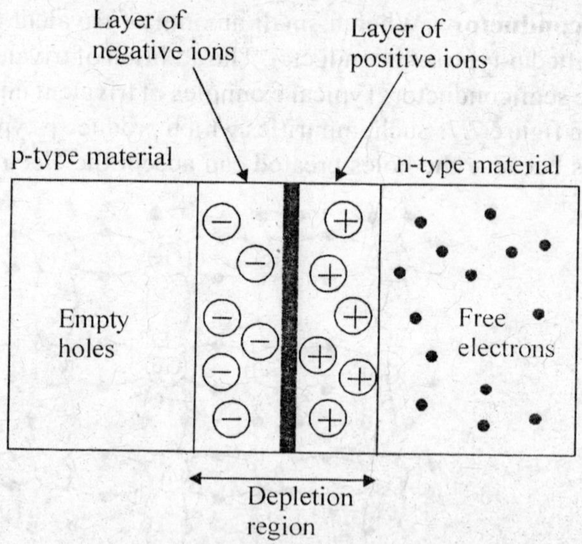

Fig.7.8 : P-N Junctions

In relation to a PN junction, there are following two bias conditions:-

(i) Forward bias : If positive terminal of battery is connected to p-region of a semiconductor and the negative battery terminal to the n-region, then it is called forward biasing (see figure7.9 (a)).

(ii)Reverse bias : If positive terminal of battery is connected to n-region of a semiconductor and the negative terminal to p-region, then it is called reverse biasing (see figure 7.9(b)).

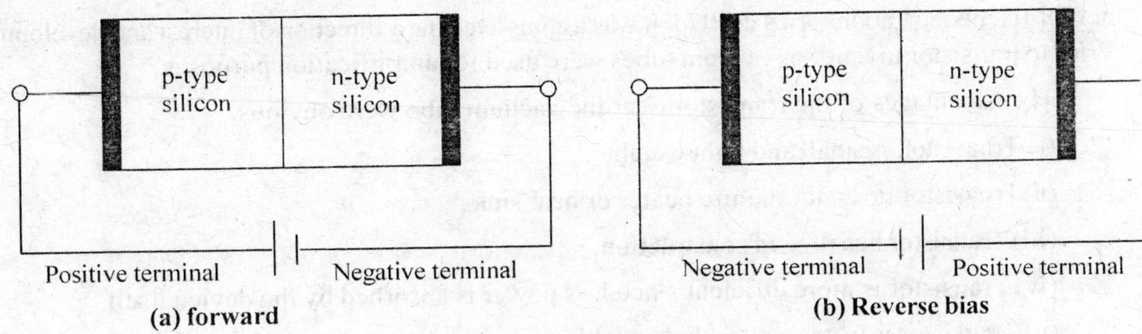

Fig.7.9 : PN Junction biasing

Note :

1. When p n junction is forward biased then current flows across junction because of low resistance offers by junction.
2. When PN junction is reversed biased then no current flows in the circuit due to the presence of high resistance.

7.4.1 P-N JUNCTION DIODE

It is a two-terminal device consisting of a p-n junction formed by semiconductor crystal. In figure 7.10, p-type region is referred to anode and n-type region is referred to cathode.

If arrowhead of diode symbol is positive w.r.t. bar of the symbol, the diode is forward biased. If the arrowhead of diode symbol is negative w.r.t bar, the diode is reverse biased.

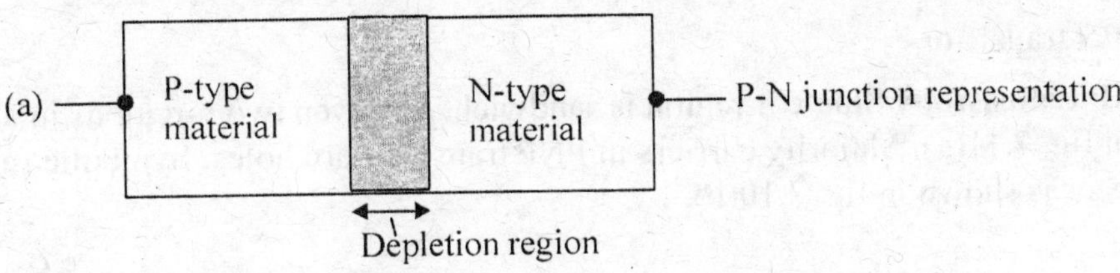

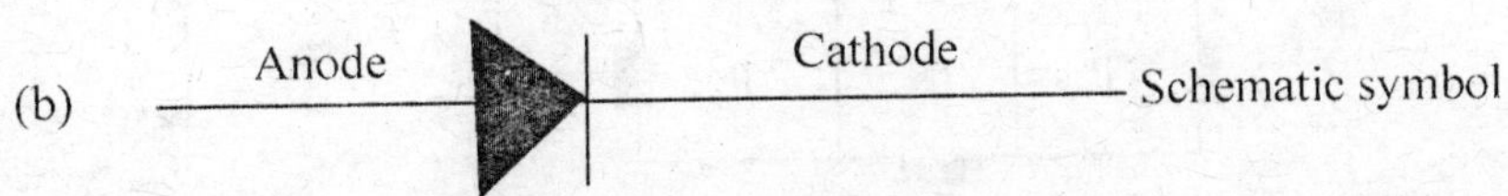

Fig.7.10 : PN Junction diode

7.5 BIPOLAR JUNCTION TRANSISTOR (BJT)

Since the diode cannot amplify a signal so it has limited application area. Bioloar Junction Transistor (BJT) is a next logical step in the development of semiconductor device after p-n junction diode.

Walter H. Brattain and John Bardeen demenstrated the amplifying action of the first transistor at Bell Telephone Laboratories in 1974. It was a completely new direction of interest and development. Prior to transistor invention, vacuum tubes were used for amplification purposes.

The advantages of the transistor over the vacuum tube were obvious.

(i) Trnasistor, is small and light weight

(ii) Trnasistor does not require heater or heat sink.

(iii) Transistor has rugged construction.

(iv) Transistor is more efficient since less power is absorbed by the device itself.

(v) Transistor is instanteneously available for use and requiring no warmup period.

(vi) Power operating voltage is possible with transistor.

Bipolar junction transistor is a two junction, bi-polar, three terminal device. Transistor is the main component in amplifiers, switching circuits, oscillators and logical circuits.

7.5.1 Type of Transistor

A bipolar junction transistor (BJT) consists of silicon (or germanium) crystal in which a layor of n-type (p type) silicon is sandwiched between two layers of p-type (n-type) silicon. In other words, a transistor may consists of a layer of n-type between two p-type layers or a layer of p-type between two n-type layers.

In a transistor, current is carried by both holes and electrons, therefore, it is called bi-polar transistor. These are two types of transistor based on the construction.

(i) PNP transistor

(ii) NPN transistor

(i) PNP transistor : Single n-region is sandwiched betwen two p-regions in a PNP transistor as shown in fig. 7.11(a). Majority carriers in PNP transister are holes. Symbolic representation of PNP transistor is shown in fig. 7.10(b).

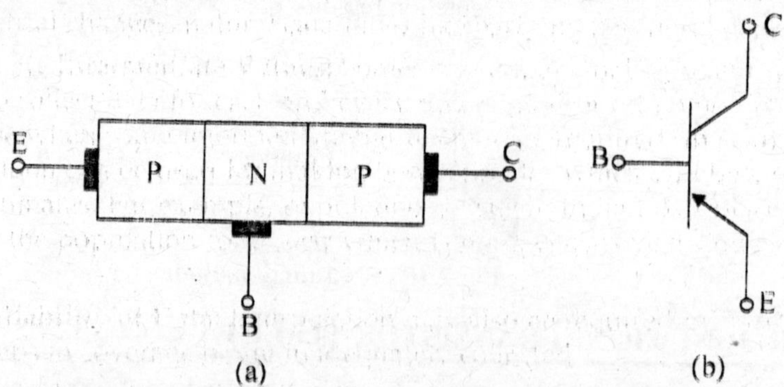

Fig. 7.11 : PNP transistor (a) Construction (b) Symbol

(ii) NPN transistor : It is consists of a lager of p-region sandwiched between two n-regions as shown in fig. 7.12(a). Electrons are majority carriers in NPN transistor. The symbol of NPN transistor is shown in fig. 7.11(b).

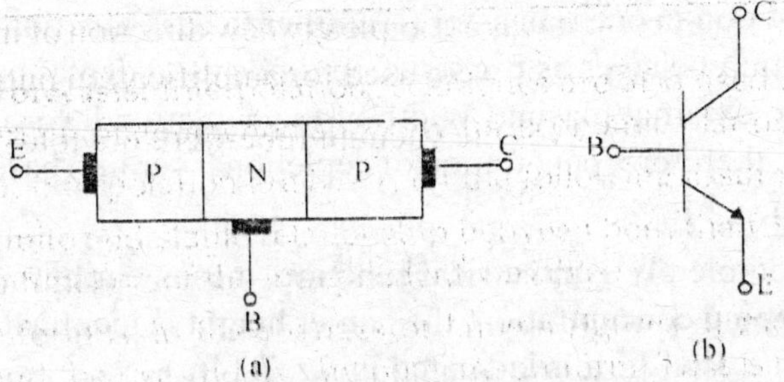

Fig. 7.12 : NPN transistor (a) Construction (b) Symbol

Direction of arrow in symbol indicates the direction of current flow in both types of transistors. These are three layers in transistor and these layers are named as emittor (E), base (B) and collector (C).

Emitter (E) : As its name suggests, it emits the majority change carriers. N-type emitter emits electrons and p-type emittor emits holes. Emitter is most heavily dopped layer. The doping layer of emittor is kept high so that it can emit the majority change carriers in a good manner.

Base (B) : Middle layer of transistor is named as base. Doping tend and size of base is kept minimum among three layers. Keeping doping level and size of basse minimum, confirms the recombination with the changes injected from emitter into base will decrease. It results that more change carriers enter into collector after passing through the base. Another advantage of thin base is that changes injected from emitter will come in contact with lesser number of opposite charge carriers of base and recombination rate will decrease.

Since the resistance is inversely proportional to the doping concentration, the resistance of base region is maximum.

Collector (C) : The function of collector is to collect the majority change carriers emitted by the emittor. Size of collector is kept largest due to that number of change carriers collected increases with size of collector and transistor amplification factor increases with increse in number of collected changes.

7.5.2 Open Circuted Transistor

If no external voltages are applied to any terminal of the transistor, all transistor currents must be zero.

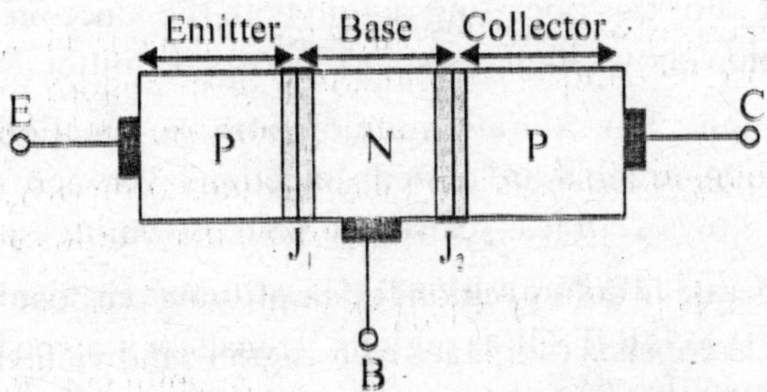

Fig. 7.13 An open circuited transistor

There are two junction in one transistor (i) emitter to base junction J_1 and (ii) base to collector junction J_2. Each junction behaves as p-n junction diode asnd depletion layer will develop across each junction under no external biasing. Width of the depletion layer is inversely proportional to doping concentrastion, therefore, penetration of depletion layer into base is more than collector and emitter as shown in fig. 7.13.

If we assume a completely symmetrical junction, means emitter and collector region having same size and same doping concentration, the banier height is identical at emitter-base junction J_1 and at collector-base function J_2. It is indicated in fig. 7.14(a).

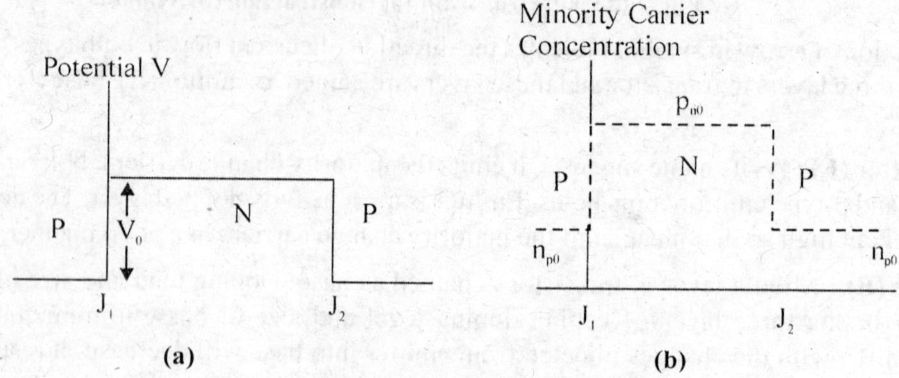

Fig. 7.14 : The potential levels at J_1 and J_2

Under open circuited conditions, the minority concentration is constant in each section and is equal to its thermal equilibrium value n_{p0} in p region (emitter and collector) and n_{n0} in n region (base). Minority carries concentration is represented in Fig. 7.14(b).

7.6 Operating Regions of Transistor

A transistor has two junctions : emitter-base junction (J_1) and base-collector junction (J_2). These two junctions can be forward biased or reverse biased according to the applied voltage across the two terminals of the transistor. Nature of biasing depends upon the type of application. Three operating regions are defined on the basis of biasing.

(i) **Active Region :** Emitter to base junction (J_1) is forward biased and base to collector junction (J_2) is reverse biased.

(ii) **Cut off Region :** For this operating region both the junction (J_1 and J_2) should be reverse biased. This region shows high resistance between emitter and collector.

(iii) **Satuartion Region :** For saturation region both the junctions should be forward biased. When both junctions are forward biased, depletion layers across the junctions are negligible and this region shows very less resistance between emitter and collector.

Transistor should operate in active region for amplifying purpose while for switching it should operate in saturation (ON) or cuttoff (OFF) regions. Transistor may be biased in different regions as explained above, is shown in fig. 7.15

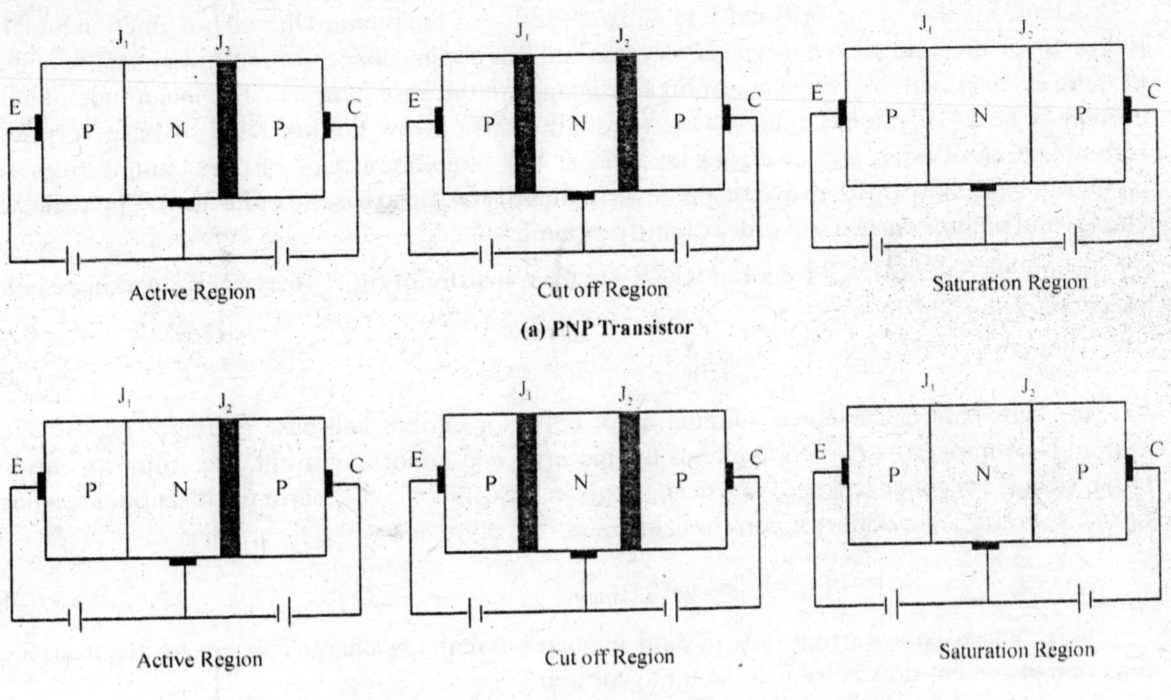

Fig. 7.15 : Transistor biased in various region

7.7 TRANSISTOR OPERATION

The basic operation of npn transistor and pnp transistor is exactly the same, if the roles of holes and electrons are interchanged. In fig.7.16 pnp transistor is redrawn in active region. It means emitter to base junction is forward biased and base to collector junction is reverse biased.

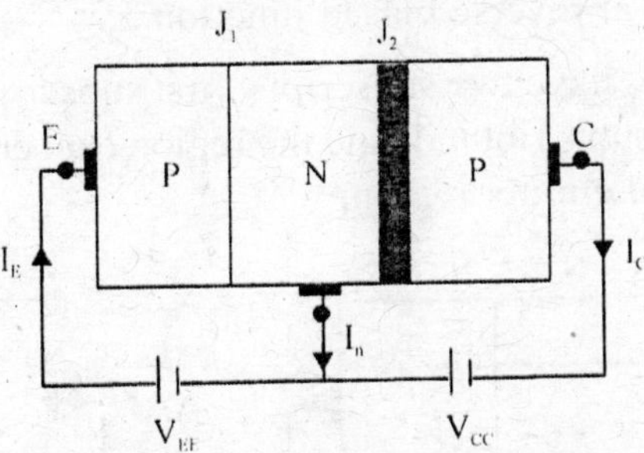

Fig. 7.16 : Transitor in active region

The junction between emitter and base is forward biased. The depletion region has been reduced in width due to the applied forward bias, resulting in a heavy flow of majority carries from p-region (emitter) to n-region (base). Another junction between base and collector is reverse biased. As the reverse baise voltage is applied across this junction, depletion layer increases. The flow of majority carries through the depletion layer is zero, resulting in only minority carrier flow.

A large number of majority carriers will diffuse across the forward biased p-n junction into the n-type. Since the sandwiched n-type is very thin and low doping concentration, a very small number of these carriers will take this path of high resistance to the base terminal. The magnitude of base current is order of microamperes. Due to small size and low doping level of base very less recombination of holes and electrons take place. The large number of carriers coming from the emitter to base, now diffuses across the reverse biased junction (base to collector.). The collector current and emitter current are order of milliamperes.

Applying Kirchoff s Current Law (KCL) to the transistor of fig, 7.16 as if it were a single hole, we obtain

$$I_E = I_C + I_B \quad ...(7.1)$$

It means emitter current is summation of collector current and base current. The collector current is comprised of two components the majority and minority current. The minority current component is called the leakage current and is given the symbol I_{CO} (I_C current with emitter terminal open). Therefore, the collector current is given as

$$I_C = I_{C,majority} + I_{CO,minority} \quad ...(7.2)$$

In NPN transistor, current flow in same manner but majority charge carriers are electrons and direction of current flow is from collector to emitter.

7.8 TRANSISTOR CURRENT COMPONENTS

Transistor is a three terminal device. Emitter current I_E, base current I_B and collector current I_c flow at emitter, base and collector terminals respectively. These currents have various components. As the transistors a bipolar device, both electron and holes contribute to current.

In active region, emitter to base junction J_1 is forward biased and base to collector junction J_2 is reverse biased. At forward biased junction J_1 current flow due to majority carriers while minority carriers conduct the current at reverse biased junction J_2.

The current components have two subscripts, first subscript denote type of carrier (either hole p or electron n) and second subscript indicate the region (i.e. emitter E, base B or collector C). For example I_{pE} is hole current of emitter region.

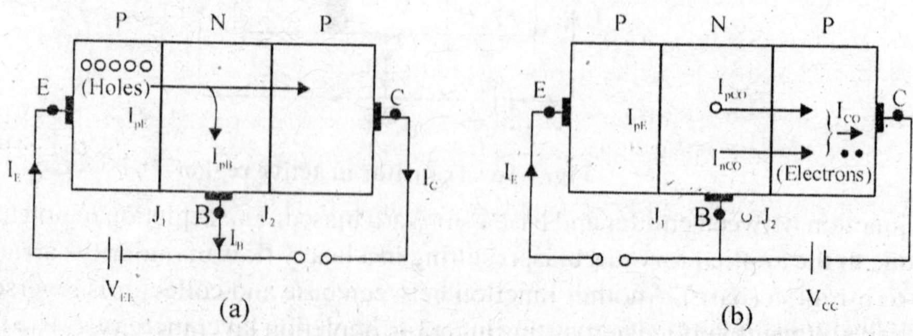

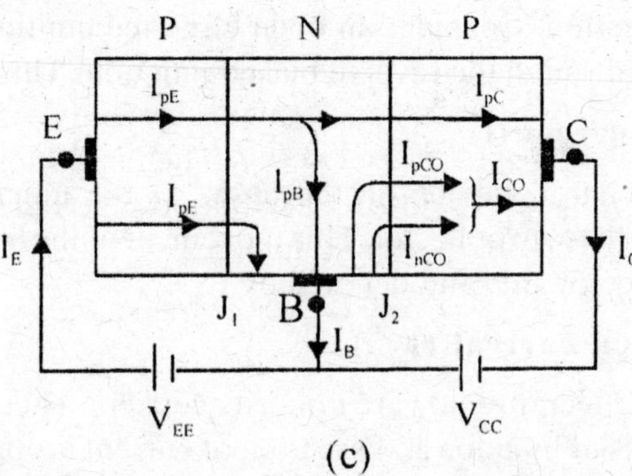

(c)

Fig. 7.17 : (a) Current components between emitter and base ((b) Current components between base and collector (c) Current components of PNP transistor

Current components at forward biased emitter base junction :

Emitter to base junction is forward biased as shown in fig. 7.17(a). Therefore majority change carriers contribute to current.

(a) Hole current : As we are discussing PNP transistor, holes are majority carrier in emitter region. Holes are emitted from emitter to n-type base. Since the base is very thin and has a very low doping concentration, most of the holes coming from emitter side do not recombine in base region. Due to this, a very small number of these carriers move towards the base terminal and contribute the current I_{pB}. The remaining large number of holes (around 98%) will cross the reverse bias junction J_2 and diffuse into collector. This is denoted by I_{pB}. Therefore, total hole current at emitter (I_{pE}) is sum of current at collector

$$I_{pB} = I_{pB} + I_{pC} \qquad ...(7.4)$$

(b) Electron current : Electrons are the majority carriers of n-type base. These majority carrier diffuses from base to emitter region. This current is denoted by I_{nE}. The direction of current I_{pE} is opposite to the motion of electrons.

The ratio of hole to electron currents, I_{pE}/I_{nE} crossing the emitter junction is proportional to the ratio of the conductivity of the p-type to that of the n-type. In a commercial transistor the doping of the emitter is made much larger than the doping of base. This feature of PNP transistor ensures that the emitter current consists almost entirely of holes.

Such a situation is desirable since the current which results from electrons crossing the emitter junction from base to emitter does not contribute. Carrier which can reach the collector.

The total emitter current is sum of current due to holes and electrons.

$$I_E + I_{pE} + I_{nE} \qquad ...(7.5)$$

Current components at reverse biased base collector junction

The base to collector junction J_2 is reverse biased, therefore, only thermally generated minority carriers can cross this junction. Consider an open circuited emitter as shown in fig. 7.17(b). The minority current I_{co} flows through the reverse biased junction. This current has two components.

(a) Minority hole current (I_{pCO})

Holes are minority charge carriers in the base. As the junction J_2 is reverse biased, these minority holes moves from base to collector. This movement of the holes contributes minority current flowing from base to collector side and denoted by I_{pCO}.

(b) Minority electron current (I_{nCO})

Electrons in p-type collector region are minority carriers. These electron will move to the base by crossing the reverse biased junction J_2. Direction of current is opposite to the direction of flow of electrons, it means current will flow from base to emitter. This current is due to electrons and denoted as I_{nCO}.

nCO

68

Electrical and Electronics Engineering

Total minority current (I_{CO}) crossing the junction J_2 is sum of minority current because of holes and electrons.

$$I_{CO} + I_{pCO} + I_{nCO} \quad \text{....(7.6)}$$

Total current at collector terminal : A large number of holes diffused from the p-type emitter also reach to collector, because of less recombination process at base. These holes constitute a current and this is named as I_{pC}. The total current I_c crossing the junction J_2 and flowing outside the collector terminal is sum of two currents I_{pC} and I_{CO}

$$I_C = I_{pC} + I_{CO} \quad \text{...(7.7)}$$

The current because of holes at collector, is a part of total emitter current that represent holes which have travelled from emitter, through the base, to collector. Therefore, we can write

$$I_{pC} = \alpha I_E \quad \text{...(7.8)}$$

Here a should be less than unity.

Thus the total current at collector terminal is given as

$$I_C = \alpha I_E + I_{CO} \quad \text{...(7.9)}$$

Total current at base terminal : Applying Kirchoff s current law (KCL) at the base terminal (B) of fig. 7.17(c)

$$I_B = I_{nE} + I_{pB} - I_{CO}$$

$$I_B = I_{nE} + I_{pB} - (I_{aCO} + I_{pCO}) \quad \text{...(7.10)}$$

Total base current at base terminal is due to both majority and minority currents.

Total current at emitter terminal : Applying Kirchoff s current law (KCL) at the emitter terminal (E) of figlk? {c)l The total emitter current (I_E) is sum of two current components.

$$I_E = I_{pE} + I_{nE} \quad ...(7.11)$$

7.9 Current Gain

Current gain is defined as the ratio of output current to input current. A transistor has two type of current gains, Alpha (α) and Beta (β).

7.9.1 Current Gain A

From the equation 7.8, current gain a is defined as the ratio of the collector current increment from cut off ($I_C = I_{CO}$) to the emitter current change from cutoff ($I_E = 0$). Mathematically

$$\alpha = \frac{I_C - I_{CO}}{I_E} \quad ...(7.12)$$

Here alpha (α) is known as the large signal current gain.

For small signals the reverse saturation current I_{CO} is negligible and may be discarded. Small signal current gain may be defined as

$$\alpha = \frac{I_C}{I_E} \quad ...(7.13)$$

For ac conditions where point of operation moves on characteristic curve, an AC alpha is defined as ratio of change in collector current (ΔI_c) to change in emitter current ((ΔI_E) at constant collector to base voltage (V_{CB}).

$$\alpha_{ac} = \left.\frac{\Delta I_C}{\Delta I_E}\right|_{V_{CB=const}} \quad ...(7.14)$$

Typical numerical values of a lie in the range 0.90 to 0.995. This is important to note that a is not a constant but varies with emitter current I_E, collector to base voltage V_{CB} and temperature.

7.9.2 Current Gain B

Ratio of collector current I_C to base current I_B is known as beta (β)

$$\beta = \frac{I_C}{I_B} \quad ...(7.15)$$

For ac conditions, the beta (P) is defined as ratio of change in collector current (ΔI_C) to change in base current (ΔI_B) at constant collector to emitter voltage (V_{CE})

$$\beta_{ac} = \left.\frac{\Delta I_C}{\Delta I_B}\right|_{V_{CE}=\text{constant}} \quad \text{...(7.16)}$$

Numerical values of β lie from 50 to more than 400. β is also called as base current amplification factor.

7.9.3 Relation Between Current Gains A and B

Small signal current gains are defined by the equations (7.13) and (7.15). From these equations we can rewrite

$$I_E = \frac{I_C}{\alpha} \text{ and } I_B = \frac{I_C}{\beta}$$

Substituted these two values in equation (7.1) *i.e.*

$$I_E = I_C + I_B$$

$$\Rightarrow \quad \frac{I_C}{\alpha} = I_C + \frac{I_C}{\beta}$$

$$\Rightarrow \quad \frac{1}{\alpha} = 1 + \frac{1}{\beta}$$

$$\Rightarrow \quad \beta = \frac{a}{1-\alpha}$$

$$\Rightarrow \quad \alpha = \frac{\beta}{1+\beta} \quad \text{...(7.16)}$$

7.10 Transistor Amplifying Action

As the relationship between emitter current I_E and collector current I_C has been established, basic amplifying action of the transistor can be introduced on surface level. For amplification, transistor is used in active region. It means emitter to base junction J_1 is forward biased and base to collector junction J_2 is reverse biased. Forward biased voltage across J_1 junction, reduces the size of depletion layer or we can say input resistance of the transistor is reduced. In the same manner when reverse bias voltage is applied to junction J_2, it increases the width of the depletion layer. It increases the output resistance of the transistor. Transferring a current from a low to a high resistance circuit, produces the basic amplification of input signal.

The basic operation of a transistor is to transfer the resister from low to high. The combination of the two forms in italics results in the label 'transistor'; that is

transfer + *resistor* → transistor.

The transistor provides power gain as well as voltage or current amplification.

7.11 Transistor Configuration

A transistor has three terminals, emitter (E) base (B) and collector (C). Out of these three one terminal remains common to input and output, remaining two terminals are used for input or output. Base terminal is preferred for input while collector is preferred for output. Depending upon the fact which terminal is chosen as common terminal, the transistor can work in three configuration namely; common emitter (CE), common base (CB) and common collector (CC). Regardless the circuit configuration, transistor can operate in any region (active, saturation or cutoff region) according to biasing of the emitter to base and base to collector junctions.

7.11.1 Common Emitter (CE) Configuration

The emitter terminals is common or reference to both input and output terminals, it is called the common emitter configuration. The input signal is applied between base and emitter while output is taken across collector and emitter.

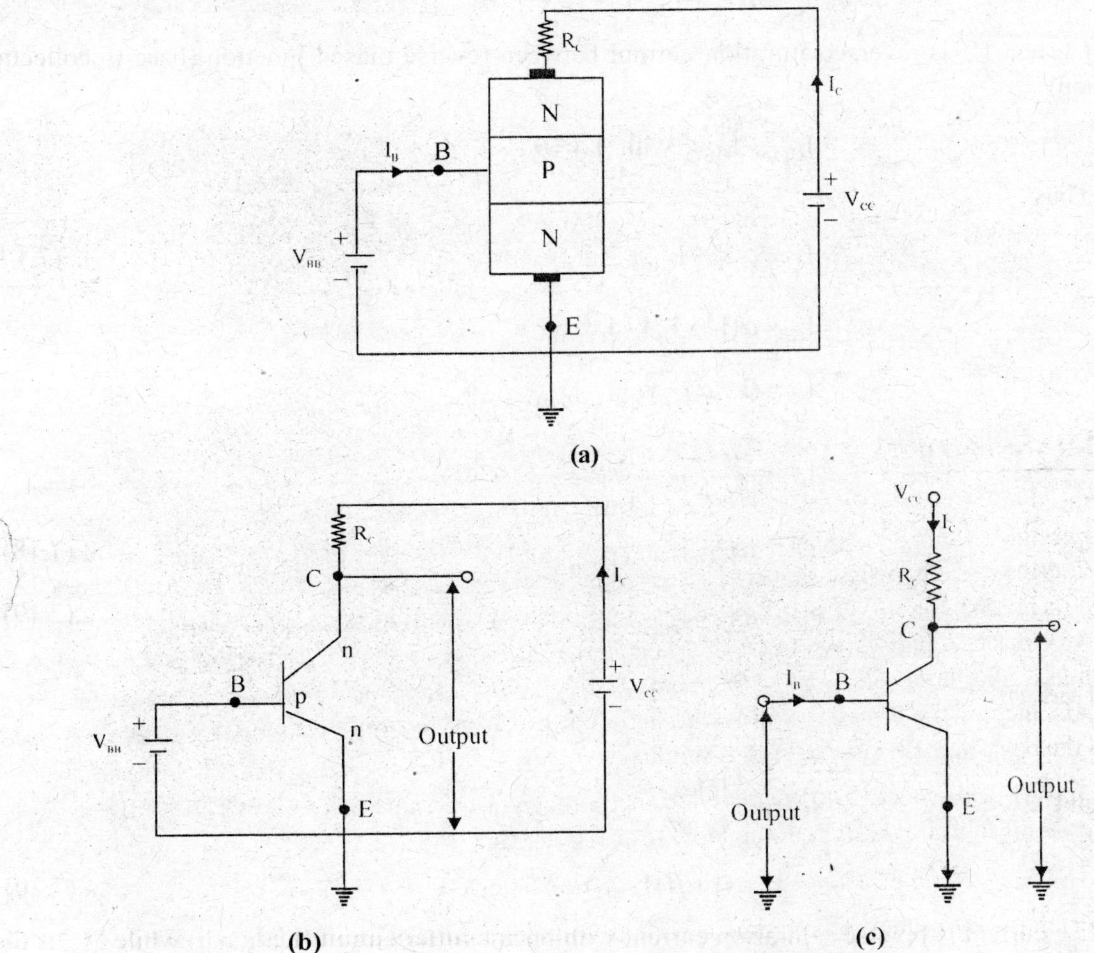

Fig. 7.18: Common Emittor (CE) Configuration.

A common emitter configuration for npn transistor is shown in fig: 7.18. To operate transistor in active region, input battery V_{BB} is connected in forward biase manner while V_{CC} is connected in such a way that base to collector function becomes reverse biased.

The current amplification factor for CE configuration is called β and it relates the output current I_C and input current I_B in following way

$$\beta = \frac{I_C}{I_B}$$

$$\Rightarrow \qquad I_C = \beta I_B$$

Analysis :

From the relation

$$I_C = \alpha I_E + I_{CO}$$

Current I_{CO} is reverse saturation current between reverse biased junction (base to collector junction)

$$I_{CO} = I_{CBO} \text{ with } I_E = 0$$

Thus

$$I_C = \alpha I_E + I_{CBO} \qquad ...(7.17$$

$$\Rightarrow \qquad I_C = \alpha(I_C + I_B) + I_{CBO}$$

$$\Rightarrow \qquad I_C = (1-\alpha) = \alpha I_B + I_{CBO}$$

$$\Rightarrow \qquad I_C = \frac{\alpha}{1-\alpha} . I_B + \frac{I_{CBO}}{1-\alpha}$$

$$\Rightarrow \qquad I_C = \beta I_B + (1+\beta) I_{CBO} \qquad ...(7.18)$$

$$\Rightarrow \qquad I_C = \beta I_B + I_{CEO} \qquad ..(7.19)$$

Here $\beta = \frac{\alpha}{1-\alpha}$

and $I_{CEO} = \frac{I_{CBO}}{1-\alpha}$

$$= (1+\beta) I_{CBO} \qquad ...(7.20)$$

I_{CBO} current is reverse saturation current with open emitter circuit (i.e.I_E = 0) while I_{CEO} is the reverse satuartion current with open base (i.e. I_B = 0) from the equation (7.20)

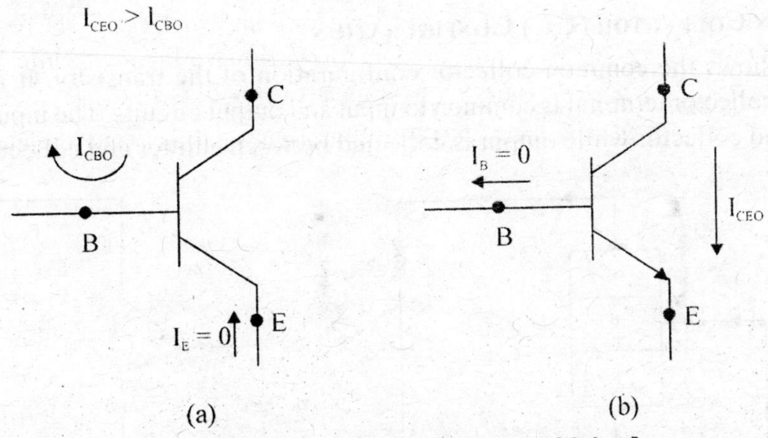

Fig, 7.19: (a) I_{CBO} with I_E =0, (b) I_{CEO} with I_B =0

Theoretically number of charge carriers contributing to I_{CEO} are more than I_{CBO}, therefore, I_{CEO} is larger than I_{CBQ}. This fact has already been proved mathematically by equations (7.20) and (7.21).

7.11.2 Common Base (CB) Configuration

For the common base configuration of the transistor, base terminal is made common between input and output. The input signal is applied between emitter and base terminals while output is taken across collector and base.

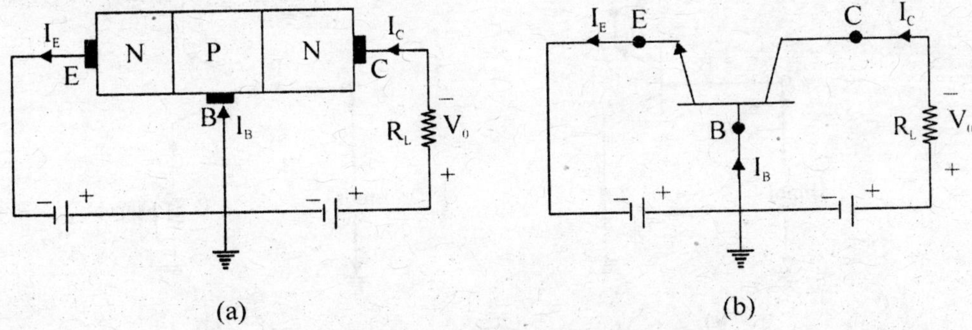

Fig. 7.20: Common Base Configuration

A common base configuration for npn transistor is shown in fig. 7.20. Emitter base junction is forward biased and base collector junction is reverse biased, therefore, transistor is in active region.

The input current is I_E and output current is I_C. These two currents are related by current gain α.

$$\alpha = \frac{I_C}{I_E}$$

$$\Rightarrow \quad I_C = \alpha I_E$$

Common base configuration is used to design current source.

7.11.3 Common Collector (CC) Configuration

Fig. 7.21 shows the common collector configuration of the transistor. It is called common collector, since collector terminal is common to input and output circuits. The input signal is applied between base and collector while output is collected between emitter and collector.

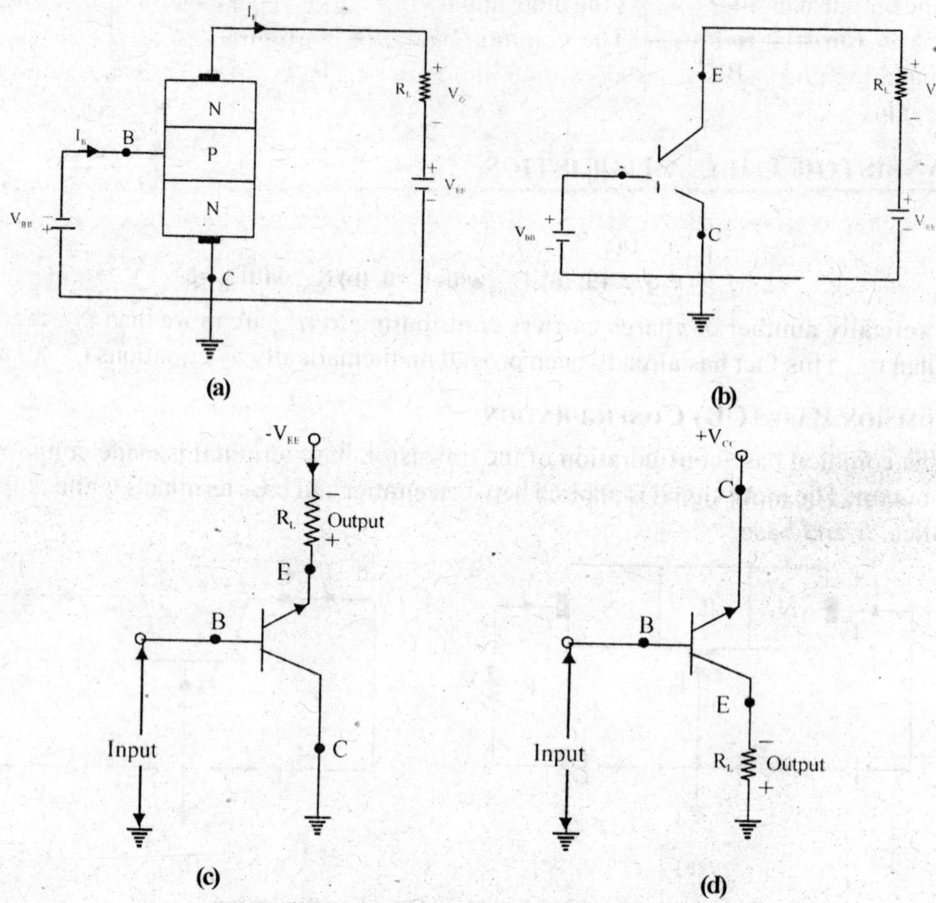

Fig. 7.21: Common Collector Configuration

In common collector configuration, the load resistor is connected from emitter to ground as \ shown in fig. 7.21 (d).

Fig. 7.21 (a) shows the common collector configuration of npn transistor. This is redrawn in fig, 7.21 (b) with symbolic representation of transistor. Fig.7.21 (c) is another way to represent CC configuration. Fig. 7.21(c) is reoriented and shown in new form in fig. 7.21(d). To operate in active region the biasing batteries are connected. This configuration resembles with common emitter configuration. There are two basic difference between CC and CE configuration.

(i) For CC configuration load is connected at emitter terminal while for CE configuration load is connected at collector terminals.

(ii) Output voltage for common collector configuration is $I_E R_L$ while for common emitter configuration is $I_C R_L$.

The voltage gain for common collector configuration is near about unity because output voltage V_E is almost equal to input voltage V_B. It means output voltage vary according to input voltage. In other words the output at emitter follows the input and for this reason common collector configuration is also named as **Emitter Follower.** The common collector configuration is used primarily for impedance matching purpose since it has a high input impedance and low output impedance with unity voltage gain.

7.12 Transistor Characteristics

Transistor characteristics are required to determine the behaviour, performance and response of the transistor. Graphically plot of interrelationship between voltage and current at input side or output side is termed as **transistor characteristic.**

There are two important characteristics of a transistor: input characteristic and output characteristic. Graphical relationship between input voltage and input current with a constant output voltage is named as input characteristic while curve that shows the variation of output voltage with output current for constant input current is termed as output characteristic.

7.12.1 Characteristics of CE Configuration

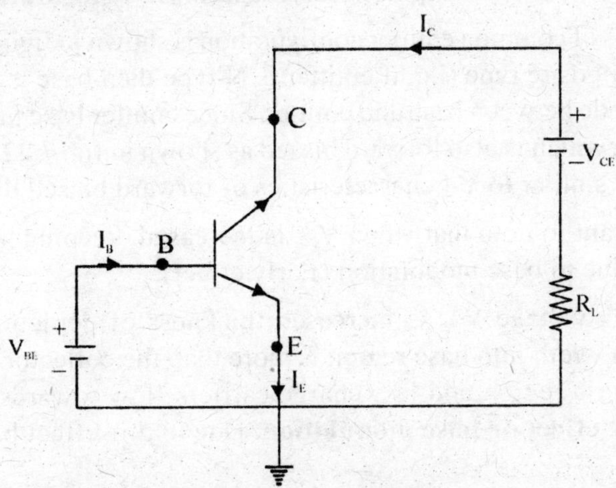

Fig. 7.22: CE configuration

Common emitter configuration of a transistor is shown in fig. 7.22, The input and output quantities of this configuration are as follows :

Input Voltage	: between base and emitter (V_{BE})
Input current	: base current (I_B)
Output Voltage	: between collector and emitter (V_{CE})
Output current	: collector current (I_C)

7.12.1.1 Input Characteristics

Input characteristics of transistor shows the relation between input current and input voltage keeping output voltage constant. For common emitter configuration input and output quantities are already defined. Therefore, input characteristics of CE configuration are the curves which show the relationship between input base current I_B and input voltage V_{BE} at constant output voltage V_{CE}.

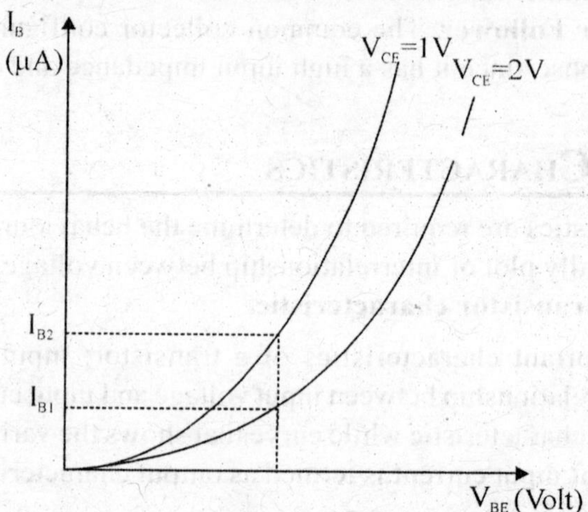

Fig. 7.23: Input characteristics of CE configuration

Input characteristics of common emitter configuration is shown in figure 7.23. Doping of emitter and base is always of opposite type (i.e. if emitter is N-type then base is P-type and vice versa) It forms a P-N junction diode between base and emitter. Since emitter base junction of transistor looks like a p-n junction diode which is also forward biased as shown in fig. 7.22, the input characteristics of CE configuration are similar to V-I characteristics of forward biased diode.

This is also important to note that when V_{CE} is increased, keeping V_{BE} constant, reduces the base current I_B. This is due to base modulation (Early effect).

As the reverse bias voltage V_{CE} is increased thickness of depletion width also increases. Penetration of depletion width into base region is more than the collector region. This reduces the effective width of the base region and less charge carriers flow towards the base terminal. This effect is known as **early effect or base modulation.** Due to this effect base current reduces with increase in V_{CE}.

7.12.1.2 Output Characteristics

Output characteristics of transistor shown the relationship between output current and output voltage with constant input current. For common emitter configuration output characteristics are drawn between V_{CE} and I_C with I_B as constant.

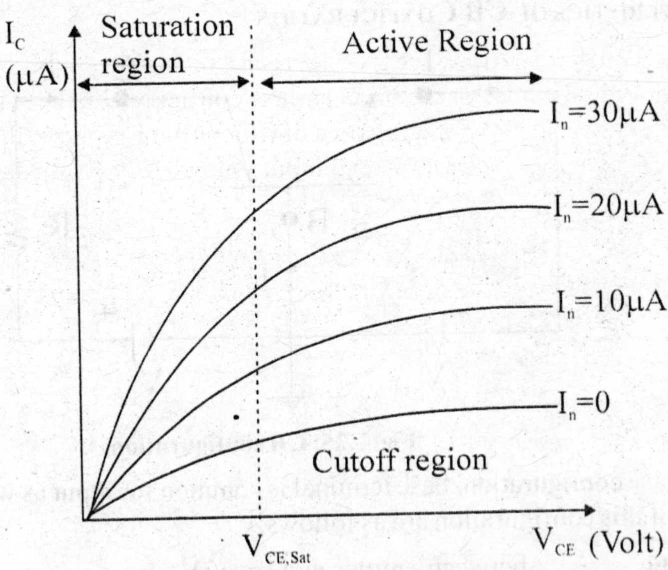

Fig. 7.24: Output characteristics of CE configuration

The output characteristic as shown in fig. 7.24, has basically three regions of operation : active, saturation and cutoff region.

Active Region : In the active region of CE configuration the collector base junction is reverse biased, while emitter base junction is forward biased. This region exists to the right of vertical dashed line at V_{CEout} and above the curve for $I_n = 0$.

In active region the percentage increase in I_C with V_{CE} is negligible, therefore, slope $\frac{D I_C}{d V_{CE}}$

negligible. Inverse for the slope $\left(\frac{d V_{CE}}{d I_C}\right)$ is output resistance. Therefore, resistance offered is very

V ^ai c)

high in active region. Slope of curve is slightly increasing with increase in V_{CE} due to early effect. Current gain β is not constant throughout the curve. It increases with increase in V_{CE}.

Saturation Region : If both emitter base junction J_1 and base collector junction J_2 are forward biased, transistor operates in saturation region. Under forward biased condition, the voltage drop V_{BE} at junction J_1 and V_{BC} at junction J_2 have very small values. The saturation value of different junction voltages in CE region is calculated as:

$$V_{CE,sat} = V_{BE,sat} - V_{BC,sat}$$

The region to the left of $V_{CE\ sat}$ in Fig. 7.24 is called saturation region.

Slope of the curve is very high in this region, therefore, the output resistance offered by the transistor is very low. Since the resistance offered to load is negligible, transistor acts like closed

Output current : collector current (I_C)

7.12.2 Characteristics of CB Configuration

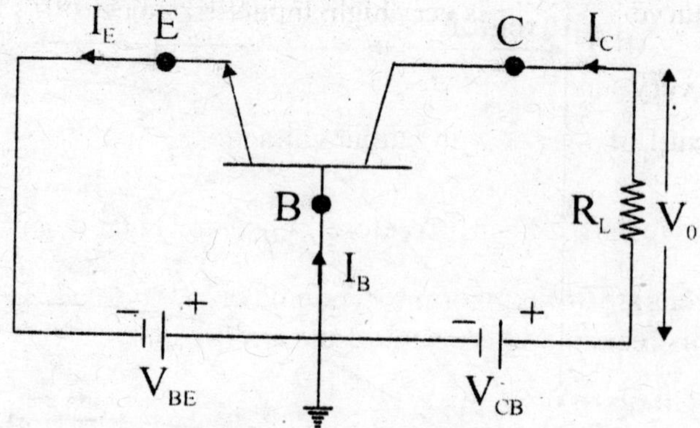

Fig. 7.25: CB configuration

In common base configuration, base terminal is common for input as well output. The input and output quantities of this configuration are as follows:

Input Voltage : between emitter and base (V_{BE})

Input current : emitter current (I_E)

Output voltage : between collector and base (V_{CB})

7.12.2.1 Input Characteristics

Input characteristics of common base configuration shows the relationship between input current I_E and input voltage V_{BE} for constant output voltage V_{CB}.

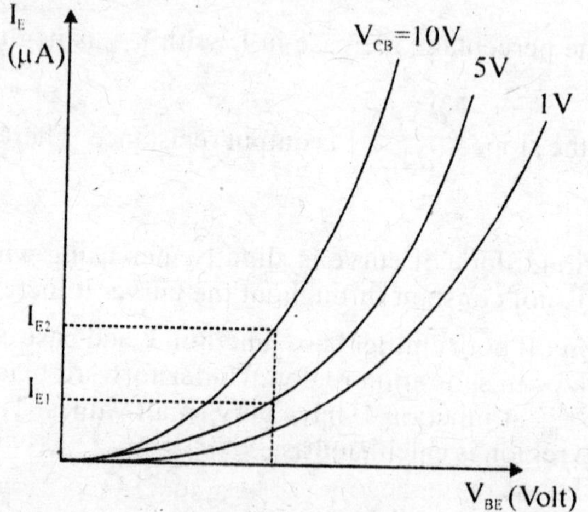

Fig. 7.26: Input characteristics of CB configuration

Input characteristics of common base configuration is shown in fig. 7.16. Input characteristics are identical to the V-I characteristics of a diode in forward region, since emitter base junction forms a p-n junction which is forward biased.

Slope of the curve $\left(\dfrac{d\,I_E}{d\,V_{BE}}\right)$ is very high. Input resistance $\left(\dfrac{d\,V_{BE}}{d\,I_E}\right)$ is inverse of the slope. So input resistance is very small.

Emitter current I_E increases with output voltage V_{CB}. With increase in V_{CB} the effective width of base reduces due to early effect. Therefore, concentration gradient $\left(\dfrac{dn}{dx}\right)$ slightly increases. As a result more electrons are diffused form N-type emitter to P-type base region and diffusion component of I_E increases. This results in increment of the emitter current I_E.

7.12.2.2 Output Characteristics

Output characteristics of common base configuration shows the variation of output current (I_C) with output voltage (V_{CB}) when input current (I_E) is kept constant.

The output characteristics of CB configuration are shown in fig.7.17. This figure contains three regions of operation : active, saturation and cutoff region.

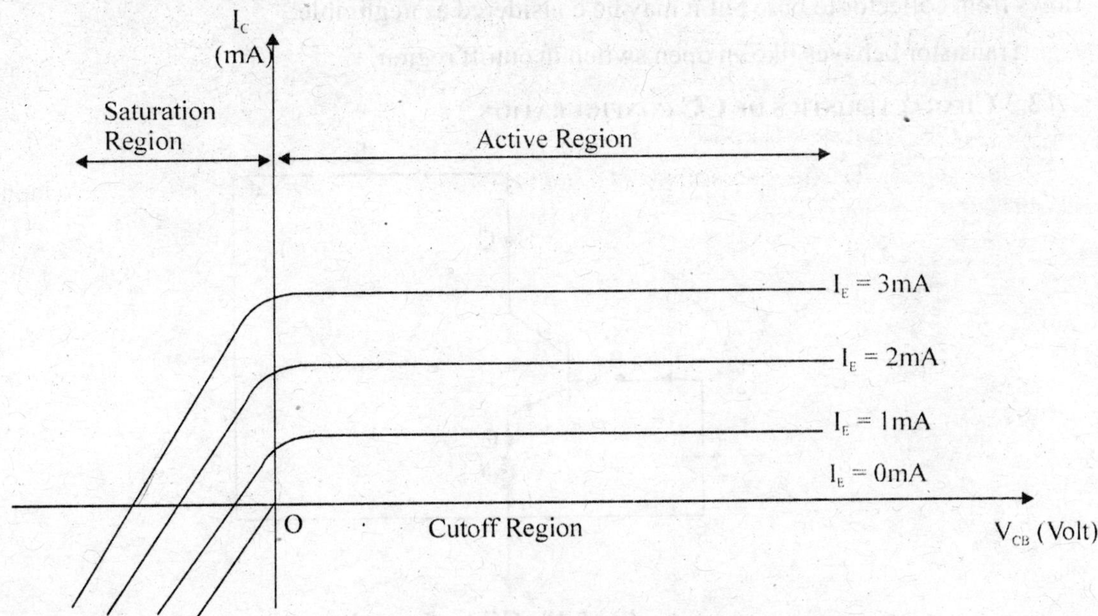

Fig. 7.27: Output characteristics of CB configuration

Active Region : In this region, emitter to base junction is forward biased and base to collector junction is reverse biased. This region exist when v_{CB} is positive and I_E is greater than zero as shown in fig. 7.27.

The output current I_C increases with increase in output voltage V_{CB} with constant input current I_E. The reason behind this is early effect as discussed earlier. This increment in I_C is marginal i.e.

approximately 1%. This may be considered as constant output current (I_C), therefore, common base configuration is used to design constant current source in active region.

The slope of the output curve is almost zero, therefore, the output resistance reaches to very high. Practically it has been observed the output resistance of CB configuration is higher than output resistance of CE configuration. The output resistance of CB configuration is of the order of mega ohms.

Saturation Region : If both the junctions of a transistor are forward biased, transistor operates in saturation region. The saturation region is defined as that region of the characteristics to left of V_{CB} = 0V. The horizontal scale in this region shows the dramatic change in characteristics in this region. Exponential increase in collector current may be noted as the voltage V_{CB} increases towards 0V.

Transistor behaves like a closed switch in saturation region.

Cut off Region : In the cutoff region the collector base and base emitter junction of a transistor are both reverse biased. As inferred by its name, the cut off region is defined as that region, where the output current (I_C) is 0 Amp as shown in fig. 7.27. Practically reverse saturation current (I_{CBO}) flows from collector to base but it may be considered as negligible.

Transistor behaves like an open switch in cutoff region.

7.12.3 CHARACTERISTICS OF CC CONFIGURATION

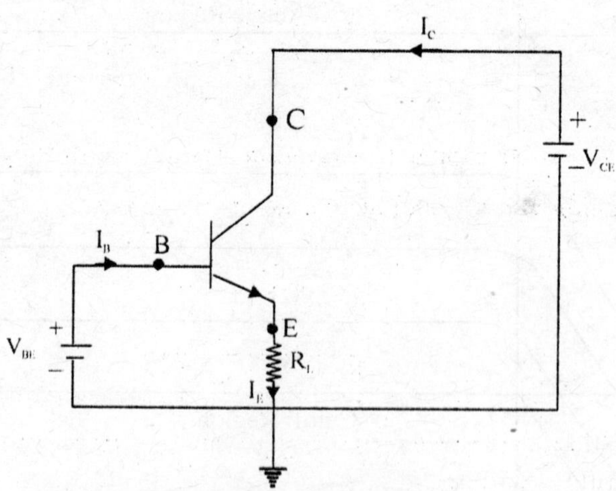

Fig. 7.18 : CC configuration

When collector terminal is made common for input and output, the transistor operates in common collector configuration. The input and output quantities are as follows :

Input Voltage : between base and emitter (V_{BE})

Input current : base current (I_B)

Output Voltage : between collector and emitter (V_{CE})

Output current : Collector current (I_C)

The basic difference of common collector configuration and common emitter configuration is that the orientation of load resistor is changed. It means load resistor is applied at emitter terminal (for CC configuration) while at collector terminal (for CE configuration). Input and output quantities for CC configuration are similar to CE configuration. Therefore, input characteristics and output characteristics of common emitter configuration can also be considered as input characteristics and output characteristics of common collector configuration respectively. Since both the these characteristics have been already discussed, therefore, it is not discussed again.

7.13 Transistor Applications

Transistor is an electronic device which may be used in different applications or in different circuits. The transistor has two basic applications as an amplifier and as a switch.

7.13.1 Transistor as a Switch

In the electronic circuits where high switching speed is compulsory, transistor is used as switching element. For switching applications, transistor is operated in saturation region (ON) and cut off region (OFF)

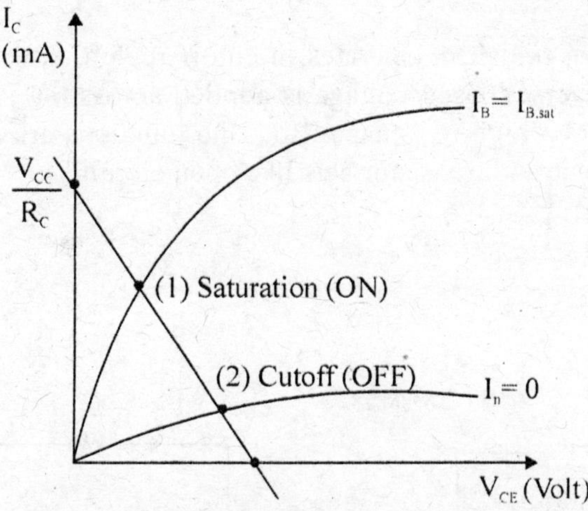

Fig. 7.28: Output characteristics of CE configuration with load line

Output characteristics of CE configuration shown in fig.(3.19 for two base currents ($I_B = 0$ and $I_B = I_{B\ sat}$). Two operating points (1) and (2) are also marked in the figure. The operating point marked as (1) is in saturation region. At this point the slope of the curve is very high.

Mathematically

$$\text{Slope(m)} = \frac{dI_C}{dV_{CE}} \rightarrow \text{Very high}$$

Therefore, output resistance of transistor

$$R_o = \frac{dV_{CE}}{DI_C} = \frac{1}{m} \rightarrow \text{Very low.}$$

Transistor offers very low or approximately zero resistance in saturation region and the transistor acts like a closed switch.

The same concept may be explained in another way that when transistor operates in saturation region, both the junctions J_1 (emitter base junction) and J_2 (base collector junction) are forward biased. The total voltage drop across the transistor is difference of voltages across the two forward biased junction. This drop is very low or negligible. So we can say in saturation region, voltage drop across the transistor is very low and collector and emitter terminals are about to short. At this time transistor behaves as closed switch.

When base current I_B is zero, the point of operation of transistor will be point (2) as shown in fig. 7.29. At this point, slope (m) of curve is very low or negligible. Therefore, the resistance offered by the transistor in cut off region is given as

$$R_o = \frac{dV_{CE}}{dI_C} = \frac{1}{m} \longrightarrow \text{Very high}$$

Thus the transistor offers a very high resistance when operated at point (2) and acts like open circuit.

In other words, when transistor operates in cutoff region, both the junctions J_1 and J_2 are reverse biased. As the reverse biased voltage is applied across the junction, the depletion layer width increases and offers a high resistance. Both the junctions are reverse biased so transistor offers a very high resistance and transistor acts like open circuit.

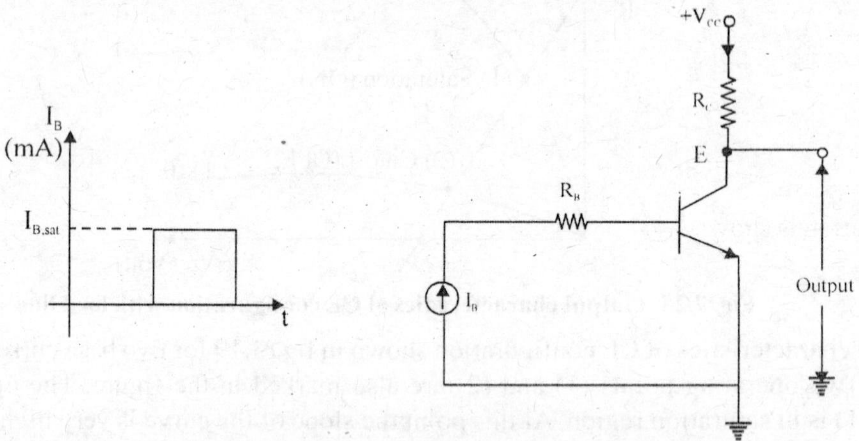

Fig.7.29 Transistor like a switching element

Base of transistor is driven by a current source of square waveshape. When base current is zero, transistor operates in cut off region. As explained earlier, in cutoff region transistor offers a high resistance and output current I_C becomes zero. This situation is shown in Fig. 7.30, where transistor is shown as open circuit.

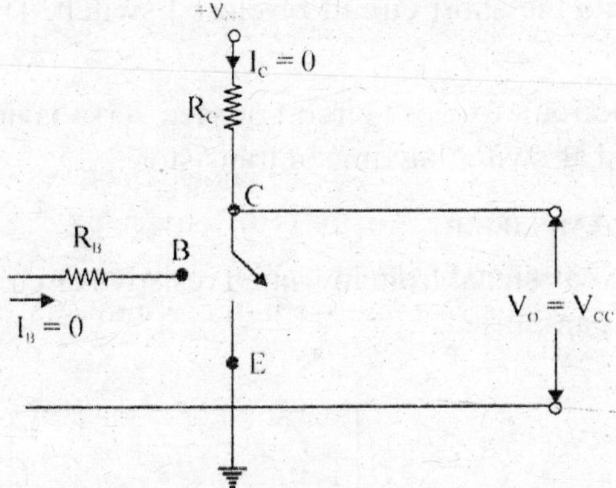

Fig.7.30 Transistor in cutoff and working like open circuit

When output current I_C is zero, the voltage drop across R_C will be zero ($I_C R_C = 0$). It results $V_0 = V_{CC}$.

When base current I_B is equal to $I_{B\ sat}$ the transistor operates in saturation region. In this region transistor offers very low or approximately zero resistance as discussed earlier. The collector current I_C is maximum in this region.

The collector current is given as

$$I_C = \frac{V_{CC}}{R_C}$$

Therefore, output voltage is given as

$$V_0 = V_{CC} - I_C R_C$$

$$V_0 = 0$$

This situation is shown in fig. 7.31

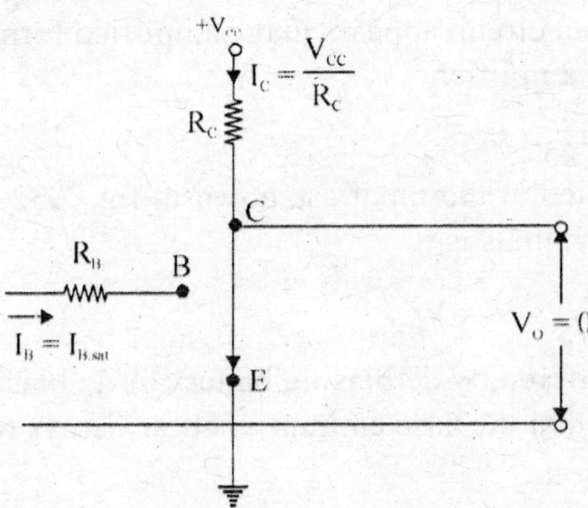

Fig. 7.31 : Transitor in saturation and working like closed circuit

Zero voltage represents the short circuit or closed switch. Transistor behaves like a closed switch.

Switching from open circuit to close circuit or vice versa is not instantly, but a finite time is required. This time is called as switching time of transistor.

7.13.2 Transistor as an Amplifier

A transistor transfers weak signal from low input resistive circuit to high output resistive circuit and therefore, signal level is amplified.

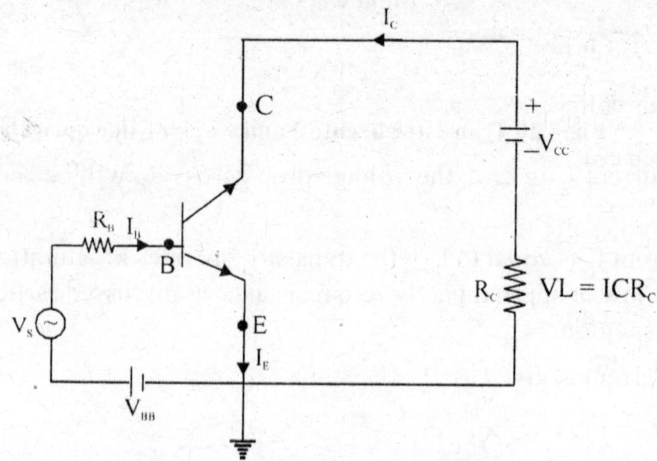

Fig.7.32 : Transistor as an amplifier

A CE configuration is shown in fig. 7.32. Transistor is kept in active region to operate as an amplifier. In active region emitter base junction is forward biased. A weak ac signal is applied at the input of the shown circuit. Output is taken across the load resistor R_C. As the input circuit has low resistance a small change in input signal voltages causes an appreciable change in base current. This causes larger change in collector current ($I_c = \beta I_B$) .

The collector current flowing through a high load resistance R_C produces a large voltage across it. Thus, a weak signal of input circuit appears in the amplified form in the output collector circuit. This way transistor act as an amplifier.

Mathematical Analysis

An ac signal V_s is applied at the input as shown in fig. 7.32. Therefore the voltage applied between base and emitter terminals is

$$V_i = V_S + V_{BB}$$

Where V_{BB} is voltage offered by dc biasing battery. A dc biasing battery with V_{BB} voltage is applied in such a manner, as that the base emitter junction always remains forward biased.

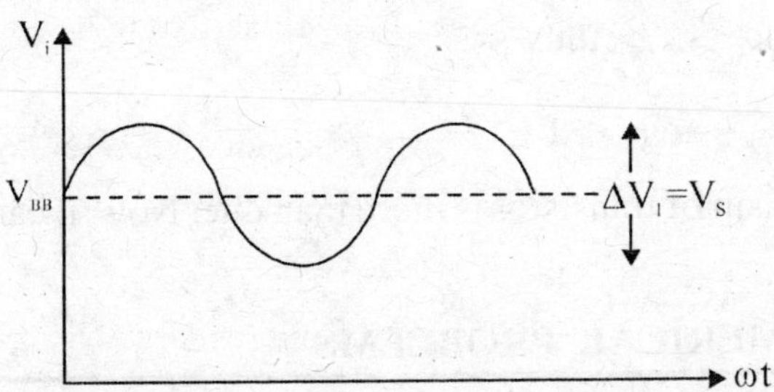

Fig. 7.33: Input voltage across base and emitter

From the fig. 7.33 it is clear that

Change in input voltage $\Delta V_i = V_S$

Similarly output collector current will be having dc component 'I_C' and ac component 'i_C'

So variation in collector current will be only due to ac component

$$\Rightarrow \quad \Delta I_C = i_C \qquad ...(7.26)$$

Collector current causes voltage drop across load resistance R_L and it varies due to variation in collector current.

$$\Delta V_L = \Delta I_C R_C$$

$$\Rightarrow \quad \Delta V_L = i_C R_C \qquad ...(7.27)$$

Voltage gain is defined as the ratio of change in output voltage to the change in input voltage

$$A_V = \frac{\Delta V_L}{\Delta V_i} \qquad ...(7.28)$$

$$A_V = \frac{i_C R_L}{V_S}$$

$$A_V = \frac{\beta i_b R_C}{V_S} \quad (\because i_c = \beta i_b)$$

$$A_V = \frac{\beta R_C}{(V_S / i_b)}$$

$$A_V = \frac{\beta R_C}{r_b}$$

Where r_b is input base resistance.

Since $\beta >> 1$ *and* $R_C >> r_b$, thus

$$A_V >> 1$$

It means voltage gasin of transistor is more than one. Now, it can be said that an transistor acts like an amplifier.

7.14 Solved Numerical Problems

Problem 1: In a CB configuration, current amplification factor is 0.9.If the emitter current is 4mA, determine the value of base current.

Solution: we know $I_C = \alpha I_E$

$I_C = 0.9 \times 4$ mA

$I_C = 3.6$ mA

And $I_B = I_E - I_C$

$I_B = 4 - 3.6 = 0.4$ mA.

Problem 2: Calculate the emitter current I_E for a transistor connected in CE configuration. Given â=60 and base current I_B=20 μA.

Solution: we know $I_C = \alpha I_B$

$I_C = 60 \times 20 \times 10^{-6}$

$I_C = 1.2$ mA

And $I_E = I_C + I_B$

$I_E = 1.2$ mA + 20 μA

$I_E = 1.22$ mA.

Problem 3: A 2N 3198 transistor has a typical $â_{dc}$ of 75. If the collector current is equal to 15 mA. Calculate (1) base current

(2) Emitter current

(3) α_{dc}.

Solution: (1) base current $I_B = \dfrac{I_C}{\beta}$

$$I_B = \frac{15mA}{75} = 200\mu A$$

(2) Emitter current $I_E = I_C + I_B$

$I_E = 15 \times 10^{-3} + 200 \times 10^{-6}$

$I_E = 15.2$ mA

(3) Alpha $\alpha_{dc} = \dfrac{\beta}{1+\beta} = \dfrac{75}{75+1} = 0.98$

Problem 4: A silicon transistor has $I_C = 1.7$ mA for $\Delta I_E = 1.79$ mA. What change in I_B will produce an equal change in I_C?

Solution: Given $I_C = 1.7$ mA and $\Delta I_E = 1.79$ mA

We know $$\alpha = \frac{\Delta I_C}{\Delta I_e} = \frac{1.7}{1.79} = 0.95$$

And $$\beta = \frac{\alpha}{1-\alpha} = \frac{0.95}{1-0.095} = 20$$

$$\Delta I_B = \frac{\Delta I_C}{\beta} = \frac{1/7mA}{20} = 0.085mA$$

Problem 5: A certain n-p-n transistor with $\beta = 100$ has a base to collector leakage current I_{CBO} of 5 µA .The transistor is connected in CE configuration. Calculate the collector current if the base current $I_B = 50$ µA.

Solution: Given $\beta = 100$, $I_{CBO} = 5$ µA, $I_B = 50$µA.

We know $I_C = \beta I_B + (1+\beta) I_{CBO}$

$I_C = (100 \times 50\ \mu A) + (101 \times 5\ \mu A)$

$I_C = 5\ mA + 0.505\ mA$

$I_C = 5.505\ mA$

Problem 6: for a transistor in CE configuration the reverse leakage current is 20.9 µA, whereas when the same transistor is connected in CB configuration, it reduces to 1.1 µA. Calculate values of á and â.

Solution: Given $I_{CEO} = 20.9$ µA , $I_{CBO} = 1.1$ µA

We know $I_{CEO} = (1+ â) I_{CBO}$

Or $$\beta = \frac{I_{ceo}}{I_{cbo}} - 1,\ \beta = \frac{20.9}{1.1} - 1 = 18$$

We know $$\alpha = \frac{\beta}{1+\beta} = \frac{18}{1+18} = 0.95$$

Problem 7: The emitter current of a transistor is 10mA. If $\alpha = 0.88$ and $I_{CBO} = 10$ µA .find the value of I_C and I_B.

Solution: Given $\alpha = 0.88$, $I_{CBO} = 10$ µA , $I_E = 10$mA

We know $I_C = \alpha I_E + I_{CBO}$

$I_C = (0.88 \times 10 \text{ mA} + 10 \mu\text{A})$

$I_C = 8.81 \text{ mA}$

And $I_B = I_E - I_C$

$I_B = 10 \text{ mA} - 8.81 \text{ mA} = 1.19 \text{ mA}$

Problem 8: Calculate the value of I_E .If $\alpha = 0.98$, $I_{CBO} = 16\ \mu\text{A}$ and $I_B = 160\ \mu\text{A}$.

Solution: We know $\beta = \dfrac{\alpha}{1-\alpha} = \dfrac{0.98}{1-0.98} = 49$

And $I_E = (1+\alpha)\, I_B + (1+\alpha)\, I_{CBO}$

$I_E = (1+49)160\ \mu\text{A} + (1+49)\ 16\mu\text{A}$

$I_E = 8.8 \text{ mA}.$

Review Questions

1. Draw the energy band diagram of conductors, insulator and semiconductors.
2. Why BJT is called a bipolar device?
3. Explain the construction and working of BJT in CB, CE and CC configurations.
4. Describe the input-output characteristics of BJT in CB, CE and CC configurations.
5. Explain the operation of n-p-n transistor.
6. Draw the circuit symbol for a pnp transistor and indicate the reference directions for the three current and the reference polarities for the three voltages.
7. Draw the circuit symbol for a npn transistor and indicate the reference directions for the three currents and the reference polarities for the three voltages.
8. For a pnp transistor biased in active region plot the potential variation in each region E, B and C.
9. Explain the shapes of the plots of potential variation in three regions E, B and C for a pnp transistor? Why is it so called?
10. What is a transistor? Why is it so called?
11. Explain the transistor's amplifying action in detail.
12. Define the current gains a and p in words and as an equation.
13. Show that alpha current gain is always less than unity.
14. Why is collector wider than emitter?
15. Why is base made so thin?

16. Can transistor action be achieved by connecting two batch to batch diode? If not, explain in detail with npn transistor.

17. What are a and p? What is their relationship?

18. Explain the transistor working in detail with the description of the different current omponents.

19. Explain the significance of current gains alpha and beta in a transistor, using various current components.

20. What are three configuration of bipolar junction transistor? Draw the input and output characteristics of transistor connection in common emitter mode.

21. What are three configuration of BJT? Compare them with regard to their input impedance, output impedance, current and voltage gain.

22. Compare the common base, common emitter and common collector configuration of a transistor for common collector configuration, prove that

$$I_C = \beta I_B + (\beta + 1) I_{CO}$$

23. Sketch and explain the typical input and output characteristic curves of common base configuration.

24. Why common collector configuration is known as Emitter Follower?

25. Sketch typical output characteristic curves of an npn transistor for common emitter configuration explain the operating regions with reference to the sketch.

26. What is base width modulation (early effect)?

27. What are the applications of transistor in different operating regions?

28. Discuss how a transistor behaves as (i) an amplifier and (ii) a switch.

Chapter 8

Thyristors

OBJECTIVE

- Silicon Controlled Rectifier (SCR)
- Operation of SCR
- SCR as two transistor analogy
- SCR as a Switch
- TRIAC (Bidirectional Thyristors)
- DIAC (Bidirectional Thyristors)
- Unijunction Transistor (UJT)
- UJT triggering of an SCR

8.1 INTRODUCTION

Thyristors, stands for thyratron transistor, because the device combines the rectification action of thyratron and control action of transistor. Thyristors is a three terminal, four layers semiconductor switching device with a control mechanism.

Thyristors basically serves two functions viz electronic switching and electronic control. The most important member of the Thyristors family is the silicon controlled rectifier (SCR).

SCR, s and triac having high voltage and current ratings are widely used in power control applications.

Some of the applications of Thyristors are given below :

(a) As static switches

(b) Relay control

(c) Phase control

(d) HVDC transmission lines

(e) Power switches

(f) Control of temperature, position and level

(g) As rectifier

(h) As invertors

Thyristors have some advantage over transistors as given below:

(1) Thyristors is a four layer device while the transistor is a three layer device.

(2) It is a possible to have Thyristors with higher voltage and current ratings.

(3) Thyristors have better power handing capacity.

(4) Internal power losses in a Thyristors are much smaller than transistor.

8.2 Silicon Controlled Rectifier (SCR)

A SCR is a semiconductor device that perform the task of an electronic switch .SCR basically serves two functions at the same time such as

(1) As rectifier for conversion of ac into dc.

(2) As electronic controller, this can control the power fed to the load.

Therefore, SCR combines the features of a rectifier and a transistor.

Note :

SCR is made of silicon only because leakage current in silicon is much smaller than germanium. Because it is basic requirement of device that acts a true electronic switch.

8.2.1 Construction

SCR is a four layer (p-n-p-n) three junction semiconductor device with three terminals, namely, the anode, the cathode and the gate. It is a unilateral device. When a PN junction is added to a BJT, the resulting three PN junction device is called a silicon controlled rectifier. According to figure 8.1 it is essentially an ordinary rectifier (PN) and a BJT (NPN) combined in one unit to form PNPN device. Outer p-type material is called anode (A) and outer n-type material is called cathode (K).The base of transistor is called gate (G).

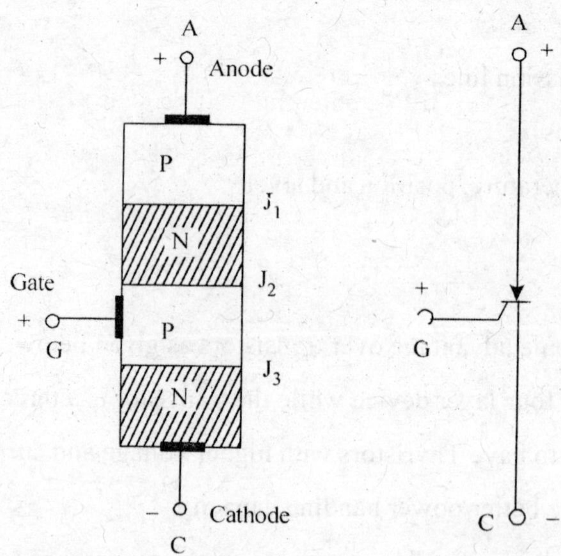

Fig.8.1 Schematic & Symbolic diagram of SCR

In SCR, conduction takes place from anode to cathode under normal operating conditions i.e. anode is held at high positive potential w.r.t cathode and gate at small positive potential w.r.t. to cathode. The schematic and symbolic diagram of SCR is shown in figure 8.1.

Mesa type construction : In this technique, the inner junction J_2 is obtained by diffusion and then the outer two layers are alloyed to it as shown in figure 8.2. The PNPN pellet is properly braced with molybdenum plates to provide mechanical strength and make it capable of handling large currents. One plate is hard soldered to a copper stud, which provides an efficient thermal path for conducting the internal losses to the surrounding medium.

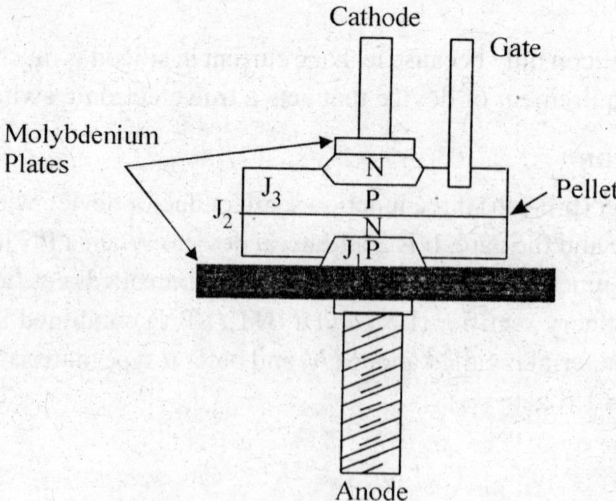

Fig.8.2 Mesa type construction of SCR

8.2.2 Operation of SCR

To understand the working of SCR, it is essential that load is connected in series with anode, and the anode is always kept at positive potential w.r.t cathode. There are two conditions.

(1) When gate is open : Figure 8.3 shows the SCR circuit with gate open i.e. no voltage applied to the gate.

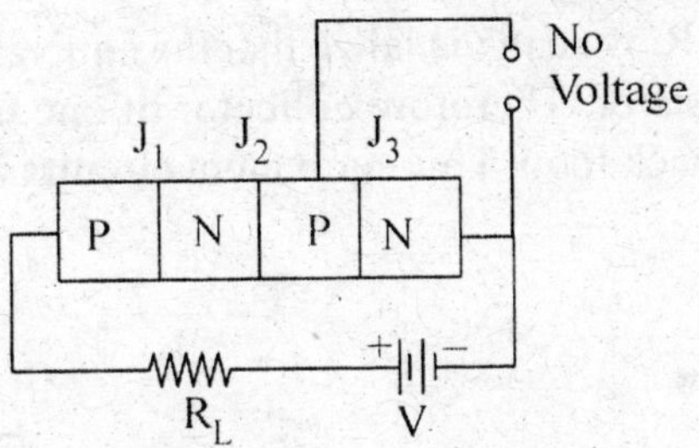

Fig.8.3 When Gate is open

Under this condition, J_2 is reverse biased and J_1, J_3 are forward biased.

Hence junction J_1 and J_3 is just as in a NPN transistor with base open. Therefore no current flows through the load R_L and SCR is cut off.

If the applied voltage is gradually increased, a stage is reached when J_2 breaks down. The SCR conducts heavily and is said to be in the ON state.

The applied voltage at which SCR conducts heavily without gate voltage is called break over voltage.

(2) When gate is positive w.r.t. cathode : By applying a small positive potential to the gate, the SCR can be made to conduct heavily as shown in figure 8.4. Now junction J_2 is reverse biased and junction J_3 is forward biased.

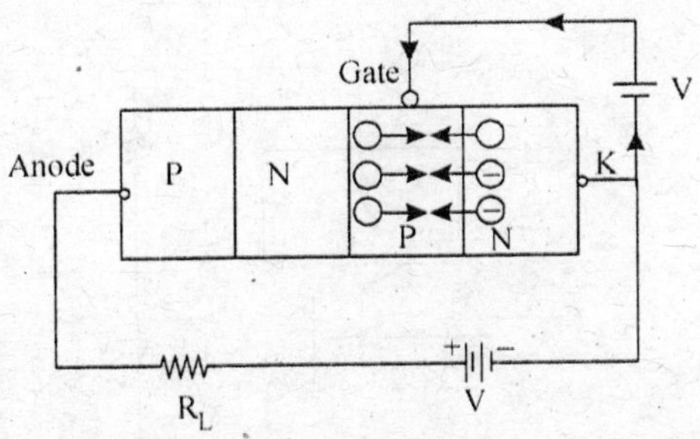

Fig.8.4 When gate is positive w.r.t. cathode

The electron from n-type material starts moving across junction J_3 towards left whereas holes from p-type towards the right. Therefore, the electrons from junction J_3 are attracted across junction J_2 and gate current starts flowing. Therefore when the gate current flows, anode current increases. The increased anode current in turn makes more electrons available at junction J_2. This process continues and J_2 breaks down and the SCR starts conducting heavily. If we remove the gate voltage, the anode current does not decrease at all. The only way to stop conduction is to reduce the applied voltage to zero.

Observation : SCR behaves like a switch because it has two states i.e. either it conducts heavily or it does not conduct.

8.2.3 Equivalent circuit of SCR

from the figure 8.5 of SCR, we can visualize that the equivalent circuit of SCR is composed of PNP transistor and NPN transistor. Therefore collector of one transistor is coupled to the base of other, making a positive feedback loop. This equivalent circuit is known as two transistor analogy.

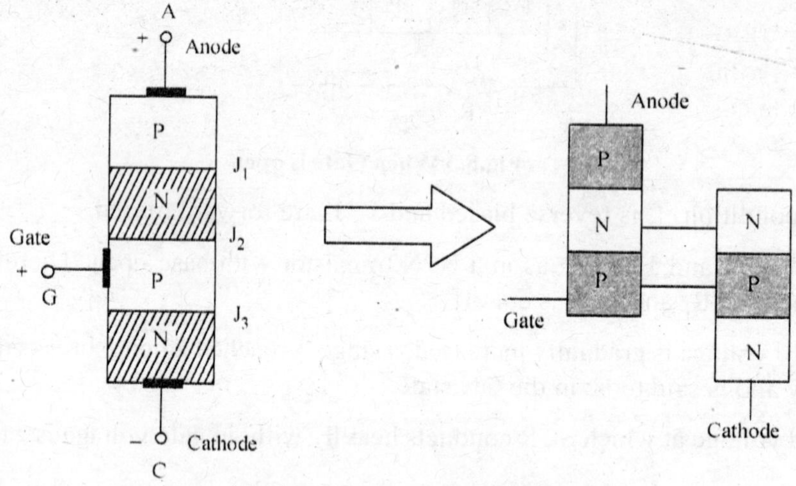

Fig.8.5 Equivalent circuit of SCR

Working of two transistor analogy : Figure 8.6 shows the equivalent circuit of SCR with supply voltage V and load resistance R_L.

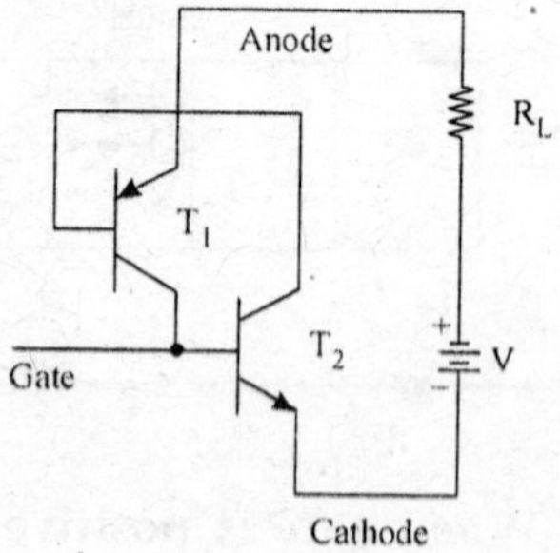

Fig.8.6 SCR as two transistors Analogy

If the supply voltage V is less than break over voltage (i.e. gate and switch open), there is no base current in transistor T_2. Hence no current flows in the collector of T_2 and T_1, .It indicates that SCR is open.

When a small gate current will flow through the base of T_2 i.e. its collector current will increase.

The collector current of T_2 is the base current of T_1. Hence collector current of T_1 increases.

Therefore an increase of current in one transistor causes an increase of current in other transistor. Thus heavy current will flow through load R_L. It indicates that SCR is close.

Break over voltage : It is the minimum forward voltage, gate being open, at which SCR starts conducting heavily i.e. turned ON.

Holding current : It is the maximum anode current, gate being open, at which SCR is turned off from ON conditions.

8.2.4 V-I CHARACTERISTICS OF SCR

Figure 8.7 shows the V-I characteristics of a SCR. It is the curve between anode-cathode voltage (V) and anode current (I) at constant value of gate current.

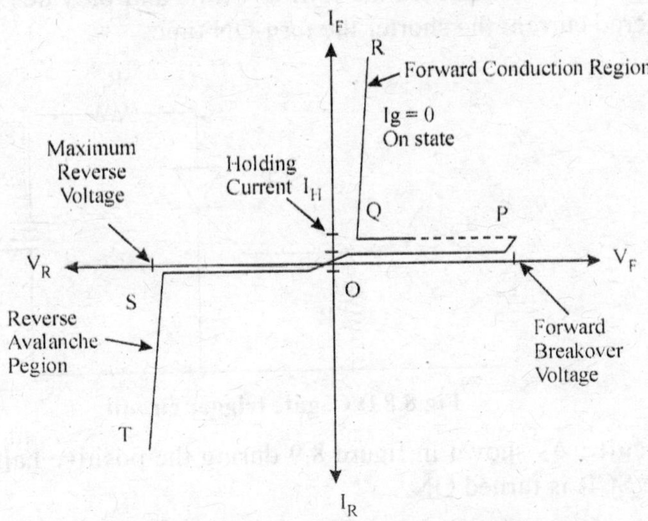

Fig.8.7 V-I characteristics of SCR

Forward characteristics : When anode is positive w.r.t. cathode, the curve between V and I is called the forward characteristics. In curve, OPQR is the forward characteristics at I_G=0. If the supply voltage is increased from zero, a point is reached (point P) when the SCR starts conducting. Therefore, the voltage across SCR suddenly drop (curve PQ) and most of supply voltage appears across the load resistance R_L.

Reverse characteristics : When anode is negative w.r.t. cathode, the curve between V and I is known as reverse characteristics. If the reverse voltage is gradually increased, at first the anode current remains small and at some reverse voltage, the SCR starts conducting heavily in the reverse direction as shown by the curve (ST). This maximum reverse voltage at which SCR starts conducting heavily is known as reverse breakdown voltage.

8.2.5 SCR AS A SWITCH

The SCR has two states (ON state and OFF state) and no state in-between. When gate current flows, it starts conducting heavily. This corresponds to the ON state. When the anode current is reduced to the holding current, SCR is turned OFF. As SCR is an electronic device, its behavior is similar to an electronic switch.

SCR turn-ON methods : In order to turn ON the SCR, the gate voltage V_G is increased up to a minimum value to start triggering.

Gate triggering voltage (V_{GT}) is the minimum value of gate voltage at which SCR is turned ON. The resulting gate current is called gate triggering current (I_{GT}).

D.C gate trigger circuit : Figure 8.8 shows an SCR with a D.C. gate bias. When the switch is closed, the gate receives sufficient positive voltage (V_{GT}) to turn the SCR ON. The resistance R connected in the circuit which improves the turn ON time and provides noise suppression. The higher the gate triggered current the shorter the turn-ON time.

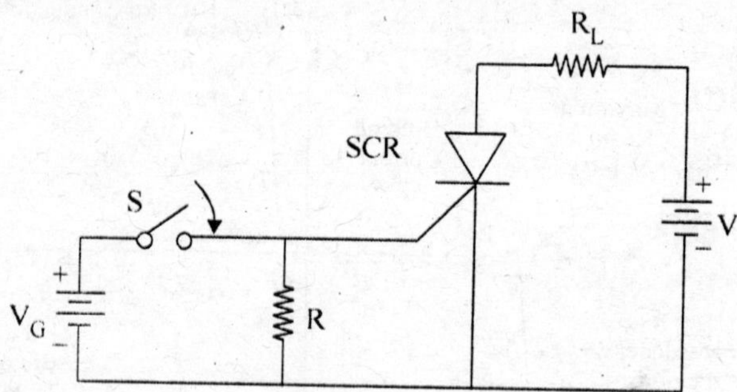

Fig.8.8 D.C. gate trigger circuit

A.C trigger circuit : As shown in figure 8.9 during the positive half cycle of the a.c. gate current, at $I_G = I_{GT}$, the SCR is turned ON.

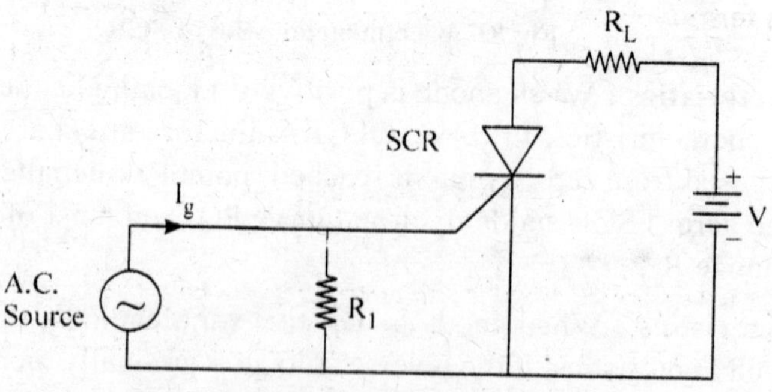

Fig.8.9 A.C. trigger circuit

SCR turn OFF methods : Anode current interruption:-The simply way to turn OFF the SCR is to open the switch S as shown in figure 8.10. When the anode current is reduced below a minimum value called holding current, the SCR turns off.

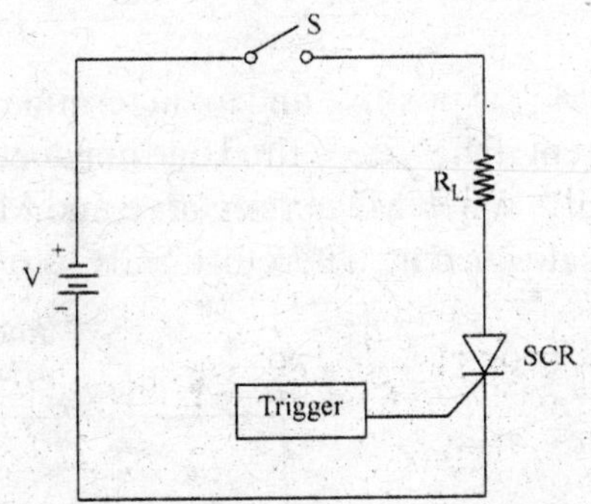

Fig.8.10 Anode current interruption

8.2.6 Application of SCR

Following are few application of SCR

(i) SCR as static contactor (ii) SCR for power control.

(iii) SCR for speed control. (iv) Over light detector

8.2.7 Advantages of SCR

Advantages of SCR are as follows :

(i) It gives noiseless operation at high efficiency because it has no moving parts.

(ii) It has small size and high switching speed.

(iii) It permits control over large current.

8.3 Triac (Bidirectional Thyristors)

We know that SCR can conduct current in one direction only. It is clear that it can control either d.c. Power or positive half cycle of a.c. in a load.

A triac (triode a.c. switch) is a three terminal semiconductor switching device which can control alternating current or can conduct current in either direction. The three terminals are main terminal MT_1, main terminal MT_2 and gate G. A triac is equivalent to two separate SCRs connected in inverse parallel with gate common as shown in figure 8.11.

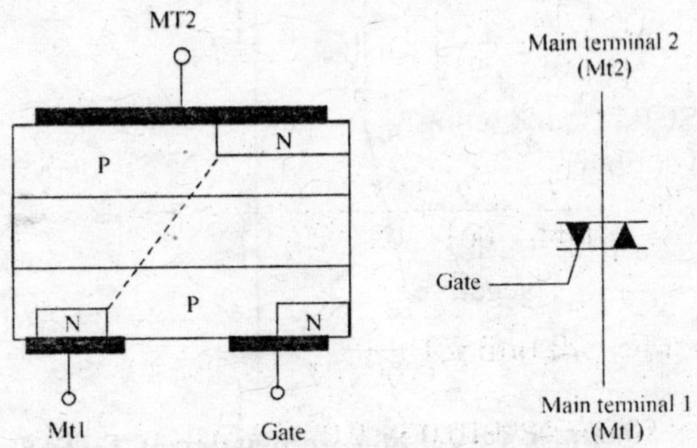

Fig.8.11 : TRIAC construction and symbol

8.3.1 Triac operation

When the main terminal MT_2 is positive and main terminal MT_1 is negative, as shown in figure 8.12, the triac will conduct current if the triac is fired into conduction by proper gate current. Therefore SCR1 is on and the SCR2 is off. When MT_2 is negative and MT_1 is positive, the triac will be fired into conduction with proper gate current. Therefore SCR_2 is on and SCR_1 is off.

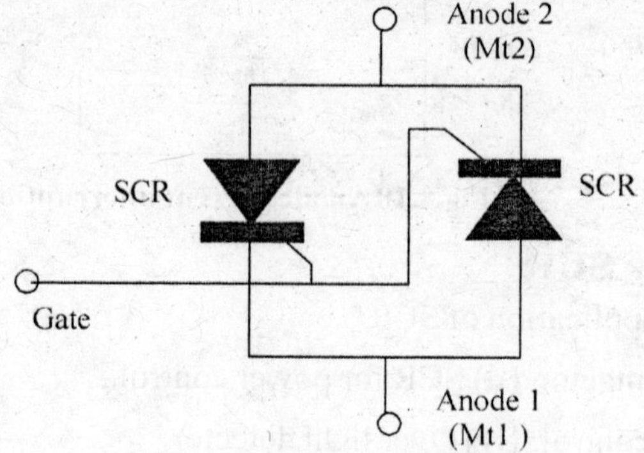

Fig.8.12 : TRIAC operation

8.3.2 Triac Characteristics

Figure 8.13 shows the V-I characteristics of a triac.

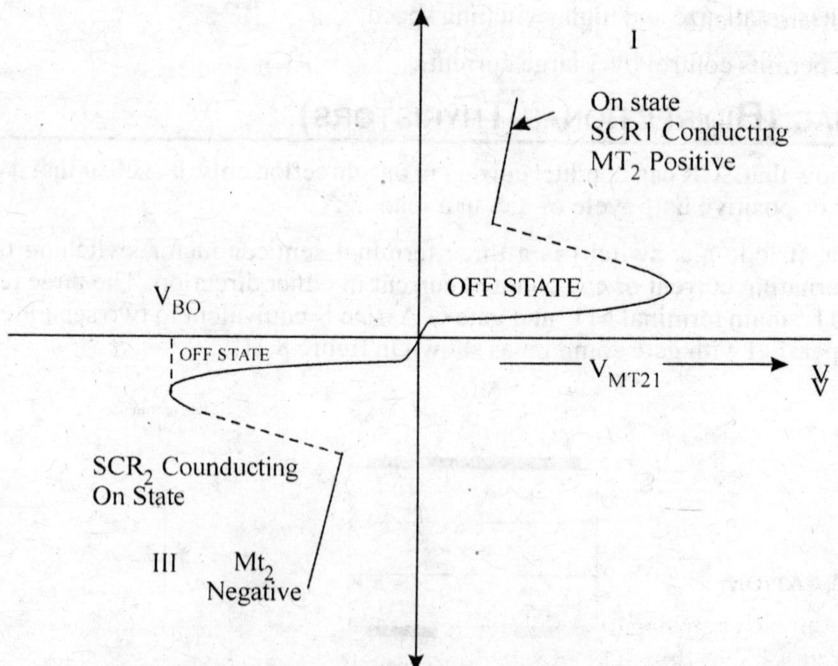

Fig.8.13 V-I characteristics of TRIAC

The triac can be operated with either positive or negative gate control voltage. The gate voltage is positive in quadrant 1 and negative in quadrant 3. The supply voltage at which the triac is turned on depends upon the gate current. The greater the gate current, the smaller the supply voltage at which the triac is turned on.

8.3.3 Application of Triac

(i) As a high power lamp switch.

(ii) Electronic change over of transformer taps.

8.3.4 Difference between Triac & SCR

Triac	SCR
It conducts in either direction.	It conducts only in one direction.
It is bidirectional Thyristors.	It is unidirectional Thyristors.
It is triggered by a small voltage pulse of either polarity.	It is triggered by a small positive voltage pulse.
Switching time is large.	Switching time is small.

8.4 Diac (Bidirectional Thyristors)

A diac is a two terminal; three layer bidirectional device which can be switched from its OFF state to ON state for either polarity of applied voltage. The structure of diac is very much similar to a transistor as shown in figure 8.14.

Fig.8.14 DIAC symbol

8.4.1 Operation

When a positive or negative voltage is applied across the terminals of a diac, only a small leakage current I_{BO} will flow. The leakage current will continue to flow until the voltage reaches the break over voltage V_{BO}. After this point current through the device increases with the decreasing values of applied voltage as shown in figure 8.15.

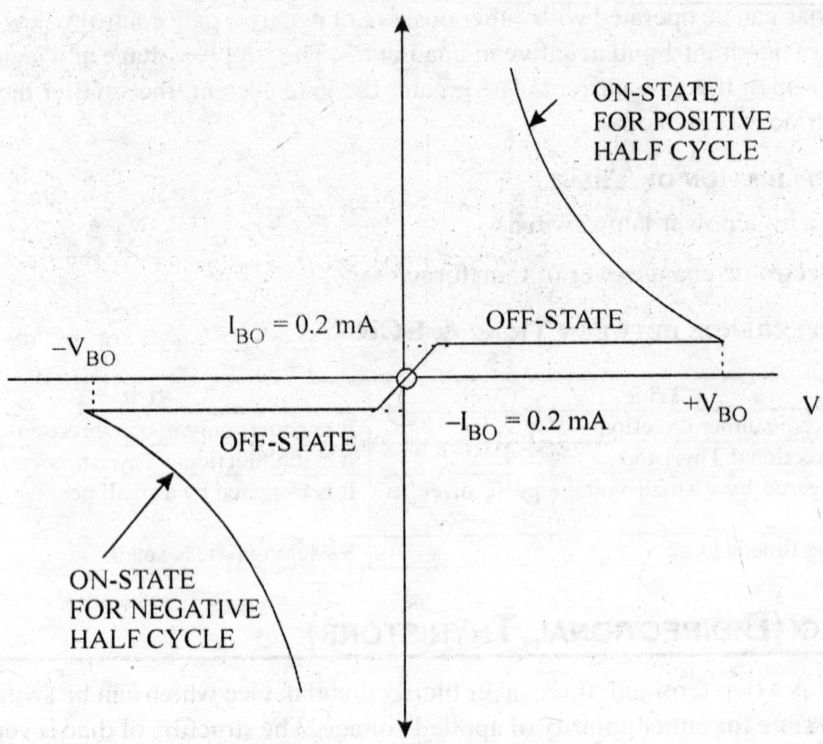

Fig.8.15 V-I characteristics of DIAC

8.4.2 Application of Diac

- lamp dimmer
- Heat control

8.5 Unijunction transistor (UJT)

The process of turning ON an SCR is called as triggering. The most popular device used for SCR triggering is UJT.

8.5.1 Construction

The UJT is a three terminal device having the basic construction as shown in figure 8.16(a). This is called as Unijunction device as there is only one junction present. In construction of SCR, a slab of lightly doped n-type silicon material has two base contacts attached to both ends of one surface.

The symbol for UJT is as shown in figure 8.16(b). The arrowhead point towards the direction of conventional current flow, when the UJT is in the forward biased or conducting state. The UJT is a three terminal device. The three terminals are Emitter(E), base1 (B_1) and base2 (B_2) as shown in figure 8.16(b).

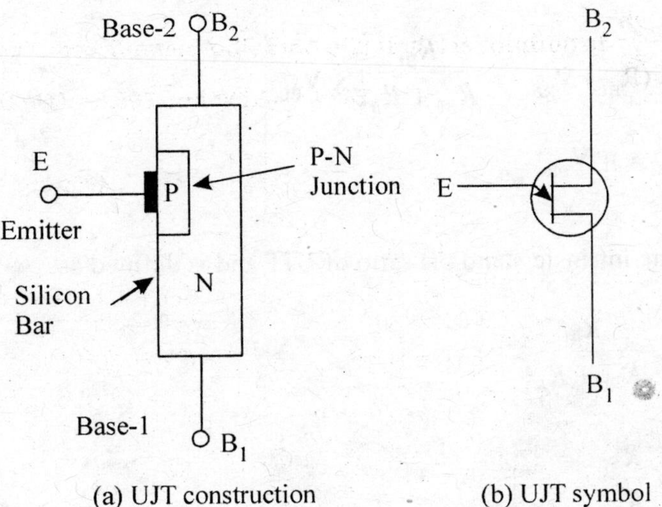

(a) UJT construction (b) UJT symbol

Fig. 8.16

8.5.2 Operation of UJT

The equivalent circuit of UJT is as shown in figure 8.17.

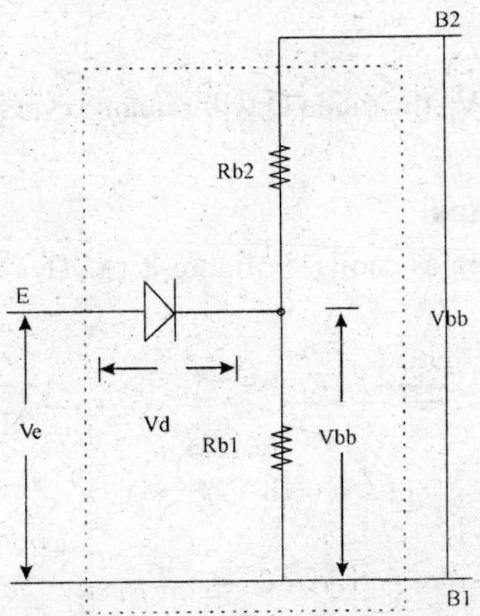

Fig.8.17 Equivalent circuit of UJT

It shows two resistances R_{B1} and R_{B2} and a diode which represents the only junction. The interbase resistance R_{BB} is the resistance between the terminals B_1 and B_2 of UJT.

$$R_{BB} = R_{B1} + R_{B2}$$

From the equivalent circuit of UJT

Voltage across (R_{B1}) $V_{RB1} = \frac{R_{B1}}{R_{B1}+R_{B2}} \times V_{BB}$

$$V_{RB1} = \eta \, V_{BB}.$$

η is called as the intrinsic stand off ratio of UJT and is defined as

$$\eta = \frac{R_{B1}}{R_{B1}+R_{B2}}$$

or $$\eta = \frac{R_{B1}}{R_{BB}}$$

As soon as the emitter voltage V_E is greater than peak voltage, the diode is forward biased and the emitter current will start flowing.

The emitter voltage at which the UJT turns ON is called as the peak voltage.

$$V_P = \eta \, V_{BB} + V_D$$

For all value of V_E below V_P the diode D will remain reverse biased .Hence UJT is in the off state.

8.5.5 UJT Characteristics

The UJT characteristics are as shown in figure 8.18. This is emitter voltage versus emitter current characteristics.

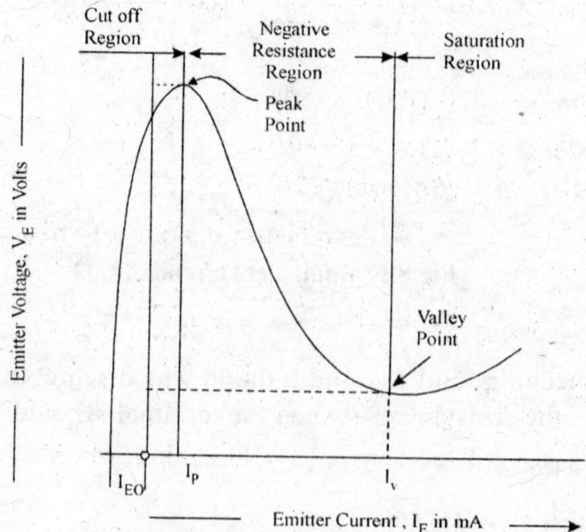

Fig.8.18 UJT characteristics

For the emitter potential less than V_p the UJT is in the OFF state and current I_E is not greater than I_{EO}. The current I_{EO} corresponds much closed with the reverse leakage current I_{CO} of a BJT. This region is known as the cutoff region. As the emitter potential increases & reaches $V_P = \eta V_{BB} + V_D$, the UJT starts conducting. Then with increase in emitter current I_E the emitter voltage decreases as show in figure. The reduction in voltage across UJT is due to the drop in resistance R_{B1} with increase in the value of I_E. The region of operation is known as the negative resistance region. Eventually the valley point will be reached and a further increase in I_E will place the device into saturation.

8.6 UJT Triggering of an SCR

The UJT has one common application for the triggering of the SCR. It is also known as the triggering circuit as shown in figure 8.19.

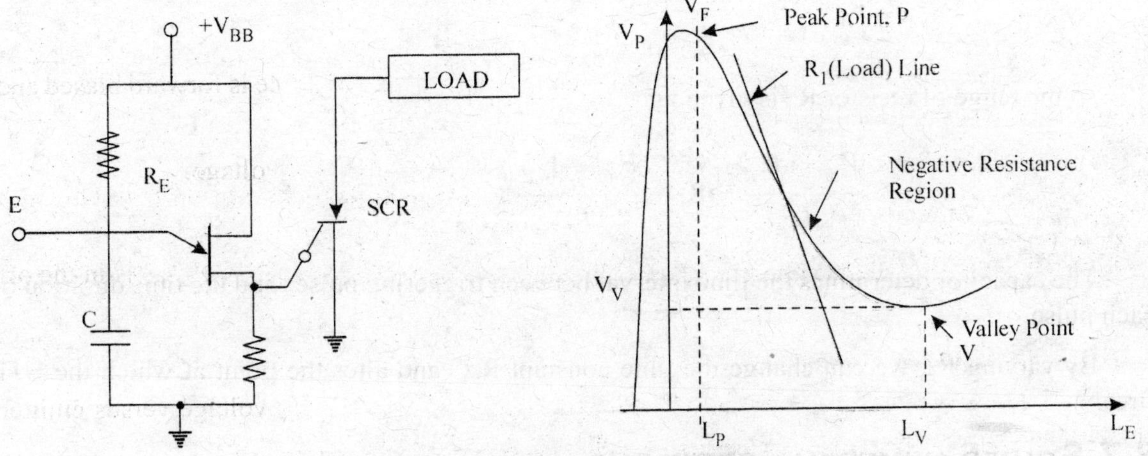

Fig.8.19 UJT triggering of an SCR

The resistor R_E is chosen so that the load line determined by R_E passes through the negative resistance region. If the load line does not pass to the right of the peak point, the device cannot turn on. For ensuring turn on of UJT

$$R_E < \frac{V_{BB} - V_P}{I_P}$$

Consider the peak point at which $I_{RE} = I_P$ & $V_E = V_P$, then

$$V_E = V_{BB} - I_{RE} R_E$$

$$\text{So } R_E(\text{max}) = \frac{V_{BB} - V_P}{I_{Re}}$$

At peak point, R_E (max) $= \frac{V_{BB} - V_P}{I_P}$ (at peak point)

Or $R_E < \frac{V_{BB} - V_P}{I_P}$

At valley point $I_E = I_V$ and $V_E = V_V$

R_E (min) = or R_E (min) =

For ensuring turn-OFF

$$R_E > \frac{V_{BB} - V_P}{I_P}$$

so the range of resistor R_E is given as

$$\frac{V_{BB} - V_P}{I_P} > R_E > \frac{V_{BB} - V_P}{I_P}$$

The capacitor determines the time interval between triggering pulses and the time duration of each pulse.

By varying R_E, we can change the time constant $R_E C$ and alter the point at which the UJT fires.

8.7 Solved Numerical Problems

Problem 1 : The intrinsic stand-off ratio for a UJT is determined to be 0.6. If the inter base resistance is 40 KΩ. What are the values of R_{B1} and R_{B2} ?

Solution: Given $R_{BB} = 40$ K&! $\eta = 0.6$

$R_{BB} = R_{B1} + R_{B2}$

$$\eta = \frac{R_{B1}}{R_{B1} + R_{B2}}$$

$0.6 = R_{B1} = 40 \times 0.6 = 24$ K&! $R_{B2} = 40 - 24 = 16$ K&!

Problem 2 : A UJT has 10V between the bases. If the intrinsic stand off ratio is 0.75, find the value of stand off voltage. What will be the peak point voltage if the forward voltage drop in the PN junction is 0.7V?

Solution : Given $V_{BB} = 10V, \eta = 0.75$ and $V_D = 0.7V$

Stand off voltage $= \eta \ V_{BB} = 0.75 \times 10 = 7.5V$

Peak point voltage $V_P = \eta \ V_{BB} + V_D = 8.2V$

8.8 Summary

1. Thyristor, stands for thyration transistor, because the devie combines the rectification action of thyratron and control action of transistor.
2. A SCR is a semiconductor device that perform the task of an electronic switch.
3. SCR is a four layer (p-n-p-n), three junction semiconductor device with three terminals, namley, the anoda, the cathode and the gate.
4. A triac (triode) is a three terminal semiconductor switching device which can control alternating current or can conduct current in either direction.
5. A diac is a two terminal , three layer bidirectional device which can be switched from its off state to ON state for either polarity of applied voltage.
6. The UJT is called as unijunction device as there is ony are junction present.
7. UJT shows the negative resistance region in the V-I characteristics. This negative resistance region is used to design firing circuit using UJT.

Review Questions

1. Explain the construction, working and characteristics of SCR.

2. Explain the construction, working and characteristics of Triac.

3. Explain the construction, working and characteristics of diac.

4. Explain the construction, working and characteristics of UJT.

5. Explain UJT triggering of an SCR.

6. (a) Describe the method to turn -on and trun-off the bidirectional thyristor.Give detail of their applcation areas.

(b) Describe the equivalent circuit and characteristics ounijunction transistor(UJT).

7. (a) Explain early effect with suitable plots of potential and minority concentration througout the base region.

(b) Disuss two possible sorces of breakdowan in a transistor as the collector to emitter voltge is increased.

❑❑❑

UNIT-5

Chapter 9

Communication Systems

Chapter 9

Communication Systems

9.1 Introduction

In its basic sense, the term communication refers to the sending, receiving and processing of information by electric ways. It started with wire telegraphy in the eighteen-forties, developing with telepony in same decades later and radio at the beginning of this century.

Radio communication was made possible by the invention of the triode tube. It became more widely used and refined through the invention of transistor, ICs and other semiconductor devices. The satellites and fiber optics has made communication even more widespread.

The modern communication systems are first concerned with the sorting, processing, storing and transmitting the information. The actual transmission then follows, with further processing and filtering of noise. Finally, at the reception side, processing steps include the decoding, storage and interpretation.

In a modern communication system number of communication forms such as radio telephony, telegraphy, broadcasting, point to point, mobile communication, radar, radio telepmertry and radio aids to navigation are used.

The basic block diagram of a communication system includes number of blocks like information source, transmitter receiver etc. Block diagram of communication system is shown in fig. 9.1

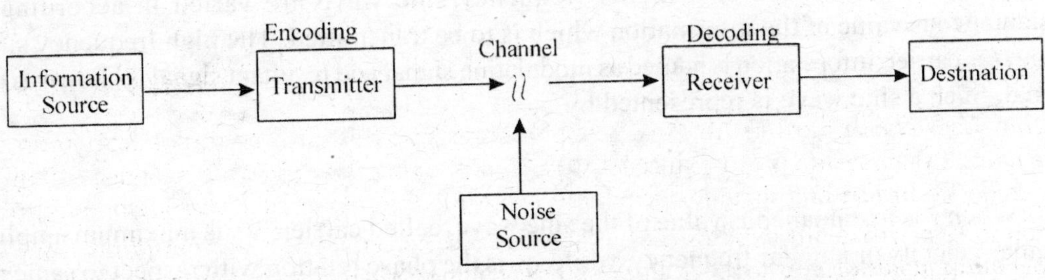

Fig. 9.1 Block diagram of Communication System

9.2 Basic Definitions

Before designing the communication system, we have to define some important terms which are mostly used in communication systems.

9.2.1 Information

The communication systems are designed to communicate a message. Information itself is that which is conveyed. The amount of information contained in any given message is measured in bits. The set, or total number of messages, consists of individual messages which may be differentiate from one to other. These may be words, groups of words, code symbols or any other prearranged units. The amount of information depends on the number of combinations made. The greater the total number of possible selections, the large amount of information conveyed.

9.2.2 Redundancy

Redundancy is additional information which is added to the actual information. The meaning of this information does not matter; only the quantity is important. It must be realized that no actual information is conveyed by a redundant message. Redundancy is not wasteful under every circumstances. Redundancy has its obvious use in teaching, entertainment and any appeal to the emotions. It also helps a message to remain unaffected under noisy environment.

9.2.3 Noise

Noise may be defined as any unwanted form of energy tending to interface with the proper and easy reception and reproduction of actual information. Noise is seen as limiting range of the systems. Noise affects the sensitivity of receivers by placing a limit on the weakest signals that can be amplified.

There are numerous ways of classifying noise. It may be subdivided according to type, source, effect or relation to the receiver, depending on circumstances. It is convenient to divide noise into two groups (i) noise whose sources are external to the receiver and (ii) noise created within the receiver itself. Examples of external noise are atmospheric noise, extra terrestrial noise and industrial noise etc. and examples of internal noise are thermal agitation noise, shot noise, transit time noise and miscellaneous noise etc.

9.3 Modulation

Same characteristics of a higher-frequency sine wave are varied in according with the instantaneous value of the information which is to be transmitted. The high frequency sine wave is known as carrier, information is named as modulating signal and resultant signal is known as modulated signal. Such a sine wave is represented by

$$v = V_m \sin(\omega t + \Phi)$$

When v is instantaneous value of the sine wave, called carrier; V_m is maximum amplitude, ω is angular velocity or angular frequency, while Φ is the phase relation with respect to same reference. Any of these three characteristics (V_m, ω and Φ) of the carrier may be varied by the modulating signal giving rise to amplitude, frequency or phase modulation respectively.

9.4 Need for Modulation

In radio transmission the audio signal is to be transmitted over long distance from broadcasting station to a receiver. To transmit the actual information, successfully and efficiented, it is necessary to convert the actual information into an electrical signal. This electrical signal may not be transmitted as it is, but is must be transmitted after the modulation. The following factors are responsible for the need of modulation.

(a) Practical Antenna Length : The length of transmitting and receiving antenna is given by

$$l = \frac{\lambda}{4} = \frac{1}{4}\frac{\text{Velocity (in meter per sec)}}{\text{frequency (in Hz)}} \qquad .....(9.1)$$

For audio range (20 Hz to 20kHz) minimum antenna length can be calculated as

$$l = \frac{3\times10^8}{4\times20\times10^3} = 3750 \text{ meters.}$$

A vertical antenna of this size is unpredictable.

If a carrier signal of 1 MHz is used to carry the signal, we need an antenna of about 75 mts. that can be easily constructed.

(b) Operating Range : Audio signals are sent to large distance in radio transmission. The energy of a wave is directly proportional to its frequency. The signal power is very small at audio frequencies (20 Hz to 20KHz). Therefore we can not feed audio signal directly to the antenna for communication purpose. The radiation of electrical energy is possible only at high frequencies (>20 KHz). The high frequency signals can be sent to thousand of kilometers even with small power.

(c) Separation of various signals : All the audio signals are concentrated with in the range from 20 Hz to 20KHz, so that all signals from the different sources would be mixed up. In order to separate the various signals, it is necessary to translate them all to different portion of the spectrum.

(d) Bandwidth requirements : The frequency range required for a given transmission should depend on the badwidth occupied by the modulation signals themselves. Before calculating the bandwidth of the communication system, it is necessary to know the bandwidth occupied by the modulating signal itself.

(e) Reduction of Noise : Noise is the main problem of any communication system. With the help of some modulation techniques, the effect of noise can be minimized.

9.5 Types of Modulation

Any sinusoidal carrier signal can be represented by the equation

$$v = V_m \sin(\omega t + \phi) \quad ....(9.2)$$

The amplitude V_m, frequency ω and phase ϕ may be known as the characteristic features of a sinusoidal signal. In the process of modulation, one of these three characteristics of the carrier is permitted to vary in accordance with the modulating signal, therefore, three basic modulation methods are possible, namely

(i) Amplitude Modulation (AM)

(ii) Frequency Modulation (FM)

and (iii) Phase Modulation (PM)

9.6 Amplitude Modulation (AM)

In amplitude modulation (AM) the amplitude of a carrier signal is varied by the modulating signal, whose frequency is lower than that of the carrier.

In other words **amplitude modulation** is defined as a system of modulation in which the amplitude of carrier is made proportional to the instantaneous amplitude of the modulating signal. This is important to note that the amplitude of the high frequency signal, named as carrier, is varied whereas frequency and phase of the carrier remains unchanged.

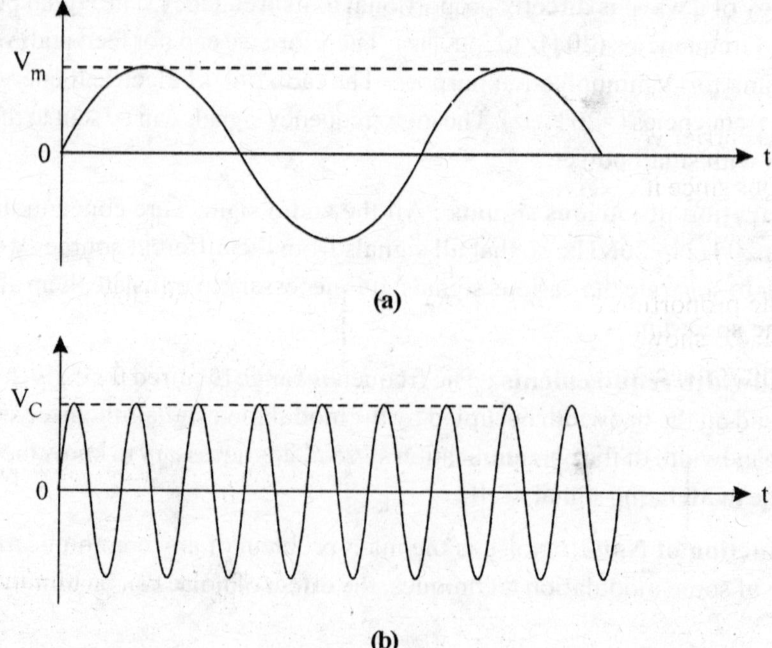

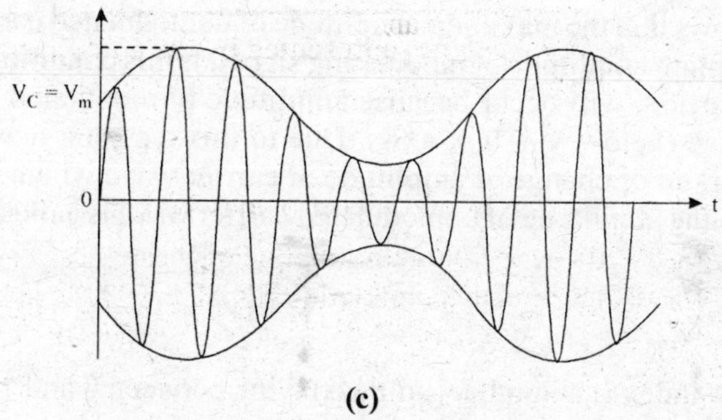

(c)

Fig. 9.2 Base band signal, (b) Carrier signal, (c) Amplitude modulated signal

One cycle of the modulated sine wave is shown in fig. 9.2(c). This lies between two limiting envelopes and has a repetition rate equal to the unmodullated carriers frequency. Amplitude of both positive and negative half cycles of carrier wave are changed in accordance with the modulating signal (Information) for the positive cylcle of modulating signal amplitude of the carrier signal increases, on the other hand, during negative cycle of modulating signal amplitude of the carrier signal decreases.

9.6.1 Modulation Index (M)

In amplitude modulation the amplitude of the carrier signal changes according to the intensity of the information (base band signal or modulating signal)

Let the carrier and modulating signals are represented by

$$v_C = V_C \sin \omega_C t \quad ....(9.3)$$

and

$$v_m = V_m \sin \omega_m t \quad .....(9.4)$$

Here v_C is carrier voltage and v_m is modulating voltage. Phase angel is ignored in both the expressions since it remains as it is by amplitude modulation process.

From the definition of amplitude modulation, the amplitude of carrier signal (V_c) will have to be made proportional to the instantaneous modulating voltage $V_m \sin \omega_m t$ when carrier is amplitude modulated. The proportionality constant is made unity for AM and superimposed onto the carrier amplitude. Fig. 4.3 shows the variation in maximum amplitude of the amplitude modulated voltage with modulating voltage.

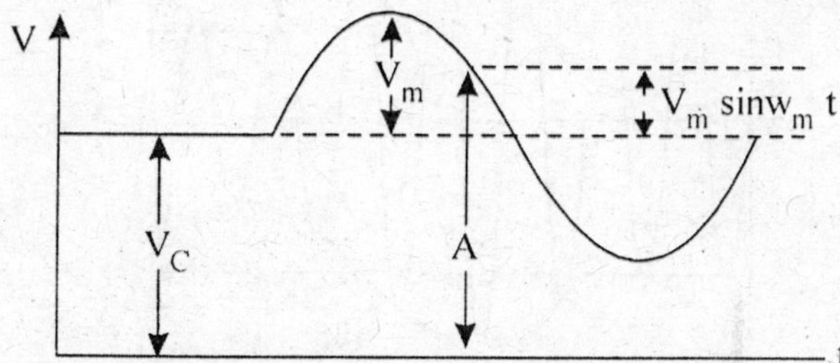

Fig. 9.3: Amplitude of AM wave

Fig. 9.3 also shows that the maximum amplitude of unmodulated carrier signal (V_c) must always be greater than maximum amplitude of modulating signal. If this condition is not satisfied, something unusual named distortion, will occur because amplitude of modulated carrier signal will move to negative side of curve (below V = 0, x-axis). Due to this reason a new term modulation index is defined. Hence, the ratio of change of amplitude of carrier wave which is equal to the amplitude of modulating signal to the amplitude of unmodulated carrier wave is called modulation index m,

$$m = \frac{V_m}{V_C} \quad \text{....(9.5)}$$

The modulation index is a number (unitless) lying between 0 and 1 and it is often expressed as a percentage and called the percentage modulation.

Mathematical Analysis of Modulation Index :

From the equations (9.3), (9.4) and Fig. 9.3 amplitude of the amplitude modulated voltage is given as

$$A = V_C + v_m = V_C + V_m \sin \omega_m t \quad \text{....(9.6)}$$

From the definiton of modulation index $V_m = m V_C$, therefore

$$A = V_C + m V_C \sin \omega_m t$$

$$= V_C (1 + m \sin \omega_m t) \quad \text{.....(9.7)}$$

The instantaneous voltage of the resulting amplitude modulated wave is given as

$$v = A \sin \omega_C t$$

$$= V_C (1 + m \sin \omega_m t) \sin \omega_C t \quad \text{....(9.8)}$$

This equation of AM wave for one cycle is pictorially represented in Fig. 9.4. This shows the maximum positive amplitude is given as $A = (V_C + V_m \sin \omega_m t)$. Similarly maximum negative amplitude as given as $-A = -(V_C + V_m \sin \omega_m t)$. The modulated wave extends between these two limiting envelops and has a repetition rate equal to the unmadulated carrier frequency.

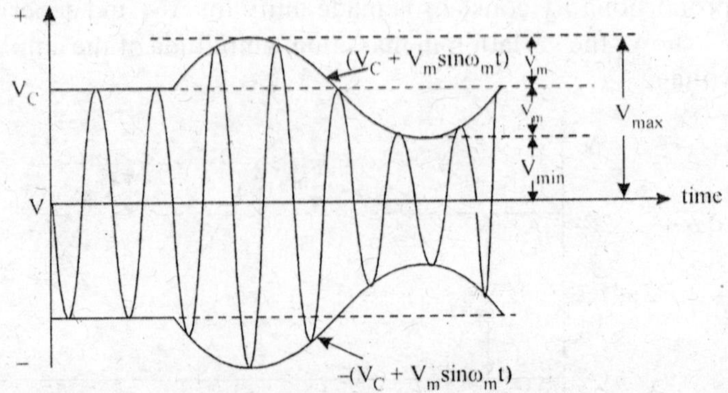

Fig. 9.4 Amplitude Modulated

Now consider the amplitude modulated waveform as shown in Fig. 9.4.

$$2V_m = V_{max} - V_{min}$$

$$\Rightarrow \qquad V_m = \frac{V_{max} - V_{min}}{2} \qquad \text{....(9.9)}$$

Similarly $V_C = V_{max} - V_m = V_{max} - \frac{V_{max} - V_{min}}{2}$

$$= \frac{V_{max} + V_{min}}{2} \qquad \text{.....(9.10)}$$

Dividing equation (9.9) by equation (9.10), we have

$$m = \frac{V_m}{V_C} = \frac{(V_{max} - V_{min})/2}{(V_{max} + V_{min})/2} = \frac{V_{max} - V_{min}}{V_{max} + V_{min}} \qquad \text{...(9.11)}$$

Equation (9.11) ensures that modulation index m is less than unity.

9.6.1 Significance of Modulation Index

The modulation index plays an important role in determining the strength and quality of signal transmitted. When modulation index has small value (m « 1), the amount of carrier amplitude variation is small, as a result of this, information transmitted at transmitter side is also week and proper (undistorted) information does not reach to the receiver side. As the degree of modulation is increased the information signal becomes stronger and clear. But if m>l i.e. the carrier wave is overmodulated, distortion occurs during reception. This is already discussed in section 9.6.1 with the help of fig. 9.3. Thus we can say modulation index m is used to describe the depth of modulation. It also shows the extent to which the amplitude of carrier wave has been vaied by the information or modulating signal.

9.6.2 Frequency Spectrum Representation of AM

The instantaneous voltage of the amplitude modulated wave as given in equation (4.8) may be rewritten as

$$v = V_C(1 + m\sin\omega_m t)\sin\omega_C t$$

$$= V_C\sin\omega_C t + mV_C\sin\omega_C t\cdot\sin\omega_m t = V_C\sin\omega_C t + \frac{mV_C}{2}\left[2\sin\omega_C t\cdot\sin\omega_m t\right]$$

Apply the trigonometrical formula

$$2\sin A\sin B = \cos(A - B) - \cos(A + B)$$

Thus $\qquad v = V_C\sin\omega_C t + \frac{mV_C}{2}\cos(\omega_C t - \omega_m t) - \frac{mV_C}{2}\cos(\omega_C t + \omega_m t) \qquad \text{.....(9.12)}$

The equation (9.12) contains three components. First component represents the unmodulated carrier signal. It is important to note here that the process of amplitude modulation has the effect of adding to the unmodulated wave, rather than changing it. Two other components produced are the two side bands. The frequency of lower side band (LSB) is $f_C - f_m$ and the frequency of upper side band (USB) is $f_C + f_m$. **For an example,** let the modulating and carrier frequency are 1 KHz and 1000 KHz respectively. Therefore, the frequency of LSB will be 999 KHz and that of USB will be 1001 KHz, each of which is very close to the carrier frequency 1000 KHz.

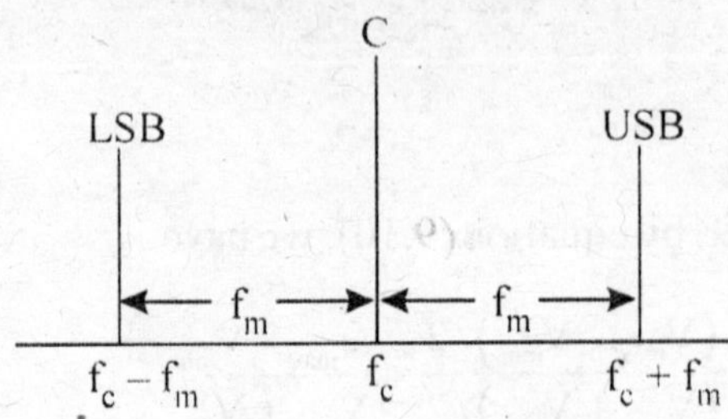

Fig. 9.5 Frequency spectrum of amplitude modulation wave

Bandwidth : The bandwidth of amplitude modulated waveform is defined as the difference between USB and LSB from the fig. 9.5, it is always equal to $2f_m$. Therefore, the bandwidth in amplitude modulation will be $2f_m$, which is twice of the modulating signal frequency.

Thus in above considered **example**, bandwidth is (1001-999) KHz or 2KHz. Which is twice the modulating signal frequency 1KHz.

9.6.3 Power Relations in the AM Wave

It has been found that the amplitude modulated signal contains three components. Modulated signal contains two additional side bands with the carrier signal. It is obvious, therefore, that the modulated signal contains more power than the carrier had before modulation. Since the amplitude of the sidebands depends on the modulation index, it is anticipated that the total power in the modulated signal will depend on the modulation index also.

The total power in the modulated signal is given by

$$P_{taks} = P_{carrier} + P_{LSB} + P_{USB} \qquad \text{....(9.13)}$$

The carrier power $(P_{carrier})$ may be calculated as

$$P_{carrier} = \frac{V_{carrier}}{R}$$

where $V_{carrier}$ is rms value and R is the resistance (e.g. antenna resistance) in which power is dissipated.

$$P_{carrier} = \frac{\left(\frac{V_c}{\sqrt{2}}\right)^2}{R} = \frac{V_C}{2R} \quad \text{....(9.14)}$$

Similarly

$$P_{LSB} = P_{USB} = \frac{V_{SB}^2}{R}$$

$$= \frac{\left[\frac{\left(\frac{m V_C}{2}\right)}{\sqrt{2}}\right]^2}{R} = \frac{m^2 V_C^2}{8R} \quad \text{....(9.15)}$$

Substitute equation (9.14) and (4.15) in equation (9.13)

$$P_{total} = \frac{V_C^2}{2R} + \frac{m^2 V_C^2}{8R} + \frac{m^2 V_C^2}{8R}$$

$$= \frac{V_C^2}{2R}\left(1 + \frac{m^2}{4} + \frac{m^2}{4}\right)$$

$$= P_{carrier}\left(1 + \frac{m^2}{2}\right)$$

$$\frac{P_{total}}{P_{carrier}} = 1 + \frac{m^2}{2} \quad \text{....(9.16)}$$

Maximum power in the AM wave is $P_{total} = 1.5\ P_{carrier}$ when m = 1. This is important, because it is the maximum power that relevant amplifiers must be capable of handling without distortion.

Modulation by several sine waves

Under the several sine waves modulation, the resulting power is to be found. The procedure consists of calculating the total modulation index and then substituted it into eq. (9.16) to calculate the total power.

These are two methods of calculating the total modulation index.

(a) Let K,, V_2, F_3 etc. be the simultaneous modulation voltages. Then the total modulating voltage V_t will be equal to the square root of the sum of the square of the individual voltages; that is

$$V_t = \sqrt{V_1^2 + V_2^2 + V_3^2 + ...}$$

Dividing both sides by V_C, we have

$$\Rightarrow \quad \frac{V_t}{V_C} = \frac{\sqrt{V_1^2 + V_2^2 + V_3^2 + ...}}{V_C}$$

$$\Rightarrow \quad \frac{V_t}{V_C} = \sqrt{\left(\frac{V_1}{V_C}\right)^2 + \left(\frac{V_2}{V_C}\right)^2 + \left(\frac{V_3}{V_C}\right)^2 + ...}$$

$$\Rightarrow \quad m_t = \sqrt{m_1^2 + m_2^2 + m_3^2} \qquad .....(9.17)$$

(b) Equation (9.16) may be rewritten on emphaisze that the total power in an AM wave consists of carrier power and sideband power.

$$P_{total} = P_{carrier}\left(1 + \frac{m^2}{2}\right)$$

$$= P_{carrier} + \frac{m^2 P_{carrier}}{2}$$

The total side band power is defined is

$$P_{SB} = \frac{m^2 P_{carrier}}{2}$$

If serval sine waves simultaneously modulate the carrier, the carrier power will be unaffected, but the total sideband power will now be the sum of the individual sideband power.

$$P_{SB,t} = P_{SB,1} + P_{SB,2} + P_{SB,3} + ...$$

$$\Rightarrow \quad \frac{m_t^2 P_{carrier}}{2} = \frac{m_1^2 P_{carrier}}{2} + \frac{m_2^2 P_{carrier}}{2} + \frac{m_3^2 P_{carrier}}{2} + ...$$

$$\Rightarrow \quad m_t^2 = m_1^2 + m_{2_1}^2 + m_3^3 .....$$

$$\Rightarrow \quad m_t = \sqrt{m_1^2 + m_2^2 + m_3^2} \qquad ....(9.18)$$

It is seen that the two approaches gives the same result. This is important to note that the total modulation index must still not exceed unity, or distortion will result as with overmodulation by a single sine wave.

9.6.4 Generation of Amplitude Modulation

The electronic circuit which can generate an amplitude modulated signal is known as AM modulatator. The square law modulator is the most popular AM modulator among the number of AM modulatrors. The circuit of square law modulator consists of a diode and a series resistor R along with a shunt band pass filter.

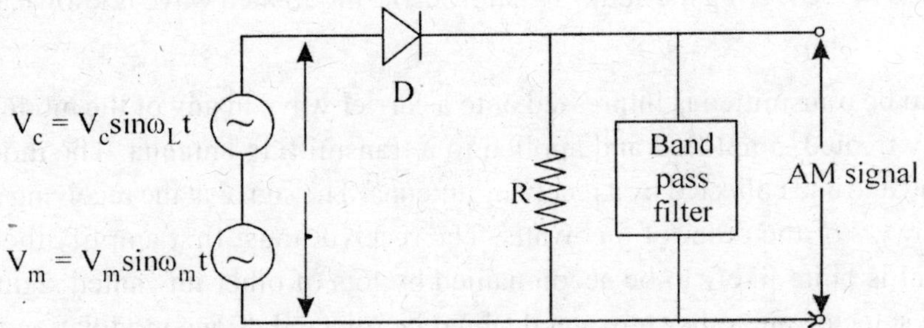

Fig. 9.6 : Square law modulator

The working principle of square law modulator is based upon the results of first steps of Taylor's series expansion. According to Taylor's series, the current and voltage may be written as

$$i = A_1 v + A_2 v^2 \quad \text{....(9.19)}$$

where $v = v_C + v_m$

$$= V_C \sin\omega_C t + V_m \sin\omega_m t \quad \text{....(9.20)}$$

Therefore,

$$i = A_1 \left(V_C \sin\omega_C t + V_m \sin\omega_m t\right) + A_2 \left(V_C \sin\omega_C t + V_m \sin\omega_m t\right)^2$$

$$= A_1 V_C \sin\omega_C t + A_1 V_m \sin\omega_m t + A_2 V_C^2 \omega_C t$$

$$+ A_2 V_m^2 \sin\omega_m t + 2A_2 V_C V_m \sin\omega_C t \sin\omega_m t$$

$$= A_1 V_C \sin\omega_C t + A_1 V_m \sin_m t + A_2 \left[V_C^2 \left(\frac{1 - \cos^2 \omega_C t}{2} \right) \right.$$

$$\left. + V_m^2 \left(\frac{1 - \cos^2 \omega_m t}{2} \right) \right] + A_2 V_C V_m \left[\cos(\omega_C - \omega_m)t - \cos(\omega_C + \omega_m)t \right]$$

$$= \frac{A_2}{2}\left(V_C^2 + V_m^2\right) + A_1 V_m \sin\omega_m t - \frac{A_2 V_C^2}{2} \cos 2\omega_C t - \frac{A_2 V_m^2}{2} \cos 2\omega_m t$$

$$+ A_1 V_C \sin\omega_C t + A_2 V_C V_m \cos(\omega_C - \omega_m)t - A_2 V_C V_m \cos(\omega_C + \omega_m)t$$

The last three terms of the equation (9.21) representing the amplitude modulated wave. The remaining four terms are of no use and can be filtered out by appropriate band passfilter. The band pass will pass the carrier, *LSB* and USB but rejects other frequencies.

9.6.5 AM Demodulator

The process of recovering the audio signal from the modulated wave is known as **demodulation or detection.**

A signal to be transmitted is impressed onto a carrier wave in any of the modulation method. It is then suitably treated, amplified and applied to a transmitting antenna. The modulated signal is radiated, propagated and collected by a receiving antenna. The signal at the receiving point is generally quite weak, powers of the order of picowatts. The receiver must first amplify the received signal. Since the signal is quite likely to be accompanied by lots of other unwanted signals. The original signal must be selected and other unwanted signal be rejected. Since modulation took place in the transmitter, the reverse process of demodulation must be performed in the receiver to recover the original modulating voltages.

Diode Detector is a commonly used detector having simplest principle of operation as shown in fig. 9.7. The detector circuit consists of tank circuit along with the diode, resistor and capacitor.

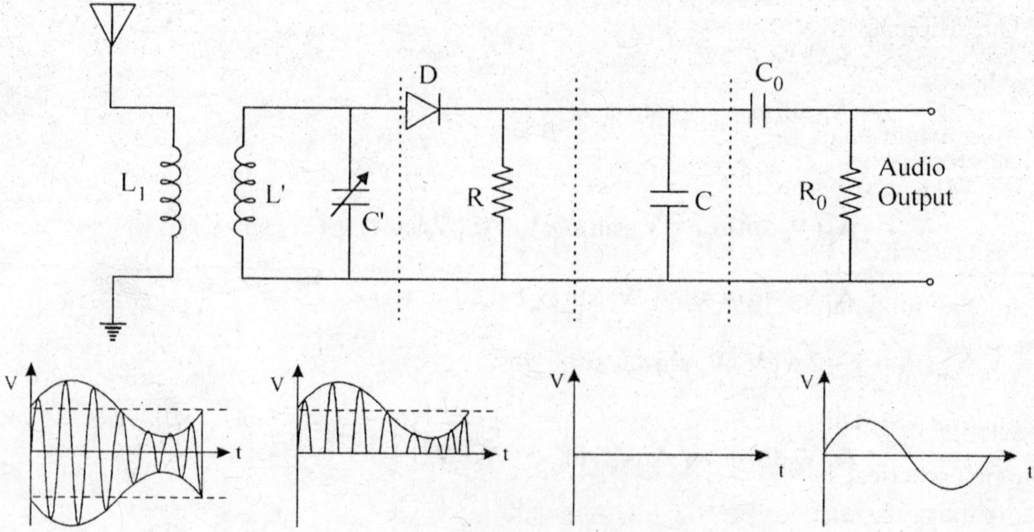

Fig. 9.7 Diode Detector

At the input stage of the detector, the AM wave of desired frequency can be selected using parallel tuned tank circuit. By varying C, the resonant frequency of L'C' circuit may be changed. With the help of this, the modulated wave of desired frequency may be selected. The received signals are amplified (amplifier not shown in fig.9.7) and then applied to the input of diode detector. During the positive half cycle, the diode (D) conducts, while during the negative half cycle, it does not conduct. Due to this rectifying action of the diode, output of the diode detector is only positive cycle of modulated signal as shown in Fig. 9.7.

The output of detector (across resistor R) contains high frequency carrier which can be filtered out using capacitor C, which bypass the high frequency carriers but provide very high reactance to the audio signal. The capacitor charges when the carrier signal amplitude increases and then discharges through resistor R. The value of R and C must be chosen in such a way that during discharge the capacitor voltage follows the wave shape of audio signal.

The output of capacitor filter has some pulsing variation due to presence of dc component. These variation can be eliminated in the output stage with the help of capacitor Co. The capacitor Co offers infinite reactance $\left(X_C = \frac{1}{2\pi f C_0} \text{ with } f = 0 \text{ for dc} \right)$ to the de component and thus stop it. Therefore, the output signal at the output stage has a wave shape very close to the original modulating signal. Now this signal is passed through a speaker, the original sound is reproduced.

9.6.6 Limitations of Amplitude Modulation

There are few limitations of amplitude modulation as follows :

(i) The efficiency of AM system is low.

(ii) AM system has limited operated range.

(iii) The reception of signals is noisy in case of AM system.

(iv) In AM troad casting station audio quality is usually poor.

9.7 Frequency Modulation (FM)

Frequency modulation is a system in which the amplitude of the modulated carrier is kept constant, while its frequency is varied by the modulating signal. In concern of practical applications, frequency modulation is preferred over the amplitude modulation. However, the generation of FM is not as simple as AM. This is mainly because FM involves minute frequency variations of the carrier.

The first practical system was come in existence in 1936 as an alternative to AM is an effort to make radio transmission more resistant to noise.

In frequency modulation, the instantaneous frequency of carrier wave is varied in accordance with the instantaneous amplitude of the modulating signal however, the amplitude of the carrier remains constant. The amount of deviation in frequency of carrier signal is proportional to the amplitude of the signal as shown in fig.(9.8) when the modulating signal amplitude is zero, the carrier frequency is at its normal frequency/When the amplitude of modulating signal increases, frequency of carrier increases and carrier frequency decreases with decrement in modulating signal amplitude.

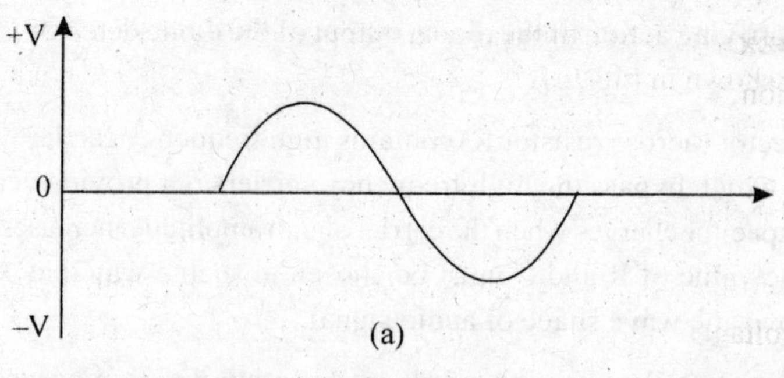

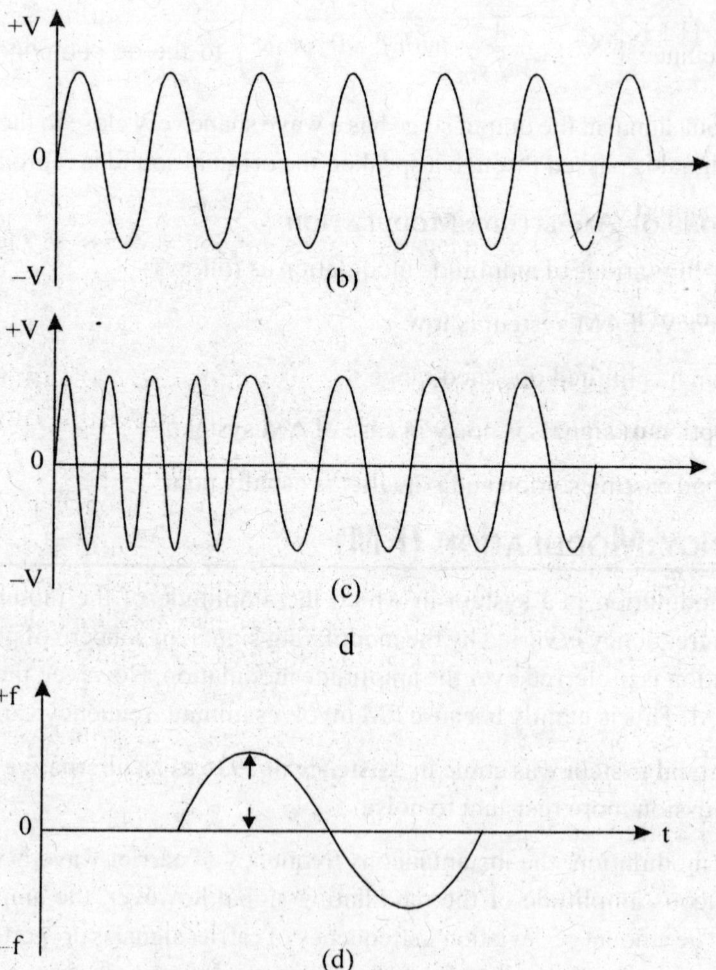

Fig. 9.8 (a) Modulating signal, (b) Carrier signal,

(c) Frequency modulated signal, (d) Frequency Vs time in FM

9.7.1 Modulation Index with Mathematical Representation of FM

In frequency modulation, the instantaneous frequency f or the frequency modulated wave is given by

$$f = f_C\left(1 + \mathrm{K}\,\mathrm{V}_m \cos\omega_m t\right) \qquad .....(9.22)$$

where f_C is unodulated carrier frequency, K is proportionality constant and $\mathrm{V}_m \cos\omega_m t$ is instantaneous modulating voltage.

The maximum deviation for this signal will occur when cosine term has its maximum value, that is ± 1. Under this condition

$$f = f_C\left(1 \pm \mathrm{K}\,\mathrm{V}_m\right) \qquad ....(9.23)$$

Therefore the maximum deviation δ will be given as

$$\delta = \mathrm{K}\,\mathrm{V}_m\, f_C$$

The instananeous amplitude of the FM signal will be given by ....(9.24)

$$v = \mathrm{A}\sin\theta$$

when A is amplitude of modulated signal which is constant for FM and θ is function of carrier signal frequency (f_C) and modulating signal frequency (f_m).

In order to find θ, ω must be integrated with respect to time

$$\theta = \int \omega\, dt = \int \omega_C\left(1 + \mathrm{K}\,\mathrm{V}_m \cos\omega_m t\right) dt$$

$$= \omega_C \int dt + \omega_C\, \mathrm{K}\,\mathrm{V}_m \int \cos\omega_m t\, dt$$

$$= \omega_C t + \frac{\mathrm{K}\,\mathrm{V}_m\, f_C \sin\omega_m t}{f_m} = \omega_C t + \frac{\delta}{f_m}\sin\omega_m t \qquad ....(9.26)$$

The ratio of maximum frequency deviation (δ) to modulating frequency (f_m) is defined as the modulation index (m_f) of frequency modulation.

$$m_f = \frac{\delta}{f_m} \qquad ....(9.27)$$

Substituting equation (9.27) into (9.26), we have

$$\theta = \omega_C t + m_f \sin\omega_m t \qquad ....(9.28)$$

Now substitute equation (9.28) into (9.25), we obtain

$$v = \mathrm{A}\sin\left(\omega_C t + m_f \sin\omega_m t\right) \qquad \text{....(9.29)}$$

It is important to note that as the modulating frequency (f_m) decreases and modulating voltage amplitude remains constant with constant carrier frequency, the modulation index (m_f) increases. Modulation which is the ratio of two frequency, is measured in radians m_f may have value more than unity.

9.7.2 Frequency Spectrum Representation of FM

The instantaneous voltage of the frequency modulated signal is given in equation (9.29) may be rewritten as

$$v = \mathrm{A}\sin\left(\omega_C t + m_f \sin\omega_m t\right)$$

Since this equations the sine of a sine, the only solution involves the use of Bassel functions. Using there equation may be expanded.

$$v = \mathrm{A}\{\mathrm{J}_0(m_f)\sin\omega_C t + \mathrm{J}_1(m_f)\left[\sin(\omega_C + \omega_m)t - \sin(\omega_C - \omega_m)t\right]$$

$$+\mathrm{J}_2(m_f)\left[\sin(\omega_C + 2\omega_m)t - \sin(\omega_C - 2\omega_m)t\right] + \mathrm{J}_3(m_f)$$

$$\left[\sin(\omega_C + 3\omega_m)t - \sin(\omega_C - 3\omega_m)t\right] + \ldots \qquad \text{......(9.30)}$$

It is seen that the output consists of a carrier and infinite number of pairs of sidebands, each preceded by J coefficients.

Bandwidth : A rule of thumb (Carson's rule) states that (as a good approximation) the bandwidth requried to pass on FM signal is twice the sum of the deviation and the highest modulating frequency. But it must be remembered that this is only an approximation. In practical cases, it gives a fair result if the modulation index is in excess of about 6.

9.7.3 Generation of Frequency Modulation

The prime requirement of a frequency modulation generator is a variable output frequency, with the variation proportional to the instantaneous amplitude of the modulating voltage. The subsidiary requirements are that the unmodulated frequency should be constant, and the deviation independent of the modulating frequency.

One method of FM generation is suggested that if either the capacitance or inductance of an LC tank circuit is varied, frequency modulation of some form will result. If this variation can be made directly proportional to the voltage applied to modulation circuits, true FM will be obtained. There are several controllable electrical and electronic phenomena which provide a variation of capacitance as a result of a voltage change. Basic reactance modulator may be designed with three terminal field effect transistor (FET) as shown in Fig.(9.9). It can be made inductive or capacitive by simple component change. The value of this reactance is proportional to the transconductance of the device, which in turn can be made to depend on the gate bias and its variations. Note that an FET is used in the explanation here for simplicity only. Identical reasoning would apply to a bipolar transistor or a vacuum tube.

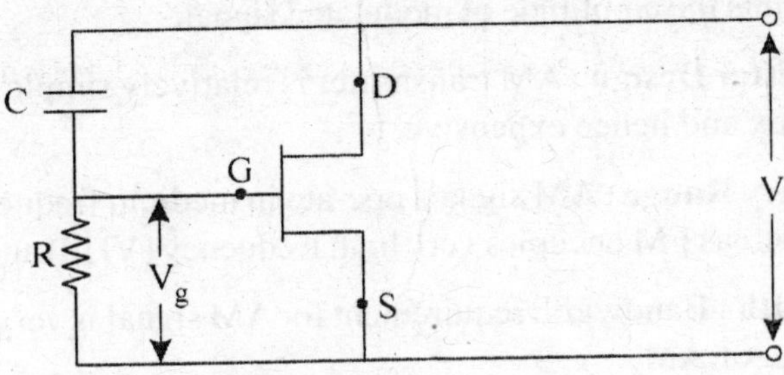

Fig. 9.9 Basic Reactance Modulator

9.7.4 Limitations of Frequency Modulation

Following are the limitations of frequency modulation.

(i) Frequency modulation equipments are complex and exapenoise

(ii) Bandwidth requirement for FM is large, hence wide channel is required.

9.8 Comparison of FM and AM

Frequency modulation and amplitude modulation are both practical systems, quite different from each other, and so the performance and characteristics of the two systems will be compared.

(i) **Amplitude :** Amplitude of the amplitude modulated signal varies in accordance with the modulating signal. This is govern by the modulation index of the system.

For FM signal amplitude remains constant and it is independent of modulation index. This is a great advantage of FM.

(ii) **Modulation Index :** For AM signal the maximum allowable value of modulation index is 1. For $m > 1$, reproduced signal will be distorted.

There is no restriction in the case of FM. It may have any value.

(iii) **Power :** In AM system, most of the power is confined in the carrier which is not useful as the modulating signal is available in sidebands. The average power in modulated wave is greater than carrier power. This added power is provided by modulating source.

In FM systems, most of the power is confined in the sidebands which contains the modulating signal and it is useful. The average power in frequency modulated wave is same as contained in the unmodulated wave.

(iv) Reception Quality : AM is more suseptible to noise interference, therefore, reception i poor. Signal to noise ratio is also low therefore it is more difficult to eliminate effects o noise.

FM reception is more immune to noise. It is also possible to reduce noise by increasin deviation δ by increasing modulation index m_f. Noise can also be minimized with th help of amplitude limits

which limits the amplitude of modulated signal.

(v) Transmitter Design : AM transmitter is relatively simple and cheap but FM transmitte is complex and hence expensive.

(vi) Frequency Range : AM signals operate in medium frequency (MF) and high frequency (HF), whereas FM occupies very high frequency (VHP) and ultra high frequency (UHF)

(vii) Bandwidth : Bandwidth requirement for AM signal is very small which is one of biggest advantage of AM.

FM requires much wider channel, due to a large number of side bands, upto 10-15 times as large as that needed by AM.

(viii) Mode of Propagation : Sky wave propagation is used for AM, therefore area of reception is large.

For FM, space wave propagation is used, therefore area of reception is limited to line of sight.

(ix) Interference : It is not possible to operate AM signals without interference while FM operates several independent transmitters on same frequency.

9.9 PHASE MODULATION (PM)

If the phase ϕ of the carrier signal is varied so that its magnitude is proportional to the instantaneous amplitude of the modulating voltage, the resulting wave is phase modulated.

The expression for the phase modulation is

$$v = A\sin\left(\omega_C t + \phi_m \sin \omega_m t\right) \qquad .....(9.31)$$

where ϕ_m is the maximum value of phase change introduced by this particular modulating signal and is proportional to the maximum amplitude of the modulation.

For the sake of uniformity, this is rewritten as

$$v = A\sin\left(\omega_C t + m_P \sin \omega_m t\right) \qquad ....(9.32)$$

where $m_P = \phi_m =$ modulation index for PM.

The expression for PM wave although that both FM and PM are closely allied, although both are distinct type of modulation system. The difference between FM and PM is quite straight forward. The modulation index is defined differently in each system.

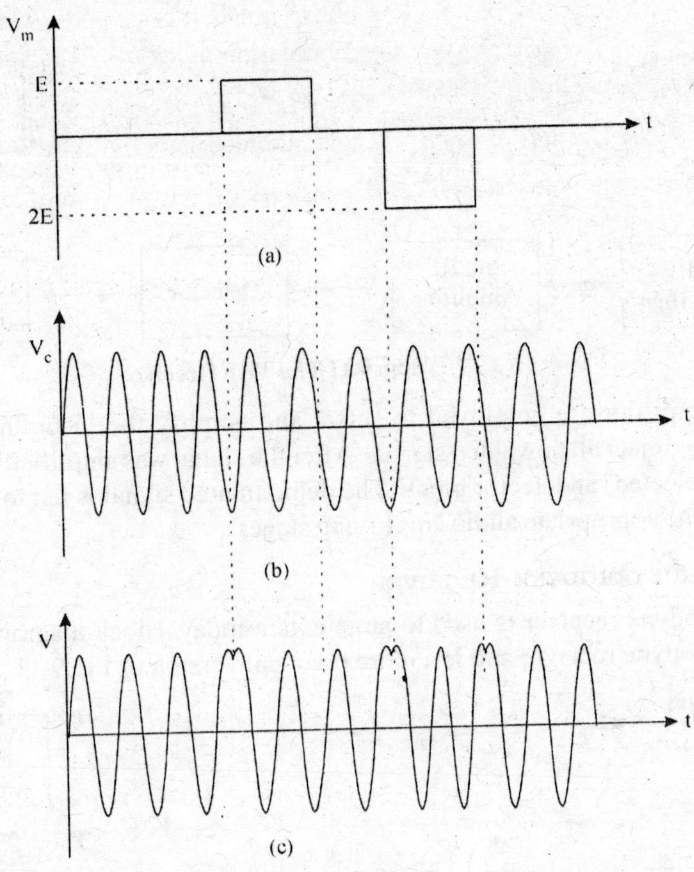

Fig. 9.10 (a) Modulating signal, (b) Carrier signal, (c) Phase modulated signal

In phase modulation, the phase deviation is proportional to the amplitude of the modulating signal and therefore independent of its frequency. In frequency modulation, the frequency deviation is proportional to the amplitude of the modulating voltage.

9.10 Radio Receiver

It is apparent that a receiver has the function of selecting the desired signal from all the other unwanted signals, amplifying and demodulating it, and displaying it in desired manner. In a very broad way receivers may be categorized as

(i) the tuned radio frequency (TRF) receiver and

(ii) the superheterodyne receiver.

9.10.1 Tuned Radio Frequency (TRF) Receiver

The TRF receiver is a simple receiver. *A.* layman with just a little knowledge of communication may expect all radio receivers to have this form. Basic block diagram of TRF receiver is shown in fig.9.11.

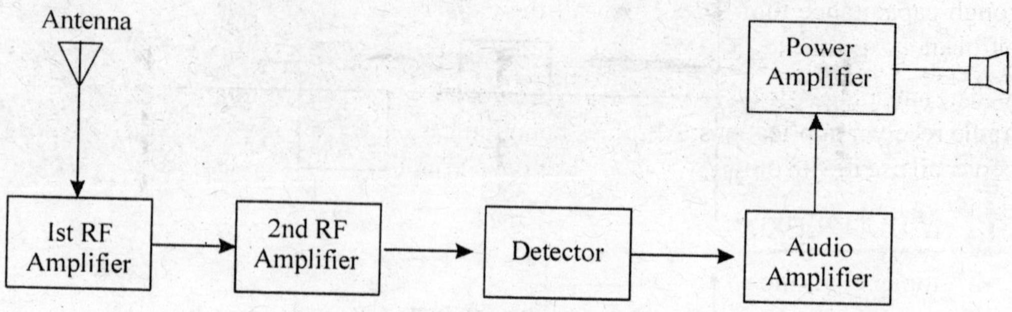

Fig. 9.11 The TRF Receiver

Two RF amplifier are employed to select and amplify the incoming signal frequency and simultaneously to reject all unwanted signals. After the signal was amplified to a suitable level, it was demodulated (detected) at detector block. The demodulated signal is fed to loudspeaker after being passed through the appropriate audio amplifying stages.

9.10.2 Superheterodyne Receiver

Superheterodyne receiver is used to large extant today. Block diagram of Fig.9.12 shows the basic superheterodyne receiver and is a more practical version of Fig.9.11

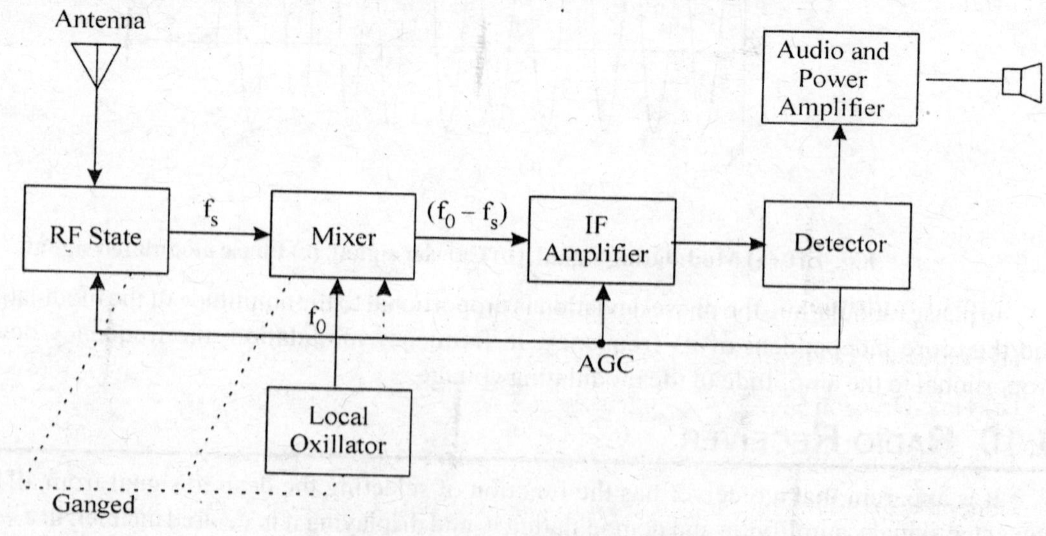

Fig. 9.12 The Superhetradyne Receiver

In the superheterodyne receiver, the signal voltage combined with the local escillator voltage and converted into a signal of a lower fixed frequency. The signal at this intermediate frequency (IF) contains the same modulation as the original carrier. It is now amplified at IF amplifier stage. Amplified signal is applied to detector to demodulate the received signal. The original signal is detected at detector stage. This signal is fed to loudspeaker after passing through the audio and power amplifier.

A constant frequency difference is maintined between the local escillator and the RF circuits through capacitance tuning in which all the capacitors are ganged together and operated by one control knob.

The advantages of superheterodyne receiver make it the most suitable type for the great majority of radio receiver applications; AM, FM, communications, signal sideband, television and even radar receiver all use it with only slight modification in principle.

9.11 MULTIPLEXING

Multiplexing is the process of simultaneously transmitting two or more individual signals over a single communication media. Due to multiplexing it is possible to increase the number of communication channel so that more information can be transmitted.

9.11.1 FUNDAMENTALS OF MULTIPLEXING

The concept of a simple multiplexer is shown in figure 9.11. It receives a large number of different input signals.Multiplexer has only one output which is connected to the single communication channel. The multiplexer combines all input signals into a single composite signal and transmit it over the communication medium.

At the receiving end of communication link, a demultiplexer is used to sort out the signal into their original form. Demultiplexing is a process which is exactly opposite to that of the multiplexing.

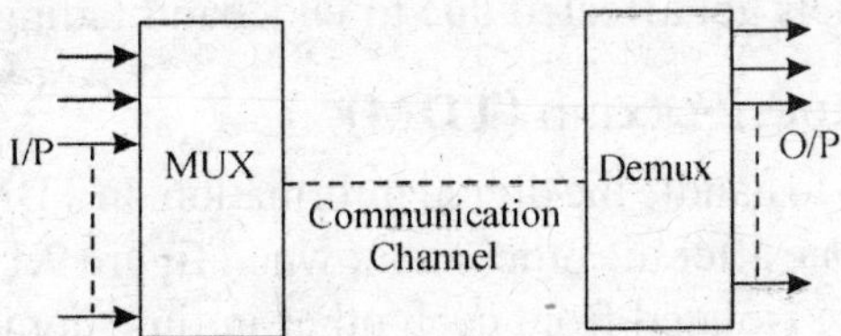

Fig. 9.13 Concept of Multiplexing

Type of multiplexing: - There are two basic types of multiplexing.

(i) Frequency Division Multiplexing (FDM)

(ii) Time Division Multiplexing (TDM)

9.11.2 FREQUENCY DIVISION MULTIPLEXING (FDM)

Generally the FDM systems are used to deal with the analog information. In FDM many signals are transmitted simultaneously where each signal occupies a different frequency slot within a common BW as shown in figure 9.14. The operation of FDM is based on sharing the available bandwidth of a communication channel among the signals to be transmitted. It means many signals are transmitted simultaneously with each signal occupying a different frequency slot within a common bandwidth.

Each signal to be transmitted modulates a different carrier. The Modulated signals are then added together to form a composite signal which is transmitted over a single channel.

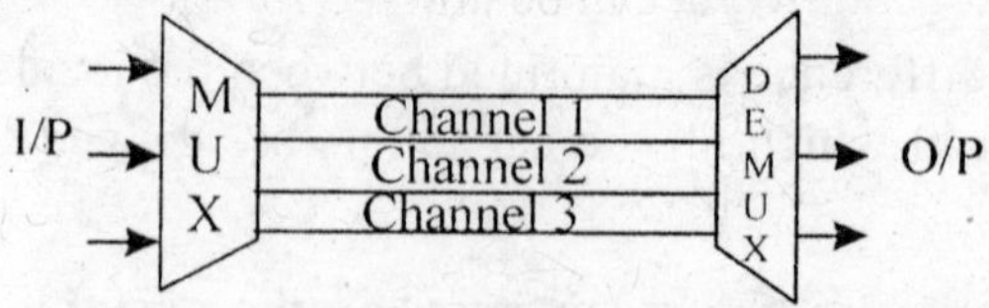

Fig. 9.14 Frequency Division Multiplexing

Advantages :

(i) A large number of signals can be transmitted simultaneously.

(ii) FDM does not need synchronization.

(iii) Demodulation of FDM is easy.

(iv) Due to slow narrow band fading only a single channel gets affected.

Disadvantages :

(i) The communication channel must have a very large bandwidth.

(ii) Inter modulation distortion takes place.

(iii) Large number of modulators and filters are required.

(iv) FDM suffers from the problem of crosstalk.

(v) All the FDM Channels get affected due to wideband fading.

9.11.3 Time Division Multiplexing (TDM)

The TDM system is used to handle the digital information. In TDM, the signals to be multiplexed are transmitted sequentially one after the other as shown in figure 9.15. Each signal occupies a short time slot. Thus the signals are isolated from each other in time domain but all of them occupy the same slot in the frequency spectrum. In TDM, the complete Bandwidth of the communication channel is available to each signal being transmitted. One frame corresponds to the time period required to transmit all the signals once on the transmission channel.

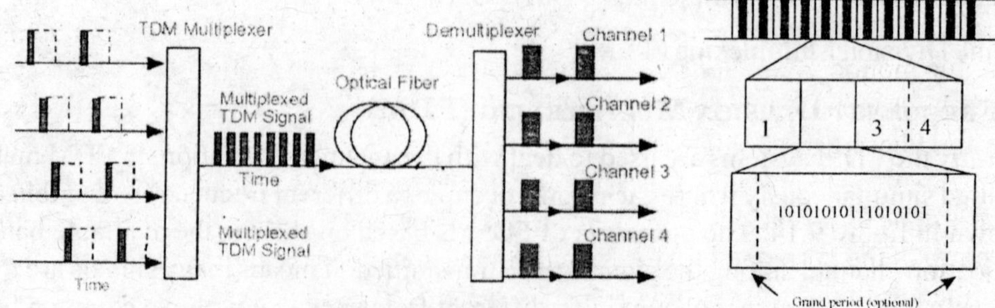

Fig. 9.15: Time Division Multiplexing

Advantages of TDM :

(i) Full available channel bandwidth can be utilized for each channel.

(ii) Inter modulation distortion is absent.

(iii) TDM circuitry is not very complex.

(iv) The problem of crosstalk is not served.

Disadvantages :

(i) Synchronization is essential for proper operation.

(ii) Due to slow narrowband fading, the entire TDM channel may get wiped out.

9.12 Introduction to Television

Television means seeing at a distance. To be successful, a television system may be required to reproduce faithfully:

(i) the shape of each object, or structural content

(ii) the relative brightness of each object, or tonal content

(iii) motion, or kinetic content

(iv) sound

(v) colour, or chromatic content

(vi) perspective, or stereoscopic content

If only the structural content of each object in a scene were shown, we would have truly black and white TV (without any shades of gray). If tonal content were added, we have black and white still pictures with shades for gray scale. Kinetic content and sound will add the motion and chromatic content sound in the picture. The chromatic content are not essential for black and white TV but required for colour televisions to fill the colors in picture. The stereoscopic content helps to make picture three dee.

A large amount of information must be transmitted by a television transmitter. Uniform standards are needed for TV transmission and reception. There are five essentially different television systems in use around the world. The two main ones are the American (Federal communications commission, **FCC** systems for monochrome and National Television Standards committee, **NTSC** system for colour) and the Europena (committee consulated if International de Radio, **CCIR,** system for monochrome and Phase Alternation by Line, **PAL,** system for colour) systems. In the two major TV systems, vestigial sideband amplitude modulation is used for video with most of the lower sideband removed. This is done to save bandwidth. Negative video modulation polarity is used for both the TV systems. The ratio of the horizontal to the vertical dimension of the receiver picture (or transmission camera) is defined as the aspect ratio. For the TV systems, it is 4 : 3.

The American system is used in the whole of North and South America with same exceptions, the European system is used by the rest of the world.

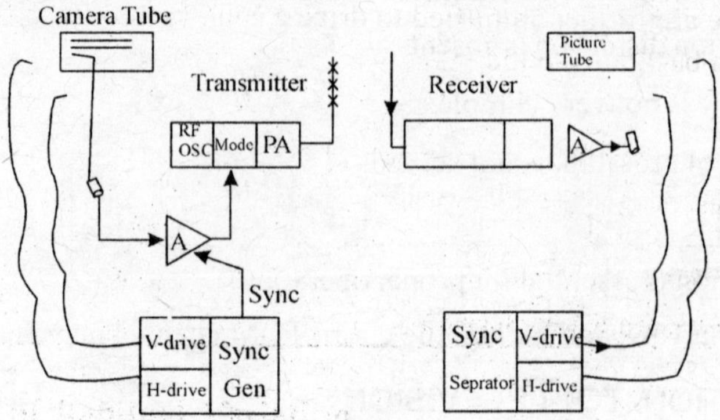

Fig. 9.16 Basic Television System

9.12.1 Transmitter (Sound and Picture)

For transmission of the sound over long distances, the sound waves are converted with the aid of a microphone into electrical signals so that they can be sent over wires as in a telephone. If these audio signals are translated into radio frequencies by modulation, they can be radiated into space by feeding the RF energy to suitable aerials.

The video signal from the camera pick-up tube, amplified and mixed with the blanking and synchronized signal from the synchronous pulse generator is supplied to the AM visual or picture transmitter.

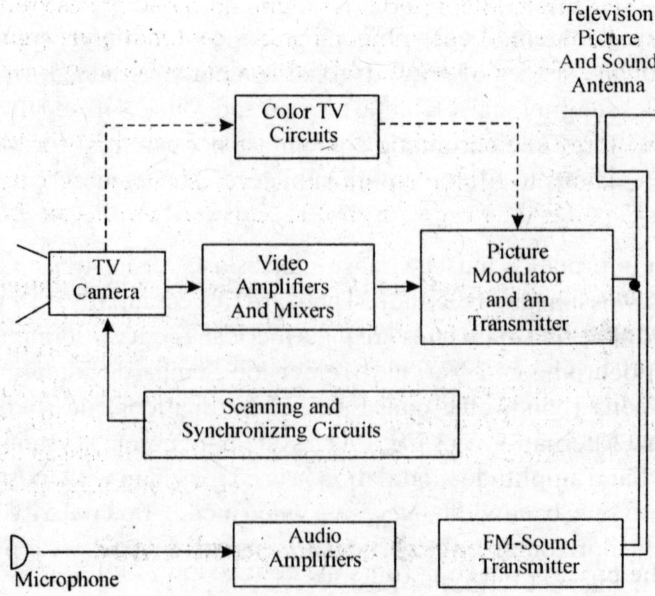

Fig. 9.17 TV Transmitter

9.12.2 Receiver (sound & picture)

At the receiver, the RF signal is amplified and the low frequency audio signals are separated by demodulation. These are further amplified to drive a sound reproducing device like a loudspeaker, which converts them back into sound.

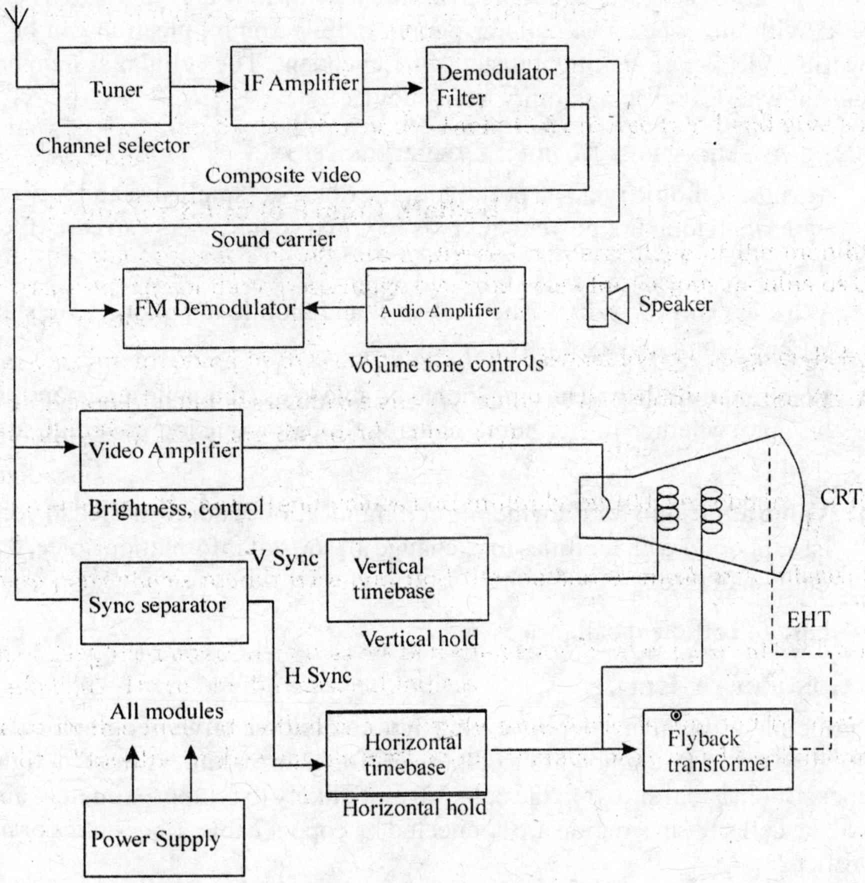

Fig. 9.18 TV receiver

The receiving antenna picks up the signal and feeds it to the tuner stage. The tuner contains a RF amplifier for amplification. The signal is converted into intermediate frequency signal by Superhetrodyning process occurring in the tuner. At the receiving end, the picture is reproduced on a cathode ray picture tube screen with the help of the focused electron beam of varying intensity falling on the screen to produce a proportional glow at the various positions on the coating of the screen.

Scanning

Scanning is the process which permits the conversion of information expressed in space and time coordinates into time variations only.

9.13 Elemenary Concept of Mobile Commuication

Cellular mobile communication is free from the constraints of location, time, distance and its kind (voice, data, video etc). In conventional landline telephone communication, telephone instrument is placed at a fixed location and caller has to go to that location for initiating a call. In mobile communication, user carries the mobile instrument with him and hence it is location and time independent. With the concept of cellular phone, mobile communication can be possible for any distance world wide, hence making it distance independent. The cellular system has the features of reachability anywhere, and accessibility to any device.

9.13.1 Requirement for Mobile Communication

(i) **Terminal mobility :** The portable subscriber set (small size and low weight) supported with small long life power battery is required so that users can carry the set anywhere he moves.

(ii) **Wireless connectivity :** The user's set and mobile telephone base station should have wireless connectivity for free mobility.

(iii) **Personal mobility :** The user should be able to use the same telephone numbers wherever he goes whether in his home coverage area or outside coverage area with roaming facility.

(iv) **Value addition :** Modern mobile communication is not restricted to voice only. It should be supported with facilities to exchange all sorts of information voice, data, image, video, multimedia etc.

(v) Fast and efficient call processing.

9.13.2 Concept of Cell

The principle of cellular telephone system is to divide a large geographical service area into small area called cell. The diameter of a cell varies from 2 to 50 km. As show in figure 9.14, the cells are hexagons. But in a real situation the cells are more likely to be approximately circular, with some overlap. All the cell sites in a region are connected by copper cable, fiber optics or microwave like to a central office.

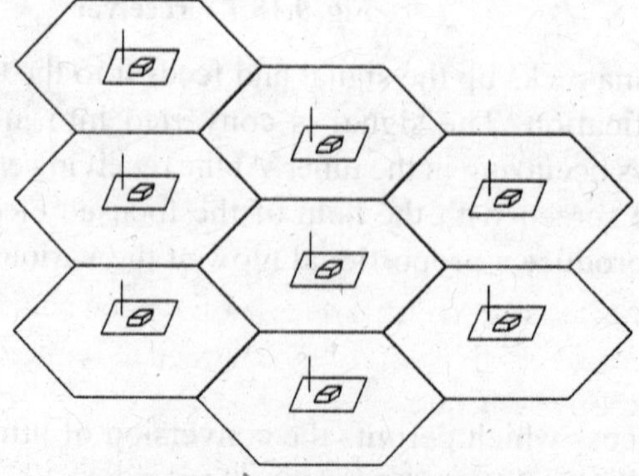

Fig.9.19 : Cell concept

9.13.3 Mobile Communication System

A basic mobile network has following components as shown in figure 9.20.

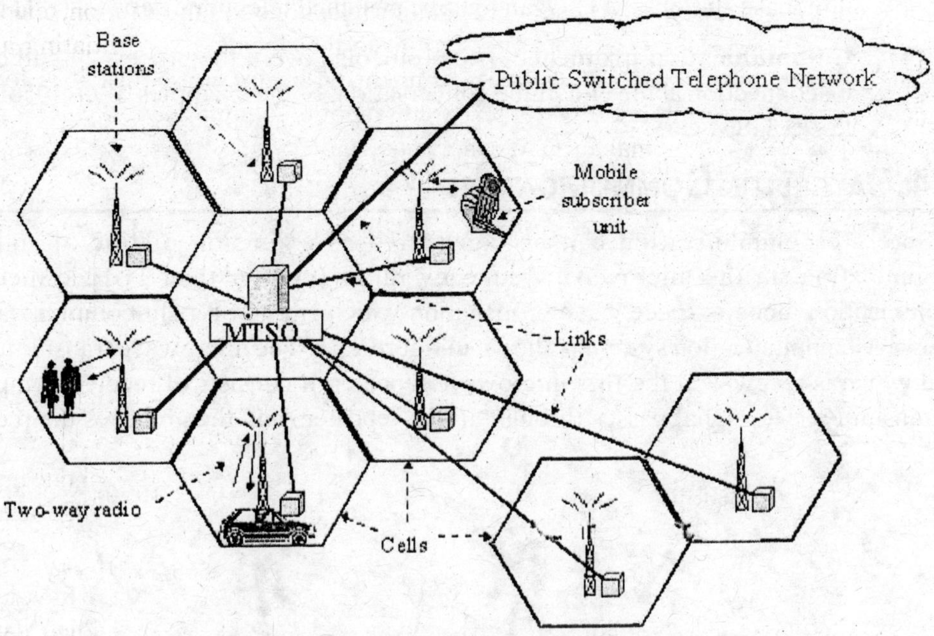

Fig.9.20 Basic Mobile Communication systems

(1) **Mobile unit :** A mobile unit is a portable low weight handset carried by the user which has wireless connectivity with nearest base station. A simple mobiles unit contains a control unit; a radio trans-receiver and an antenna system. The trans-receiver consists of a transmitter and receiver for two way telephonic conversation.

(2) **Cell site :** It provides an interface between the mobile unit and the MTSO .It is also known as base station and consists of control unit, radio cabinets, antennas, a power plant and data terminals. It is connected with MTSO on one side using wide band media, and on the other side has wireless connectivity with mobile unit.

(3) **MTSO (Mobile telephone switching office) :** It is the heart of the mobile system. Its processer provides cellular administration and central co-ordination .It may also be connected with landline telephone networks. It contains the cellular processor and the cellular switch.

The cellular switch is an analog or digital telephone exchange which controls switching between landline subscriber unit to base station for landline to mobile connectivity and vice versa. It controls switching between a base station to another base station for mobile to mobile conversation

The cellular processor processes the data received from base station controllers regarding the status of mobile unit. It also processes the diagnostic data and billing information.

(4) **System interconnect :** The radio connectivity, voice grade four wire line connectivity, optical fiber connectivity, microwave link, data link etc. can be used to connect mobile unit, base stations, MTSO, and public switched telephone network(PSTN).

(5) **Communication protocols:-**The protocols govern the process of call connection and disconnection at the end of the conversation. Some examples of protocols are IS-54 and IS-95, GSM etc.

9.14 Satellite Communication

Satellite communication can be considered as extension of line-of-sight microwave communication. In the microwave frequency range (greater than 1GHz), the only possible communication mode is space wave propagation which is line-of-sight communication. A typical microwave communication system is shown in figure 9.21. The microwave transmitter transmits the signal which is received at the first microwave repeater. It consists of receiver, amplifier-reshaper and transmitter. The signal passes through all the repeaters and then reaches the receiver.

Fig.9.21 : Microwave communication system

9.14.1 Satellite Communication System

Basically, a communication satellite is a radio repeater in the sky. A satellite communication system is shown in figure 9.22. It consists of a satellite that links earth station transmitter and earth station receiver. Usually, one satellite is connected to many earth stations. The user is connected to the earth station. A user interface between the user and the terrestrial system, and a terrestrial interface between terrestrial system and the earth station are to be used at the earth stations. The satellite consists of transponder (the word being derived from TRANS mitter-res-PONDER), power supply system, arrangements to control its position, fuel etc. The earth station transmitter is referred as uplink section and the earth station receiver is referred as down link section.

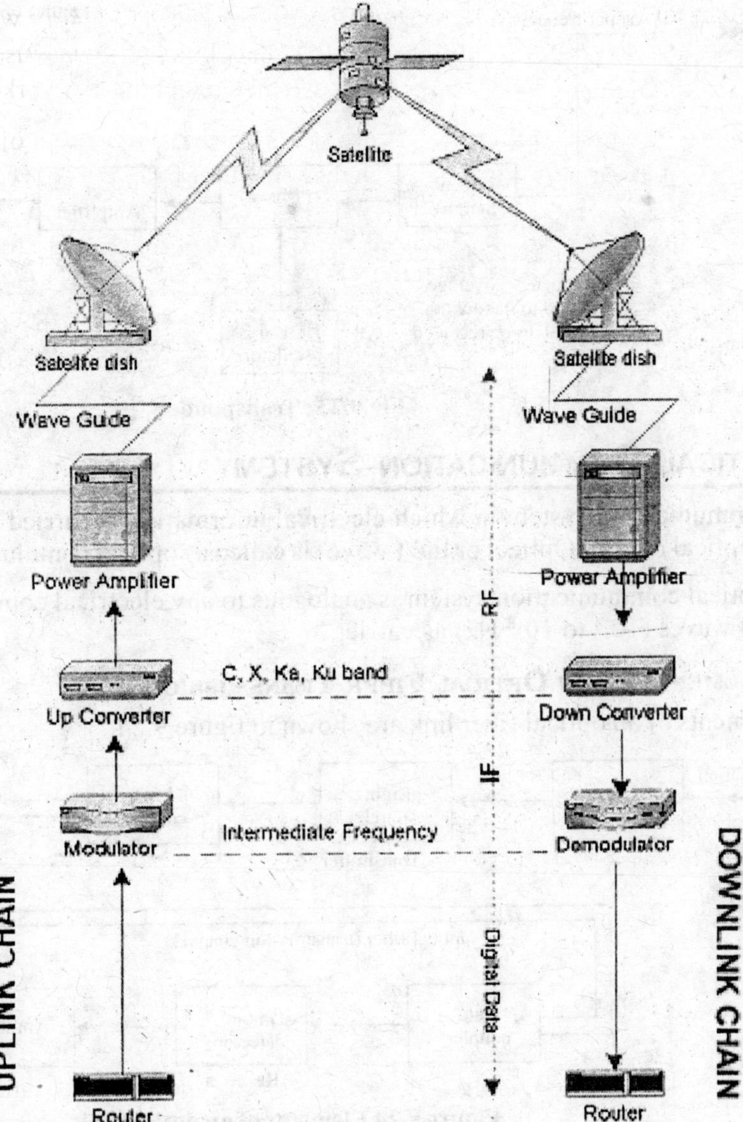

Fig. 9.22 Satellite Communication System

Up link section : The up link section as shown in figure 9.22 consists of an IF modulator, an IF to RF up-convertor, a high power amplifier and a transmitting antenna.

Transponder : The transponder as shown in figure 9.23, consists of a receiving antenna, an input low noise amplifier, a frequency translator, an output low power amplifier and a transmitting antenna. The frequency translator is basically an RF to RF repeater.

Down link section : The downlink section as shown in figure 9.22 consist of a receiving antenna, a low noise amplifier, an RF to IF down convertor and IF demodulator.

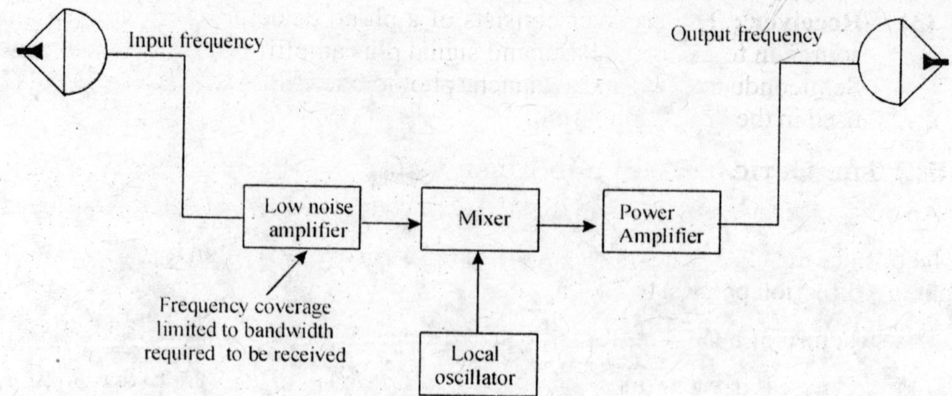

Fig.9.23 : Transponder

9.15 Optical Communication System

The communication system in which electrical information is carried by the electromagnetic wave in the optical region (Optical or light wave) is called as optical communication.

Thus optical communication system is analogous to any electrical communication system but using optical waves (10^{14} to 10^{18} Hz) as carrier.

9.15.1 Elements of an Optical Fiber Transmission Link

The elements of an optical fiber link are shown in figure 9.24.

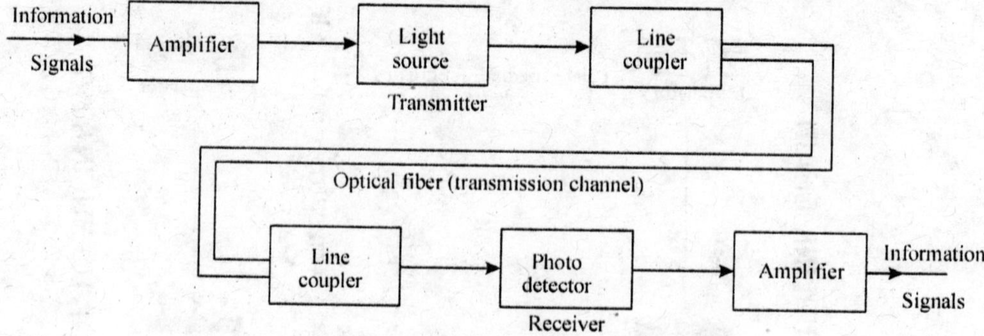

Figure 9.24 Elements of an optical link

(1) **Transmitter :** It consists of an optical source and associated drive circuitry. The light source is either a laser or a light emitting diode (LED). This optical source is modulated by electrical base band signal with the help of associated drive circuitry. The optical source acts as a carrier of baseband signal.

(2) **Optical fiber :** This is hair thin glass fiber transmission media of non magnetic material to pass the light through it with minimum possible loss. The optical fiber is placed inside a cable offering environmental and mechanical protection.

(3) **Receiver :** The receiver consists of a photo detector which converts modulated light source in to electrical baseband signal plus amplification and signal restoring circuitry. Semiconductor PIN and avalanche photodiode (APD) is the two principle photo detectors used in the optical fiber link.

9.15.2 The Optical Fiber and Fiber Cable

An optical fiber is a cylindrical dielectric waveguide which operates at optical frequency.

It confines the electromagnetic energy in the form of light within its surface and guides the flow of light in a direction parallel to its axis.

The structure of a single fiber is shown in figure 9.25. It has basically three parts.

(1) **Core :** It is a single solid dielectric of refractive index n_1. The core material used for low and medium loss fibers is glass. The diameter of core varies from 8 to 200μm.

(2) **Cladding :** The core is surrounded by a solid dielectric cladding having refractive index n_2 that is smaller than n_1. The cladding material is also either glass or plastic. It adds mechanical strength to the fiber. It protects the core from absorbing surface contaminants.

(3) **Buffer coating :** An elastic plastic buffer encapsulates the fiber. This adds further strength to the fiber and mechanically isolates the fiber from small geometrical irregularities or roughness of the adjacent surface used for supporting the structure while placing the fiber in a cable.

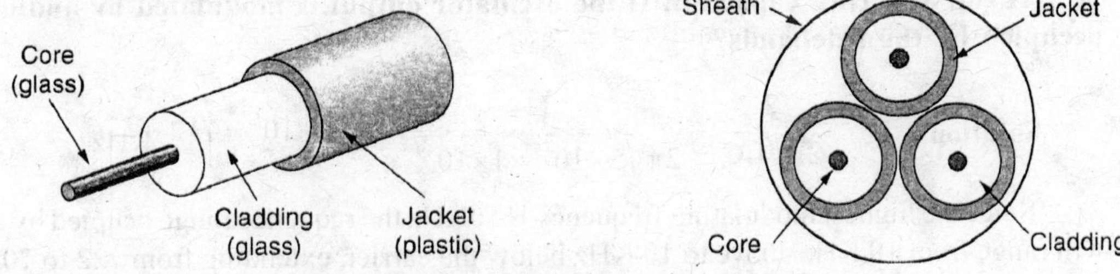

Fig. 9.25 Optical fiber structure

Advantages of Optical fiber communication

(1) Tremendous Bandwidth

(2) Light weight and small diameter

(3) Low loss

(4) Less number of repeaters

(5) High security

Disadvantages

(1) Installation cost is high.

(2) Maintenance and repair is costly.

9.16 SOLVED NUMERICAL PROBLEMS

Example 1. Determine the length of an antenna designed to transimit a signal of 30 KHz frequency.

Solution : The length of transmitting antenna is given by

$$l = \frac{\lambda}{4} = \frac{1}{4} \cdot \frac{C}{f}$$

$$= \frac{3 \times 10^8}{4 \times 30 \times 10^3} = 2500 \text{ m}$$

Example 2. The maximum peak of peak voltage of an AM wave is 20 mV and minimum peak to peak voltage is 5mV. Determine the modulation index.

Solution : The modulation index is given as

$$m = \frac{V_{max} - V_{min}}{V_{max} + V_{min}} = \frac{20 - 5}{20 + 5} = 0.6$$

Example 3. The tuned circuit of the oscillator in a simple AM transmitter employs a 50 μH coil and 1 nF capacitor. If the oscillator output is modulated by audio frequencies occupied by the sidebands?

Solution : $f_C = \frac{2}{2\pi\sqrt{LC}} = \frac{1}{2\pi\sqrt{5 \times 10^{-5} \times 1 \times 10^{-9}}} = 7.12 \times 10^5 = 712$ KHz

Since the highest modulating frequency is 10KHz the requency range ocupied by the sidebands will range from 10 KHz above to 10 KHz below the carrier, extending from 722 to 702 KHz.

Example 4 A sinusoidal carrier wave of frequency 2 MHz and amplitude 20mV is amplitude modulated by a sinusoidal wave of frequency 5 KHz. Determine the frequency and amplitudes of lower and upper side bands. Consider modulation index as 0.8.

Solution: Carrier frequency $f_C = 2 \times 10^6$ Hz.

modulating frequency $f_m = 5 \times 10^3$ Hz.

carrier amplitude $V_C = 20m\,V$

LSB frequency $= f_C - f_m = 2 \times 10^6 - 5 \times 10^3 = 1.995$ MHz

USB frequency $= f_C + f_m = 2 \times 10^6 + 5 \times 10^3 = 2.005$ MHz

Amplitude of each side band $= \frac{m V_C}{2} = \frac{0.8 \times 20 \times 10^{-3}}{2} = 8$ mV

Example 5. A 500 watt carrier is modulated to a depth of 75% calculate the total power in the modulated wave.

Solution : $P_{total} = P_{carrier}\left(1+\dfrac{m^2}{2}\right) = 500\left(1+\dfrac{0.75^2}{2}\right)$

$= 500 \times 1.281 = 640.50\ \text{W}$

Example 6. A broadcast radio transmitter radiates 10 kW when the modulation percentage is 60. How much of this is carrier power?

Solution : $P_{total} = P_{carrier}\left(1+\dfrac{m^2}{2}\right)$

$$P_{carrier} = \frac{P_{total}}{1+\dfrac{m^2}{2}} = \frac{10}{1+(0.6)^2/2} = \frac{10}{1.18} = 8.47 \quad \text{KW}$$

Example 7. A carrier wave of 100V and 1000KHz is modulated by a 50V, 900 KHz sine wave signal. Find the modulation factor.

Solution : Modulation index is defined as

$$m = \frac{V_m}{V_C} = \frac{50}{100} = 0.5$$

Example 8. A sinusoidal carrier voltage of requecny 1.5 MHz and amplitude 30 Volts in amplitude modulated by sinusoidal voltage of requecny 20 KHz producign 50% modulation. Calculate the frequency and amplitude of lower and upper sideband terms.

Solution : Given that

Carrier frequency f_C =1.5 MHz

modulating frequency f_m = 20 KHz

modulation index m = 50% = 0.5

carrier amplitude V_C = 30 V

Frequency of LSB $= f_C - f_m = 1.5\times10^6 - 20\times10^3 = 1.48\ \text{MHZ}$

Frequency of MSB $= f_C + f_m = 1.5\times10^6 + 20\times10^3 = 1.52\ \text{MHz.}$

and amplitude of each side band $= \dfrac{m V_C}{2} = \dfrac{0.5\times30}{2} = 7.5\,\text{V}$

Example 9. A sinsuodial carrier voltage of frequency 1 MHz and ampliude 120V is ampliude modulated by sinusodial voltage of requency 8 KHz producing 60% modulation. Determine the frequency and amplitude of lower and upper side band terms.

Solution : Carrier frequency $f_C = 1 \times 10^6$ Hz

Modulating frequency $f_m = 8 \times 10^3$ Hz

Modulation index $m = 60\% = 0.6$

Amplitude of carrier $V_C = 120\,V$

Lower side band frequency

$$f_{LSB} = f_C - f_m$$

Upper side band frequency

$$f_{USB} = f_C + f_m$$

$$= 1 \times 10^6 + 8 \times 10^3 = 1008\ \text{KHz}$$

Amplitude of each side band term $= \dfrac{mV_C}{2} = \dfrac{0.6 \times 120}{2} = 36\,V$

Example 10. The antenna current of an AM transmitter is 10A when only carrier is sent, but it increases to 10.5A when the carrier is modulated by a single sine wave. Find the percentage modulation. Determine the antenna current when the depth of modulation changes to 0.6

Solution : (a) Total current $I_t = 10.5$ A

carrier current $I_C = 10$ A

$$\left(\frac{I_t}{I_C}\right)^2 = 1 + \frac{m^2}{2}$$

$$m = \sqrt{2\left[\left(\frac{I_t}{I_C}\right)^2 - 1\right]} = \sqrt{2\left[\left(\frac{10.5}{10}\right)^2 - 1\right]} = 0.45 = 45\%$$

(b) Now if modulation index $m = 0.6$

and carrier current $I_C = 10$ A

$$I_t = I_C\sqrt{1 + \frac{m^2}{2}} = 10\sqrt{1 + \frac{(0.6)^2}{2}} = 10.86\,A$$

Example 11. A carrier wave of power 150W is subjected to a modulation index of 0.70 for amplitude modulation. Determine the total power in sidebands and the power of the modulated wave.

Solution : Total power in sideband $P_{SB} = \dfrac{m^2 V_C^2}{4R}$

and carrier power $P_C = \dfrac{V_C^2}{2R}$

therefore $\dfrac{P_{SB}}{P_C} = \dfrac{m^2}{2}$

$$\Rightarrow \qquad P_{SB} = \frac{m^2}{2} P_C = \frac{(0.70)^2}{2} \times 150 = 36.75 \text{ W}$$

Example 12. A certain transmitter radiates 10 KW with the carrier signal, and 12.5 KW when the carrier is sinusoidally modulated. Calculate the modulation index. If another sine wave, corresponding to 40 percent modulation is transmitted simultaneously, determine the total power.

Solution : Carrier power $P_C = 9$ KW

Total power $P_t = 10$ KW

$$\frac{P_t}{P_C} = 1 + \frac{m^2}{2}$$

$$\Rightarrow m = \sqrt{2\left[\frac{P_t}{P_C} - 1\right]} = \sqrt{2\left[\frac{12.5}{10} - 1\right]} = \sqrt{0.5} = 0.707$$

For second part, the total modulation index will be $m_t = \sqrt{m_1^2 + m_2^2}$

$$= \sqrt{(0.707)^2 + (0.4)^2} = \sqrt{0.5 + 0.16} = 0.812 = 81.2\%$$

therefore, the total power

$$P_t = P_C\left(1 + \frac{m_t^2}{2}\right) = 10\left[1 + \frac{(0.812)^2}{2}\right] = 13.29 \text{ kW}$$

Example 13. The antenna current of an AM broadcast transmitter, modulatated to a depth of 40 percent by an audio sine wave is 11A. It increases to 12A as a result of simultaneous modulation by another audio sine wave. What is modulation index due to this second wave?

Solution : Carrier current is given as

$$I_C = \frac{I_t}{\sqrt{1+m^2/2}} = \frac{11}{\sqrt{1+(0.4)^2/2}} = 10.58\,\text{A}$$

If total current is increased from 11 A to 12 A,

$$m_t = \sqrt{2\left[\left(\frac{I_t}{I_C}\right)^2 - 1\right]} = \sqrt{2\left[\left(\frac{12}{10.58}\right)^2 - 1\right]} = 0.757$$

Now we have relation $m_t^2 = m_1^2 + m_2^2$

$$m_2 = \sqrt{m_t^2 - m_1^2} = \sqrt{(0.757)^2 - (0.4)^2} = 0.643$$

Example 14. A carrier wave of power 80W is to be amplitude modulated with a modulation index of 0.7. What will be the carrier power after modulation? What must be the audio power required if the modulator has an efficiency of 75%.

Solution : The total power of AM signal is confined in carrier and sidebands. The carrier power is not affected by modulation and remains as it is. Therefore the power of modulating signal is confined in the sideband power provided 100% efficiency of the modulator.

Since the carrier power remains unaltered by modulation, the carrier power after modulation is 80W.

Sideband power is given as

$$P_{SB} = \frac{m^2}{2} P_C = \frac{(0.7)^2}{2} \times 80 = 19.6 \text{ W}$$

Modulator efficiency $\eta = \frac{P_{SB}}{P_a}$

where P_a is the power in audio signal, therefore

$$P_a = \frac{P_{SB}}{\eta} = \frac{19.6}{0.75} = 26.13\,\text{W}$$

Example 15. An amplitude modulated wave consists of the following frequency components :

Carrier	**: 12V peak**
Lower sideband	**: 5V peak**
Upper sideband	**: 5V peak**

If the am litude m odulated w ave drives a load resistance of 3.7 $k\Omega$ **, determine the power delivered to the load resistor by**

(a) Carrier (b) LSB and USB (c) AM signal

Solution : (a) Power delivered to the load resistor by carrier is given as

$$P_C = \frac{V_C^2}{2R} = \frac{(12)^2}{2\times3.7\times10^3} = 19.46 \text{ mW}$$

(b) Power delivered to the load resistor by either sideband is given by

$$P_{LSB} = P_{USB} = \frac{V_{SB}}{2R} = \frac{(5)^2}{2\times3.7\times10^3} = 3.3 \text{ mW}$$

(c) Total power delivered

$$P_{total} = P_C + P_{LSB} + P_{USB} = (19.46 + 3.3 + 3.3) \text{ mW} = 26.06 \text{ mW}$$

Example 16. What is the modulation index of an FM carrier having a carrier swing of 300 KHz and a modulating signal of 15 KHz.

Solution : Carrier swing CS = 2δ

$$\Rightarrow \quad \delta = \frac{CS}{2} = \frac{300\times10^3}{2} = 150 \text{ KHz}$$

The modulation index $m_f = \frac{\delta}{f_m} = \frac{150}{15} = 10$

Example 17. A 110 MHz carrier wave frequency modulated by a 10 KHz simusoidal modulating signal. If the maximum frequency deviation is 20 KHz find the modulation index.

Solution : Modulatin index is given as

$$m_f = \frac{\delta}{f_m} = \frac{20\times10^3}{10\times10^3} = 2$$

Example 18. An FM signal has a resting frequency of 100 MHz and highest frequency of 100.5 MHz when modulated by a signal of 10 KHz frequency. Determine (1) frequency deviation (ii) carrier swing (iii) modulation index (iv) percentage modulation.

Solution : (i) Frequency deviation $\delta = 100.5\text{-}100 = 0.5\ MHz$

(ii) Carrier swing $CS = 2 \times \delta = 2 \times 0.5 = 1$ MHz

(iii) Modulation index $$m_f = \frac{\delta}{f_m} = \frac{0.5 \times 10^6}{10 \times 10^3} = 50$$

(iv) Percentage modulation $$m = \frac{\delta_{actual}}{\delta_{max}}$$

Here $\delta_{max} = 75$ KHz (allowed from FM broadcast)

$$m = \frac{0.5 \times 10^6}{75 \times 10^3} = 666\%$$

Example 19. In an FM system, when the modulating frequency is 600 Hz and the modulating voltage is 2.6V, the deviation is 5 KHz. If the modulating voltage is now increased to 7V, what is new deviation? If the modulating voltage is raised to 10.2V while modulating frequency is dropped to 300 Hz. What is the deviation. Obtain the modulation index in each case.

Solution : In frequency modulation, deviation is proportional to amplitude of modulating signal.

Therefore

$$\delta \propto V_m$$

Therefore

$$\frac{\delta_1}{\delta_2} = \frac{V_{m1}}{V_{m2}}$$

Since $\delta_1 = 5$ KHz, $V_{m1} = 2.6$ V and $V_{m2} = 7$ V, then

$$\delta_2 = \frac{V_{m2}}{V_{m1}} \times \delta_1 = \frac{7}{2.6} \times 5 \times 10^3 = 13.46 \text{ KHz.}$$

Similarly when $A_{m3} = 10.2$ V then $\delta_3 = 19.61$ KHz.

Note that the change in modulating frequency has no effect on the deviation since it is independnet of modulating frequency ($\delta = K V_m f_C$)

Now the modulation index can be calculated as

$$m_{f1} = \frac{\delta_1}{f_{m1}} = \frac{5}{0.6} = 8.334$$

$$m_{f2} = \frac{\delta_2}{f_{m2}} = \frac{13.46}{0.6} = 22.434$$

$$m_{f3} = \frac{\delta_3}{f_{m3}} = \frac{19.61}{0.3} = 64.367$$

Thus, the change in modulating frequency changes the modulation index sharply.

Example 20. A frequency modulated signal is expressed as

$$v = 15 \sin\left(5 \times 10^8 t + 5 \sin 1000t\right)$$

Determine the carrier and modulating frequencies, the modulation index and the maximum deviation. Also determine the amount of power that will dissipates by this FM signal in a resistor of 10 Ω.

Solution : The FM signal is given as

$$v = A \sin\left(\omega_C t + m_f \sin \omega_m t\right)$$

camparing the numerical equation, we have A= 15V

$$A = 15\,V$$

$$f_C = \frac{5 \times 10^8}{2\pi} Hz = 79.55 \text{ MHz}$$

$$f_m = \frac{1000}{2\pi} Hz = 159.1 Hz$$

$$m_f = 5$$

$$\delta = m_f f_m = 5 \times 159.1 = 795.5 Hz$$

and $$P = \frac{\left(A/\sqrt{2}\right)^2}{R} = \frac{(15)^2}{2 \times 10} = 11.25 \text{ W}$$

Example 21. A 25 MHz carrier is modulated by a 400 Hz audio sine wave. If the carrier voltage is 4V and the maximum deviation is 10KHz, write the equation of this modulated wave for (a) FM and (b) PM. If the modulating frequency is now changed to 2KHz all else remaining constant, write a new equation for (c) FM and (d) PM. Calculate the frequencies in radian.

Solution : Calculating the frequencies in radians, we have

$$\omega_C = 2\pi \times 25 \times 10 = 1.57 \times 10^8 \text{ rad/sec}$$

$$\omega_m = 2\pi \times 400 = 2513 \text{ rad/sec.}$$

The modulation index will be

$$m = m_f = m_p = \frac{\delta}{f_m} = \frac{10000}{400} = 25$$

This yields the equations

(a) $v = 4\sin\left(1.57 \times 10^8 t + 25\sin 2513t\right)$ (FM)

(b) $v = 4\sin\left(1.57 \times 10^8 t + 25\sin 2513t\right)$ (PM)

Note that two expressions are identical. Now when the modulating frequency is multiplied by 5, the equation will show a fivefold increase inthe modulating frequency. However while the modulation index in FM is reduced fivefold, for PM the modulation index remains constnat. Hence

(c) $v = 4\sin\left(1.57 \times 10^8 t + 5\sin 1256t\right)$ (FM)

(d) $v = 4\sin\left(1.57 \times 10^8 t + 25\sin 1256t\right)$ (PM)

Note, once again, that the difference between FM and PM is not apparent at a single modulating frequency. Rather, it reveals itself in the differing behaviour of the two systems when the modulating frequency is varied.

9.17 SUMMARY

1. Basic block diagram of a communication system incudes information source, transmitter and receiver sections.

2. Some characteristics of a high frequency sine wave are varied in accordance with the instantaneous value of modulating signal.

3. Any of three characteristics (V_m , ω and ϕ) of the carrier may be varied by the modulating signal giving rise to amplitude, frequency or phase modulation respectively.

4. Modulation is requried to reduce the antenna length, to increases the operating range, to separate various signals and to determine the bandwidth of the system.

5. The modulation index plays an important role in determining the strength and quality of signal transmitted.

6. For AM system, modulation index lying between 0 and 1. If m>l, the carrier wave is overmodulated and distortion occurs during reception.

7. The process of amplitude modulation, adds two side bands to the carrier signal instead of changing it.

8. Bandwidth of amplitude modulated signal is defined as the difference between USB and LSB and equal to $2f_m$.

9. Maximum power in the AM is 1 .5 $P_{carrier}$ when modulation index is equal to unity.

10. The process of recovering the audio signal from the modulated wave is known as demodulation or detection.

11. Frequency modulation (FM) is a system in which the amplitude of the modulated carrier is kept constant, while its frequency is varied by the modulating signal.

12. The ratio of maximum frequency deviation (δ) to the modulating frequency (f_m) is defined as the modulation index (m_f) of frequency modulation. m_f may have value more than unity.

13. Frequency modulated signal consists of a carrier and infinite number of pairs for sidebands.

14. If the phase ϕ of the carrier signal is varied so that its magnitude is proportional to the instantaneous amplitude of modulating voltage, the resulting wave is phase modulated.

15. Multiplexing is a technique of combining channels to ensure that a large number of them can be carried on the one bearer without interference.

16. There are two types of multiplexing (i) Frequency division multiplexing (FDM) and (ii) Time division multiplexing (TDM)

17. Receivers may be categorized as (i) Tune radio frequency (TRF) and (ii) Superheterodyne receiver.

Review Questions

1. What do you mean by modulation and demodulation? Why the modulation is neccessary in communication system?
2. Define the following terms

   (a) Information (b) Redundancy (c) Noise (d) Bandwidth
3. Define amplitude modulation and modulation index. Use a sketch of a sinusoidally modulated AM waveform to help explain the definiti..
4. Derive the relatin between the output power of an A A transmitter and the depth of modulation.
5. From the expression for the instantaneous voltage of AM wave, derive a formula for the rms value of this wave.
6. Explain the meaning of the following terms as referred to amplitude modulation,

   (i) Lower sideband frequency

   (ii) Upper sideband frequency
7. Derive an expression for the fraction of total power carried by the sidebands in AM.
8. Describe how a square law electronic device can be used to generate an amplitude modulated signal.
9. Explain the operation of diode demodulator.
10. Explain the significance of modulation index in context of amplitude modulation.
11. If the maximum and minimum voltage of an AM wave are V_{max} and V_{min} respectively, then show that modulation indese (m) is given by

$$m = \frac{V_{max} - V_{min}}{V_{max} + V_{min}}$$

12. Describe frequency and phase modulation.
13. Derive the formula for the instantaneous value of an frequency modulated voltage and define the modulation index.
14. In an FM system, if m_f is doubled by halving the modulating frequency, what will be the effect on the maximum deviation?
15. Explain different type of modulations. Obtain an expressionfor a frequency modulated carrier with sinusoidal modulation.
16. What are the advantages of frequency modulation?

17. Compare amplitude modulation and frequency modulation.

18. Define the term multiplexing. Also explain the types of multiplexing.

19. With the aid of the bock diagram of a simple receiver, explain the superheterodyne principle.

20. What are the advantes that the superheterodyne receiver has over the TRF receiver? Are there any disadvantages?

21. How is the constnat intermediate frequency (IF) achieved in the superheterodyne reciever?

22. Write shrot note on Television.

23. Explain how television is capable of displaying complete moving pictures, despite the fact that at any instant of time only a tiny portion of picture tube screen is active.

24. Draw the basic block diagram of television transmitter and receiver.

Index

A

B

C

D

E

R

S

T

U

V

W